Environmental Science

Series editors: R. Allan · U. Förstner · W. Salomons

Springer

Berlin
Heidelberg
New York
Barcelona
Hong Kong
London
Milan
Paris
Singapore
Tokyo

Marcelo Juanicó · Inka Dor (Eds)

Hypertrophic Reservoirs for Wastewater Storage and Reuse

Ecology, Performance and Engineering Design

With 166 Figures and 75 Tables

 Springer

Editors

Dr. Marcelo Juanicó
Juanicó & Friedler Mediterranean Ltd.
Ram On
M.P. Megido 19205, Israel
Phone: +972-6-6499241
Fax: +972-6-6499237

Prof. Inka Dor
Division of Environmental Sciences
The Hebrew University of Jerusalem
Jerusalem 91904, Israel
Phone: +972-2-6585554
Fax: +972-2-6586155

The maps included in this book are presented to illustrate technical issues. They do not imply any statement by the editors, publisher or authors regarding the legal status of any territory or the delimitation of its frontiers or boundaries.

The cover figure shows a schematic lay out of a wastewater reservoir preceded by two anaerobic ponds.

ISSN 1431-6250
ISBN 3-540-65598-0 Springer-Verlag Berlin Heidelberg New York

Library of Congress Cataloging-in-Publication Data
CIP applied for

Cataloging-in-Publication Data applied for

Die Deutsche Bibliothek - CIP-Einheitsaufnahme
Hypertrophic reservoirs for wastewater storage and reuse :
ecology, performance and engineering design ; with 75 tables /
Marcelo Juanicó ; Inka Dor (eds.). - Berlin ; Heidelberg ; New York ;
Barcelona ; Hong Kong ; London ; Milan ; Paris ; Singapore ; Tokyo ;
Springer, 1999
    (Environmental science)
    ISBN 3-540-65598-0

Cover Design: Struve & Partner, Heidelberg
Dataconversion: Büro Stasch, Bayreuth

SPIN: 10569129   32/3020 – 5 4 3 2 1 0 – Printed on acid-free paper

*The editors wish to dedicate this book to the Israeli farmers,
great individuals who anonymously contributed most
to the development of wastewater reservoirs.*

Preface

Wastewater reservoirs are unique limnological ecosystems sharing with waste stabilisation ponds the highest extreme of the hypertrophic range. They are also a practical, efficient, cost effective wastewater treatment and storage technology.

Wastewater reservoirs have been operated in Israel for almost 30 years for the storage and treatment of wastewater effluents during the rainy winter months, in order to use them for agriculture irrigation during the dry summer months. Today there are more than 200 of these reservoirs operating in the country.

The use of reservoirs to store and treat wastewater is an old practice, but the development of deep reservoirs to cope with the evaporation rates of semi-arid and arid regions, their massive use, and the huge R&D effort invested to better understand their behaviour and improve design and operational criteria are relatively new.

Wastewater Reservoirs as a Unique Hypertrophic Ecosystem

The habitat of wastewater reservoirs resembles that of waste stabilisation ponds, but the increased depth (average between 6 and 8 m, maximum up to 20 m), greater volumes (up to several million m³) and the non-steady-state hydraulic regime, introduce several differences in the both the hydrology and the biological community.

Many of the characteristics quoted in the literature as typical of hypertrophic ecosystems can be found in wastewater reservoirs: mass growth of a few dominant species, suppression of plankton periodicity, lack of inorganic nutrient depletion, oxygen oversaturation during the day followed by low dissolved oxygen during the night, high oxygen production over the year, high bio-turbidity, etc.

A unique limnological characteristic of wastewater reservoirs is that the operational regime (which sharply changes during the year), and not the climatic seasons, is the main factor affecting the changes of the biological community and water quality during the year. The frequent short-term external changes in input flows cause the system to be continuously in a state of transition between multiple steady states. The result is an ecosystem with high regulatory potential due to its environmentally robust biota. For practical purposes, this means that the system can be regulated through design and operational parameters, depending less on climatic seasonal changes and unpredictable 'natural' plankton succession.

Besides, in spite of their 'sewage treatment' character, wastewater reservoirs are 'green':

- The processes occurring within the reservoirs are natural. They utilize solar energy (mechanical sewage treatment plants use electricity). Algae within the reser-

voirs produce most of the oxygen required by the depuration processes (mechanical plants take oxygen from the atmosphere, with high energy consumption).
- Aquatic birds and other water related animals find the reservoirs a good refuge: Israel's wastewater reservoirs are commonly visited by thousands of aquatic birds every year. This is important in semi-arid and arid areas where the few 'wet' natural habitats of the birds have been invaded by tourism or urban or agricultural development.

Wastewater Reservoirs as an Emergent Wastewater Treatment Technology

These reservoirs perform simultaneously as storage volumes for better flow management, and as deep stabilisation ponds with a nonsteady-state hydraulic regime. They have a high treatment capacity for both high-rate decaying pollutants such as pathogens, and low-rate decaying pollutants such as refractory compounds. The reservoirs are a must in almost any wastewater reuse project where there is always a need for regulation between the almost constant sewage flow and the irregular water demand for irrigation. Besides, they can be used in numerous other situations:

- Beach protection: Wastewater may be stored in coastal areas during the summer in order to avoid the contamination of beaches during the tourism season. By the end of summer – when the last tourist has gone – wastewater can be released from the reservoirs into the sea. Meanwhile, the effluents will reach excellent quality due to long residence time within the reservoirs during the summer months.
- River/stream recovery (I): Wastewater is stored during the dry season when the rivers run with minimum flow. Wastewater of high quality will be released from the reservoirs to the river when river-flow is at maximum, thus obtaining maximum dilution and minimum negative ecological impact.
- River/stream recovery (II): Wastewater is stored when river flow is at maximum. Wastewater of very high quality is then released from the reservoirs to the rivers during the dry period as a substitute for freshwater, in order to avoid total drying and ecosystem destruction in rivers whose dry-season freshwater sources have been diverted for other purposes.
- High quality effluents are required: Wastewater contains not only organic matter but also significant concentrations of pathogens, heavy metals, hard detergents, pesticides, organic micro-pollutants, and other refractory pollutants which are not removed by the classic sewage treatment plants. Stabilisation reservoirs are able to remove most of them.

This Book

The present publication has two main objectives:

- To summarize in a single book the huge amount of interdisciplinary experience gained in Israel during the last 30 years and that is spread in numerous publications, most of them internal reports and local journals in Hebrew
- To fulfil the difficult task of transforming limnological and R&D studies into engineering design and operational criteria

The book is divided into two sections:

- A series of reviews on different aspects of wastewater reservoirs. The authors are the researchers, agronomists, engineers, biologists, etc., who actively worked on the specific subject of each chapter.
- A series of case studies summarizing some of the most interesting field work performed on these systems.

Meanwhile, R&D on deep wastewater reservoirs continues both in Israel and abroad, and we expect to publish an updated second edition of this book some years from now.

Dr. Marcelo Juanicó
Prof. Inka Dor   Jerusalem, 1999

Contents

Part I Limnology and Technology 1

1 Research and Development Policy 3
1.1 Introduction 3
1.2 Background Data 3
  1.2.1 Water Scarcity and Wastewater Reuse 3
  1.2.2 Wastewater Reclamation Plan 4
  1.2.3 Surface Wastewater Reservoirs 5
1.3 R&D Support 5
  1.3.1 The R&D Policy 5
  1.3.2 Overview of the Ongoing R&D Programme 6
1.4 Results Dissemination 9
1.5 Discussion 9
  References 10

2 Wastewater Storage and Reuse for Irrigation in Israel 13
2.1 Introduction 13
2.2 Wastewater Treatment 14
  2.2.1 Pretreatment 14
  2.2.2 Treatment: Conventional Processes 14
  2.2.3 Advanced Treatment 15
2.3 Sewage Treatment and Reuse in Israel 15
2.4 Main Water Reclamation Projects in Israel 18
2.5 Effect of Irrigation with Treated Sewage Effluent on Soil, Crops and Environment 19
  References 20

3 Health and Treatment Requirements for Wastewater Irrigation 23
3.1 Introduction 23
3.2 The Development of Wastewater Recycling and Reuse 23
  3.2.1 Early Major Wastewater Irrigation Projects 23
  3.2.2 Present Status of Interest in Wastewater Recycling and Reuse 24
3.3 Public Health Risks Associated with Wastewater Irrigation 26
  3.3.1 Pathogenic Microorganisms in Wastewater 26
  3.3.2 Survival of Pathogens in the Environment 26
  3.3.3 Proposed Model to Predict the Relative Effectiveness of Pathogens in Causing Infections Through Wastewater Irrigation 27

3.4   Evaluation of the Epidemiological Evidence of Human Health Effects
       Associated with Wastewater Irrigation 28
       3.4.1   Early Microbial Guidelines for Wastewater Irrigation
               Lack Epidemiological Basis 28
       3.4.2   Population Groups Consuming Vegetables and Salad Crops
               Irrigated with Raw Wastewater 29
       3.4.3   Health Effects among Sewage Farm Workers 31
       3.4.4   Health Effects among Population Groups Residing
               Near Wastewater-Irrigated Fields 32
       3.4.5   Reduction in Negative Health Effects by Wastewater Treatment 32
       3.4.6   Implications for Developing Countries 33
3.5   Evaluating Health Guidelines for Recycling Wastewater in Agriculture 34
       3.5.1   Scientific, Historical, and Social Influences 34
       3.5.2   The Engelberg Report and the New WHO Guidelines 36
       3.5.3   The USEPA/USAID Recommended Wastewater Reuse Guidelines 37
       3.5.4   A Risk Assessment/Cost-Effectiveness Approach
               for Comparing the Various Guidelines 37
3.6   Reduction of Health Risks by Various Agronomic Technics 38
       3.6.1   Restricting Crops 39
       3.6.2   Modification and Control of Irrigation Techniques 39
3.7   Appropriate Low-Cost Methods of Wastewater Treatment for Irrigation 40
       3.7.1   Goals of Wastewater Treatment for Recycling and Reuse in Irrigation ... 40
       3.7.2   Interseasonal Wastewater Stabilisation/Storage Reservoirs 42
       3.7.3   Advantages of Centrally Managed, Engineered
               Environmental Interventions 44
       References 44

4   **Empirical Data for Monitoring and Control** 47
4.1   Introduction 47
4.2   Materials and Methods 47
4.3   Results 49
4.4   Discussion 57
4.5   Conclusions 59
       Acknowledgements 60
       References 60

5   **Process Design and Operation** 61
5.1   Introduction 61
       5.1.1   Water Demand for Irrigation in Israel and the
               Hydrological Cycle in the Reservoirs 61
       5.1.2   Basic Concepts in Designing Wastewater Reservoirs 61
5.2   Basic Operational Regimes and Water Demand Curves for Irrigation 63
5.3   The 'Old' Continuous-Flow Single Reservoirs 65
       5.3.1   Volume and Depth 65
       5.3.2   Outlet and Inlet Location 66
       5.3.3   The Hydraulics of Continuous-Flow Wastewater Reservoirs
               as Reactors 66

5.3.4   *MRT* and *PFE* in Continuous-Flow Wastewater Reservoirs 69

5.3.5   Calculating Surface Organic Loading 69

5.3.6   Performance and Limitations of the Continuous-Flow Reservoirs 71

5.3.7   The Removal of Coliforms 73

5.4   The New Sequential Batch Reservoirs 74

5.4.1   Example I. Batch Operation During Winter 74

5.4.2   Example II. Batch Operation During Spring 75

5.4.3   Several Reservoirs Working in Sequential Batch 76

5.5   Maximum Organic Loading 78

5.5.1   Mean Surface Organic Loading 78

5.5.2   Increasing the Surface Organic Loading 79

5.5.3   Shocks of High Organic Loading 80

5.6   The Tools for Design 80

5.7   Control and Monitoring 80

5.7.1   Sampling 81

5.7.2   Data Analysis 81

     References 82

6   Hydraulic Age Distribution 85

6.1   Introduction 85

6.2   Concepts: The Hydraulic Age Distribution 87

6.2.1   *PFE*: Percentage of Fresh Effluents 87

6.3   The Leslie Matrix Model 90

6.3.1   Background 90

6.3.2   The Leslie Matrix Model Applied to Age Class Structure
         of Effluents in a Reactor 91

6.3.3   Age Distribution in Different Types of Reactors 95

6.3.4   Computer Algorithms 98

6.4   A Case Study: The Geta'ot Reservoir in 1989 100

6.5   Conclusions 102

6.6   Notation 102

     Acknowledgements 103

     References 103

7   Modelling 105

7.1   Introduction 105

7.1.1   Milestones in the Development of Models of Aquatic Ecosystems ... 105

7.2   Models of Stabilisation Reservoirs 106

7.2.1   Models for Predicting Maximal Permitted Dissolved *BOD*
         in Stabilisation Reservoirs 106

7.2.2   Model for the Prediction of the Accumulation of Organic Matter
         in the Sediment of Stabilisation Reservoirs 109

7.2.3   Statistical models for Predicting *BOD* and *COD* Removal
         as a Function of the Operational Regime 110

7.2.4   Model of Hydraulic Age Distribution in Stabilisation Reservoirs 112

7.3   Simulation Model of Stabilisation Reservoirs 113

7.3.1   Introduction 113

7.3.2		Assumptions	113
7.3.3		Forcing Functions	116
7.3.4		State Variables	118
7.3.5		The Logical Structure of the Model and the Computer Programme	126
7.3.6		Results	126
7.4		Summary and Conclusions	139
7.5		Notation	141
		References	143

8   Sediment-Water Interrelationship 145

8.1   Introduction 145
8.2   Accumulation of Nutrients in Bottom Soils 145
8.3   Release of Nutrients from the Bottom 146
8.4   Nutrient Balance of Reservoirs 148
8.5   Modelling of Soil Bottom Processes 150
        References 150

9   Specific Construction Details 153

9.1   Earthen Reservoirs 153
9.2   The Spillway 154
9.3   The Floating Outlet Pipe 154
        9.3.1   Elements of the Outlet Pipe 154
        9.3.2   Details of the Design 155
        9.3.3   Details of the Joints 157
9.4   The Road at the Top of the Embankment 158

10   Nitrogen and Nitrification 159

10.1   Introduction 159
10.2   Ammonia in Wastewater Reservoirs 160
        10.2.1 General Aspects 160
        10.2.2 Inhibition by Ammonia of Algal Photosynthesis and Growth 160
10.3   Nitrifying Bacteria in Reservoirs 163
        10.3.1 Abundance of Nitrifying Bacteria 163
        10.3.2 Detection of Nitrifying Bacteria in Wastewater Reservoirs 163
        10.3.3 Nitrifying Bacteria in Anaerobic Environments 166
10.4   Unbalanced Nitrification of Ammonia 167
        10.4.1 Nitrification of Ammonia in Wastewater Reservoirs 167
10.5   Summary 169
        Acknowledgements 170
        References 171

11   Phytoplankton 173

11.1   Introduction 173
11.2   Wastewater Reservoir as an Algal Habitat 173
11.3   Composition and Seasonality of Phytoplankton 174
11.4   Algae-Bacteria Relationships 177
        References 179

12 Fauna 181
12.1 Introduction 181
12.2 Limnological Parameters 181
12.3 Faunal Limits and Definitions 182
12.4 Faunal Composition and Frequency 182
12.5 Aquatic Birds 185
12.6 Spatial and Temporal Distribution 185
12.7 Food Chains 188
12.8 The Effect of Grazing 188
12.9 The Reservoirs as an Environmental Opportunity 189
12.10 Unwanted Effects of the Fauna of the Hypertrophic Reservoirs ... 190
12.11 Fish Introduction for Manipulative Purposes 190
12.12 Faunistic Succession and Typology of the Hypertrophic Reservoirs ... 191
12.13 Conclusions 191
        Acknowledgements 192
        References 192

13 Odorous Compound 195
13.1 Introduction 195
13.2 Materials and Methods 196
        13.2.1 Sampling at the Na'an Reservoir 196
        13.2.2 Materials 197
        13.2.3 Analytical Methods 197
13.3 Results and Discussion 198
13.4 Conclusions 203
        Acknowledgements 203
        References 203

14 Degradation of Organosynthetic Pollutants 205
14.1 Introduction 205
14.2 Results and Discussion 206
        14.2.1 Organic Pollutants Identified 206
        14.2.2 Reservoirs in the Judean Hills 208
        14.2.3 Na'an Reservoir 212
        14.2.4 Givat Brenner Wastewater Treatment Facility 212
        14.2.5 Distribution of Organosynthetic Chemicals in Agricultural Soils
                Irrigated with Effluents from the Reservoirs 213
        14.2.6 Degradation Mechanisms 214
14.3 Conclusions 216
        Acknowledgements 216
        References 217

15 Trace Metals 219
15.1 Introduction 219
15.2 Materials and Methods 219
        15.2.1 The Studied Reservoirs and Water Balance 219
        15.2.2 Sampling 220

15.2.3 Analytical Methods 221
15.2.4 Potential Sources of Errors 222
15.3 Results and Discussion 222
15.3.1 Operation of the Reservoirs and Water Balance 222
15.3.2 Concentration of Metals in Inflow and Outflow 224
15.3.3 Input/Output Amounts 225
15.3.4 Losses by Seepage 227
15.3.5 Removal and Metal Budget 228
15.3.6 Sedimentation Rates 228
15.3.7 Sedimentation Quantities 230
15.3.8 Parameters Affecting the Removal of Metals in the Reservoirs 230
15.4 Conclusions and Recommendations 231
Acknowledgements 232
References 232

16   The Clogging Capacity of Effluents 233
16.1 Introduction 233
16.2 Drip Irrigation 233
16.3 Measurements of the Clogging Capacity of Effluents on Screen Filters 234
16.3.1 Filterability Index and Filter Specific Resistance 234
16.3.2 "ATMMIN" and "ATMLIT" 234
16.3.3 Clogging Time 236
16.4 Causes for Screen Filter Clogging 236
16.4.1 Causes Related to Filter Operation 238
16.4.2 Causes Related to the Effluent 238
16.5 Reservoir Management Alternatives 239
16.5.1 The Biological Approach 239
16.5.2 The Chemical Approach 241
16.5.3 The Operational Approach 243
16.6 Concluding Comments 244
References 245

17   Particle Characterization and Filtration 247
17.1 Introduction 247
17.2 Particle Characterization 247
17.3 Mechanical Filtration 253
17.4 Granular Media Filtration 256
17.5 General Conclusions 260
References 261

18   Satellite Remote Sensing of Water Quality 263
18.1 Introduction 263
18.2 SPOT Images of Wastewater Reservoirs 264
18.2.1 Deep Wastewater Reservoirs 264
18.2.2 Ground Truth 264
18.2.3 Reservoir Classification 264
18.2.4 Radiance Values 265

18.2.5 Chromatic Coordinates 266
18.2.6 Atmospheric Effects 268
18.3 The Spectral Radiance Model 269
18.3.1 The Optical Properties of Hypertrophic Wastewater Reservoirs 269
18.3.2 Model Structure and Inputs 269
18.3.3 The Optical Properties of the Water Constituents 271
18.3.4 Model Results 272
18.4 Principal Components Analysis 275
18.4.1 Principal Components 275
18.4.2 Spectral Interpretation of the PCA 276
18.4.3 Interpretation of Model Results by Principal Components 277
18.5 Summary and Conclusions 280
References 281

19   Experiences Outside Israel 283
19.1 Independent Experience 284
19.1.1 Canada and the United States 284
19.1.2 Spain 287
19.1.3 China 290
19.1.4 Germany 291
19.1.5 Italy 292
19.2 Wastewater Reservoirs Designed Abroad by Israeli Consultants 293
19.2.1 Chile – The Santiago Poniente Experimental Treatment Plant 293
19.2.2 Morocco – The Ben Slimane System 294
19.2.3 India – Wastewater Reservoirs in Gujarat State 295
References 300

Part II Case Studies 303

20   The Na'an Reservoir 305
20.1 Introduction 305
20.2 Materials and Methods 306
20.2.1 Structure and Operation of the Reservoir 306
20.2.2 Field Sampling and Measurements 307
20.2.3 Laboratory Analyses 308
20.3 Results 308
20.4 Discussion 313
Acknowledgements 325
References 325

21   The Geta'ot Reservoir 327
21.1 The Physical Environment 327
21.1.1 Introduction 327
21.1.2 Methods 327
21.1.3 Results and Discussion 330
21.1.4 Conclusions 338
21.2 Changes in Water Quality 339

21.2.1 Introduction 339
21.2.2 Methods 339
21.2.3 Results and Discussion 339
21.3 Conclusions 351
Acknowledgements 351
References 352

22   The Ma'aleh HaKishon Reservoir 355
22.1 Introduction 355
22.2 Description of the Reservoir 356
22.3 Operation of the Reservoir 356
22.4 The Biological Cycle in the Reservoir 357
22.5 Removal of Organic Material 357
22.6 Removal of Heavy Metals and Detergents 358
22.7 Nitrogen, Phosphorus and Potassium 358
22.8 Suspended Solids 359
22.9 Enteric Microorganisms 359
22.10 Discussion 360
References 360

23   The Negev Desert Reservoirs 361
23.1 Introduction 361
23.2 The Treatment Plant 361
23.3 The Soil-Aquifer Treatment (SAT) 362
23.4 Reservoirs of Reclaimed Water 363
23.5 Growth of Algae in the Reservoirs 364
23.6 Control of Algae by Grazing Fish 364
23.7 Control of Algae by Lowering Water Level and Temporary Desiccation
of the Reservoir Walls 366
References 367

24   The Enan Reservoir 369
24.1 Introduction 369
24.2 Methods 370
24.2.1 Study Site 370
24.2.2 Sampling Methods 372
24.3 Results 372
24.3.1 Water Temperature and Dissolved Oxygen 372
24.3.2 Phytoplankton and Primary Productivity 376
24.3.3 Effect of Impoundment on Water Quality 378
24.3.4 Fish Farming 380
24.4 Discussion 382
Acknowledgements 385
References 385

Index 389

Contributors

Prof. Aharon Abeliovich
Laboratory of Environmental Microbiology
Ben Gurion University, Sde Boker Campus
Sde Boker 84990, Israel
Phone: +972-7-6596830, Fax: +972-7-6596831

Prof. Avner Adin
Division of Environmental Sciences
The Hebrew University of Jerusalem
Jerusalem 91904, Israel
Phone: +972-2-6585550, Fax: +972-2-5635266

Prof. Yoram Avnimelech
Laboratory of Environmental Systems Management
Technion – Israel Institute of Technology
Haifa 32000, Israel
Phone: +972-4-8292480, Fax: +972-4-8221529

Dr. Yossi Azov
Department of Environmental Engineering
Technion – Israel Institute of Technology
Haifa 32000, Israel
Phone: +972-4-8228737, Fax: +972-4-8228737

Prof. Nissim Ben Yosef
Applied Physics Division
The Hebrew University of Jerusalem
Jerusalem 91904, Israel
Phone: +972-2-6584204, Fax: +972-2-5663878

M.Sc. Chaim Braude
Applied Physics Division
The Hebrew University of Jerusalem
Jerusalem 91904, Israel
Phone: +972-2-6584204, Fax: +972-2-566 3878

Dr. Hether Bromley
Division of Environmental Sciences
The Hebrew University of Jerusalem
Jerusalem 91904, Israel
Phone: +972-2-6585554, Fax: +972-2-6586155

Dr. Chanan Dimentman
Department of Evolution, Systematics, and Ecology
The Hebrew University of Jerusalem
Jerusalem 91904, Israel
Phone: +972-2-6585990, Fax: +972-2-6584574

Prof. Inka Dor
Division of Environmental Sciences
The Hebrew University of Jerusalem
Jerusalem 91904, Israel
Phone: +972-2-6585554, Fax: +972-2-6586155

Dr. Gabi Eitan
Division of Pollution Prevention and Wastewater Reuse
The Water Comission, P.O. Box 184
Jerusalem 91001, Israel
Phone: +972-2-6290152, Fax: +972-2-6290142

Dr. Jacob Eren († June 1998)
Mekorot Water Company of Israel
Tel Aviv 62100, Israel

Prof. Badri Fattal
Division of Environmental Sciences
The Hebrew University of Jerusalem
Jerusalem 91904, Israel
Phone: +972-2-6586161, Fax: +972-2-5635886

Dr. Eran Friedler
Juanicó & Friedler Mediterranean Ltd.
Ram On
M.P. Megido 19205, Israel
Phone: +972-6-6499241, Fax: +972-6-6499237

Dr. Sarig Gafny
Institute for Nature Conservation Research
Tel-Aviv University
Tel Aviv 69978, Israel
Phone: +972-3-6409813, Fax: +972-3-6407304

Prof. Avital Gasith
Institute for Nature Conservation Research
Tel-Aviv University
Tel Aviv 69978, Israel
Phone: +972-3-6409813, Fax: +972-3-6407304

Dr. Boris Ginzburg
Division of Environmental Sciences
The Hebrew University of Jerusalem
Jerusalem 91904, Israel
Phone: +972-2-6584191, Fax: +972-2-6586155

Dr. Jenia Gun
Division of Environmental Sciences
The Hebrew University of Jerusalem
Jerusalem 91904, Israel
Phone: +972-2-6584191, Fax: +972-2-6586155

Dr. Marcelo Juanicó
Juanicó & Friedler Mediterranean Ltd.
Ram On
M.P. Megido 19205, Israel
Phone: +972-6-6499241, Fax: +972-6-6499237

Dr. Adam Kanarek
Mekorot Water Company of Israel
Lincoln 9
Tel Aviv 62100, Israel
Phone: +972-3-6230568, Fax: +972-3-6230866

Dr. Drora Kaplan
Laboratory of Environmental Microbiology
Ben Gurion University, Sde Boker Campus
Sde Boker 84990, Israel
Phone: +972-7-6596830, Fax: +972-7-6596831

Prof. Ovadia Lev
Division of Environmental Sciences
The Hebrew University of Jerusalem
Jerusalem 91904, Israel
Phone: +972-2-6584191, Fax: +972-2-6586155

M.Sc. Anat Maoz
Division of Environmental Sciences
The Hebrew University of Jerusalem
Jerusalem 91904, Israel
Phone: +972-2-6585554, Fax: +972-2-6586155

Dr. Lea Muszkat
Department of Chemical Agroecology
Agricultural Research Organization, Volcani Center
Bet-Dagan 50250, Israel
Phone: +972-3-9683440, Fax: +972-3-968 3647

Prof. Francis Por
Department of Evolution, Systematics, and Ecology
The Hebrew University of Jerusalem
Jerusalem 91904, Israel
Phone: +972-2-6585990, Fax: +972-2-6584574

Dr. Michal Raber
Division of Environmental Sciences
The Hebrew University of Jerusalem
Jerusalem 91904, Israel
Phone: +972-2-6585554, Fax: +972-2-6586155

Chemist Roza Ravid
Department of Environmental Engineering
Technion – Israel Institute of Technology
Haifa 32000, Israel
Phone: +972-4-8292778, Fax: +972-4-8228898

Ing. Meir Romem
Agir Projects (Sealing) Ltd.
Kibbutz Gazit
M.P. Jeezrael 19340 Doar Na Yizreel, Israel
Phone: +972-6-6768916, Fax: +972-6-6768916

Dr. Hanna Schechter
Division of Environmental Sciences
The Hebrew University of Jerusalem
Jerusalem 91904, Israel
Phone: +972-2-6585554, Fax: +972-2-6586155

Dr. Yehuda Shevah
Tahal Consulting Engineers Ltd.
P.O. Box 11170
Tel Aviv 61111, Israel
Phone: +972-3-6924279, Fax: +972-3-4283

Prof. Hillel Shuval
Division of Environmental Sciences
The Hebrew University of Jerusalem
Jerusalem 91904, Israel
Phone: +972-2-6586161, Fax: +972-2-5635886

Dr. Benjamin Teltsch
Mekorot Water Company of Israel
The Central Laboratory
P.O. Box 610
Nazaret Illit 17105, Israel
Phone: +972-6-6500648, Fax: +972-6-6461883

Dr. Myriam Waldman
The Ministry of Science
P.O. Box 18195
Jerusalem 91181, Israel
Phone: +972-2-5819098, Fax: +972-2-5815595

Part I
Limnology and Technology

Chapter 1

Research and Development Policy

Yehuda Shevah · Myriam Waldman

1.1
Introduction

Wastewater is considered a sanitary and health issue dealt with normally by local and health authorities. The degree of treatment is usually dictated by these authorities who lay down the criteria for treatment and discharge of effluents to water bodies, according to quality standards and the diluting effect of the estuaries and other recipient water bodies. Current standards in Israel are for a biological treatment which reduces the *BOD* level to 20 mg l^{-1} and *TSS* to 30 mg l^{-1}. In addition, the local authorities are required to dispose of the effluents in a manner not harmful to the recipient water body. Treating wastewater to this standard is very costly and the Israeli Government has to assist local governments by providing grants and soft loans amounting to more than US$150 million yr^{-1}. In Israel the situation is further complicated because, due to water scarcity, wastewater is highly regarded as potential water that can be treated and recycled, while considering the associated problem of environmental impact and cumulative pollution processes. Thus, the treatment and safe disposal of treated effluents is combined with a policy for reclamation and reuse of effluents as an integral part of the available water resources.

This policy is supported by an R&D programme which is directed toward the following objectives:

- Treatment and storage of wastewater
- Securing a reliable and safe source of water for irrigation
- Safeguard of the environment, soil and groundwater resources
- Cost allocation and economic justification

1.2
Background Data

1.2.1
Water Scarcity and Wastewater Reuse

Israel has a surface area of about 20 000 km^2, a population of about 5.5 million (1995) and an annual water demand of about 2 000 million cubic meters (MCM), compared with a natural potential of about 1 600 MCM. This pressing imbalance has been the focus for the development and conservation of water resources, including wastewater reuse for irrigation (agriculture requires about 1 300 MCM yr^{-1} of water of varying quality; Table 1.1).

The total agricultural consumption has not changed significantly over the last 20 years, although there was a steady increase in the agricultural production output (Gelb and

Table 1.1. Water supply and demand in Israel – 1995 and projection for 2010 (*Source:* Water Commission and authors' estimates)

Water supply and sources	1995	Sector demand	1995	2010
Natural fresh	1 600	Urban and industrial	700	900
Natural brackish	180	Agriculture		
Domestic effluents	220	Fresh water	900	650
		Marginal resources	400	600
Total	2 000	Total	2 000	2 150

Table 1.2. Irrigated crops and irrigation water requirements, 1995 (*Source:* Ministry of Agriculture, Central Planning Authority, November 1996)

Crop	Area ($\times$ 1000 ha)	Water requirements (MCM yr^{-1})
Citrus and other tree plantations	71.3	576
Field crops	51.7	151
Cotton	24.5	120
Vegetables and flowers	39.0	212
Others	12.8	194
Total	199.3	1 253

Kislev 1982). The achievements made in water saving have been accompanied by a gradual change to crops which can be irrigated with low-quality effluents (Table 1.2).

Of the current water demand for irrigation, 70% can be supplied by water of non-potable quality, making the substitution of fresh water with domestic effluents a viable option that attracted farmers, economists and politicians alike. Accordingly, a national wastewater and reuse policy was formulated, assuming that it could be achieved without damaging the environment.

1.2.2
Wastewater Reclamation Plan

The major principles of the adopted policy include the increase of reclaimed wastewater used for irrigation from the current 17% to 34%, and increasing the utilized effluent quantities from the current 220 MCM to 420 MCM by the year 2010 (Table 1.3).

Two major courses have been adopted for the treatment and disposal of urban wastewaters, as shown in Table 1.4. For Greater Tel Aviv the treatment includes advanced tertiary treatment to the level of potable water by Soil Aquifer Treatment, while for other cities and towns, secondary treatment is advocated in a system which combines the treatment of wastewater with controlled irrigation schemes (Waldman and Shevah 1985). A major element in these systems is the large storage bodies which regu-

Table 1.3. Water supply for irrigation in 1995 and projection for 2010

Source	1995 (MCM)	(%)	2010 (MCM)	(%)
Fresh water	900	69	650	52
Brackish water	180	14	180	14
Domestic effluents	220	17	420	34
Total	1 300	100	1 250	100

Table 1.4. Wastewater Treatment and Reuse Plan (MCM yr^{-1}) 1995–2010 (*Source:* Shevah 1996)

Year	Potential wastewater	Reuse plan Tertiary effluents	Secondary effluents	Total
1995	440	100	120	220
2010	560	160	260	420

late between the relatively constant flow of wastewater during the whole year and the seasonal demand for irrigation (only six to eight months). The storage body is either a surface water body or an underground confined aquifer. Both types of storage provide an additional treatment step.

1.2.3
Surface Wastewater Reservoirs

These reservoirs are the key element in the sewage treatment system, employing conventional biological treatments in anaerobic sedimentation ponds, followed by a series of aerobic ponds with or without mechanical aeration means, and the reservoir which receives the secondary effluents all the year round, thus regulating between sewage production and irrigation demand (Dor 1986; Juanicó and Shelef 1994). Such treatment and reuse schemes proved to be a feasible technology and more than 200 reservoirs with a capacity ranging between 0.1 MCM and 12 MCM are already available turning in more than 150 MCM of sewage effluents.

1.3
R&D Support

1.3.1
The R&D Policy

The experience gained with the use of these systems required the initiation of a comprehensive R&D system that enabled the expansion of the programme without detrimental effects to the environment. Stabilization ponds which are relatively cheap to construct and operate are clearly more efficient in protecting the environment than the conventional treatment systems designed principally to reduce the content of organic matter and suspended solids. On the other hand, effluents contain many pathogenic microorganisms and toxic compounds of industrial origin which may have direct health hazards (WHO 1979), although no epidemiological evidence indicating a link between sickness and the use of effluents for irrigation was reported (Hillman 1985). However, because of the associated health and environmental risks, the policy that emphasizes reuse of wastewater on an unconventional scale could have enormous implications on public health and as such it had to be accompanied by a well-defined research and monitoring programme, built in as an integral part of the national wastewater reclamation programme (Waldman and Shevah 1985).

Several interdisciplinary research groups were established to study the biological, chemical and engineering aspects of reservoirs impounding sewage effluents. An elaborated research programme was established to standardize sampling procedures, analytical methods and data banks. Over the years, the research programme was expanded to include: a) oxygen balance and the development of odours; b) survival and regrowth of enteric micro-organisms; c) behaviour and characteristics of suspended solids; d) accumulation and release of nitrogen, phosphorous and organic carbon sediments at the bottom of the reservoirs; e) nitrogen balance, isolation of nitrifying organisms and determination of optimal conditions for nitrification; f) epidemiological and population studies; and g) mathematical modelling for the description of the processes which incur in the reservoirs.

1.3.2
Overview of the Ongoing R&D Programme

Despite their heterogeneity, wastewater reservoirs constitute a well defined category of aquatic ecosystem. They represent a unique hypertrophic aquatic ecosystem due to a combination of high organic loading, which makes possible the development of active biomass, and relatively deep water that lends importance to the limnological processes taking place. Being seasonal reservoirs, they function under non steady-state conditions which induce great fluctuations in the quality of effluents as evolved from the ongoing comprehensive research programme reviewed in the following.

1.3.2.1
Limnology

Several research studies have been conducted in order to study the effect of the long detention period on the quality of the effluents as expressed by various indicators, both chemical (Abeliovich 1980; Dor 1982) and biological (Kott 1980). The emphasis is on the concentration of organic matter expressed as *BOD* and *COD* and the removal of heavy metals and detergents (Eren 1987). Similarly, the effects of the low rate reactions which take place during the long retention time on the removal of the hard pollutants not affected by the treatment were studied (Abeliovich 1985), as well as the large populations of phytoplankton and zooplankton which extremely fluctuate and vary over the year (Dor et al. 1987a,b) and the considerable impact of large planktonic algae and metazoan species on the performance of the micro-irrigation systems which are highly susceptible to clogging (Teltch et al. 1991).

1.3.2.2
Health Aspects

The elimination of pathogenic bacteria, coliform, faecal streptococci and enteroviruses after long detention period followed by disinfecting at the end of the storage period was studied by Kott (1980). This study prompted the development of a research programme on viruses in water (Shuval 1984). This programme has been sustained through nearly two decades, and has generated much valuable information regarding the detection of cultivable enteroviruses and the common adenoviruses, and the need to examine their significance in the context of waterborne virus diseases (Marzuk et al. 1980) and exposure to aerosolized viruses from wastewater used for sprinkler irrigation (Fattal and Teltch 1982).

1.3.2.3
Nutrients Balance

Stored effluents contain a high concentration of ammonia which triggers the nitrification-denitrification process. The effects of a variety of environmental, physical, and chemical factors on the nitrogen cycle, and the transformation of ammonia to nitrite, nitrate, and free nitrogen was studied by Abeliovich (1993). Nitrogen and other elements' concentrations are also affected by the sedimentation process and the storage capacity of the particles comprising the bottom sediment. The nutrient fluxes to and from the sediment, and nutrients' balance was studied by Avnimelech and Wodka (1988) and Avnimelech (1989).

1.3.2.4
Heavy Metals

The pathways of metals found in municipal wastewater through the treatment and reuse process was studied by Kaplan et al. (1986), who assessed the heavy metal pollution hazard by measuring the concentration of dissolved and bound fractions, and the chelation process of zinc, cadmium, lead and copper. Similarly, the presence of micro pollutants, which cannot be totally removed and can therefore accumulate in the food chain were intensively investigated, using advanced instruments that facilitate their measurement in water (Rebhun et al. 1987).

1.3.2.5
Organic Micropollutants

The positive effect of solar photocatalytic mineralization of pesticides such as alkyl benzene and alkyl phenolic constituents, phathalates alkyl phosphates and di- and triethoxylates was assessed in eight reservoirs receiving domestic effluents (Muskat et al. 1995; see chapter in this book).

1.3.2.6
Monitoring Programme

For effective management of wastewater reservoirs, an effective monitoring programme is essential in order to control water quality, identify problematic areas from the pollution perspective, and to build a database which could be used at a later stage to measure the success or failure of the system. Data normally collected include *BOD*, *COD*, suspended solids, total solids, and chemical elements, namely: sodium, chloride, potassium and boron. Current monitoring programmes are however compounded by the large number and the complexity of the measured parameters and the related cost. Various attempts are made to achieve the maximum value from the available sampling and analytical facilities, while new monitoring techniques are being developed.

Standard analytical methods rely on sampling and laboratory analysis using specialized and highly expensive equipment. These cumbersome field sampling and laboratory testing methods can be replaced by rapid tests using accelerated immunofluorescence assays (Nasser et al. 1993; Bustyak et al. 1995). Automatic sampling is being introduced in order to overcome the deficiencies of grab samples which represent the quality of water at the time of sampling. To monitor the influx of solutes into the water body, a Multi-Level Sampler (MLS) device was developed to monitor water quality and to study the process of solutes transport (Ronen et al. 1987). A selective and highly sensitive chemical sensor system that will continuously monitor pollutants in water bodies is also being investigated using fiberoptic evanescent wave spectroscopy technique for continuous monitoring in real time and *in situ* (Schnitzer et al. 1990). An inexpensive portable system based on a spectrometer comprising a tunable diode laser and a sensitive infrared detector will be developed in the near future.

Remote sensing techniques are also investigated using reflected light measurements by satellite multispectral imaging (Dor and Ben-Yosef 1995; see chapter in this book), while mathematical modelling was developed for the description of the processes which incur in the reservoirs (Friedler 1993; see chapter in this book).

1.3.2.7
Socio-Economic Research

Early studies of the economic aspects of treatment and reuse systems were conducted by Horowitz and Shevah (1985) who analysed the economic benefits accruing to each of the three sectors involved in the treatment and utilization of domestic effluents for irrigation, namely: the local authorities, the government, and the agricultural sector. Reuse of effluents as an ultimate disposal system, results in significant savings to the local authorities, the party primarily responsible for sewage treatment and disposal, who can do without the energy-intensive treatment processes needed to eliminate the organic content of sewage. Similarly, the central government benefits from the generation of additional water sources, protection of the environment, and the postponement of large-scale desalination of sea water, which is the ultimate solution for water scarcity.

A high treatment level is generally favored by local authorities, but the additional costs should be considered as an input in the decision-making process, taking into consideration that reuse for irrigation results in adequate treatment while reducing both treatment and fertilizer cost. In addition to the quantity of effluents allocated to the farmer, further benefits derive from the nutrient content of wastewater which has an economic value (Haruvi and Sadan 1994).

The composite benefits of treatment and reuse of effluents for irrigation are obvious. However, the allocation of costs between the sectors involved presents certain difficulties (Selbst 1980). The determination of fair sharing of investment costs and equitable water pricing policies requires estimation of the direct benefits to each of the beneficiaries. Horowitz and Shevah (1985) proposed suitable cost allocation and pricing mechanisms, while also discussing the need for water rate adjustments, so that agriculture is subjected to realistic competition with other users.

1.4
Results Dissemination

The use of wastewater reservoirs as a central element in wastewater treatment and reuse has become a major feature of environmental importance, attracting a wide interest from researchers, engineers and administration personnel. The necessity and the importance of these reservoirs is reflected by their number, which increased from almost nil in 1970 to about 200 units in 1996. The reservoirs, initially a controversial environmental and public health issue, are today the subject of extensive investigations and routine monitoring programmes by scientists and environmental protection agencies.

The results of these activities are disseminated through international journals and reports submitted to the financing agencies. Workshops and seminars dedicated to wastewater reservoirs are frequently organized and attract a large audience. Furthermore, a management information system has been developed for the collection, data analysis, evaluation and dissemination of data on the biological, chemical, microbial and public health aspects of the wastewater reservoirs. Periodical reports, in the form of annual reports and up-to-date data when available, are produced by the Water Commission (Eitan 1995), providing a vital link between the supervisory bodies and the users.

1.5
Discussion

In Israel, the national policy proclaims reuse of effluents for irrigation as an inevitable outcome of the acute shortage of water and the need to conserve existing water resources. Further utilization of domestic effluents is recognized as the most economic course for developing the additional 200 MCM of water required by the turn of the next century (Shevah 1996). These needs are compounded by economic benefits associated with the use of effluents for irrigation, including release of freshwater for domestic use, increase of the irrigated area, recycling of nutrients and trace elements in the effluent, and reduction in the cost of water for irrigation (Haruvi and Sadan 1994).

Secondary effluents from sewage works have, normally, low concentrations of suspended solids and *BOD*. On the other hand, these effluents contain many pathogenic microorganisms and toxic compounds of industrial origin and as such they may have direct health hazards for field workers and the local population, and indirectly hazards for the general public (Hillman 1985). Health hazards derive from viruses, bacteria, and various protozoa (particularly those that are cyst-like), worms, and *Ascaris* larva which can withstand treatment processes and reach the soil to continue their life cycle (WHO 1973, 1979; Shuval 1984).

Therefore, the dilemma remains as to whether we should opt for the costly tertiary treatment and disinfection process, which reduces the health hazard to zero, while at the same time eliminating those very factors that make the use of treated effluents for irrigation economically attractive. It is to be noted that despite the potential health hazards, there is not yet sufficient epidemiological evidence indicating a link between sickness and the use of effluents for irrigation (Hillman 1985). Moreover, the positive effects of long detention periods in stabilization ponds was demonstrated indicating the high efficiency of stabilization ponds in producing safe effluents to be used for irrigation, especially with regard to the removal of pathogenic worms, intestinal nematodes, bacteria, and viruses (Feachem 1984).

Under these circumstances, the introduction of wastewater reservoirs – as extensively discussed in this volume – provides a public health safety measure and a low cost method for preventing environmental pollution. In addition, the utilization of secondary effluents results in energy saving because there is no need to eliminate the organic content of the effluent as discussed by Horowitz and Shevah (1985). Suitable cropping systems and organizational structures of the agricultural sector could have added advantages which effectively permit the integration of effluents in the supply of water for agriculture.

The inclusion of long detention reservoirs and the associated processes as an intermediate step between treatment and reuse is of vital importance in breaking the direct link between the unsafe raw wastewater and the treated effluents suitable for irrigation use. The large volume of the reservoirs compensates the short and long-term differences between water supply and demand, while providing a simple, robust and cost effective method.

Despite the many advantages of wastewater reservoirs and the extensive supporting research which was conducted, it is concluded that further supporting research and stringent monitoring are still a basic requirement, whereby the epidemiological and environmental research and monitoring are essential.

References

Abeliovich A (1980) Biology of wastewater effluent reservoirs. Report Biology Dept., Ben Gurion Univ., (in Hebrew)

Abeliovich A (1985) Biological treatment of chemical industry effluents by stabilization ponds. Wat Res 19(12):1497–1503

Abeliovich A (1993) The transformations of ammonia and the environmental impact of nitrifying bacteria. Biodegradation 3

Avnimelech Y (1989) Modelling the accumulation of organic matter in the sediment of a newly constructed reservoir. Wat Res 23(10):1327–1329

Avnimelech Y, Wodka M (1988) Accumulation of nutrients in the sediments of Ma'aleh HaKishon Reservoir. Wat Res 22(11):1437–1442

Bustyak S, Goldiner I, Nasser A, Goldstein L (1995) A homogeneous immunoflorescence assay based on dye-sensitized photobleaching. Anal Biochemisytry 225(1):127–134

Dor I (1986) Performance of a deep wastewater reservoir storing effluents for agriculture. Proc. Third Int. Conf. on Wastewater Reuse. Larnaca, Cyprus, pp 1–14

Dor I, Berend Y (1982) Research on Na'an reservoir. Limnological, sanitary and engineering character-istics of a deep effluent reservoir. Report Hebrew Univ., Jerusalem, (in Hebrew)

Dor I, Ben-Yosef N (1995) Monitoring effluent quality in the hypertrophic wastewater reservoirs using remote sensing. 3rd Int. Conf. on Appropriate Waste management Tech. for Dev. Countries. NEERI, Nagpur, India, pp 199–227

Dor I, Schechter H, Bromley H (1987a) Limnology of a hypertrophic reservoir storing wastewater efflu-ent for irrigation in Israel. Hydrobiologia 150:225–241

Dor I, Kalinsky I, Eren J, Diementman C (1987b) Deep wastewater reservoirs in Israel. 1. Limnological changes following self-purification. Wat Sci Tech 9:317–322

Eitan G (1995) Wastewater collection, treatment and reclamation survey, Water Commission, Ministry of Agriculture, Jerusalem (in Hebrew)

Eren J (1987) Changes in wastewater quality during long term storage. Proc. Water Reuse Symp. III, pp 1291–1300

Fattal B, Teltch B (1982) Viruses in wastewater aerosols. Env Int 7:35–38

Feachem R (1984) Sanitation and diseases. Health aspects of excreta and wastewater management. Pub-lished for the World Bank by John Wiley and Sons

Gelb E, Kislev Y (1982) Farmers' financing of agricultural research in Israel. Research Policy 11:321–327

Haruvi N, Sadan E (1994) Cost benefit analysis of wastewater treatment in the water scarce economy of Israel: A Case Study, Journal of Financial Management and Analysis 7/1:44–51

Hillman P (1985) Health aspects of reuse of treated wastewater for irrigation. FAO Regional Seminar on the Treatment and Use of Sewage Effluent for Irrigation. Nicosia, Cyprus

Friedler E (1993) Mathematical model of stabilization reservoirs. Dsc Thesis, Technion Haifa (in Hebrew)

Horovitz U, Shevah Y (1985) The supplementing and substitution of irrigation water with treated sew-age effluents. In: Scientific Basis for Water Resources Management. International Association of Hydrological Science, Public. N° 153, pp 185–193

Juanicó M, Shelef G (1994) Design operation and performance of stabilization reservoirs for wastewater irrigation in Israel. Wat Res 28(1):175–186

Kaplan D, Abeliovitz A, Ben-Yaakov S (1986) Chelating of heavy metals by organic compounds. In: Pro-ceeding of Chemicals in the Environment Conf., Lisbon

Kott Y (1980) The inactivation of microorganisms by ozonation. In: Current Research in Water Technol Report. National Council for Research and Development, Jerusalem

Marzuk Y, Goyal S, Gerba C (1980) Relationship of viruses and indicator bacteria in wastewater in Is-rael. Wat Res 14:1585–1592

Muskat L, Bir L, Feigelson L (1995) Solar photocatalytic mineralization of pesticides in polluted water. J Potochem Photobiol 87:85–88

Nasser A, Elkana Y, Goldstein L (1993) A nylon filter A-ELISA for detecting viruses in water. Wat Sci Tecnol 27:135–141

Rebhun M, Bracha C, Arvin E (1987) Behaviour and fate of organic micropollutants in the subsurface. Environmental Engineering and Water Resources Centre, Technion Report, Haifa

Ronen D, Magaritz M, Levy I (1987) In situ multi level sampler for preventive monitoring and study of hydrochemical profiles in aquifers. Groundwater Monit Rev 7:69–74

Selbst N (1980) Economic, social and administrative considerations in the use of wastewater. In: Shuval H (ed) Water quality management under conditions of scarcity: Israel as a case study. Academic Press, New York, USA, pp 243–269

Schnitzer I, Katzir A, Schiessl U, Riedl W, Tacke M (1990) Evanescent Field R spectroscopy using optical fibers and tunable diode lasers. Material Sci Eng B5:333

Shevah Y (1996) Israel national policy for wastewater treatment and reuse. Proc. Int. Conf. on Water Resources Management under Scarcity. Tel-Aviv, Israel

Shuval H (1984) Studies on viruses in the water environment in Israel. In: Melnick J L (ed) Enteric vi-ruses in water. Virol Monogram, 15, Kargar, Basel, Switzerland, pp 98–104

Teltsch B, Juanicó M, Azov Y, Ben-Harim I, Shelef G (1991) The clogging capacity of reclaimed wastewater: A new quality criterion for drip irrigation. Wat Sci Tec 24(9):123–131

Waldman M and Shevah Y (1985) Research and development as integral elements in water resour-ces management in Israel. In: Scientific Basis for Water Resources Management. IAHS Publ N° 153, pp 433–446

WHO Technical Report Series N° 517 (1973) Reuse of effluents: Methods of wastewater treatment and health safeguards. Report of a WHO Meeting of Experts, Geneva

WHO Technical Report Series N° 639 (1979) Human viruses in water, wastewater and soil. Report of a WHO Scientific Group, Geneva

Chapter 2

Wastewater Storage and Reuse for Irrigation in Israel

Gabi Eitan

2.1
Introduction

Water scarcity is a common concern in many arid and semi-arid countries. Shortage of water might be accompanied by groundwater contamination, public health threats, aquifer depletion and well abandonment. Solutions are expensive and may take a long time to achieve, but more immediate actions may also be undertaken: reduce water consumption through conservation, and increase the availability of water supply through wastewater reuse. The efforts invested in water reuse will preserve freshwater for potable uses.

There are several aspects that should be considered while discussing effluents reuse for irrigation:

1. Urban and industrial areas produce massive quantities of wastewater. In order to reuse effluents they should be conveyed from the urban conglomerates where it is produced, to the agricultural regions. In some places pipelines will stretch over long distances to get there.
2. The continuous flow of effluents all the year round must be regulated/stored in order to reuse it only during the irrigation season. Storage can be done by:
   – Open reservoirs to hold and keep the wastewater until it is used. Reservoirs are built along the pipe-lines, in the rural regions where land is available.
   – The recharge of an aquifer as an underground wastewater storage.
3. Wastewater must be treated before any disposal to the environment and irrigation is only one way of effluent disposal. Wastewater contains organic materials that are degraded to stable components while passing through the treatment plant. The required quality of the final effluents can be achieved by using the appropriate technology. Quality and quantity of industrial sewage should be considered when a treatment plant is designed for a city or region. Some components in industrial wastes can be toxic and should be eliminated in a pre-treatment installation inside the industrial area.
4. Pathogenic microorganisms (bacteria, viruses, protozoa and helminths) are present in domestic sewage and special actions are required to reduce their potential threat to public health.
5. The use of water invariably causes an increase in the concentration of soluble salts. The main ions are Sodium, Chloride and bicarbonates. These ions increase the total salt content (salinity) and the sodicity of the water. They are not removed during the sewage treatment process.

6. Macronutrient elements are found in reclaimed effluents especially phosphorus and nitrogen. The presence of these nutrients should be taken into consideration as a source of plant nutrients as well as groundwater contaminants.
7. Certain trace elements required by plants (B, Cu, Mn, Mo, Zn) may be present in effluents in excessive levels. Other elements may also contribute to toxicity hazards (Cd, Pb, Hg). Their origin is mainly from industrial wastes.

2.2
Wastewater Treatment

Wastewater is a mixture of water and other components (organic and inorganic, solubles and solids) carried in the flow. These components can be separated into six groups:

a   Coarse organic materials: plastics, paper, rugs etc.
b   Suspended, colloidal and soluble organic compounds: fats, proteins, carbohydrates, etc.
c   Inorganic suspended components: sand, gravel, etc. (grit)
d   Soluble inorganic nutrients, minerals and salts
e   Toxic compounds: heavy metals, micro-organic pollutants, solvents, detergents, etc.
f   Various types of microorganisms, part of them are pathogenic

Disposal of untreated wastewater to the environment can cause nuisances, health problems and contamination of surface and groundwater. As the main component of urban wastewater is water (99.9%) the purpose of treatment is to reclaim the water while removing all the mentioned waste materials that have been added to the water through domestic and industrial use. Wastewater that flows into a treatment plant passes several stages till the final effluents leave the installations with the required quality. Environmental and economic considerations dictate what quality level is reached. The main components of sewage treatment plants are pretreatment, treatment, and advanced treatment.

2.2.1
Pretreatment

Coarse screens are normally employed at the first treatment unit and are used to protect plant equipment from large solids and debris that may get into pumps and valves, mechanical aerators, etc., causing physical damage.

Grit chambers are designed to remove sand, gravel, cinders or other heavy solid materials that have a specific gravity substantially greater than those of organic matter in wastewater.

2.2.2
Treatment: Conventional Processes

Activated sludge is a biological treatment process in which biodegradable organic materials are utilized as food and degraded by micro-organisms. Treatment is accomplished by agitating and aerating a mixture of wastewater and biological sludge (micro-organisms). The mixed liquor is settled in settling tanks, the clear final effluents flows out while the sludge is returned to the process. There are many variations to the basic conventional activated sludge process, which is very flexible and can be adopted to almost any type of biologically biodegradable waste problem.

Trickling filters consist of a deep rock or plastic media bed overlying an underdrain system. Wastewater is applied over the media and allowed to trickle through it. A microbial biofilm grows on the media surface consuming organic matter from wastewater as food. The effluents that pass through the media flows to a settling tank that removes organic slime that fell of the media. The media need adequate ventilation in order to keep the aerobic population active.

Stabilization ponds are a popular technology for secondary treatment of wastewater. They are shallow, lined earthen basins and can be classified as aerobic, anaerobic or facultative, according to the nature of biological activity. Stabilization ponds may serve only small and medium-sized communities due to the excessive land requirements for large populations. The main advantage of these systems is the low cost of construction (when land is cheap and available) and operation.

2.2.3
Advanced Treatment

Advanced wastewater treatment and disinfection – if applied – will produce effluents of highest quality standards.

2.3
Sewage Treatment and Reuse in Israel

During the last twenty years, many water infrastructure projects have been constructed and developed all over the country: new water sources, wastewater treatment plants, open reservoirs and dams. The annual flow of sewage in Israel is about 390 MCM (Table 2.1). About 77% of this volume is treated using different types of technologies (Table 2.2). About 65% of the sewage flow is reused; part of it (23%) is treated by Soil Aquifer Treat-

Table 2.1. Annual sewage flow, treatment and reuse in Israel

	$m^3 \times 1\,000$ CM yr^{-1}	Total volume (%)
Domestic sewage	313 000	80
Industrial waste	68 000	18
Dairy farm waste	8 000	2
Total raw sewage	389 000	100
Sewage treated	298 100	77
Stored in reservoirs for irrigation	164 000	42
Stored in aquifer for irrigation	90 000	23
Total reclamation and reuse	254 000	65

Table 2.2. Population served and sewage volumes treated by different technologies

Treatment technology	N° of treatment plants	N° of cities and settle-ments connected	Population connected (millions)	Volume of treated waste-water ($\times 1000\ m^3\ yr^{-1}$)	Volume of ef-fluents reused and recharged ($\times 1000\ m^3\ yr^{-1}$)
Activated sludge or similar techniques	97	180	3.15	197 400	187 000
Stabilization ponds and other extensive plants	495	676	1.50	100 700	67 000
Percolation pits and others and untreated	–	458	1.05	90 900	0
Total	592	1 314	5.70	389 000	254 000

Table 2.3. Quality of raw sewage quality in Israel (based on samples from several different places)

Parameter	EC^a	SAR	Na^b	K^b	Cl^b	B^b	$N\ Kjel^b$	$N\text{-}NH_4{}^b$	P^b	BOD^b	COD^b	TDS^b	TSS^b
Minimum	16	0.7	30	1	50	0.1	4	10	1	55	64	472	30
Maximum	9 400	50	600	200	1 000	1.6	160	150	70	2 900	5 600	6 800	2 800
Average	1 750	4	165	25	240	0.5	60	45	15	360	650	1 000	320

[a] $(\mu mhos\ cm^{-1})$;
[b] $(mg\ l^{-1})$.

ment (SAT) in an isolated aquifer in the dunes south of Tel Aviv and pumped south for irrigation in the Neveg desert; the other part (42%) is kept in open reservoirs for several months before being used for irrigation. Table 2.3 presents typical sewage quality in Israel, based on data from different places all over the country (Eitan 1995).

In Israel, 380 open reservoirs with a total storage capacity of about 250 MCM (million cubic meters) have been constructed over the past 50 years (Figs 2.1 and 2.2). These reservoirs increased the water availability for the intensive agriculture that developed in the country including the arid regions. Most reservoirs (60%) hold wastewater effluents or a combination of effluents with winter floods and other marginal waters. The rest (40%) are filled only with rain water where there is enough flow in the river beds. Only a few are used to regulate the supply of potable water. Reservoir volumes vary from less than 30 000 m³ to more than 10 million m³, but most of them are between 200 000 and 800 000 m³ (Fig. 2.3). The most effort on reservoir construction was applied between 1970 and 1990. The total capacity constructed during that period was more than 190 MCM, and new reservoirs continue to be constructed nowadays. Holding water in open reservoirs creates a unique ecological system that enables flora and fauna to flourish. The fact that these reservoirs are fed with wastewater effluents – or a combination of several sources of marginal waters – makes them very special. This ecosystem and several phenomena related to it are discussed later in this book.

Fig. 2.1. Distribution of both wastewater and freshwater reservoirs in Israel

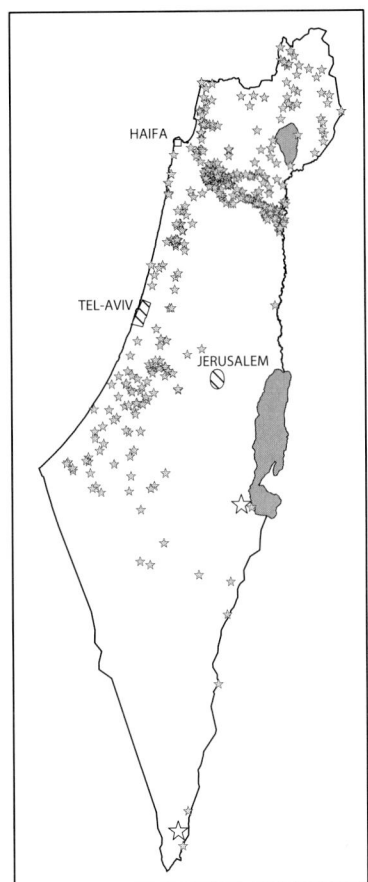

Fig. 2.2. Open reservoirs that have been built during the last 50 years

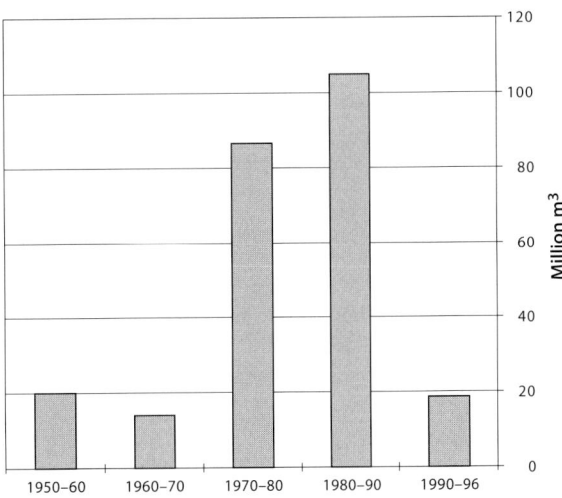

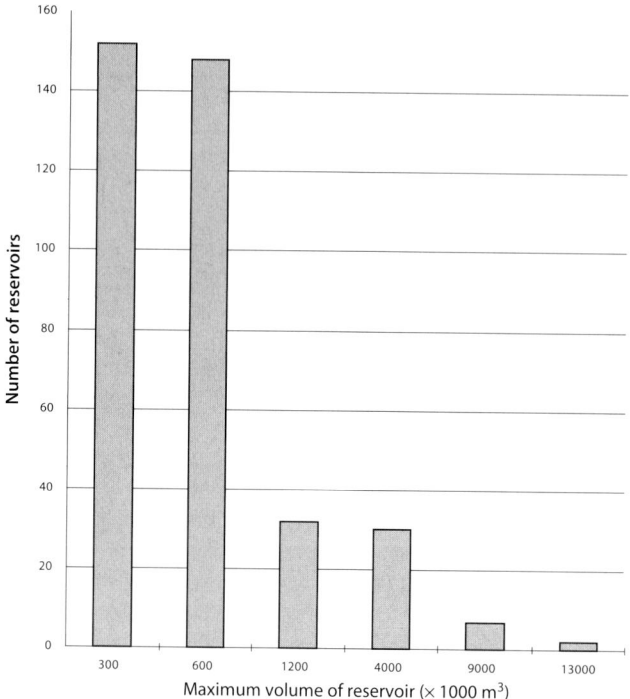

Fig. 2.3. Open reservoirs by size

2.4
Main Water Reclamation Projects in Israel

The Greater Tel Aviv wastewater treatment plant and reclamation project is located south of the Tel Aviv urban area, connecting more than 20 cities and townships and serving a population of about 1.4 million. The project receives more than 100 MCM yr^{-1} of sewage which is treated according to high standards. Effluents are pumped to sand basins not far from the treatment plant and percolate to an isolated aquifer below for SAT. There are several wells that draw water from that aquifer to use it for unrestricted irrigation in agricultural fields located about 100 km south of the site.

The Haifa wastewater treatment plant and reclamation project serves a population of about 0.5 million living in Haifa and 18 nearby towns, producing about 30 MCM yr^{-1} of wastewater. The treated effluents are conveyed by pumping through a 28 km main to the Kishon open reservoirs with a capacity of 12 MCM which is located in the centre of a prosperous agricultural valley.

The Jeezrael Valley Project is a unique system that uses semi-intensive technology for sewage treatment (anaerobic ponds followed by low-energy aerated lagoons) combined with long detention in numerous open reservoirs. There is a central control for the operation, treatment and reuse of the effluents. The project serves a number of medium and small towns and agricultural settlements with a total population of about 300 000.

Most small rural settlements – less than 1 000 people – have extensive sewage treatment systems. In some places effluents are used directly from the oxidation ponds, while in other places effluents are stored in a reservoir that receives also floods during the winter.

2.5
Effect of Irrigation with Treated Sewage Effluent on Soil, Crops and Environment

Pollution of soil, crops, and groundwater is a major concern under modern intensive agricultural conditions. Water flowing below the root zone carries various contaminants to the unsaturated zone and eventually further to the aquifer. Migration of chemicals through the soil depends not only on conventional subsurface fluid flow but also upon a number of interactive processes by which their concentrations may be attenuated (e.g., adsorption, exchange, precipitation, complexation and chelations, microbial activity and plant uptake). An understanding of the transport of nutrients in the soil is important for a good irrigation and fertilization management, particularly when using sewage effluents (Feigin et al. 1991).

Salts are added to water during domestic and industrial use. Thus, sewage has a salt content higher than that of the original freshwater. Salts accumulate in soil because whereas the water is taken up by the plant roots or evaporates, most of the salts remain in the soil (Bouwer 1982). Excessive salt accumulation in the soil can only be prevented through leaching with surplus irrigation or by rainfall. Thus, leaching may be considered the key to the successful use of irrigation water containing high solute levels (Boaz et al. 1976).

Chloride causes specific damage in some crops, mainly in woody perennial fruit trees and ornamentals. The common initial symptom is bronzing of the leaves of citrus trees. One of the causes of chloride injury is thought to be nutritional imbalance which is most likely to be of importance for very sensitive crops such as avocado.

Sodium can cause nutritional imbalance through impaired Ca or K nutrition, but specific toxicities are usually limited to fruit trees. Sodium also affects the physical condition of the soil through clay particle swelling and dispersion. This in turn causes a reduction in soil porosity, hydraulic permeability, infiltration rate, and aeration. When dispersion results in migration of clay particles to deeper soil horizons, the damage may be irreversible (Bolt 1978).

Boron is an essential nutrient for most crop plants, but toxicity appears when its concentration in the irrigation water exceeds 0.5–1.0 mg l^{-1}. The range of optimum Boron concentration between deficiency and toxicity is thus rather narrow (Keren and Bingham 1985).

Nitrogen is added in considerable amounts to the soil in irrigation with secondary treated effluents, depending on the content of N in the effluent and the volume of water applied. Consequently, the actual quantity added to crops varies widely and is greatly affected by climatic conditions, soil and effluent properties, cropping and irrigation management (Broadbent and Reisenauer 1985). Since effluent N eventually enters the N cycle in soil, irrigation with sewage effluent affects nitrogen uptake by the crop, the level of available N in the soil, gaseous N losses and N leaching below the rooting depth. Irrigation management determines the distribution of N application over time. It is clear that appropriate fertilizer management in effluent-irrigated soils is essential to obtain the maximum benefit from the presence of N in the effluents, and to prevent the application of excessive amount of N (Feigin et al. 1978, 1979). Excessive N affects crop growth and yield (Bielorai et al. 1984). Enhanced vegetative growth as indicated by plant height and reduction in yield was found in cotton plants to which excessive amounts of N were added through sewage effluents or fresh water amended with fertilizer. The nitrogen level in plant tissue also affects carbohydrate metabolism: its high content reduces the carbohydrate concentration in the plant and its storage organs.

Phosphorus is also added in large quantities to obtain enhanced crop yields. Secondary sewage effluent often has a high phosphorus content, making it an important source of P for wastewater irrigated soils (Hook 1983). Effluent can partially – or even completely – replace other sources for P, and using effluent for irrigation results in great savings and has an important economic role. The amount of P added to the soil through sewage effluent irrigation is usually not excessive, however excessive levels of P may result in nutrient imbalance such as Cu, Fe and Zn deficiencies (Kardos and Hook 1976).

Potassium is taken up by crops in large quantities. Since the level of available K in the soil is often below that necessary for production of high crop yield, fertilizers and organic material (e.g., manure) are often added. K is also found in significant concentrations in treated wastewater.

Trace elements are important components of sewage and found even in higher concentrations in sewage sludge: Silver, Arsenic, Barium, Cadmium, Cobalt, Chromium, Copper, Mercury, Manganese, Molybdenum, Nickel, Lead, Selenium, Tin, Vanadium and Zinc. Trace elements participate in many of the biochemical processes that affect plants, animals and humans. Some of them are essential for plant development and growth, some are required by animals. Some trace elements are hazardous to organisms, while the toxicity of others is low (Chaney 1989). The quantity of trace elements added to soil by means of sewage effluent irrigation is generally low. Despite this low content, the presence of trace elements is an important characteristic of secondary effluent when massive use of wastewater for irrigation and groundwater recharge is performed. Although it may take a century till levels of trace elements in wastewater irrigated soils reach the currently proposed upper limit, the potential harmful effect in the future should not be ignored. The movement of trace elements in the soil profiles has received considerable attention, since even a slow transport through soil and subsoil material may result in an increased content of trace elements in groundwater.

References

Bielorai H, Vaisman I, Feigin A (1984) Drip irrigation of cotton with treated municipal effluents. 1: Yield response. J Environ Qual 13:231–234

Boaz M, Hausenberg I, Pozin Y (1976) Salinity survey in Israel. In: Dregne HE (ed) Managing saline water for irrigation. Proc Int Salinity Conf, Texas Tech College, Lubbock, pp 388–399

Bolt G (1978) Transport and accumulation of soluble soil components. In: Bolt GH, Bruggenwert MGM (eds) Soil chemistry. A: Basic elements. Elsevier, Amsterdam, pp 126–140

Bouwer H (1982) Wastewater reuse in arid areas. In: Middlebrooks EJ (ed) Water reuse. Science Publisher, Ann Arbor, pp 137–180

Bouwer H, Idelovitch E (1987) Quality requirements for irrigation with sewage water. J Irrig Drain Div ASCE 113:516–535

Broadbent F, Reisenauer H (1985) Fate of wastewater constituents in soil and groundwater: nitrogen and phosphorus. In: Pettygrove G, Asano T (eds) Irrigation with reclaimed municipal wastewater – a guidance manual. Lewis, Chelsea, pp 12.1–12.16

Chaney R (1989) Toxic element accumulation in soils and crops: Protecting soil fertility and agricultural food-chains. In: Bar-Yosef B, Barrow N, Goldshmid J (eds) Inorganic contamination of the vadose zone. Springer Ecol Stud 74:140–158

Eitan G (1995) Wastewater treatment and reuse in Israel – 1994. Survey published by the Water Commission – Jerusalem, 70 pp

Feigin A, Bielorai H, Dag Y, Kipnis T, Giskin M (1978) The nitrogen factor in the management of effluent-irrigated soils. Soil Sci 125:248–254

Feigin A, Bielorai H, Shalhevet J, Kipnis T, Dag J (1979) The effectiveness of some crops in removing minerals from soils irrigated with sewage effluent. Prog Wat Technol 11:151–162

Feigin A, Ravina I, Shalhevet J (1991) Irrigation with treated sewage effluent. Springer-Verlag. Adv Ser Agr Sci 17:224

Hook J (1983) Movement of phosphates and nitrogen in soil following application of municipal wastewater. In: Neldon DW, Erick DE, Tanji KK (eds) Chemical mobility and reactivity in soil systems. Soil Sci Soc Am Madison Spec Publ 11:241–255

Kardos L, Hook J (1976) Phosphorus balance in sewage effluent treated soils. J Environ Qual 5:87–90

Keren R, Bingham F (1985) Boron in water, soils, and plants. Advances in soil science. Springer, Berlin Heidelberg New York, pp 229–276

Chapter 3

Health and Treatment Requirements for Wastewater Irrigation

Hillel Shuval · Badri Fattal

This chapter draws partially on material from the author's studies for the UNDP/World Bank Water and Sanitation Program (Shuval 1990; Shuval et al. 1986).

3.1
Introduction

Recycling and reuse of wastewater in agriculture can be a highly effective strategy for developing a sustainable water resource in water short areas, nutrient conservation, and environmental protection. However, it is essential to understand the health risks involved and to develop appropriate strategies for the control of those risks, including microbial guidelines for effluent quality, regulation of the types of crops to be irrigated by various irrigation techniques, and the treatment of the wastewater to an appropriate degree so as to control potential health risks both to the farmers and the consumers of crops from pathogenic microorganisms in the wastewater stream. This chapter will deal with the above questions with particular reference to the role of low cost hypertrophic stabilization ponds and reservoirs for wastewater storage and reuse.

3.2
The Development of Wastewater Recycling and Reuse

3.2.1
Early Major Wastewater Irrigation Projects

With the publication of a report of the First Royal Commission in England in 1865, land treatment (or what is now known as recycling and reuse) became one of the principal means of sewage disposal. Sewage farms were established in Edinburgh, London, Manchester, and other major cities of the United Kingdom. By 1875 there were approximately 50 land treatment sites in Britain. Widespread wastewater irrigation also became popular in other parts of Europe during the late 1800s and early 1900s. Paris, for example, had sewage farms as early as 1868, and by 1904 the great intercepting sewers of Paris had stopped discharging into the Seine altogether. All the dry weather wastewater flow was applied to sewage farms, which by then had a total area of 5 300 ha.

The city of Berlin established its first sewage farm in 1876. By 1910 Berlin had about 17 200 ha devoted to sewage farming and was treating about 310 000 m³ of wastewater per day. Melbourne, Australia established its first large sewage farm – Werribbee Farm – in 1897 and grazed sheep and cattle on the grass grown in the sewage-irrigated plots.

This well-managed farm is still in operation and today irrigates large areas of pasture with the effluent from its stabilization pond system. In 1904 planned sewage farming was established in Mexico City. An irrigation district was organized in the end of the Valley of Mexico where the city's untreated wastewater was used to irrigate large areas. The programme area has since been expanded under careful government control, and in 1997 approximately 100 000 ha were irrigated with wastewater. Sewage irrigation was also under way in the United States in those early years. It was practised as early as 1871 in Lenox and Worcester, Massachusetts, and in 1876 near Augusta, Maine. According to Fuller (1912), by 1904 the country had 14 municipal sewage farms or broad irrigation projects serving a population of about 200 000, and a number of institutional plants were in operation. Early municipal sewage irrigation projects near Chicago and Los Angeles had to be abandoned, however, because of the rapid growth of the two cities and their suburbs in the direction of sewage-irrigated lands. Apparently the health authorities intervened when the odour from these sites became a nuisance (Fuller 1912).

As in the United States, many of the early large irrigation and sewage farm projects in Europe were abandoned because urban development encroached upon the sewage farm areas. The problems with odour and concerns about public health – particularly fears about the possible transmission of disease from vegetable crops irrigated with raw sewage – were largely responsible for the decline of sewage farming. Another disadvantage in temperate areas with plentiful rainfall was that, with the cessation of sewage irrigation during heavy rainy seasons, raw or partially treated wastewater was frequently discharged into neighbouring streams, or crops were harmed by oversaturation of the irrigated land areas. This was only a minor problem in the more arid and western areas of the United States, however, and thus sewage farming has continued there up to the present.

By 1912 the trend away from sewage farming was already evident. According to Fuller (1912):

"…the present outlook is that broad irrigation or sewage farming is decidedly on the wane with little prospects of adoption even in the arid districts, except perhaps for an occasional project where local conditions are unusually favourable."

Eventually, sewage farming was almost completely abandoned in most areas of the highly urbanized industrial countries of the Western world. All this changed after World War II, however, when agricultural, scientific, and engineering interest in wastewater reuse was revived in both the industrialized and the developing countries.

3.2.2
Present Status of Interest in Wastewater Recycling and Reuse

After 1945 wastewater treatment and disposal through land application gained increasing attention as a mean of preventing river pollution and increasing water resources in arid and semiarid areas. The arid developing countries were particularly interested in utilizing wastewater for agricultural development. Such countries have few flowing streams with sufficient capacity to serve as natural repositories, even for well-treated wastewater effluent. Thus, wastewater reuse in agriculture provided almost the only feasible, relatively low-cost method for sanitary disposal of municipal wastewater that minimized pollution of the region's waterways. These factors, coupled with rapid urban growth and the need to increase agricultural production, made sewage farms attractive to the agricultural community and municipal planners.

Furthermore, the health regulations developed by the State of California (to be discussed in detail later in this chapter) helped to reestablish the feasibility of wastewater reuse in agriculture in the western part of the United States. Soon thereafter a similar trend began in many of the rapidly developing countries faced with water shortages and with insufficient waterways to properly dilute and dispose of municipal wastewater. The early strict California standards that were drafted to provide essentially a "zero risk" basis for reuse were copied in other areas of the United States and in countries throughout the world. However, those standards were not based on sound epidemiological criteria and have been accepted more on faith than on scientific evidence.

A survey of wastewater reuse practices in developing countries carried out by the World Bank-UNDP Program on Water and Sanitation has estimated that some 80% of the wastewater flow from urban areas in developing countries is currently used for permanent or seasonal irrigation (Gunnerson et al. 1985). In many areas (i.e., Santiago, Lima, Teheran, Bombay, and Kabul) untreated wastewater flows through channels and/or rivers to adjacent areas where it is diverted by subsistence farmers to small plots of unregulated vegetable and salad crops grown for nearby urban markets. The public health risks in such uncontrolled use of raw wastewater for irrigation are obvious.

In other areas, government-controlled irrigation projects divert partly or fully treated wastewater to farming operations organized for the irrigation of controlled crops. Examples of such operations include a 2 800 ha greenbelt irrigated with treated wastewater at Khartoum; a 100 000 ha area of restricted grain and fodder crops irrigated with wastewater near Mexico City; 10 000 ha Werribbee Farm at Melbourne, where 50 000 sheep and 20 000 cattle graze on pasture irrigated with well-treated stabilization pond effluent; and a carefully controlled 9 000 ha farm near Kuwait City which is irrigated with well-treated wastewater effluent. In Israel by 1996 some 65% of the urban wastewater was being recycled in conformity with strict health regulations. The Tel-Aviv "Dan Region Water Reclamation Project" treats and recycles the wastewater of some one million persons to the standard required for unrestricted crop irrigation in the southern semi-arid Negev Region of Israel (Eitan 1994).

Most other cities in developing countries fall somewhere between being totally unregulated and being carefully and effectively controlled. Many governments have recognized the importance of wastewater recycling through irrigation and have developed national wastewater reuse programmes as part of their water resources management policy. Such programmes have been established in Tunisia, Saudi Arabia, Israel, India, the Republic of South Africa, and some states in the United States (i.e., California, Arizona, and Florida).

Although wastewater reuse has been practised more widely in developing countries over the past forty years, much of it is unplanned and uncontrolled and poses a threat to public health. These risks must be fully understood and appropriate measures taken to provide technically feasible and economically attractive solutions so that the public can reap the full benefits of wastewater reuse without suffering harmful effects. Only then can such a practice become a truly successful development policy.

3.3
Public Health Risks Associated with Wastewater Irrigation

3.3.1
Pathogenic Microorganisms in Wastewater

Enteric diseases of the human intestinal tract are caused by many of the pathogenic microorganisms, including bacteria, viruses, protozoa, and helminths. These diseases are transmitted when pathogenic microorganisms are excreted to the environment by an infected person (the initial "host"), transported by a suitable vector, such as contaminated water or food, and ingested by another susceptible human "host". Large numbers of the disease-causing pathogens are excreted in the urine and faeces of infected individuals, and these pathogens contaminate the wastewater which is disposed of into the environment.

The concentration of pathogenic bacteria and viruses in the faeces of an infected person usually ranges from 1 million to 100 million (10^6–10^8) organisms per gram of faeces. The concentration of protozoa is about 10–100 thousand (10–10^5) per gram of faeces, and the concentration of encysted helminth eggs ranges from 100 to 10 000 (10^2–10^4) per gram of faeces. The wastewater stream of a community carries the full spectrum of pathogenic microorganisms excreted by the diseased and infected individuals living in that community. The calculated concentration of pathogenic microorganisms in the wastewater stream is many millions per litre for bacteria, thousands per litre for viruses, and a few hundred per litre for some of the helminth eggs (Feachem et al. 1983; Shuval et al. 1986).

3.3.2
Survival of Pathogens in the Environment

In order to infect a susceptible individual, pathogens must be able to survive in the environment (i.e., in water, soil, or food) for a period of time and they must be ingested in a sufficiently high number. Factors that affect the survival of pathogens in soil include antagonism from soil bacteria, moisture content, organic matter, pH, sunlight, and temperature. Excreted enteric pathogens such as bacteria, viruses, protozoa, and helminth eggs do not usually penetrate undamaged vegetables but can survive for long periods in the root zone, in protected leafy folds, in deep stem depressions, and in cracks or flaws in the skin.

Data from numerous field and laboratory studies have made it possible to estimate the persistence of certain enteric pathogens in water, wastewater, soil, and on crops (Fig. 3.1). For example, it appears that *Campylobacter* may survive in soil or on crops for only a few days, whereas most bacterial and viral pathogens can survive from weeks to months. The highly resistant eggs of helminths such as *Trichuris*, *Taenia*, and *Ascaris* can survive for nine to twelve months, but their numbers are greatly reduced during exposure to the environment.

Field studies in Israel have demonstrated that enteric bacteria and viruses can be dispersed for up to 730 m in aerosolized droplets generated by spray (sprinkler) irrigation, but their concentration is greatly reduced by detrimental environmental factors such as sunlight and drying (Teltsch et al. 1980; Applebaum et al. 1984; Shuval

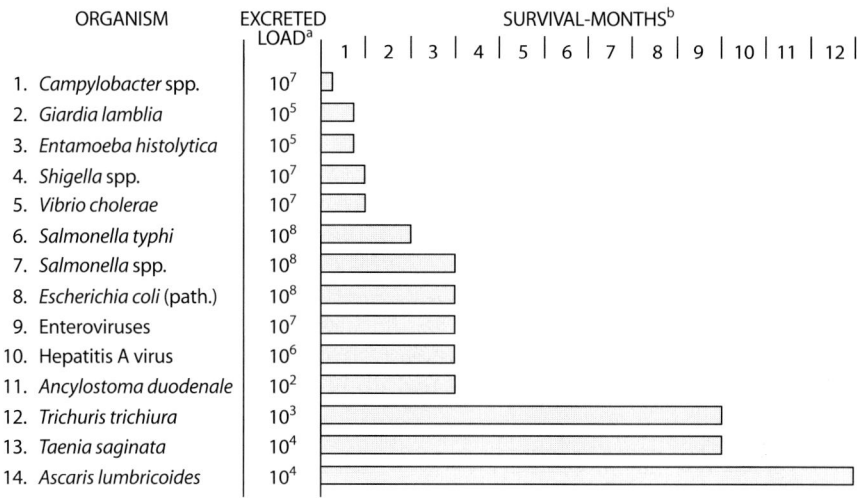

ORGANISM	EXCRETED LOAD[a]	SURVIVAL-MONTHS[b]
1. *Campylobacter* spp.	10^7	
2. *Giardia lamblia*	10^5	
3. *Entamoeba histolytica*	10^5	
4. *Shigella* spp.	10^7	
5. *Vibrio cholerae*	10^7	
6. *Salmonella typhi*	10^8	
7. *Salmonella* spp.	10^8	
8. *Escherichia coli* (path.)	10^8	
9. Enteroviruses	10^7	
10. Hepatitis A virus	10^6	
11. *Ancylostoma duodenale*	10^2	
12. *Trichuris trichiura*	10^3	
13. *Taenia saginata*	10^4	
14. *Ascaris lumbricoides*	10^4	

[a] Typical avg. number of organism/gm feces
[b] Estimated average life of infective stage at 20°–30°C

Fig. 3.1. Persistence of selected enteric pathogens in water, wastewater, soil, and on crops (based on data from Feachem et al. 1983)

et al. 1988; Shuval et al. 1989a). Thus, most excreted pathogens can survive in the environment long enough to be transported by the wastewater to the fields and to the irrigated crops. The contaminated crops eventually reach the consumer, although by then the concentration of pathogens is greatly reduced. The rapid natural die-away of pathogens in the environment is discussed in a later section as it is an important factor in reducing the health risks associated with wastewater reuse.

3.3.3
Proposed Model to Predict the Relative Effectiveness of Pathogens in Causing Infections Through Wastewater Irrigation

Theoretical analysis suggests that a number of epidemiological factors determine whether various groups of pathogens will cause infections in humans through wastewater irrigation. We have developed a model to evaluate the empirical epidemiological data and to formulate control strategies.

Table 3.1 summarizes the epidemiological characteristics of the main groups of enteric pathogens as they relate to the five factors that influence the transmission and degree of illness disease resulting from wastewater irrigation. This summary provides a simplified theoretical basis for ranking the groups of pathogens according to their potential for transmitting disease through wastewater irrigation. On this basis, it appears that the helminth (worm) diseases are the most effectively transmitted by irrigation with raw wastewater because they persist in the environment for relatively long periods; their minimum infective dose is small; there is little or no immunity against

Table 3.1. Epidemiological characteristics of enteric pathogens vis-a-vis their effectiveness in causing disease through wastewater irrigation

Pathogen	Persistence in environment	Minimum infective dose	Immunity	Concurrent routes of infection	Latency/soil development stage
Viruses	Medium	Low	Long	Mainly home contact, food and water	No
Bacteria	Short/medium	Medium/high	Short/medium	Mainly home contact, food and water	No
Protozoa	Short	Low/medium	None/little	Mainly home contact, food and water	No
Helminths	Long	Low	None/little	Mainly soil contact outside home and food	Yes

them; concurrent infection in the home is often limited; and latency is long and a soil development stage is required for transmission.

In contrast, the enteric viral diseases should be least effectively transmitted by irrigation with raw wastewater in developing countries with low levels of sanitation in the home, despite their small minimum infective doses and ability to survive for long periods in the environment. Due to poor hygiene in the home, and the prevalence of concurrent routes of infection in some areas, most of the population has been exposed, and acquired immunity, to most of the enteric viral diseases as infants. Most enteric viral diseases impart immunity for life, or at least for very long periods, so that they are not likely to reinfect individuals exposed to them again, for example, through wastewater irrigation. The transmission of bacterial and protozoan diseases through wastewater irrigation lies between these two extremes.

3.4
Evaluation of the Epidemiological Evidence of Human Health Effects Associated with Wastewater Irrigation

3.4.1
Early Microbial Guidelines for Wastewater Irrigation Lack Epidemiological Basis

The strict health regulations governing wastewater reuse that have been developed in the industrial countries over the past 60 years, such as those of the Department of Health of the State of California, which requires an effluent standard of 2 coliforms/100 ml for irrigation of crops eaten uncooked, have been based to a great extent on scientific data indicating that most enteric pathogens can be detected in wastewater and that they can survive for extended periods in wastewater-irrigated soil and crops (Fig. 3.1). Most health authorities have concluded that, because pathogens can survive long enough to contaminate crops, even if their numbers are very low, they pose a serious risk to public health. However, these regulations were formulated at a time when sound epidemiological evidence was rather scanty. As a result, policy makers used the cautious "zero risk" approach and introduced very strict regulations that they hoped would protect the public against the potential risks thought to be associated with wastewa-ter reuse. Most industrial countries were not concerned that these regulations were overly restrictive because the economic and social benefits of wastewater reuse were of only marginal interest.

One of the goals of our studies for the UNDP/World Bank (Shuval et al. 1986; Shuval 1990) as described in this chapter was to reevaluate all the credible, scientifically valid, and quantifiable epidemiological evidence of the real human health effects associated with wastewater irrigation. Such evidence is needed to determine the validity of cur-rent regulations and to develop appropriate technical solutions for existing problems.

The following evaluation is based on available scientific papers published in rec-ognized journals and on numerous unpublished government reports, university the-ses, and private papers obtained during an intensive worldwide search carried out with the help of international and national agencies and individuals. Over 1 000 documents, some more than 100 years old, were examined in the course of this study, but few of-fered concrete epidemiological evidence of health effects. Most of them based their conclusions on inference and extrapolation. Nonetheless, about 50 of these reports provided enough credible evidence based on sound epidemiological procedures to make a detailed analysis useful. Those studies are reviewed in detail in the UNDP-World Bank report on which this chapter is partially based (Shuval et al. 1986). Our general conclusions of some about the more pertinent studies are presented below.

3.4.2
Population Groups Consuming Vegetables and Salad Crops Irrigated with Raw Wastewater

In areas of the world where the helminths (worm) diseases caused by *Ascaris* and *Tri-churis* are endemic in the population, and where raw, untreated wastewater is used to irrigate salad crops and/or other vegetables generally eaten uncooked, the consump-tion of such wastewater-irrigated salad and vegetable crops may lead to infection. Khalil (1931) demonstrated the importance of this route of transmission in his pio-neering studies in Egypt. Similarly, a study in Jerusalem (Shuval et al. 1984) provided strong evidence that massive infections of both *Ascaris* and *Trichuris* may occur when salad and vegetable crops are irrigated with raw wastewater. The disease almost to-tally disappeared from the community when raw wastewater irrigation was stopped (Fig. 3.2). Two studies from Darmstadt, Germany (Krey 1949; Baumhogger 1949) pro-vided additional support for this conclusion.

These studies also indicate that regardless of the level of municipal sanitation and personal hygiene, irrigation of vegetables and salad crops with raw wastewater can serve as a major pathway for continuing and long-term exposure to *Ascaris* and *Trichuris* infections. Both of these infections are of a cumulative and chronic nature, so that repeated long-term reinfection may result in a higher worm load and increased negative health effects, particularly among children.

Cholera can also be disseminated by vegetable and salad crops irrigated with raw wastewater if it is carrying cholera vibrio. This possibility is of particular concern in nonendemic areas where sanitation levels are relatively high, and the common routes of cholera transmission, such as contaminated drinking water and poor personal hy-giene, are closed. Under such conditions, the introduction of a few cholera carriers

Fig. 3.2. Relationship between *Ascaris*-positive stool samples in population of western Jerusalem and supply of vegetables and salad crops irrigated with raw wastewater in Jerusalem, 1935–1982. (*Sources:* Ben-Ari 1962; Jjumba-Mukabu and Gunders 1971; Shuval et al. 1984)

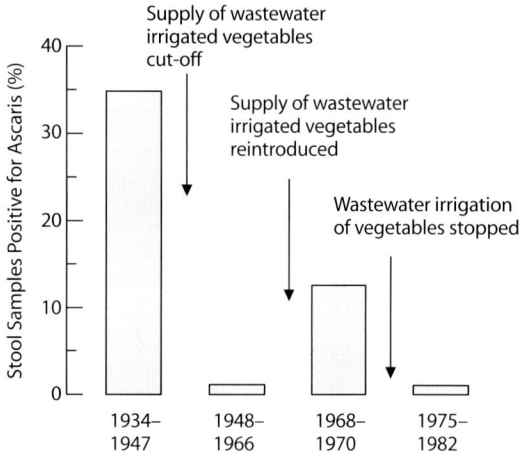

Fig. 3.3. Hypothesized cycle of transmission of *Vibrio cholerae* from first cholera carriers introduced from outside the city, through wastewater-irrigated vegetables, back to residents in the city (*Source:* Fattal et al. 1986a)

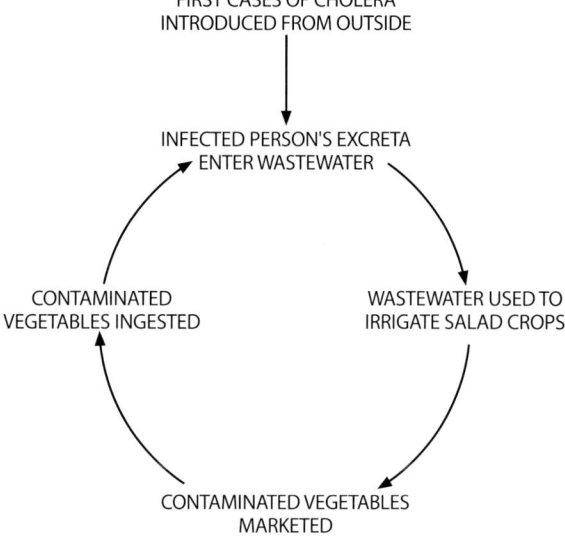

(or subclinical cases) into a community could lead to massive infection of the wastewater stream and subsequent transmission of the disease to the consumers of the vegetable crops irrigated with the raw wastewater, as occurred in Jerusalem in 1970 (Fig. 3.3 shows the cycle of transmission of *Vibrio cholerae* through wastewater-irrigated vegetables; Fattal et al. 1986b).

Similarly, evidence from Santiago, Chile strongly suggests that typhoid fever can be transmitted by fresh salad crops irrigated with raw wastewater. The number of typhoid fever cases in Santiago rose rapidly at the beginning of the irrigation season, after 16 000 ha of vegetables and salad crops (usually eaten uncooked) had been irrigated with raw wastewater (see Fig. 3.4; Shuval 1984). The relatively high socioeco-

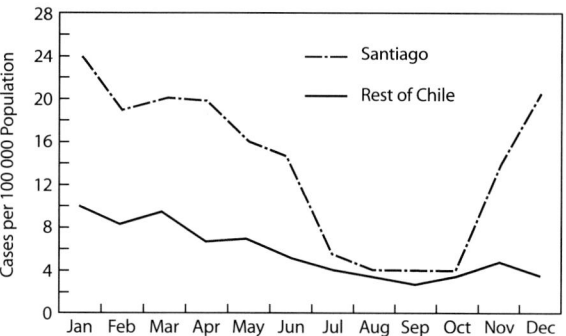

Fig. 3.4. Seasonal variation in typhoid fever cases in Santiago and the rest of Chile average rates, 1977–1981 (*Source:* Field investigation carried out for the World Bank, Shuval 1984)

nomic level, good water supply, and good general sanitation in the city supports the hypothesis that wastewater irrigation can become a major route for the transmission of such bacterial disease.

There is only limited epidemiological evidence to indicate that beef tapeworms (*Taenia saginata*) have been transmitted to populations consuming the meat of cattle grazing on wastewater-irrigated fields or fed crops from such fields. However, there is strong evidence from Melbourne, Australia (Penfold and Phillips 1937) and from Denmark (Jepson and Roth 1949) that cattle grazing on fields freshly irrigated with raw wastewater or drinking from raw wastewater canals or ponds can become heavily infected with the disease and develop cysticercosis. This condition can become serious enough to require veterinary attention and may lead to economic loss. Irrigation of pastures with raw wastewater from communities infected with tapeworm disease may provide a major pathway for the continuing cycle of transmission of the disease to animals and humans.

3.4.3
Health Effects among Sewage Farm Workers

Sewage farm workers exposed to raw wastewater in areas of India where *Ancylostoma* (hookworm) and *Ascaris* infections are endemic have much higher levels of infection than other agricultural workers (Krishnamoorthi et al. 1973). The risk of hookworm infection is particularly high in areas where farmers customarily work barefoot because the broken skin of their feet is readily penetrated by the motile hookworm larva. Sewage farm workers in this study also suffered more from anemia (a symptom of severe hookworm infestation) than the controls. Thus, there is evidence that continuing occupational exposure to irrigation with raw wastewater can have a direct effect on human health and productivity, and on the economy.

Sewage farm workers are also liable to become infected with cholera if the raw wastewater being used for irrigation is from an urban area experiencing a cholera epidemic. This situation is particularly likely to arise in areas where cholera is not normally endemic and where the level of immunity among the sewage farm workers is low or nonexistent. This proved to be the case in the 1970 cholera outbreak in Jerusalem (Fattal et al. 1986b).

Studies from industrialized countries have thus far produced only limited and often conflicting evidence of the incidence of bacterial and viral diseases among wastewater irrigation workers exposed to partly or fully treated effluent, or among workers in wastewater treatment plants exposed directly to wastewater or wastewater aerosols. Most morbidity and serological studies have been unable to give a clear indication of the prevalence of viral diseases among such occupationally exposed groups.

It is hypothesized that many sewage farmers or treatment plant workers have acquired relatively high levels of permanent immunity to most of the common enteric viruses endemic in their communities at a much younger age. Thus, by the time they are exposed occupationally, the number of susceptible workers is small and not statistically significant. Presumably this is also the case among infants and children in developing countries because they are exposed to most endemic enteric viral diseases by the time they reach working age. Although this is not the case for some bacterial and protozoan pathogens, multiple routes of concurrent infection with these diseases may well mask any excess among wastewater irrigation workers in developing countries.

3.4.4
Health Effects among Population Groups Residing Near Wastewater-Irrigated Fields

There is little evidence linking disease and/or infection among population groups living near wastewater treatment plants or wastewater irrigation sites with pathogens contained in aerosolized wastewater. Most studies have shown no demonstrable disease resulting from such aerosolized wastewater, which is caused by sprinkler irrigation and aeration processes. Researchers agree, however, that most of the earlier studies have been inadequate.

Recent studies in Israel suggest that aerosols from sprinkler irrigation with poor microbial quality wastewater can, under certain circumstances, cause limited infections among infants living near wastewater-irrigated fields. The studies, however, also concluded that these were negligible and could be controlled by better treatment (Fattal et al. 1986a; Fattal et al. 1987; Shuval et al. 1988; Shuval et al. 1989b).

These findings support the conclusion that, in general, relatively high levels of immunity against most viruses endemic in the community block additional environmental transmission by wastewater irrigation. Therefore, the additional health burden is not measurable. The primary route of transmission of such enteroviruses, even under good hygienic conditions, is through contact infection in the home at a relatively young age. As already mentioned, such contact infections are even more common in developing countries, so that a town's wastewater would not normally be expected to transmit a viral disease to rural areas using it for irrigation.

3.4.5
Reduction in Negative Health Effects by Wastewater Treatment

Some epidemiological studies have provided evidence that negative health effects can be reduced when wastewater is treated for the removal of pathogens. For example, Baumhogger (1949) reported that in 1944 residents of Darmstadt who consumed salad crops and vegetables irrigated with raw wastewater experienced a massive infection of *Ascaris*; but the residents of Berlin, where biological treatment and sedimentation were applied to the wastewater prior to the irrigation of similar crops, did not.

Another study on intestinal parasites was conducted on school children near Mexico City (Sanchez Levya 1976). The prevalence of intestinal parasites in children from villages that used wastewater irrigation did not differ significantly from that in children from the control villages, which did not irrigate with wastewater. The lack of significant difference between the two groups may have resulted from long-term storage of the wastewater in a large reservoir for weeks or months prior to its use for irrigation. Presumably, sedimentation and pathogen die-away during long-term storage were effective in removing the large, easily settleable protozoa and helminths, which were the pathogens of interest in this study. This study provides the first strong epidemiological evidence of the health protection provided by microbial reductions achieved in wastewater storage reservoirs.

Furthermore, the absence of negative health effects in Lubbock, Texas (Camann et al. 1983) and in Muskegon, Michigan (Clark et al. 1981), appears to be associated with the fact that well-treated effluents from areas of low endemicity were used for irrigation.

Data from these field studies strongly suggest that pathogen reduction by wastewater treatment, including long term storage in wastewater reservoirs, can have a positive effect on human health. In all the above studies, this positive effect was achieved despite the use of effluent which had not been disinfected and which contained a few thousand of faecal coliform bacteria per 100 ml. These data agree with water quality data on pathogen removal and suggest that appropriate wastewater treatment resulting in effective reduction of coliforms to the level of a few thousand/100 ml, but not total removal, can provide a high level of health protection.

3.4.6
Implications for Developing Countries

In summary, epidemiological studies on the health effects of wastewater reuse in agriculture from both developed and developing countries indicate that the following diseases are occasionally transmitted via raw or very poorly treated wastewater:

1. The general public may develop ascariasis, trichuriasis, typhoid fever, or cholera by consuming salad or vegetable crops irrigated with raw sewage, and probably tapeworm by eating the meat of cattle grazed on raw sewage irrigated pasture. There may also be limited transmission of other enteric bacteria and protozoa.
2. Wastewater irrigation workers may develop ancylostomiasis (hookworm), ascariasis, possibly cholera, and, to a much lesser extent, infection caused by other enteric bacteria and viruses, if exposed to raw wastewater.
3. Although there is no demonstrated risk to the general public residing in areas where wastewater is used in sprinkler irrigation, there may be minor transmission of enteric viruses to infants and children living in these areas, especially when the viruses are not endemic to the area and raw wastewater or very poor-quality effluent is used.

Thus, the empirical evidence on disease transmission associated with wastewater irrigation in developing countries strongly suggests that helminths are the number one problem, with some limited transmission of bacterial and viral disease. The above ranking, based on empirical data, agrees with that predicted in our model.

In interpreting the above conclusions, one must remember that the vast majority of developing countries are in areas where helminthic and protozoan diseases such as hookworm, ascariasis, trichuriasis, amoebiasis and tapeworm are endemic. In some of these areas, cholera is endemic as well. It can be assumed that in most developing countries, in populations with low levels of personal and domestic hygiene, the children will become immune to the endemic enteric viral diseases when very young through contact infection in the home.

In conclusion, epidemiological evidence of disease transmission associated with the use of raw wastewater in agriculture in developing countries indicates that the pathogenic agents may be ranked in the following declining order of importance:

1. High risk: helminths (*Ancylostoma*, *Ascaris*, *Trichuris*, and *Taenia*)
2. Lower risk: enteric bacteria (cholera, typhoid, shigellosis, and possibly others); protozoa (amebiasis and giardiasis)
3. Least risk: enteric viruses (viral gastroenteritis and infectious hepatitis)

As pointed out earlier, these negative health effects were all detected in association with the use of raw or primarily treated wastewater. Therefore, wastewater treatment processes that effectively remove all, or most, of these pathogens, according to their rank in the above list, could reduce the negative health effects caused by the utilization of raw wastewater. While helminths are very stable in the environment, bacteria and viruses rapidly decrease in numbers in the soil and on crops.

Thus, the ideal treatment process should be particularly effective in removing helminths, even if it is somewhat less efficient in removing bacteria and viruses. Wastewater treatment technologies that can be used to achieve this goal are discussed later in this chapter. In general, the above ranking of pathogens will not apply to the more developed countries or other areas in which helminth diseases are not endemic. In those areas the negative health effects, if any, resulting from irrigation with raw or partly treated wastewater will probably be associated mainly with bacterial and protozoan diseases and, in a few cases, with viral diseases. Whatever the country or the conditions, however, the basic strategies for control are the same: the pathogen concentration in the wastewater stream must be reduced and/or the type of crops irrigated must be restricted.

Overall, our studies have demonstrated that the extent to which disease is transmitted by wastewater irrigation is much less than was widely believed to be the case by public health officials in the past. Moreover, this study does not provide epidemiological support for the use of the much-copied California standard requiring a coliform count of 2/100 ml for effluent to be used in the irrigation of edible crops. No detrimental health effects were detected when well-treated wastewater of much higher coliform counts was used.

3.5
Evaluating Health Guidelines for Recycling Wastewater in Agriculture

3.5.1
Scientific, Historical, and Social Influences

This section will review the scientific basis and historical and social forces that influenced the evolution of microbial standards and guidelines for wastewater reuse for agricultural purposes. This analysis will draw extensively on recent World Bank and World Health Organization studies and reports whose goal was a cautious reevaluation of the credible scientific evidence which could provide a sound basis for establishing safe and feasible health guidelines for wastewater reuse (Engelberg Report 1985; Shuval et al. 1986; WHO 1989; Shuval 1990).

Many of the current standards restrict the types of crops to be irrigated with conventional wastewater effluent to those not eaten raw. Regulations like those in California, requiring the effluent used for the irrigation of edible crops to have a bacterial standard approaching that of drinking water (2 coliforms/100 ml), are usually not practical, even for developed countries. Such a standard is even less feasible for developing countries. In reality, a standard of 2 coliforms/100 ml for irrigation is superior to the quality of drinking water for the majority of urban and rural poor in developing countries (where faecal coliforms are generally in excess of 10/100 ml).

In developed countries, where these crop restrictions can normally be enforced, vegetable and salad crops are not usually irrigated with wastewater. In the developing countries, many of which have adopted the same strict regulations, public health officials do not approve of the use of wastewater for irrigation of vegetable and salad crops eaten raw. However, when water is in short supply such crops are widely irrigated, illegally, with raw or poorly treated wastewater. This usually occurs in the vicinity of major cities, particularly in semiarid regions.

Since the official effluent standards for vegetable irrigation are not within the obtainable range of common engineering practice and economic considerations, new projects to improve the quality of effluent are not usually approved. With the authorities insisting on unattainable, expensive, and unjustifiable standards, farmers are practising uncontrolled, unsafe widespread irrigation of salad crops with raw wastewater. The highly contaminated vegetables are supplied directly to the nearby urban markets, where such truck garden products can command high prices. This is a classic case in which official insistence on the "best" prevents cities and farmers from achieving the "good."

Some inconsistency exits between the strict California standards, which require edible crops to be irrigated with wastewater of drinking water quality, and the actual agricultural irrigation with normal surface water as practised in the United States and other industrialized countries with high levels of hygiene and public health. There are few, if any, microbiological limits on irrigation with surface water from rivers or lakes – which may be polluted with raw or treated wastewater. For example, the US Environmental Protection Agency's water quality criteria for unrestricted irrigation with surface water is 1 000 faecal coliforms/100 ml (USEPA 1972). A WHO world survey of river water quality has indicated that most rivers in Europe have mean faecal coliform counts of 1 000–10 000/100 ml. And yet none of these industrialized countries have restrictions on the use of such river water for irrigation.

A number of microbial guidelines have been developed for recreational waters considered acceptable for human contact and swimming. In the United States, for example, microbial guidelines for recreational water have in the past ranged from 200–1 000 faecal coliforms/100 ml, although there is currently a move to reduce those numbers. In Europe, guidelines vary from 100 coliforms/100 ml in Italy to 20 000 coliforms/100 ml in Yugoslavia. The European Economic Community has recommended a guideline of 10 000 faecal coliforms/100 ml for recreational waters (Shuval et al. 1986).

It is difficult to explain the logic of a 2 coliforms/100 ml standard for effluent irrigation when farmers all over the United States and Europe can legally irrigate any crops they choose with surface water from free-flowing rivers and lakes which often have faecal coliform levels of over 1 000/100 ml. It is even more difficult to explain the epidemiological rationale of the 2 coliforms/100 ml standard for effluent irrigation, while in Europe recreational water for bathing is considered acceptable at 1 000–10 000 faecal coliforms/100 ml.

3.5.2
The Engelberg Report and the New WHO Guidelines

In July 1985 a group of environmental experts including engineers and epidemiologists meeting at Engelberg, Switzerland, under the auspices of the UNDP, World Bank, WHO, UNEP, and the International Reference Center for Waste Disposal formulated new proposed microbiological guidelines for treated wastewater reuse in agricultural irrigation (Engelberg Report 1985). The group reviewed the epidemiological evidence gathered in the UNDP-World Bank study (Shuval et al. 1986) and the epidemiological analysis prepared by Blum and Feachem (1985). The group accepted the main findings and recommendations of the UNDP-World Bank study and concluded that "current guidelines and standards for human waste use are overly conservative and unduly restrict project development, thereby encouraging unregulated human waste use." The new guidelines recommended in the Engelberg Report and later approved by and

Table 3.2. Recommended microbiological quality guidelines for wastewater use in agriculture[a] (*Source:* WHO 1989)

Category	Reuse conditions	Exposed group	Intestinal nematodes[b]; (arithmetic mean number of eggs per liter)	Faecal coliforms (geometric mean number per 100 ml[c])	Wastewater treatment expected to achieve the required microbiological quality
A	Irrigation of crops likely to be eaten uncooked, sports fields, public parks[d]	Workers, consumers, public	≤1	≤1 000[d]	A series of stabilization ponds designed to achieve the microbiological quality indicated, or equivalent treatment
B	Irrigation of cereal crops, industrial crops, fodder crops, pasture and trees[e]	Workers	≤1	No standard recommended	Retention in stabilization ponds for 8–10 days or equivalent helminth and faecal coliform removal
C	Localized irrigation of crops in category B if exposure of workers and the public does not occur	None	Not applicable	Not applicable	Pretreatment as required by the irrigation technology, but not less than primary sedimentation

[a] In specific cases, local epidemiological, socio-cultural and environmental factors should be taken into account, and the guidelines modified accordingly;
[b] *Ascaris* and *Trichuris* species and hookworms;
[c] During the irrigation period;
[d] A more stringent guideline (<200 faecal coliforms/100 ml) is appropriate for public lawns, such as hotel lawns, with which the public may come into direct contact;
[e] In the case of fruit trees, irrigation should cease two weeks before fruit is picked, and no fruit should be picked off the ground. Sprinkler irrigation should not be used.

recommended by the WHO Meeting of Experts (WHO 1989) are presented in Table 3.2. Since the possibility of transmitting helminth disease by wastewater irrigation of even nonedible crops was identified as the top health problem, a new, stricter approach to the use of raw wastewater was developed. The new WHO guidelines recommend effective water treatment in all cases to remove helminths to a level of one or fewer helminth eggs per litre.

The main innovation of the WHO guidelines is: for crops eaten uncooked, an effluent must contain one or fewer helminth eggs per litre, with a geometric mean of faecal coliforms not exceeding 1 000/100 ml. This is a much more liberal coliform standard than the early California requirement of 2 total coliforms/100 ml.

An attractive feature of the new WHO (1989) effluent guidelines is that they can be readily achieved with low cost, robust stabilization pond systems and hypertrophic wastewater storage reservoirs. The high levels of pathogen removal that can be achieved by such low-cost stabilization pond systems are shown in Fig. 3.4. Even higher degrees of treatment with an added safety factor can be achieved in the sequential batch operated wastewater reservoirs discussed in this book.

3.5.3
The USEPA/USAID Recommended Wastewater Reuse Guidelines

The USAID together with the USEPA established their own rigorous guidelines in 1992 at zero faecal coliforms/100 ml, a *BOD* of 10 ppm, a turbidity of 2 NTU and a chlorine residual of 1 mg l^{-1}. This quality of wastewater effluent can only be achieved in very costly high-tech wastewater treatment plants that require a high level of technological infrastructure to operate and maintain, so that they can continuously meet such very rigorous standards. These guidelines were drafted by one of the leading American consulting engineering firms under contract to USAID. Such consulting engineering firms often tend to favour such high-tech treatment processes. Again the new American guidelines are essentially as strict as those required for drinking water. The fact that little if any natural river water or water at approved bathing beaches in the United States or elsewhere could meet these recommended irrigation guidelines did not seem to bother those who drafted and approved the new American guidelines. No one has suggested that such river water should not be allowed for irrigation purposes, neither has any health risk from such irrigation been reported. Israel has followed the American fail-safe approach and has established a strict standard of 10 faecal coliforms/100 ml for unrestricted irrigation of crops eaten raw.

3.5.4
A Risk Assessment/Cost-Effectiveness Approach for Comparing the Various Guidelines

The debate over the appropriateness of the various guidelines has so far been on a qualitative level. We have carried out a related study aimed at developing a quantitative risk assessment and cost-effectiveness approach based on a mathematical model and experimental data, to arrive at a comparative risk analysis of the various recommended wastewater irrigation microbial health guidelines for unrestricted irrigation of vegetables normally eaten uncooked (Shuval et al. 1997; Shuval and Fattal 1996). The guidelines that were compared are those of the World Health Organization (1 000 faecal coliforms/100 ml), those recommended by the USEPA/USAID (0 faecal coliforms/100 ml) and those of the Ministry of Health of Israel (10 faecal coliforms/100 ml) which closely follow those of the United States. First, through laboratory experiments we determined estimates of the risk of ingesting enteric bacteria or viruses from the consumption of wastewater irrigated vegetables. This allows us to estimate the risk of infection and disease based on the risk of infection and disease model developed for drinking water by Haas et al. (1993). Thus, the annual risk of virus disease – infectious hepatitis for example – from regularly eating vegetables irrigated with raw wastewater is shown to be as high as 10^{-2} to 10^{-3} (one person per 100 or per 1 000 per year might become ill). The study indicates that the annual risk of succumbing to a virus disease (such as infectious hepatitis) from regularly eating vegetables irrigated with treated wastewater effluent meeting the World Health Organization Guidelines of 1 000 faecal coliforms/100 ml is negligible and of the order of 10^{-6} to 10^{-7} (one person per million or 10 million).

The USEPA considers an annual risk of 10^{-4} (one person per 10 000) to be acceptable for microbial contamination of drinking water (Regli 1991). Thus according to this study the new recommended WHO Guidelines for Wastewater Reuse in Agriculture are some 100 to 1 000 times safer than what the USEPA itself recommends as the degree of safety required for drinking water. The additional benefit that might result from a further reduction of risk gained by adhering to the 1992 USEPA/USAID Wastewater Reuse Guidelines, which require no detectable faecal coliforms/100 ml or the Israeli standard of 10 faecal coliforms/100 ml appears to be insignificant in relation to the major additional costs associated with the expensive technology required to treat effluent to such a rigorous standard. According to our preliminary cost-effectiveness estimate, treating wastewater to meet the USAID/USEPA guidelines for a city of one million persons would result in an additional cost, over the cost of treatment to meet the WHO guideline levels, of some US$30 million per case of virus disease prevented. To meet the Israeli standard might cost some US$15 million per case of disease prevented. This preliminary study suggests that it is questionable if such a high level of wastewater treatment is justified from an economic and/or public health point of view.

Our conclusion is that the new WHO guidelines, which are based on extensive epidemiological evidence and can be achieved with relatively cost effective wastewater treatment technology provide a high degree of public health protection at a reasonable cost. There is little or no justification, in our opinion, in the restrictive "zero risk" USEPA/USAID recommended guidelines or even to those of the Ministry of Health of Israel which are a bit less restrictive but still exceedingly expensive to achieve and require costly high-tech wastewater treatment technology which provide little if any measurable increase in health protection.

3.6
Reduction of Health Risks by Various Agronomic Technics

The risk of transmission of communicable disease to the general public by irrigation with raw or settled wastewater can be reduced by a number of agronomic techniques. Some of these restrict the types of crops grown, and others – through modification and/or control of irrigation techniques – prevent or limit the exposure of health-related crops to pathogens in the wastewater.

3.6.1
Restricting Crops

One of the earliest and still most widely practised remedial measures is to restrict the type of crops irrigated with raw wastewater or with the effluent of primary sedimentation. Since there is ample evidence that salad crops and other vegetables normally eaten uncooked are the primary vehicles for the transmission of disease associated with raw wastewater irrigation, forbidding the use of raw effluent to irrigate such crops can be an effective remedial public health measure. Although such regulations have been effective in countries with a tradition of civic discipline and an effective means for inspection and enforcement of pollution control laws, they will likely be of less value in situations where those preconditions are absent.

In many arid and semiarid areas, where subsistence farmers near major urban centres irrigate with raw wastewater, the market demand for salad crops and fresh vegetables is great. Thus, governmental regulations forbidding farmers to grow such crops would be little more than a symbolic gesture. Even under the best of circumstances, it is difficult to enforce regulations that work counter to market pressures; to enforce regulations that prevent farmers from obtaining the maximum benefit from their efforts under conditions of limited land and water resources would be impossible.

3.6.2
Modification and Control of Irrigation Techniques

Basin or sprinkler irrigation of salad and vegetable crops usually results in direct contact of the crops with wastewater, thus introducing a high level of contamination. Well-controlled ridge-and-furrow irrigation reduces the amount of direct contact and contamination, and drip irrigation causes even less (Fig. 3.5). Although none of these methods can completely eliminate direct contact of the wastewater with leafy salad crops and root crops, ridge-and-furrow and drip irrigation methods cause less contamination than the basin and sprinkler methods.

Many vegetables that grow on vines (i.e., tomatoes, cucumbers, squash, and the like) can be partially protected from wastewater contact if properly staked and/or grown hanging from wires that keep them off the ground, although some of these vegetables will inevitably touch the ground.

Fig. 3.5. Cross section of irrigation furrows showing flow path of water into ridges

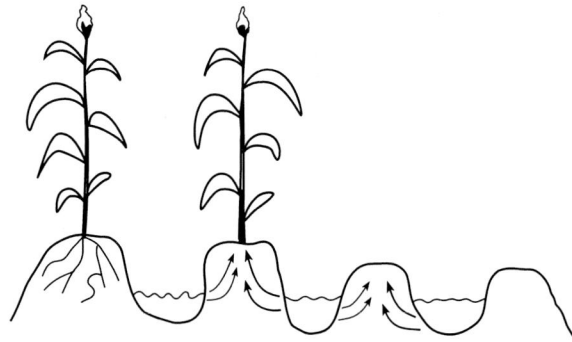

Fruit groves do well with basin or ridge-and-furrow irrigation, but normal overhead sprinkler irrigation leads to direct contamination of the fruit. With low-level, low-pressure sprinkler irrigation, however, the main spray is below the level of the branches, and the fruit is less likely to be contaminated. In all cases, windfall picked from the ground will be in contact with wastewater-contaminated soil. Another possible control measure is to discontinue irrigation with wastewater at a specified period before harvesting the crop. This option is feasible for some crops, but the timing of a vegetable harvest is difficult to control. In addition, some types of vegetables are harvested over long periods of time from the same plot.

All of the above irrigation control techniques can help reduce the danger of crop contamination, but they are feasible only in fairly advanced and organized agricultural economies. Health regulations dependent upon any of the above procedures to protect certain high-risk crops from contamination must be enforced by legal sanctions and frequent inspections. If well-organized inspection and law enforcement systems are not present, as in some developing countries, the value of these options as a major remedial strategy may be limited. However, in the case of large, centrally operated sewage farms managed by the government or under its control, such procedures can be of value.

3.7
Appropriate Low-Cost Methods of Wastewater Treatment for Irrigation

A review of all possible wastewater treatment technologies suitable for wastewater recycling and reuse is beyond the scope of this chapter. However, we shall attempt to review some general principles and to emphasize low cost treatment technologies particularly suited to warm countries.

3.7.1
Goals of Wastewater Treatment for Recycling and Reuse in Irrigation

In areas with plentiful rainfall, wastewater has traditionally been disposed of or diluted in large bodies of water, such as rivers and lakes. High priority has been given to maintaining the oxygen balance of these bodies of water to prevent serious wastewater pollution. Most of the conventional processes used to treat wastewater in industrial countries have been designed primarily to remove the suspended and dissolved organic fractions which decompose rapidly in natural bodies of water. The organic matter in wastewater, usually measured as biochemical oxygen demand (BOD), provides rich nutrients to the natural microorganisms of the stream, which multiply rapidly and consume the limited reserves of dissolved oxygen (DO) in the streams. If oxygen levels drop too far, anaerobic conditions may develop, serious odours may evolve, and fish may die.

A secondary goal of conventional wastewater treatment has been to reduce pathogenic microorganisms in order to protect the quality of the sources of drinking water used by downstream communities. However, the conventional wastewater treatment systems are not particularly efficient in removing pathogens. Thus, communities that draw their drinking water from surface sources cannot depend upon upstream wastewater treatment plant systems to reduce pathogens to a safe level. Therefore, they must remove the pathogens with their own drinking water treatment plants using a series of highly efficient, technical, and costly processes (i.e., coagulation, sedimentation, filtration, and chemical disinfection). The most effective conventional wastewater treatment system is activated sludge, which removes only 90–99% of the viruses, protozoa, and helminths, and 90–99.9% of the bacteria. Conventional processes cannot achieve higher levels of pathogen removal without great additional expense for chemical disinfection, such as chlorination, and the additional sand filtration. Further research and development are needed to improve the removal of helminths by conventional methods. As yet, little effort has been made to develop new and more effective methods.

In contrast to conventional treatment systems, studies have shown that well-designed multi-cell stabilization ponds allowing 20–30 days of retention can remove almost 100 percent of the helminth eggs (Yanez et al. 1980; Feachem et al. 1983; Mara and Silva 1986; Fattal et al. 1997). Bacteria, viruses, and protozoa are often attached to larger faecal particles that settle out in pond systems. At best, however, only 90% can be removed by sedimentation. The most effective process for removing bacteria and viruses in stabilization ponds is die-off, which increases with time, pH, and temperature. Many developing countries have hot climates in which stabilization ponds are exposed to the direct rays of the sun and may reach temperatures up to 40° C. The pH at midday is commonly 9 or higher owing to the photosynthetic activity of the algae. Predatory or competing microorganisms may also affect die-away by attacking or damaging pathogens directly or indirectly. Exposure to the ultraviolet rays of the sun may also play a role in killing pathogens in ponds. Long retention times, however, appear to be the most important factor in reducing bacterial concentrations in pond systems.

In warm climates with temperatures in excess of 20° C, a pond system with 4–5 cells and a 20- to 30-day retention time usually reduces the faecal coliform concentration by 4–6 log orders of magnitude – that is, by 99.99 to 99.9999%. Thus, if the initial concentration of faecal coliform bacteria in the raw effluent is approximately 10^7/100 ml, the effluent will contain 10^3 or 1 000/100 ml. The same pond system will reduce enteric viruses by 2–4 log orders of magnitude (i.e., from an initial concentration of about 1 000/100 ml to 10 or fewer/100 ml). Helminths will be removed almost completely, while the *BOD* will be reduced by about 80%. Figure 3.6 shows the generalized removal curves for *BOD*, helminth eggs, bacteria, and viruses in a multicell stabilization pond system in a warm climate.

Our cooperative studies with Egyptian colleagues at the city of Suez on stabilizations pond treatment of wastewater for irrigation and aquaculture have demonstrated that with well designed and well operated multi-cell ponds with 30 days of detention an effluent of 10–100 faecal coliforms/100 ml was consistently achieved (Mancy 1996). In our studies at a stabilization pond pilot plant system treating primary effluent from the Jerusalem Municipal treatment plant an effluent of 1 000 faecal coliforms/100 ml was achieved with 25–30 days of detention (Fattal et al. 1997).

Stabilization ponds are therefore highly suitable for treating wastewater for irrigation. They are more efficient in removing pathogens, particularly helminths, than are conventional wastewater treatment systems. In addition, they produce a biologically stable, odourless, nuisance-free effluent without removing too many of the nutrients.

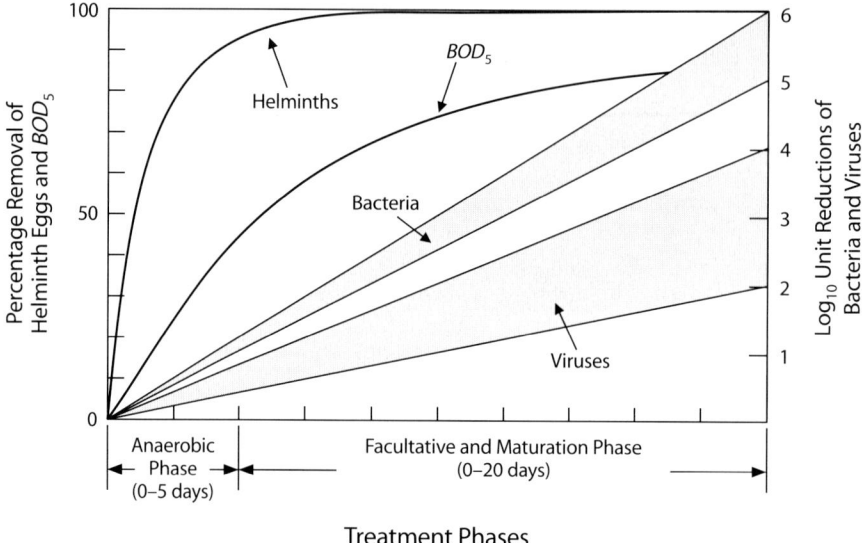

Treatment Phases

Fig 3.6. Generalized removal curves for *BOD*, helminth eggs, excreted bacteria, and viruses in wastewater stabilization ponds at temperatures above 20° C

Thus, ponds should be the system of choice for wastewater irrigation in warm climates, especially if land is available at a reasonable price. Ponds are particularly attractive for developing countries because they cost little to maintain and are robust and fail-safe. They should never be considered a cheap substitute. In reality they are superior conventional methods of treatment in almost all respects. Although ponds require relatively large land areas, land costs are, in many cases, not a serious obstacle.

The goals of sewage treatment for irrigation of crops for human consumption are not the same as goals of conventional sewage treatment. The primary goal for treatment of wastewater to be used for irrigation must be removal of pathogenic microorganisms in order to protect the health of the farmers and consumers. Removal of the organic material, however, which contains valuable agricultural nutrients is neither necessary nor desirable, although aerobic conditions should be maintained because a black, highly odorous, anaerobic wastewater effluent would probably be an environmental nuisance to farmers and nearby residents.

3.7.2
Interseasonal Wastewater Stabilization/Storage Reservoirs

Since wastewater is generated by the community 365 days a year and the irrigation season in most areas is limited to a number of months per year, a means must be found to handle wastewater flows during non-irrigation periods. If allowed to flow unrestricted, the effluent will contaminate the region's natural bodies of water.

A suitable solution is the large inter-seasonal storage reservoirs pioneered in Israel, described in detail in this book. These reservoirs are designed to upgrade the quality of the effluent during the long residence time and to store up to 8 months of wastewater flow to be used for irrigation during the dry summer season. They are often preceded by settling pond and/or by conventional stabilization ponds, and may also be designed to catch surface runoff. The operational regime of such reservoirs vary. Some are operated changing between nonsteady-state flow and batch while others are operated sequentially, receiving and storing influent for extended periods after which the inflow is stopped for a period of stabilization and bacterial decay prior to the discharge for irrigation. The ranges of the reservoirs' key parameters are: mean hydraulic residence time, 50 to 200 days; maximum depth, 5 to 13 m; temperature, 14–17° C in winter and 23–30° C in summer. Log mean concentration of microorganisms in often partially treated typical wastewater influent at the entrance of the reservoir per 100 ml are: heterotrophic bacteria 10^7–10^8, faecal coliforms 10^6–10^7, enterococci 10^4–10^6, and F^+ bacteriophages 10^5–10^6; mean BOD, COD and TSS at the entrance are: 225, 275 and 250 mg l^{-1} respectively. The results at outlet in such reservoirs showed that:

1. In general, the quality of the effluents is much improved and best at the beginning of the irrigation season when the reservoir is full of old effluent (long detention time), but sharply deteriorates when water level drops and new wastewater continues to be pumped into the reservoir (Juanicó and Shelef 1994; Pearson et al. 1996).
2. The die-off rate of microorganisms in the summer is higher than in the winter. We have shown that there is a significant correlation between the mean hours of sunshine in the month and the rate of bacterial die-away. Thus, with greater sunshine duration in the summer months, the bacterial die-away is highest (Fattal et al. 1996). However, in single reservoirs during the summer, the relative volume of the influent, raw or partially treated, is leading to poorer bacterial quality of the effluent since the pond levels are at their lowest during the intensive irrigation season.
3. Coliform removal is high in the epilimnion where high pH values occur due to algal activity, and low in the hypolimnion where pH values are low (Liran et al. 1994).
4. The logs mean reduction of hetrotrophic bacteria, *Escherichia coli*, *Enterococci* and *F. bacteriophages* at outlet of the reservoir are: 1, 3, 2.5 and almost 4 logs respectively (Fattal et al. 1993).
5. Significant correlation coefficients were found between the concentrations of the various microbiological indicators tested, and also a high correlation were found between the number of indicator microorganisms (excluding heterotrophic bacteria) and pollution parameters (BOD, COD and TSS); good correlation was found between BOD vs. COD, BOD vs. TSS and TSS vs. COD (Fattal et al. 1993).

In order to improve the quality of the effluent, innovative changes were made in the design of such stabilization/storage reservoirs in Israel (Liran et al. 1994; Juanicó and Shelef 1994; Juanicó 1996; Friedler and Juanicó 1996). The storage of the effluent was changed from seasonal to multi-seasonal and/or from single reservoir to two or more reservoirs used sequentially as batch reservoirs, supplying effluents for irrigation only from reservoirs which no longer receive fresh influent. By doing so, the input of the effluent from treatment units into the reservoir is stopped before reservoir effluent is released for irrigation. Another improvement was made by implementing better treatment of the wastewater before entering into the reservoir. The performance of the improved batch stabilization reservoirs, when properly designed and operated, showed that they are able to remove up to 90% of the *BOD*, *COD* and *TSS* and detergents, and faecal coliforms by up to five orders of magnitude (i.e. 99.999%). A significant removal of heavy metals, bacteriophages and other pollutants is also obtained. Such improved alternating reservoir systems can normally produce an effluent which can easily meet the WHO guidelines for unrestricted irrigation of all crop of 1 000 faecal coliforms/100 ml or less.

3.7.3
Advantages of Centrally Managed, Engineered Environmental Interventions

Immediate health benefits can be obtained from remedial measures taken by a central authority and involving environmental interventions that lower the level of exposure of large populations to environmentally transmitted disease. Such measures as central plants for the purification of drinking water supplies, pasteurization of milk, and area-wide campaigns for reducing breeding sites of malaria-carrying mosquitoes are well-known examples of success using this strategy. Any remedial action based on changing the personal behaviour and lifestyle in the history of public health progress has proven that the broadest and most thorough public education, law enforcement, or both, is a much slower process and, in general, has succeeded only in areas with relatively high educational and living standards.

The wastewater treatment/storage reservoir option reviewed above offers this type of centrally managed and engineered form of environmental intervention. It is the only remedial measure that will simultaneously reduce the negative health effects for sewage farm workers and for the public that consumes wastewater-irrigated vegetables. It is also the only measure that can bring about health benefits in a short time without massive changes in personal behaviour or restrictive regulations that depend on complex inspection and law enforcement procedures. However, it does require central organizational and management capacity, financial resources, and availability of land.

Although it may be appropriate in some situations to restrict the type of crops grown or to control wastewater irrigation practices, such regulations are difficult to enforce where there is great demand for salad crops and garden vegetables. In arid and semiarid zones (as well as some humid areas) where irrigation is highly desirable, many economists and agricultural authorities consider it economically prudent to allow unrestricted wastewater irrigation of cash crops in high demand. That goal can only be achieved with an effective high level of wastewater treatment as suggested in this chapter.

References

Applebaum J, Guttman-Bass N, Lugten M, Teltsch B, Fattal B, Shuval HI (1984) Dispersion of aerosolized enteric viruses and bacteria by sprinkler irrigation with wastewater. In: Melnik JL (ed) Enteric viruses in water. Karger Basel, Monogr Virol 15:193–201

Baumhogger W (1949) Ascariasis in Darmstadt and Hessen as seen by a wastewater engineer. Zeitschrift für Hygiene und Infektionskrankheiten 129:488–506 (in German)

Ben-Ari J (1962) The incidence of *Ascaris limbriocoides* and *Trichuris trichuria* in Jerusalem during the period of 1934–1960. Am J Trop Med Hyg 11:336–368

Blum D, Feachem RG (1985) Health aspect of night soil and sludge use in agriculture and aquaculture: An epidemiological perspective. Dubendorf, Switzerland, International Reference Centre for Wastes Disposal

Camann DE, Northrop RL, Graham PJ, Guentzel MN, Harding HJ, Yimball KT, Mason RL, Moore BF, Sorber CA, Becker CM, Jakobowsky W (1983) An evaluation of potential infectious health effects from sprinkler application of wastewater to land. Lubbock, Texas, second interim report to the US-Environmental Protection Agency, Cincinnati

Clark CS, Bjornson HS, Holland JW, Elia VJ, Majeti VA (1981) Evaluation of the health risks associated with the treatment and disposal of municipal wastewater and sludge. EPA-600/S-1 81-030. Cincinnati, US Environmental Protection Agency

Engelberg Report (1985) Health aspects of wastewater and excreta use in agriculture and aquaculture. Report of review meeting of environmental specialists and epidemiologists, Engelberg, Switzerland. Sponsored by the World Bank/WHO. Dubendorf, Switzerland, International Reference Centre for Wastes Disposal

Eitan G (1994) Survey on wastewater collection, treatment and reuse in Israel. Ministry of Agriculture, Water Commission, Jerusalem, 175 pp (in Hebrew)

Fattal B, Wax Y, Davies M, Shuval HI (1986a) Health risks associated with wastewater irrigation. An epidemiological study. Am J Public Health 76:977–979

Fattal B, Yekutiel P, Shuval HI (1986b) Cholera outbreak in Jerusalem 1970 revisited: The case for transmission by wastewater irrigated vegetables. In: Goldsmith JR (ed) Environmental epidemiology: Epidemiological investigation of community environmental disease. Boca Raton, LA: CRC Press, pp 49–59

Fattal B, Margalith M, Shuval HI, Wax Y, Morag A (1987) Viral antibodies in agricultural populations exposed to aerosols from wastewater irrigation during a viral disease outbreak. Amer J Epidem 125:899–906

Fattal B, Puyesky Y, Eitan G, Dor I (1993) Removal of indicator microorganisms in wastewater reservoir in relation to physico-chemical variables. Wat Sci Tech 27:321–329

Fattal B, Goldberg T, Dor I (1996) Model for measuring the effluent quality and the microbial die-off rate in Na'an wastewater reservoir. J Handasat Maim 26:26–30 (in Hebrew)

Fattal B, Berkovitz A, Shuval HI (1997) Wastewater treatment efficiency in experimental stabilization pond: The effect of detention time, longitudinal and cross-sectional baffles on system function. J Maim and Hashkaya 367:7–20 (in Hebrew)

Feachem RG, Bradley DJ, Garelick H, Mara DD (1983) Sanitation and disease: Health aspects of excreta and wastewater management. Chichester, John Wiley and Sons

Friedler E, Juanicó M (1996) Treatment and storage of wastewater for agricultural irrigation. Int Wat Irrig Rev 16:26–30

Fuller GW (1912) Broad Irrigation in sewage disposal. New York, McGraw-Hill

Gunnerson CG, Shuval HI, Arlosoroff S (1985) Health effects of wastewater irrigation and their control in developing countries. In: Future of Water Reuse, Denver, AWWA Research

Haas CN, Rose JB, Gerba CP, Regli S (1993) Risk assessment of viruses in drinking water. Risk Analysis 13:545–552

Jepson A, Roth H (1949) Epizootiology of cyiticercus bovis-resistance of the eggs of *Taenia saginata*. In: Report of the 14th international veterinaiy congress, vol 2, London, His Majesty's Stationery Office

Jjumba-Mukabu R, Gunders E (1971) Changing patterns of intestinal helminth infections in Jerusalem. Am J Trop Med Hyg 20:109–116

Juanicó M (1996) The performance of batch stabilization reservoirs for wastewater treatment, storage and reuse in Israel. Wat Sci Tech 33:149–159

Juanicó M, Shelef G (1994) Design operation and performance of stabilization reservoirs for wastewater irrigation in Israel. Wat Res 28:175–186

Khalil M (1931) The pail closet as an efficient means of controlling human helminth infections as observed in Tura prison, Egypt, with a discussion on the source of *Ascaris* infection. Annals of Tropical Medicine and Parasitology 25:35–62

Krey W (1949) The Darmstadt ascariasis epidemic and its control. Zeitschrift für Hygiene und Infektionskrankheiten 129:507–518 (in German)

Krishnamoorthi RP, Abdulappa MR, Anwikar AR (1973) Intestinal parasitic infections associated with sewage farm workers with special reference to helminths and protozoa. In: Proceeding of symposium on environmental pollution. Central Public Health Engineering Research Institute. Nagpur, India

Liran A, Juanicó M and Shelef G (1994) Coliform removal in a stabilization reservoir for wastewater irrigation in Israel. Wat Res 28:1305–1314

Mancy HK (1996) Comparative research on environmental health in pursuit of peace in the Middle East. In: Steinberger Y (ed) The wake of change. ISEEQS, Jerusalem, Israel, pp 736–741

Mara DD, Silva SA (1986) Removal of intestinal nematode eggs in tropical waste stabilization ponds. J Tropical Medicine and Hygiene 99:71–74

Pearson HW, Mara DD, Cawley LR, Oragui JI, Silva SA (1996) Pathogen removal in experimental deep effluent storage reseroirs. Wat Sci Tech 33:251–260

Penfold WJ, Phillips M (1937) *Taenia saginata* and *Cysticercosis bovis.* J Helminthology 15

Regli S, Rose JB, Haas CN, Gerba CP (1991) Modelling the risk from Giardia and viruses in drinking water. J Am Water Works Assoc 83:76–84

Sanchez Levya R (1976) Use of wastewater for irrigation in district 03 and 88 and its impacts on human health. Master's thesis, School of Public Health, Mexico City (in Spanish)

Shuval HI (1984) Health aspects of irrigation with sewage and justification of sewerage construction plan for Santiago, Chile. Unpublished Internal Report, World Bank, processed

Shuval HI (1990) Wastewater irrigation in developing countries – health effects and technical solutions. Summary of World Bank Technical Paper N° 5, UNDP-World Bank wastewater and sanitation discussion paper series DP N° 2, Washington, DC

Shuval HI, Fattal B (1996) Which health guidelines are appropriate for wastewater recycling to agriculture? A risk-assessment/cost-effectiveness approach. In: Potable water and wastewater. AQUATEC Cairo, Egypt

Shuval HI, Yekutiel P, Fattal B (1984) Epidemiological evidence for helminth and cholera transmission by vegetables irrigated with wastewater. Jerusalem as a case study. Wat Sci Tech 17:433–442

Shuval HI, Adin A, Fattal B, Rawitz E, Yekutiel P (1986) Wastewater irrigation in developing countries: Health effects and technical solutions. World Bank Technical Paper N° 51, Washington, DC, World Bank

Shuval HI, Wax Y, Yekutiel P, Fattal B (1988) Prospective epidemiological study of enteric disease transmission associated with sprinkler irrigation with wastewater: An overview. In: Implementing water reuse. Proceedings of Water Reuse Symposium IV, Denver, Am Water Works Associ pp 765–781

Shuval HI, Guttman-Bass N, Applebaum J, Fattal B (1989a) Aerosolized enteric bacteria and viruses generated by spray irrigation of wastewater. Wat Sci Tech 21:131–135

Shuval HI, Wax Y, Yekutiel P, Fattal B (1989b) Transmission of disease associated with wastewater irrigation: A prospective epidemiological study. Am J of Public Health 79:850–852

Shuval HI, Lampert Y, Fattal B (1997) Development of a risk assessment approach for evaluating wastewater recycling and reuse standards for agriculture. Wat Sci Tech, p 35 (in press)

Teltsch B, Shuval HI, Tadmor J (1980) Die-away kinetics of aerosolized bacteria from sprinkler irrigation of wastewater. Appl Environ Microbiol 39:1191–1197

USEPA (1972) Water quality criteria – 1972. A report of the committee on water quality criteria, National Academy of Sciences USEPA, Washington, DC

USEPA/USAID (1992) Guidelines for water reuse. United States Environmental Protection Agency, Washington. Wash Tech Report, Sept. 1992 81:252

World Health Organization (1989) Health guidelines for the use of wastewater in agriculture and aquaculture. Report of a WHO Scientific Group, WHO Technical Report Series 778, Geneva

Yanez F, Rojas R, Castro MI, Mayo C (1980) Evaluation of the San Juan stabilization ponds: Final research report of the first phase. Pan American Center for Sanitary Engineering and Environmental Sciences. Lima, Peru

Chapter 4

Empirical Data for Monitoring and Control

Inka Dor · Michal Raber

4.1
Introduction

During the past few decades, deep wastewater reservoirs have become increasingly popular in Israel. They provide a means for sewage disposal and advanced treatment while supplying nutrient rich water for irrigation of industrial crops. The raw effluents undergo primary and secondary treatment, after which they are collected starting from September until June and reused for irrigation of industrial crops during the later summer months.

One hundred and twenty such reservoirs are presently operational in Israel. Despite their considerable local success these systems are poorly documented and the only data available on the subject are several publications (Eren 1978; Abeliovich 1982; Dor 1986; Dor et al. 1987a,b; Shelef et al. 1987) and local reports. Health risks of irrigation with the stored wastewater were recently evaluated in the studies of Shuval et al. (1988) and Fattal et al. (1986).

In order to achieve the water quality required for irrigation (Shelef 1977) and to avoid environmental nuisances, the reservoirs must be properly operated. A failure of the wastewater reservoir is usually recognized late, when nuisance smells appear. As theoretical principles and standard procedures for the operation are not yet established for this kind of deep, highly polluted reservoir, an attempt to optimize performance is usually undertaken by individual managers by trial and error.

The present study analyses data collected from twelve representative reservoirs during a four-year period. The aims of this study are: 1) to evaluate a rapid method for reservoir monitoring using light penetration measurements which will allow early recognition of arising problems, and 2) to provide an empirical basis for reservoir control through regulation of the entering organic load.

4.2
Materials and Methods

The reservoirs investigated and their locations are given in Map 1. Twenty-six field trips and corresponding data about the reservoirs are listed and numbered according to the increasing *BOD* of the reservoir water at the time of sampling (Table 4.1). For comparison some oxidation ponds were also included.

The Na'an Reservoir and both basins of the Ma'aleh HaKishon Reservoir were investigated seasonally in detail (local reports in Hebrew) and are included several times. Two reservoirs, Netiv Halamed He and Tel Adashim, as well as the Na'an oxidation pond, are included twice: for the winter and the summer seasons, respectively. Other reservoirs were tested only once.

Table 4.1. Data on the reservoirs (R) and oxidation ponds (O) investigated

N°	Name	Maximal depth (m)	Maximal volume[a]	Effluent origin	BOD (mg l^{-1})	Tempera- ture (°C)[b]	Date of sampling
1	Ma'aleh HaKishon South (R)	11	12 000	D + F	1	22.6	Apr. 1985
2	Ma'aleh HaKishon North (R)	11	12 000	D + F	3	20.8	Apr. 1985
3	Ma'aleh HaKishon North (R)	11	12 000	D + F	3.5	20.1	Apr 1984
4	Tzora (R)	5	600	D + M	4.0	10.7	Jan. 1986
5	Na'an (R)	10	700	D	4.0	21.4	May 1985
6	Ma'aleh HaKishon North (R)	11	12 000	D + F	7.0	28.5	Aug. 1985
7	Ma'aleh HaKishon South (R)	11	12 000	D + F	7.0	28.9	Aug. 1985
8	Ma'aleh HaKishon North (R)	11	12 000	D + F	7.9	26.2	Jun. 1984
9	Yodefat (R)	6	200	D + I	11	21.8	May 1985
10	Ginnegar (R)	6	200	D + M	20	21.5	May 1986
11	Ma'aleh HaKishon North (R)	11	12 000	D + F	21	17.4	Oct. 1984
12	Maisha (R)	14	1000	D + I	25	21.5	May 1986
13	Or Haner (R)	11	650	D + I	28	14.3	Jan. 1986
14	Hafez Hayyim (R)	9	800	D+F+M+I	30	22.6	Jun. 1985
15	Giva'at Hayyim (R)	7	360	D + I + M	37	25.7	Jun. 1986
16	Netiv Halamed He (R)	5	300	D + I	37.5	17.8	Feb. 1984
17	Na'an (R)	10	700	D	42	29.0	Jul. 1982
18	Netiv Halamed He (R)	5	300	D + I	54	25.2	May 1984
19	Na'an (O)	1.6	10	D	58	24.0	May 1985
20	Eyal (R)	7	260	D + M	60	27.8	Jun. 1985
21	Tel-Adashim (R)	10	500	D + I	61	21.0	May 1985
22	Tel-Adashim (R)	10	500	D + I	68	13.6	Dec. 1985
23	Na'an (R)	10	700	D	79	25.0	May 1982
24	Na'an (R)	10	700	D	120	14.0	Feb. 1982
25	Na'an (O)	1.6	10	D	150	17.0	Mar. 1986
26	Ashdod (O)	2	16	D + I	370	17.4	Feb. 1986

D: domestic; F: flood or freshwater added; I: industrial; M: manure;
[a] × 1000 m^3;
[b] mean for the upper m^3 of the water column.

Sampling and measurements were performed from an inflated boat, about 20 m from the shore, between 10:00 and 12:00 a.m. Profiles of dissolved oxygen and temperature (YSI Model 58 Oxygen Meter) and light penetration (Lambda Li – Cor 185 Quantum Meter) were measured on site.

Water was sampled near the surface and at depths of 0.5 and 1.0 m using a transparent Wildco 1120 Alpha Bottle. Part of each sample was used for the measurements of respiration (oxygen consumption) rate in the dark bottles suspended from a floating bar at the corresponding depths. Exposure duration was 1–7 h, according to need.

The initial and final oxygen content in the bottles was measured using a YSI Model 54 Oxygen Meter (Dor et al. 1987a). Remaining water was kept overnight at 4° C and analysed the next day. Organic matter (as $BOD 5$ total) and total suspended solids (TSS) were determined according to APHA et al. (1980). Chlorophyll a content was estimated after centrifugation and extraction with hot methanol (Talling and Driver 1963).

To calculate correlation coefficients involving the above variables, the profile measurements were summarized as a single, representative value: a) light penetration data were expressed as a depth of 1% of surface light, the value of which is accepted as an approximate compensation level, i.e., limit of the euphotic zone, where net oxygen is produced by algae; b) dissolved oxygen was given as mean concentration in the upper meter, calculated from the readings taken every 10 cm in the water column; c) chlorophyll a, respiration BOD and TSS were averaged from three estimates: near the surface, at 0.5 and 1.0 m depth.

The BOD values of the entering wastewater and the organic loads for the various reservoirs were obtained by the courtesy of the Water Commissioner, Ministry of Agriculture. Organic load data of N°s 3, 8, 11, 19, 25 and entering BOD data of N°s 3, 8, 11 for the period investigated were not available. In addition chlorophyll data of N° 16, light data of N° 10 and TSS data of N°s 8 and 26 are missing because of technical failures. Accordingly, various correlation coefficients were calculated from a number of values ranging from 20–24. Some unpublished results of our former studies of the Na'an and Netiv Halamed He Reservoirs were added in Fig. 4.12 in order to better underline the obtained trend.

In the plotting of values on the scatter diagrams, the semi-log or log-log scales were selected empirically, to obtain the best-fitting linear relationships. Several deviating reservoirs – marked on the scatter diagrams by dotted lines – were not included in the calculation of the correlation coefficients, which is discussed later.

4.3
Results

Besides having similar operational schemes (see Introduction), the reservoirs differ in their locations (Fig. 4.1), volume, depth and source of collected wastewater (Table 4.1). Volumes ranged from 200 000 to 12 000 000 m³ in the reservoirs and were only 10 000–16 000 m³ in the oxidation ponds. The depths were 5–14 m in the reser-

Table 4.2. Range of the variables investigated in the various reservoirs

Organic load (BOD_5 kg ha^{-1} d^{-1})	2 –	150[a]
BOD_5 total of reservoir effluent	1 –	120[b]
TSS (mg l^{-1})	6 –	150
Dissolved oxygen (mg l^{-1})	0.1 –	19
Chlorophyll a (µg l^{-1})	3 –	2700
Respiration rate (mg l^{-1} h^{-1})	0.07 –	5
Compensation level (cm)	20 –	350

[a] Up to 350 kg ha^{-1} d^{-1} in oxidation ponds;
[b] Up to 370 mg l^{-1} in oxidation ponds.

Fig. 4.1. Location of the wastewater reservoirs and oxidation
ponds investigated

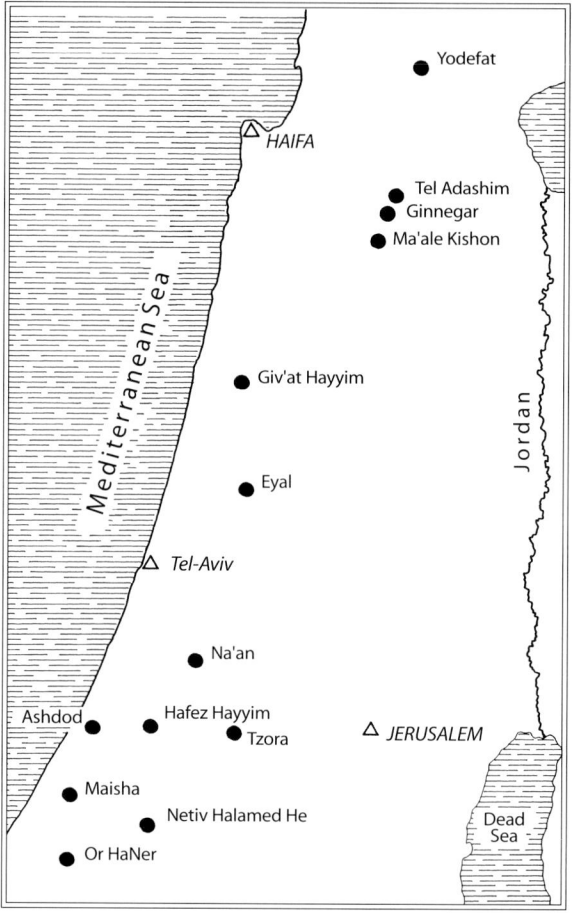

voirs and only 1.6–2 m in the oxidation ponds. Temperatures of 10–17° C were
represenative of winter and 20–29° C of summer. Additional elements of heterogeneity were introduced by the time of sampling, windy or calm weather, seasonal light
intensity and presence or absence of grazers.

The variables investigated differed widely in the various reservoirs and oxidation
ponds (Table 4.2). The lowest values of organic load, *BOD*, *TSS* and respiration rate
were recorded in the Ma'aleh HaKishon Reservoir (N° 1), the largest system of this kind
in Israel, which receives a highly pretreated effluent; the highest values were found in
the heavily loaded Ashdod oxidation pond.

The maximal and the minimal compensation levels were again recorded in the
above N°s 1 and 26, respectively. The highest concentrations of dissolved oxygen were
measured in the stratified reservoirs Na'an (N° 17) and Nativ Halamed He in the summer (N° 18). The lowest oxygen contents were recorded during winter mixing in the
Nativ Halamed He (N° 16) and Na'an (N° 24) Reservoirs and in Na'an oxidation pond
(N° 19). The highest concentrations of chlorophyll a were connected with the popula-

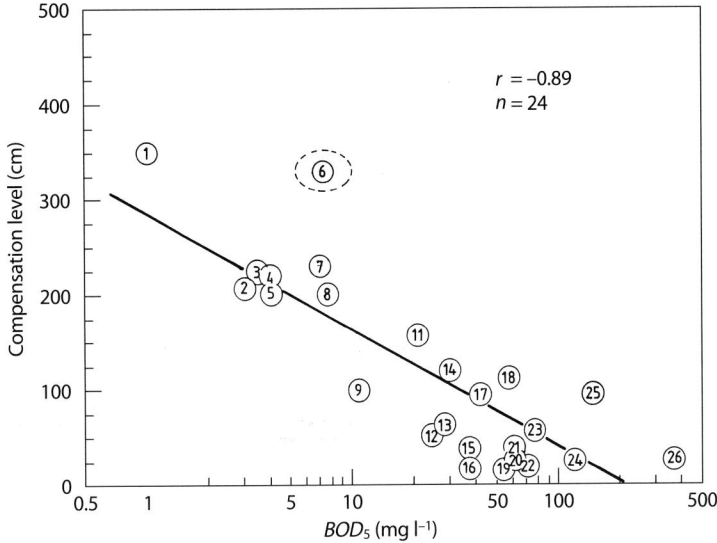

Fig. 4.2. *BOD* and compensation level relations; $P = 0.001$

Fig. 4.3. *TSS* and compensation
level relations; $P = 0.001$

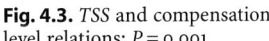

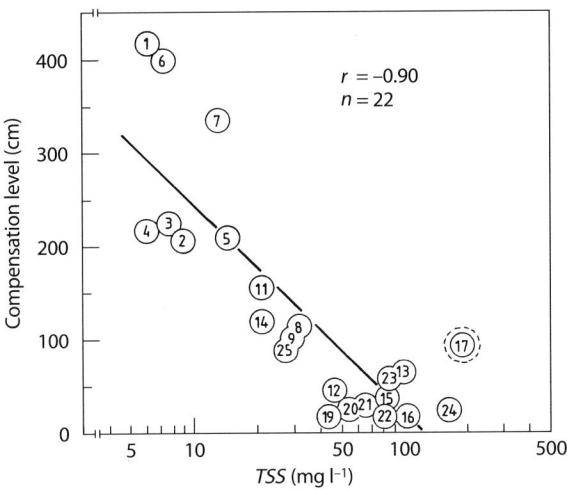

tions of large-celled algae belonging to the genera *Euglena, Chlamydomonas, Oocystis, Micractinium* and *Closteridium* in Givat Hayyim (N° 15), Eyal (N° 20) and Tel-Adashim (N° 22). Reservoirs populated by small-celled algae belonging to the genera *Chlorella* and *Selenastrum*, e.g., reservoir Na'an (N° 23), had a comparatively lower content of chlorophyll a under similar concentrations of *BOD*.

Light penetration, as reflected in compensation levels, appears negatively correlated with the concentration of *BOD* (Fig. 4.2), *TSS* (Fig. 4.3) and chlorophyll a (Fig. 4.4),

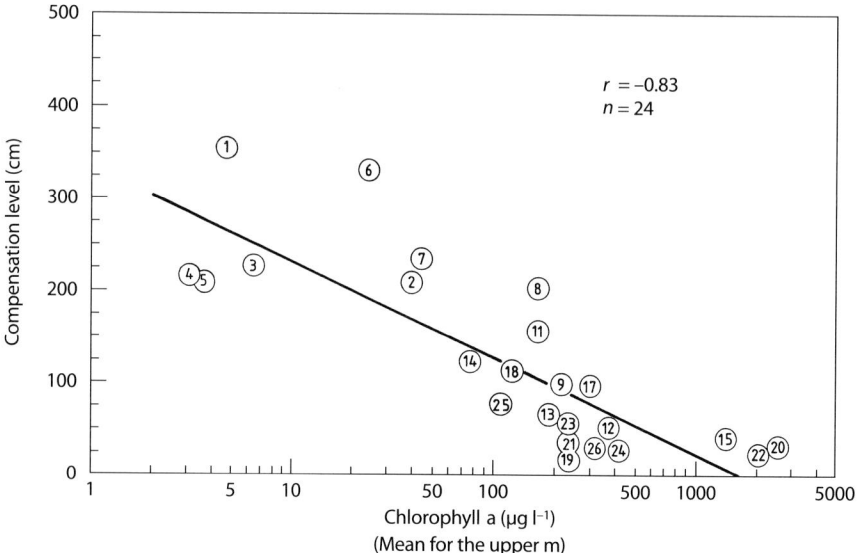

Fig. 4.4. Chlorophyll a and compensation level relations; $P = 0.001$

Fig. 4.5. Compensation level and dissolved oxygen relations; $P = 0.001$

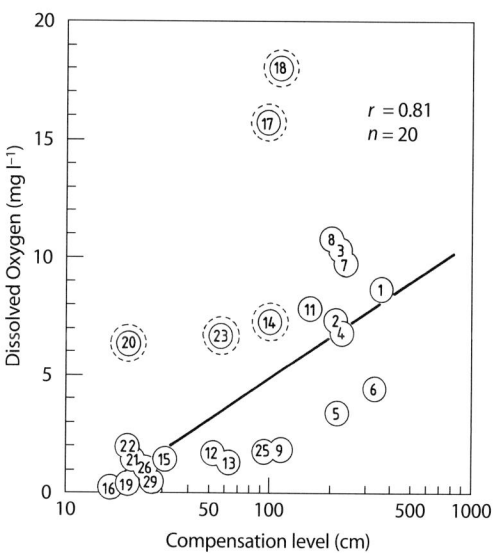

while it is positively correlated with dissolved oxygen (Fig. 4.5). *BOD* is positively correlated with *TSS* (Fig. 4.6) and with the concentration of chlorophyll a (Fig. 4.7) while having negative correlation with dissolved oxygen (Fig. 4.8). *TSS* is positively correlated with chlorophyll a (Fig. 4.9). *BOD* of the reservoirs is positively correlated with the organic load (Fig. 4.10). Figure 4.11 summarizes relationships between the above vari-

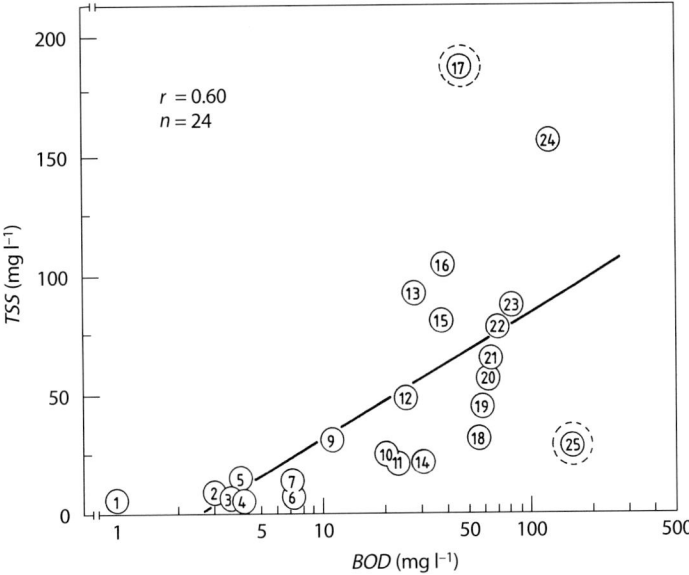

Fig. 4.6. *BOD* and *TSS* relations; $P = 0.001$

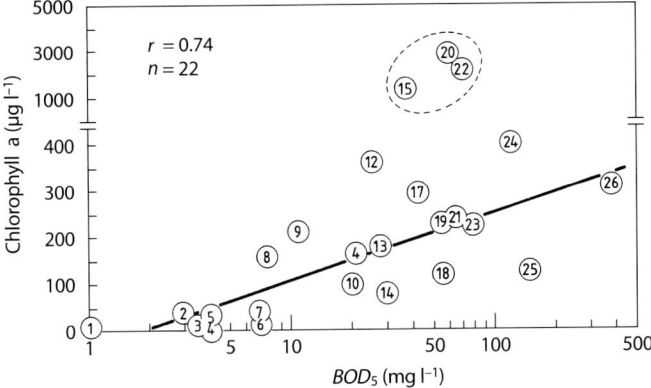

Fig. 4.7. *BOD* and chlorophyll a relations; $P = 0.001$

ables. No correlation was found between chlorophyll a and dissolved oxygen ($r = -0.26$) and graphical representation is not given in this case.

In Fig. 4.12 percentages of *BOD* reduction in various reservoirs in relation to the entering *BOD* are plotted. Up to *BOD* values not exceeding 40–50 mg l^{-1}, the reservoirs reduce, on average, some 80% of the entering *BOD*, irrespective of season. Under the existing design, operation and local climatic conditions, this seems to be the maximal efficiency of the *BOD* removal in these systems. With higher concentration of

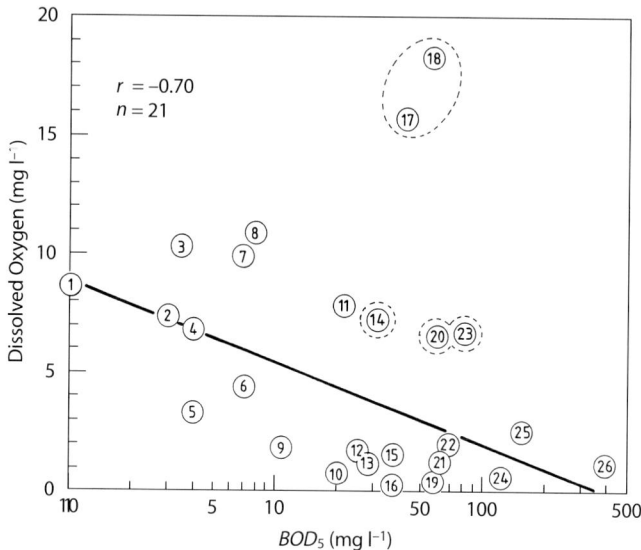

Fig. 4.8. *BOD* and dissolved oxygen relations; $P = 0.001$

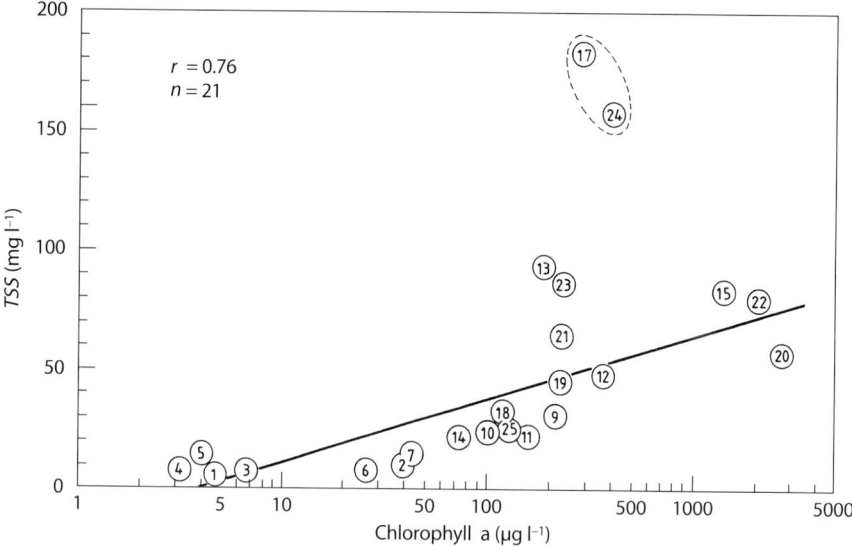

Fig. 4.9. *TSS* and chlorophyll a relations; $P = 0.001$

organic matter, the efficiency of *BOD* reduction in the reservoirs declines steeply, indicating inhibition of the aerobic processes. Similar decline in oxidation ponds (N[os] 19, 25 and 26), has a much gentler slope, which is connected with their shallowness, a more

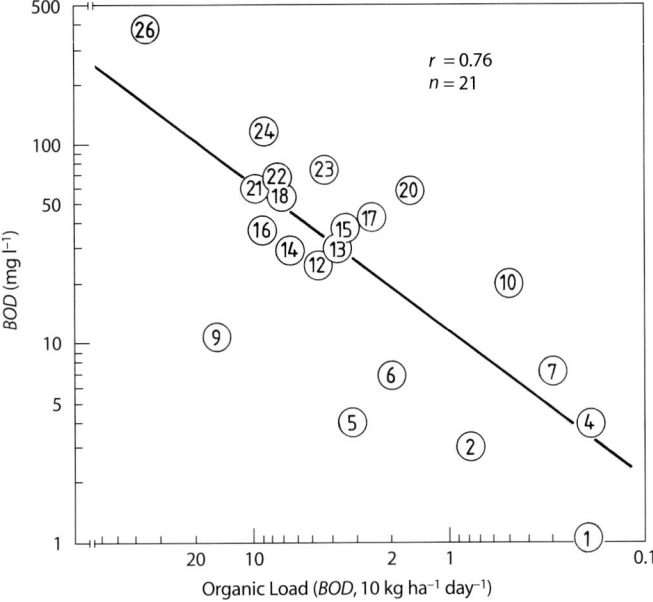

Fig. 4.10. Organic load and reservoir effluent *BOD* relation; $P = 0.001$

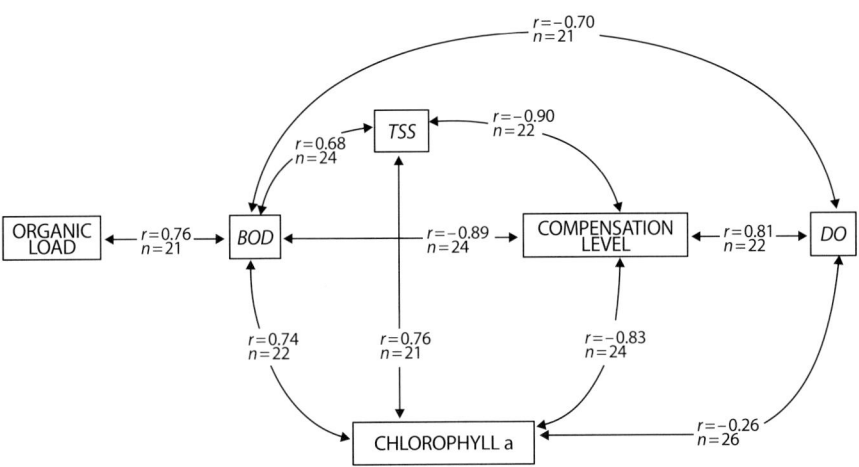

Fig. 4.11. Summary of correlations between variables investigated

efficient oxygenation and a resulting higher carrying capacity. The same figure indicates that reservoirs having higher *BOD* values tend to be stratified (marked by black rings), which obviously results from their higher turbidity and trapping solar energy within the upper water layer.

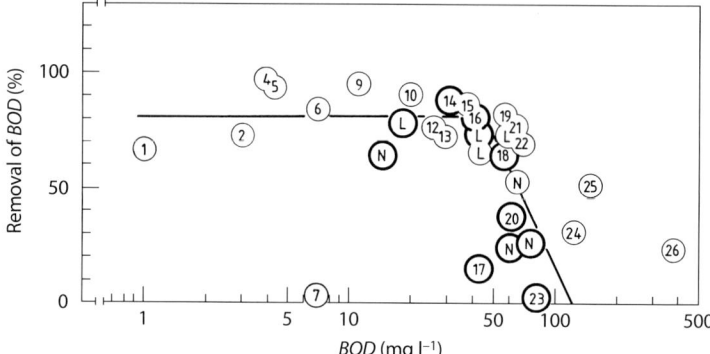

Fig. 4.12. Percentage of *BOD* removal as a function of reservoir effluent *BOD*. Numbers *19, 25* and *26* are oxidation ponds; *black rings* mark reservoirs thermally stratified at the time of sampling; letters *L* and *N* mark unpublished data obtained previously in reservoirs Netiv Halamed and Na'an respectively and not listed in Table 4.1

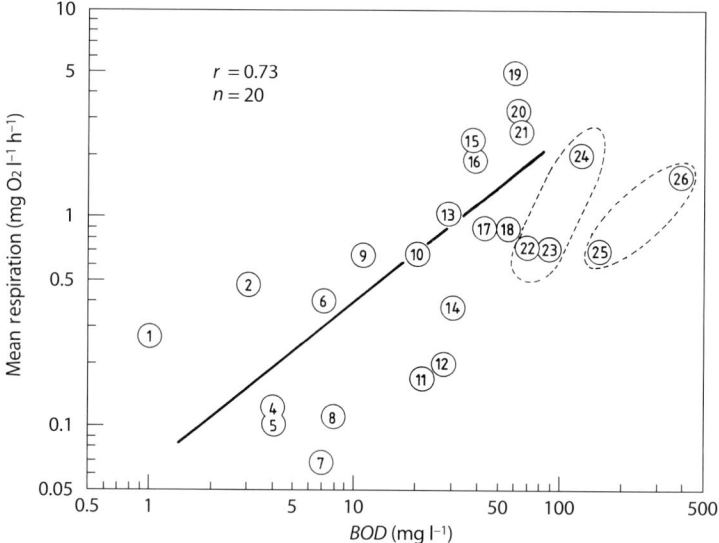

Fig. 4.13. *BOD* and rate of respiration relations. Numbers *22, 23, 25* and *26*, which show decline of respiration rate at *BOD* values exceeding 50 mg l^{-1}, were not included in the calculation of correlation coefficients; $P = 0.001$

Figure 4.13 shows that the rate of respiration is significantly correlated with the increasing content of organic matter (as *BOD*), but only up to a *BOD* of 60 mg l^{-1}. Above this value a decline of respiration per unit volume is recorded, which indicates an inhibition of aerobic bacterial activities.

Odour nuisances were repeatedly recorded in oxidation ponds N°s 19 and 26 and in reservoirs N°s 16, 21, 22, 24. In all the above cases the compensation depth was less than 20 cm and, in all except N° 20, oxygen concentration at midday was very low (Fig. 4). The comparatively high mean oxygen concentration in N° 20, receiving a considerable amount of manure, resulted from supersaturation close to the surface where blooming of the flagellate algae, chiefly *Euglena* sp., was concentrated; however, under the bloom the water column was anaerobic. In all the above cases, again except N° 20, organic load exceeded 70 kg *BOD* ha^{-1} d^{-1}.

4.4
Discussion

Water qualities in the reservoirs differ seasonally and according to the source of effluent, organic load, local morphometry and sampling circumstances. Considering such wide heterogeneity, the correlation coefficients between the variables tested were indeed surprisingly high. These results confirm that, in accordance with the former studies (Abeliovich 1982; Dor et al. 1987a,b), the wastewater reservoirs constitute a well-defined category of aquatic ecosystems and exhibit a predictable behaviour over a wide range of conditions.

Light penetration, chlorophyll a, rate of respiration, and dissolved oxygen content change consistently, in accordance with the content of organic matter (as *BOD*), which appears to be a powerful independent factor, affecting all other variables. *BOD* reduction in the reservoirs remains around 60–90% within a wide range of organic matter content, indicating an efficient decomposing activity of bacteria up to a *BOD* concentration of about 40–50 mg l^{-1}. At higher concentrations of *BOD* a step drop in the efficiency of removal is recorded. Also the decline of respiration rate occurs approximately at the same limit, thus providing an independent line of evidence that a *BOD* of 40–50 mg l^{-1} may constitute a critical value for the operation of these reservoirs. At a higher concentration of organic matter and a concomitant increase of bacterial population, the lack of oxygen presumably inhibits the aerobic, decomposing activities. Indeed, the reservoirs operated at the higher *BOD* levels cause odour nuisances, indicating anaerobic incidents. However such nuisances can also appear with a lower content of organic matter if the entering effluent contains manure. In this case mostly flagellate algae develop, forming a dense surface bloom and limiting the photic zone to a few tens of centimetres. Under the algal bloom, in the lack of photosynthetic activity, anaerobic conditions arise and objectionable gases are produced.

Twelve reservoirs selected for investigation constitute 10% of the total number of the wastewater reservoirs existing in Israel, thus constituting a reliable representative sample of these systems.

BOD and compensation level showed a close and highly significant association ($r = -0.89$; $P = 0.001$; Fig. 4.2), which requires explanation. The organic matter content of domestic wastewater, expressed in terms of *BOD*, is normally composed of suspended and dissolved materials which scatter and absorb light so limiting its penetration. Besides this direct effect, the organic matter itself, as well as the mineral nutrients released from its decomposition, stimulates growth of bacteria and algae, respectively, resulting in biological turbidity and additionally decreasing the transparency of the water column. The fact that all the concerning factors act in the same direction can explain the above high correlation obtained. However, the *BOD* and compensation level relations obviously lack a direct causative dependence, and therefore precautions must be undertaken in extrapolating from the present findings to the other situations where the source of *BOD* is different (for example if it is composed of soluble organic matter from industry).

High correlation coefficients between total suspended solids (*TSS*), chlorophyll and compensation level (Figs 4.3 and 4.4) are self-explanatory. Association between *TSS* and chlorophyll a (Fig. 4.9) provides evidence that a significant part of the suspensions are algal cells.

Sanitary and environmental performance of the wastewater reservoirs ultimately depend on the availability of dissolved oxygen for biodegradation of residual organic waste. This study demonstrates that *DO* is negatively correlated with the concentration of *BOD* (Fig. 4.8) and shows strong, positive association with the level of compensation (Fig. 4.5). This suggests that the oxygen content in deep, highly enriched water bodies is closely correlated with the metabolic activities responsible for oxygen production through photosynthesis and oxygen consumption by respiration. Evidently, wind mixing, which intensifies diffusion and distribution of the oxygen throughout the water column, is not strong enough in these reservoirs to mask the predominant effect of the metabolic activities.

The central position of compensation level, mediating between the *BOD*, *TSS*, chlorophyll and dissolved oxygen content, predestinates underwater light measurements to serve as a useful parameter for a rapid assessment of reservoir water quality.

Moreover, compensation level must be closely related to the optical properties of the water body and these can be detected by infra-red or visible spectrum photography from an aeroplane or satellite (Buttner et al. 1987). Present results therefore provide a background for the future application of the modern remote sensing techniques to the monitoring of the wastewater reservoirs.

The somewhat less significant association between *BOD* and *DO* ($r = -0.70, P = 0.01$, Fig. 4.8) suggests that certain non-synergistic factors mediate between these variables. *BOD*, representing the biodegradable organic matter, affects dissolved oxygen through bacterial heterotrophs, which exert an immediate respiratory demand, thus decreasing concentrations of oxygen. However, *BOD* in domestic wastewater is normally coupled with mineral nutrients (N and P) (Dor et al. 1987a), which promote algae, thus stimulating photosynthetic oxygen production. This contradictory effect undoubtedly affects the relationship between the above variables. Another disturbing influence may be caused by zooplankton displaying a seasonal mass growth which is connected with an intense local consumption of oxygen and drop in its content. The last effect appears chiefly in reservoirs within the lower range of *BOD* where normally near-saturation values of *DO* should be expected.

An increasing chlorophyll a content affects the mean concentrations of dissolved oxygen in conflicting ways: it promotes oxygen production through photosynthesis but exerts a self-shading effect, which limits light penetration, restricting photosynthetic activity to the uppermost water layer. As a result these variables do not correlate (Fig. 4.11).

In Figs 4.2–4.3, 4.5–4.9 and 4.13 all the available values were plotted, but some reservoirs, forming deviating points or sub-groups marked on the scatter diagrams, were excluded from the calculation of the correlation coefficients. It must be pointed out that including all the plotted values only slightly decreases the correlation coefficients and in all cases they remain significant. However, the above exclusions are very important because they mark the sub-groups of differently behaving reservoirs. For example, N° 6 in Fig. 4.2 (representing Ma'aleh HaKishon Reservoir), deviated from the regression line showing a compensation level higher than predicted from the calculated slope. Light penetration in the Ma'aleh HaKishon Reservoir was evidently more efficient than could be expected from the respective *BOD*. This may be caused by a long-term occurrence of filter feeders (unpublished), harvesting unicellular algae and bacteria and thus improving transparency of the water column in this reservoir.

Other deviating reservoirs were N[os] 15, 20 and 22 (Fig. 4.7), in which chlorophyll a was much higher than in other reservoirs, and higher than should be expected from the respective *BOD*. In all the above reservoirs large-celled algae prevailed *Oocystis* sp. in N° 15, *Euglena* sp., *Chlamydomonas* sp. and *Micractinium* sp. in N° 20 and *Micractinium* sp. with *Closteridium* sp. in N° 22. These algae are evidently rich in chlorophyll. By comparison most of the reservoirs having a lower content of chlorophyll were predominately populated by small algae, such as *Chlorella vulgaris* and *Selenastrum minutum* (unpublished information).

Numbers 14, 17, 18, 20 and 23 (Figs 4.5 and 4.8) are characterized by a higher oxygen content than may be extrapolated from the regression obtained for the *BOD-DO* relations in the other 20–21 cases. The above reservoirs were tested at a time when a sharp thermocline cut the water column into the epi- and hypolimnion. In such conditions dissolved oxygen, produced by algae in the photic zone, accumulates in the upper water mass and frequently reaches high levels of supersaturation.

Seasonal differences in behaviour of the various reservoirs can be demonstrated in several cases. Comparing the winter and summer season in Ma'aleh HaKishon (N[os]2 and 6, respectively) and Netiv Halamed He (N[os] 16 and 18, respectively), an increase in compensation level is seen in the summer (Fig. 4.2) which can be related to the filtering activity of the zooplankton (as explained above) and/or to the decreased turbulence during the stagnation period. The Tel-Adashim Reservoir had a much lower chlorophyll a content in summer as compared to winter (N[os] 21 and 22, Fig. 4.4). However, in this case the expected difference in compensation level was negligible. This lack of difference may be caused by a heavy bloom of red photosynthetic bacteria present in this reservoir almost continuously (unpublished information), which limits light penetration and mask the effect of the algae. In the Na'an Reservoir, *BOD*, *TSS* and chlorophyll a increased in winter (N° 24) as compared to summer (N[os] 5 and 23), which was connected with the concomitant decrease of compensation level (Figs 4.2, 4.3 and 4.4).

Oxidation ponds (N[os] 19, 25 and 26), although constituting extreme points in some graphs (Figs 4.2, 4.5 and 4.8), are mostly situated close to the calculated regression slopes proving that their behaviour, at least with respect to the variables investigated here, forms a continuum with the wastewater reservoirs.

The studies presently initiated deal with the microbial qualities of the wastewater reservoir effluents and with the rate of removal of enteric bacteria and viruses in these systems.

4.5
Conclusions

Despite wide variability, wastewater reservoirs constitute a well-definied category of water bodies where a chain of processes is induced by the content of organic matter.

The scatter diagrams, regressions and correlation coefficients, obtained from the analysis of the selected physico-chemical and biological variables in a number of reservoirs, contribute to the understanding rules of behaviour of these systems and allow an approximate prediction of their performance.

Based on the present results, the following operational conclusions are reached: a) light penetration readings can be used for the rapid assessment of the reservoir water quality; b) BOD levels of 40–50 mg l^{-1} are an upper limit for the safe performance of the reservoirs under the present design and local climatic conditions; and c) in order to support the above level of the BOD in the reservoir, the organic load should not exceed 60 kg BOD ha^{-1}d^{-1}.

With these requirements fulfilled, the wastewater reservoirs are at their best, i.e., not only supplying water for irrigation without causing odour nuisances but also reducing BOD by some 80%, close to the maximal efficiency of these systems.

Acknowledgements

We thank Mr. A. Aharoni for the unpublished results concerning the Ma'aleh HaKishon and Netiv Halamed He Reservoirs from 1984. Mr. G. Eitan helped in selecting the representative reservoirs for this study, provided background operational data, and participated in most of the research trips. Some BOD data were obtained by the courtesy of J. Eren and G. Eitan. Discussion of the results with H. Shuval is much appreciated. The study was supported by a grant from the Water Commissioner, Ministry of Agriculture, Israel.

This chapter was reproduced with permission from Water Research, vol. 24, N° 9:1077–1084, 1990.

References

Abeliovich A (1982) Biological equilibrium in a wastewater reservoir. Wat Res 16:1135–1138
APHA, AWWA, WPCF (1980) Standard methods for the examination of water and wastewater. Washington, DC
Buttner G, Korandi M, Gyomorei A, Kote Z, Szabo G (1987) Satellite remote sensing of inland waters: Lake Balaton and Reservoir Kiskore. Acta Astronautica 15:305–311
Dor I (1986) Performance of a deep wastewater reservoir storing effluent for agriculture. Proc. Third Int. Conf. Wastewater Reuse, Larnaca, Cyprus, pp 1–14
Dor I, Schechter H, Bromley HJ (1987a) Limnology of a hypertrophic reservoir storing wastewater effluent for agriculture at Kibbutz Na'an, Israel. Hydrobiologia 150:225–241
Dor I, Kalinsky I and Eren J (1987b) Deep wastewater reservoirs in Israel: Limnological changes following selfpurification. Wat Sci Technol 19:317–322
Eren J (1978) Succession of phyto- and zooplankton in a wastewater storage reservoir. Verh Int Verein Limnol 20:1926–1929
Fattal B, Yekutiel P, Wax Y (1986) Prospective epidemiological study of health risks associated with wastewater utilization in agriculture. Wat Sci Technol 18:199–209
Shelef G (1977) Commission for the proposal of standards of wastewater quality for agricultural irrigation. Final Report (in Hebrew)
Shelef G, Juanicó M, Vikinski M (1987) Reuse of stabilization pond effluent for agricultural irrigation in Israel. Wat Sci Technol 19:299–305
Shuval HI, Wax Y, Yekutiel P, Fattal B (1988) Prospective epidemiological study of enteric disease transmission associated with sprinkler irrigation with wastewater: An overview. In Implementing Water Reuse, pp 765–791. Proc. Water Reuse Symp. IV, AWWA Research Foundation, Denver, Colorado
Talling GF, Driver D (1963) Some problems in the estimation of chlorophyll a in phytoplankton. Proc. Conf. on Primary Production Measurements, Marine and Freshwater, Hawaii, 1961, pp 142–146. US Atomic Energy Commission, TID-7633

Chapter 5

Process Design and Operation

Marcelo Juanicó

5.1
Introduction

5.1.1
Water Demand for Irrigation in Israel and the Hydrological Cycle in the Reservoirs

Mediterranean climate has a rainy winter and a dry summer. Thus, seasonal storage of wastewater in reservoirs is imperative if all wastewater is to be reused in irrigation. The combination of continuous inflow of wastewater all year round and discontinuous outflow of reclaimed effluents for irrigation determines the annual cycle of this kind of reservoir: filling during the winter, emptying during the summer (Fig. 5.1).

5.1.2
Basic Concepts in Designing Wastewater Reservoirs

There are two main differences between the design of wastewater reservoirs and the design of classic wastewater treatment units such as waste stabilization ponds or activated sludge:

1. *Wastewater reservoirs are nonsteady-state reactors.* Most biological treatment units are steady-state reactors with a fixed volume, the same amount of sewage enters the reactor every day, and the same amount of treated effluents leaves the reactor every day (generally by simple overflow). There are small differences in the quantity and quality of sewage that enters the reactor from day to day and weekends, but these differences are so small that the performance of the reactor can be analysed by assuming steady-state conditions. The assumption of steady-state conditions allows the use of simple analytical solutions for kinetic rates in the process design of the reactor. On the contrary, wastewater reservoirs share with sequential batch reactors (e.g., SBR activated sludge; Fig. 5.2) the category of nonsteady-state reactors whose process design requires special equations which can not be solved by analytical procedures. Modelling and numerical solutions are required (Juanicó and Friedler 1994).
2. *The design of wastewater reservoirs includes 'limnological' elements.* Wastewater reservoirs belong to the category of 'extensive' or 'natural' wastewater treatment systems, together with wastewater stabilization ponds, constructed wetlands, and other. All these systems are large, but wastewater reservoirs are especially large with volumes from 50 000 m^3 to 12 million m^3 (typical values ranging between half to two million m^3). As a consequence of their large size, reservoirs behave not only as chemical reactors but also as limnological units (such as lakes, water supply reser-

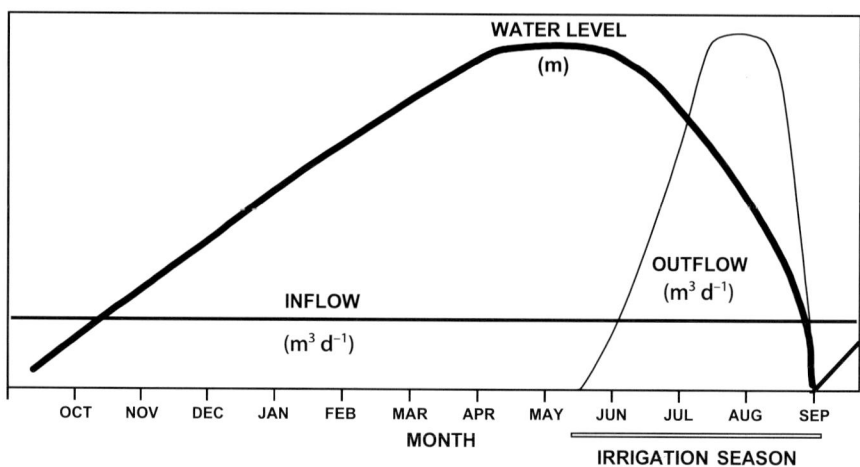

Fig. 5.1. Typical annual cycle of a continuous-flow wastewater reservoir in Israel

Fig. 5.2. Schematic cycle of sequential batch activated sludge and wastewater reservoirs

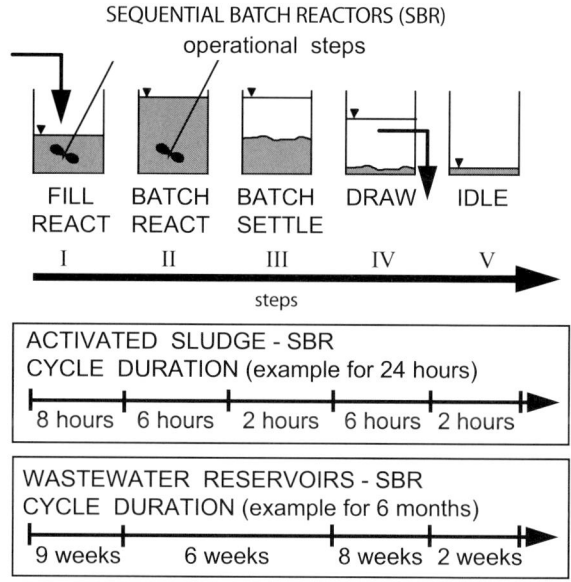

voirs, and other large water bodies). Some of the limnological elements which affect the behaviour of the reservoirs are evaporation, solar radiation, stratification (Juanicó 1994c), winds, waves, currents (Figs 5.6 and 5.10), tides, settling of particles and living organisms (Avnimelech and Wodka 1988; Avnimelech 1989), succession of plankton communities (Eren 1978), etc. Water quality management in 'limnological units' is based on different concepts and tools than the ones used for the control of chemical reactors; this point must be taken into consideration by the designer.

5.2
Basic Operational Regimes and Water Demand Curves for Irrigation

Continuous or discontinuous *input* of wastewater to the reservoir, and continuous or discontinuous *discharge* of effluents from the reservoir are two main operational parameters affecting general performance. Figure 5.3 describes some basic operational

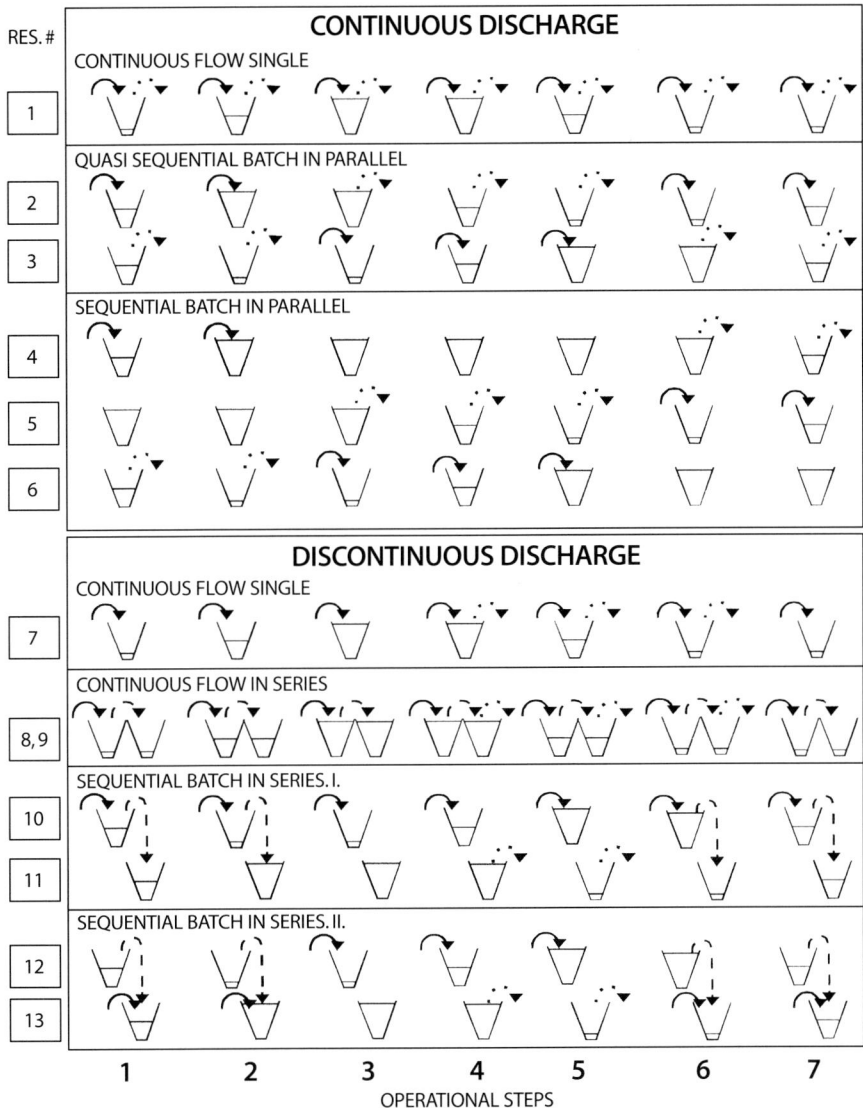

Fig. 5.3. Some (13) basic operational regimes of stabilization reservoirs. Inflow (*plain arrows*) and outflow (*dotted arrows*) rates are not necessarily equal

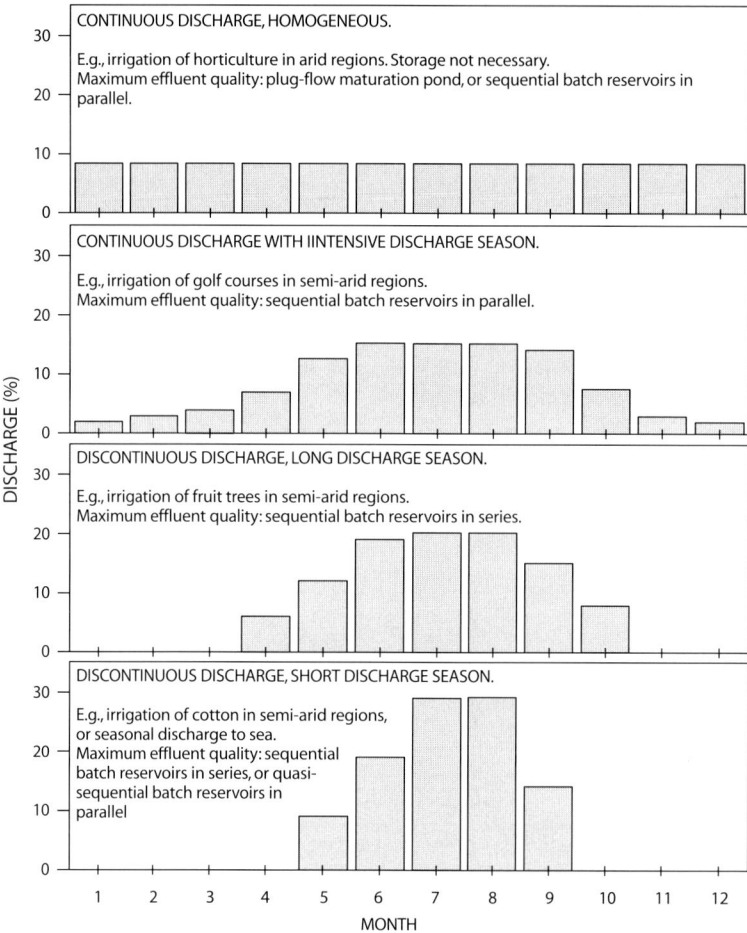

Fig. 5.4. Some typical effluent discharge curves from stabilization reservoirs (shadowed area = 100% of annual discharge)

regimes combining different *input* and *discharge* possibilities (there are more alternatives than the few ones showed in Fig. 5.3). Continuous-flow (or continuous input) regimes (reservoirs N°s 1, 7, 8, 9) receive wastewater all the year round. Sequential batch regimes are operated in such a way that only one of the several reservoirs receives wastewater, while the reservoir which releases effluents stops to receive wastewater before its outlet is opened. In the quasi-sequential batch regime (N°s 2, 3) the input of wastewater to the reservoir is stopped simultaneously with the opening of the outlet. In the sequential batch in series regimes, the input may enter always to the same reservoirs of the series (N° 10) or alternatively to both of them (N°s 12, 13).

Different discharge curves are typical of different situations (Fig. 5.4). For example, horticulture irrigation in arid zones may have an almost homogeneous demand for water all the year round. In semi-arid regions such as those of Mediterranean climate, golf courses (shallow roots) may demand irrigation all the year round but much more intensively during the summer, while some fruit trees (deeper roots) may demand irrigation only during spring-summer-fall but not in winter. Under the same conditions, quick growing crops such as cotton may need large amounts of water concentrated during a short irrigation season of 3–4 months. Wastewater storage reservoirs constructed to avoid the discharge of effluents to the sea, lakes or rivers during some periods of the year may have a short discharge season such as that of cotton irrigation. The operational regime and active volume of wastewater reservoirs is a function of the requirements of the water discharge curves.

5.3
The 'Old' Continuous-Flow Single Reservoirs

Most of the reservoirs constructed in Israel during the seventies (Pano 1975; Berend and Pano 1976) and part of eighties followed the operational regime N° 1: a single reservoir was constructed, wastewater entered the reservoir all the year round, and wastewater was released from the reservoir only during the irrigation season (generally May to September).

The criteria to design these reservoirs are empirical and were developed by both researchers and farmers, by trial and error and common sense.

5.3.1
Volume and Depth

The *active* volume of the reservoir is computed by making a balance between the gains (inflow and rain) and the losses (evaporation and seepage) during the non-irrigation season. A *dead* volume of about 1 m depth must be added to account for the effluents which remain in the reservoir at the end of the irrigation season due to the impossibility of pumping them out without dragging the bottom sediments. It is still necessary to add the freeboard or *dry* volume (Fig. 5.5).

Some reservoirs may receive flood waters and other residual waters besides wastewater. If the determination of the active volume is based on the average flow of these water sources, it will happen that in rainy years the reservoir will be full by the end of the winter before the beginning of the irrigation season. As a consequence, the opera-

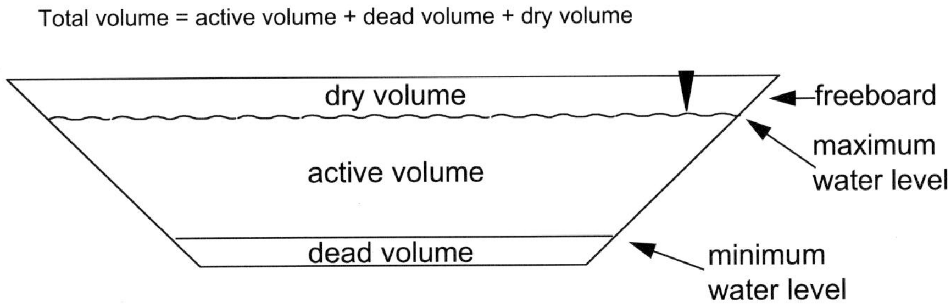

Total volume = active volume + dead volume + dry volume

Fig. 5.5. Division of the total volume of a reservoir

tor will be obliged to release some effluents to the closest water body. If the goal of the project is that *all* the wastewater must go to irrigation with no release to water bodies, the calculation of the active volume must be based on maximum flows and not on the average ones.

Deep reservoirs with a small *area/volume* relationship are recommended in semi-arid regions where evaporation losses may account for more than 15% of inputs to the reservoir and increase the salinity of the remaining effluents (Meron and Eren 1985). Reservoirs with an active depth of 7 to 15 m are normally used in Israel. Deeper reservoirs up to 20 m active depth are being designed nowadays. Dissolved oxygen in reservoirs comes from photosynthetic activity of algae and from diffusion of atmo-spheric oxygen. Thus, deeper reservoirs with a small area/volume ratio have a poorer supply of oxygen than shallow water bodies. Deeper reservoirs require lower surface organic loading or the use of aerator and/or mixers to maintain aerobic conditions.

5.3.2
Outlet and Inlet Location

In most reservoirs the outlet is made of a pipe suspended from a raft, with the open-ing at about 1 m below the water surface in order to take the effluents from the oxy-gen rich epilimnion which has effluents of better quality than the anaerobic hypolim-nion (see Chapter 9).

Both field data and simulation modelling show that continuous-flow reservoirs with the inlet located at the bottom of the reservoir perform better, because this location improves the oxygen balance of the reservoir and avoids hydraulic short-circuiting between inlet and outlet.

The outlet must not be located at the leeward side of the reservoir. The wind in-duced currents (Fig. 5.6) concentrate suspended solids, resuspended pollutants (e.g., heavy metals) and resuspended pathogens at the leeward side of the reservoir, so the outlet must not be located there.

Inlet must be located as far as possible from outlet in continuous-flow reservoirs. Inlet and outlet may be designed as a single unit in sequential batch reservoirs due to the close batch period which eliminates the need to separate them.

In well designed continuous-flow reservoirs the inlet-outlet axis is perpendicular to the direction of dominant winds in order to avoid inlet-outlet hydraulic short-cir-cuiting due to wind induced surface currents and deep counter-currents.

5.3.3
The Hydraulics of Continuous-Flow Wastewater Reservoirs as Reactors

Waste stabilization *ponds* are designed as steady-state flow reactors, with almost con-stant inflow, outflow, volume and depth. Physico-chemical and biological processes and pond performance follow an annual cycle related to temperature and/or solar radiation; in tropical areas the performance of the pond is homogeneous during the year. Deep stabilization ponds also belong to this category (Wachs and Berend 1968) because they are steady-state reactors in spite of their depth.

On the contrary, continuous-flow stabilization *reservoirs* have an annual empty-full-empty cycle (Fig. 5.1). Inflow is almost constant throughout the year while outflow is

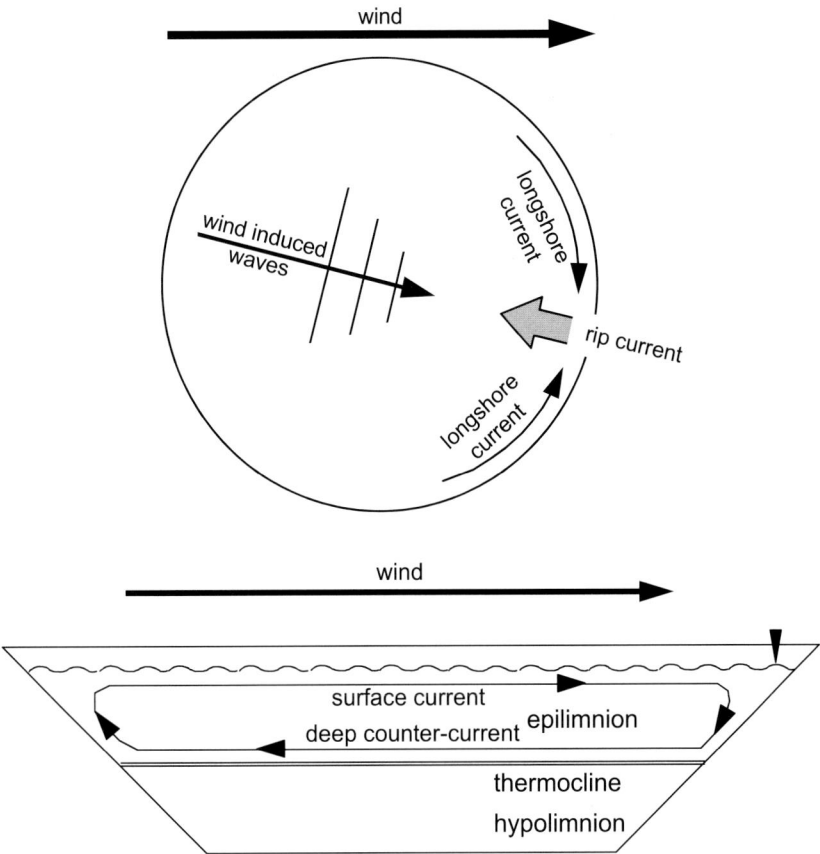

Fig. 5.6. Schematic representation of wind induced currents in a reservoir

zero during the winter and very high during the irrigation season. Stabilization reservoirs can be described as accumulative batch systems with a relatively short discharge, belonging to the wide category that Levenspiel (1972) denominates as 'semi-
batch reactors'. The residence time distribution, volumetric hydraulic loading, and
surface organic loading change during the year as a function of changes in inflow-
outflow rates, water level, area and volume of the reservoir.

The hydraulic mixing of three stabilization reservoirs has been studied by using a
tracer pulse-injection technique while the reservoirs were maintained in steady-state
flow: Argaman et al. (1988) used Rhodamine B to study the Northern and Southern
Kishon Reservoirs (6 million m³ each), and Moreno et al. (1988) used tritiated water
in a 50 000 m³ reservoir. In both cases it was concluded that reservoirs were perfectly
mixed or close to that.

The tools for the analysis of the hydraulic age distribution of effluents within
nonsteady-state reactors with perfect mixing were developed by Juanicó and Friedler
(1994) (see chapter on the subject in this book).

The *Mean Residence Time* (*MRT*) of effluents within a steady-state flow reactor (such as a stabilization pond) is computed as follows under the assumption of perfect mixing:

$$MRT(\text{in days}) = \overline{V} / \overline{Q} \quad , \tag{5.1}$$

where $\overline{V}$ is mean volume (m³) and $\overline{Q}$ is mean flow (m³ d⁻¹).

Equation (5.1) is not applicable to a nonsteady-state flow reactor. Thus, Eq. (5.2) was developed to calculate *MRT* in stabilization reservoirs. At the end of a given day *d*, when there is no outflow from the reservoir, under the assumption of perfect mixing:

$$MRT_d = \frac{\left[(MRT_{d-1} + 1)VOL_{d-1}\right] + (0.5\,IN_d)}{(VOL_{d-1} + IN_d)} \quad , \tag{5.2}$$

where:
- d = end of a given day d
- MRT = mean residence time of effluents within the reservoir (d)
- VOL = volume of effluents within the reservoir (m³)
- IN_d = amount of inflow (m³) that enters the reservoir during day d

The *MRT* of the previous day (d–1) is needed in Eq. 5.2, so the calculation must be started when the reservoir is empty and MRT_d = 0 at the end of the irrigation season of the previous year, and computed forward on a daily basis.

As the system is not steady state, daily values of volume and inflow are needed. When outflow occurs, Eq. 5.2 introduces a small error in the calculation of *MRT* due to the one day time-step used while changes in inflow, outflow and volume are continuous. The difference between the values of *MRT* obtained when using a step of one day or a step of one hour is less than 4% over a one-year simulation.

The *Percentage of Fresh Effluents* (*PFE*) within a reservoir is the small fraction of effluents with a relatively short residence time within the reservoir (fresh efflu-ents) expressed as a percentage of the whole volume of effluents within the reser-voir. The analysis of several reservoirs indicated that the hydraulic operation of the reservoir is the main factor affecting the reservoir's behaviour and water quality (more important than seasonal changes in temperature and solar radiation) and that *PFE* is the parameter which best represents the hydraulic operation of the reser-voir (Juanicó and Shelef 1991, 1994; Liran et al. 1994; Friedler 1993; Juanicó 1995; see Chapter 21). The percentage of fresh effluents having 30 or less than 30 days within the reservoirs is called PFE_{30}; *PFE* of effluents having 5 or less than 5 days within the reservoirs is called PFE_5, etc. The concentration of organic matter in the effluents of the reservoir is closely related to PFE_{30} and PFE_{10} (i.e., the *BOD* in the effluents from the reservoir is affected by the quantity and quality of wastewater in-troduced in the reservoir during the last 30 or 10 days). On the contrary, the con-centration of faecal coliforms in the effluents from the reservoir is more related to PFE_5 and PFE_1 (i.e., to the quantity and quality of wastewater introduced to the reser-voir during the last 5 days or the last 1 day. The reason for the difference is that the rate of faecal coliforms die-off is much higher than the rate of organic matter degra-dation.

5.3.4
MRT and *PFE* in Continuous-Flow Wastewater Reservoirs

Figure 5.7 and Table 5.1 present typical values of *MRT* and PFE_{30} in continuous-flow reservoirs operated in Central-Northern Israel. *MRT* varies from 30–40 days in early winter to 130–160 days in midsummer. *PFE* values at the end of the irrigation season are four or more times higher than at the beginning. The four reservoirs present very similar *MRT* and PFE_{30} values in spite of the big differences in volume. These two parameters depend on the relationships between volume, inflow and outflow of the reservoir and not on the absolute values themselves.

5.3.5
Calculating Surface Organic Loading

Surface organic loading (expressed as kg $BOD\,ha^{-1}\,d^{-1}$) is a parameter normally used for the design of stabilization *Ponds* which have a constant water surface area. In stabilization *Reservoirs* the water surface area is not constant but varies with water level during the year. The reduction in water surface area during the irrigation season (a function of the profile of the reservoir) may double the surface organic loading even in situations where the inflow amount and *BOD* are constant during the whole year

Fig. 5.7. The percentage of fresh effluents (PFE_{30}) and mean hydraulic retention time (*MRT*) in four wastewater reservoirs, during the hydrological year

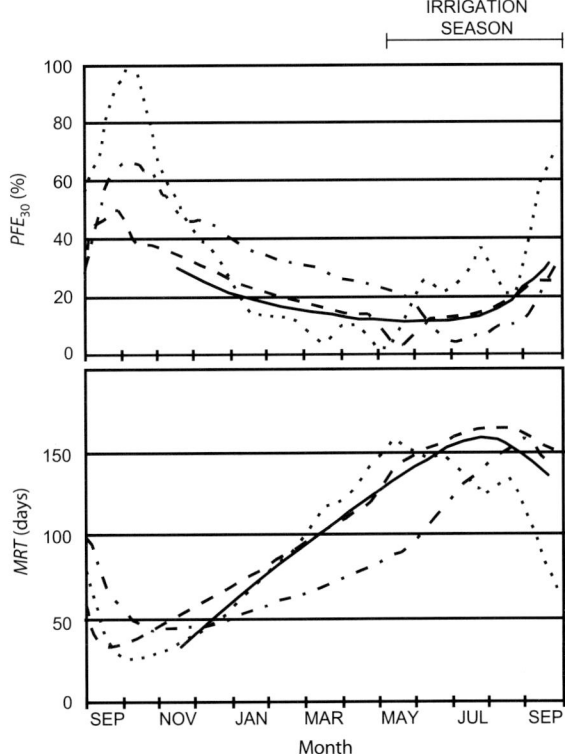

Table 5.1. Annual mean of main design and operation parameters, surface organic loading and removal percentages of some effluent quality parameters, in six wastewater reservoirs in Israel[a]

Reservoir	Max volume (m³)	Max depth (m)	MRT[b] (days)	Inflow BOD (mg l⁻¹)	Oxygen condition (summer)	BOD Load kg/(ha d)	BOD Removal (%) mean	BOD max[c]	COD Load kg/(ha d)	COD Removal (%) mean	COD max[c]	TSS Load kg/(ha d)	TSS Removal (%) mean	TSS max[c]	MBAS (detergents) Load kg/(ha d)	MBAS Removal (%) mean	MBAS max[c]	Total coliforms Inflow (MPN/100 ml)	Total coliforms Removal (%) mean	max[c]
Kishon S	6.5 E6	10.5	100		aerobic	6	30	90	40	8	90	10	−6	95	0.45	30	99			
Kishon N	6.4 E6	10.5	80	40	facult.	20	80	95	70	50	75	20	40	90	0.75	50	70	10^5	89	99.98
Geta'ot	500000	5.5	105	80	facult.	30	70	95	90	60	95	110	80	98	0.001	60	90	10^6	90	99.99
Eliahu	260000	9	130	150	facult.	40	70	85	90	50	60	90	65	90						
Adashim[d]	560000	7	80	250	facult./anaerob.	170	50	70	400	35	50	110	1	40						
Genigar	850000	8	125	1000	anaerob.	310	85	95	800	85	95	140	70	95				10^8	92	99.90

[a] Mean water temperature in wastewater reservoirs in Israel: 13°C in winter, 30°C in summer;
[b] MRT: mean residence time of effluents within the reservoir;
[c] Maximum removal values refer to those at the beginning of the irrigation season (May);
[d] Reservoir with permanent short-circuiting of effluents between inlet and outlet (which are closely located) and large dead areas.

(Fig. 5.3). Thus, the calculation of surface organic loading in stabilization reservoirs should be made for each day of the year. The annual mean of these daily organic loading values can be then computed for the whole hydrological year for comparison between reservoirs with different loading (Table 5.1).

Another difficulty arising from the nonsteady-state condition of these systems is that surface organic loading values have not the same meaning when the reservoir is full and when the reservoir is almost empty, due to the dilution of the new incoming wastewater into different volumes of old effluents.

5.3.6
Performance and Limitations of the Continuous-Flow Reservoirs

In reservoirs with an irrigation season of 4–5 months, the annual mean of *MRT* is 80–130 days (Table 5.1); the annual mean *BOD* removal percentage is 70–85%, *COD* removal is 50–85%, *TSS* removal is 40–80% and detergents removal is 50–60%. Removal of N and P in two stabilization reservoirs in series with a joint annual *MRT* of 180 days (Kishon North and Kishon South) is 70% and 60% respectively (Fig. 5.8).

The low removal values of the Southern Kishon Reservoir are due to the high quality of the inflow entering the reservoir (*BOD* = 10 mg l^{-1}): growth of algae and zooplankton within the reservoir adds to the total *BOD*, *COD* and *TSS* values. The poor performance of the Adashim Reservoir is due to a design error: inlet and outlet were closely located in this reservoir, resulting in a permanent hydraulic short-circuiting between them.

The main limitation of continuous-flow reservoirs is the degradation of effluent quality during the irrigation season (Fig. 5.9). The quality of the effluents is optimal at the beginning of the irrigation season when the reservoir is full of effluents with long residence time, and the organic loading and the percentage of fresh effluents

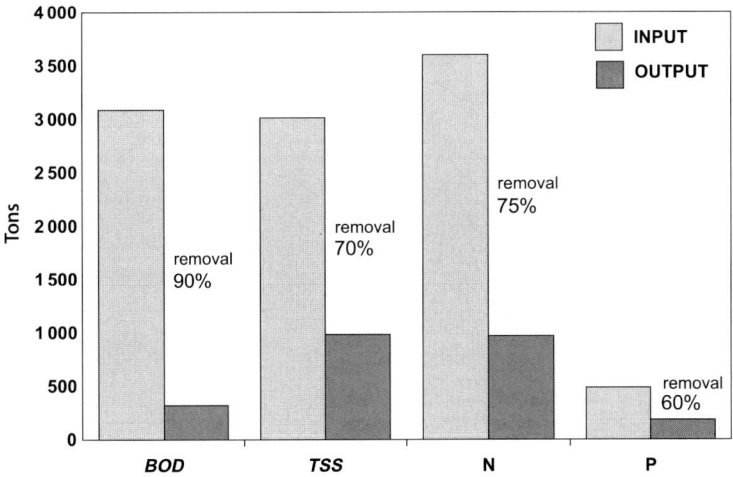

Fig. 5.8. Accumulative amounts of *BOD*, *TSS*, N and P that entered and left two wastewater reservoirs in series (Kishon N and S, Table 5.1), during 6 years (1984 to 1990)

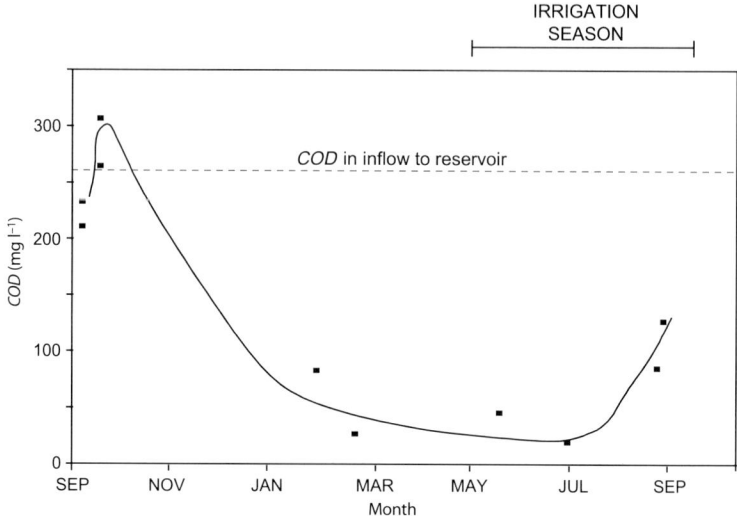

Fig. 5.9. The concentration of *COD* in the epilimnion of the Geta'ot Reservoir during the year

within the reservoir are minimal. Then water quality begins to deteriorate as water level drops while new effluents continue to enter the reservoir. Effluent quality is worst by the end of the irrigation season when the reservoir is almost empty and 30–50% of the effluents have 30 days or less within it. Figure 5.9 shows that the concentration of *COD* within the reservoir may be even higher than in the effluents that enter the reservoir; this is due to the contribution of organic matter to the water column from the bottom sediments (mainly as soluble *COD*). Other parameters, including pathogens, behave in the same way as *COD* (Juanicó and Shelef 1991, 1994; Liran et al. 1994; Bar-Or and Keshet 1996; Nature Reserves Agency 1997). The differences in the quality of the outflow between the beginning and the end of the irrigation season are so conspicuous, that it is impossible to assign the quality of the outflow from a particular reservoir to any standard of water quality for irrigation. Under these conditions, the continuous-flow reservoirs can be considered as storage units only, because their treatment capacity is so irregular.

Maximum removal percentages were obtained at the beginning of the irrigation season, reaching one order of magnitude for *BOD*, *COD*, *TSS* and detergents, and 3–4 orders of magnitude for total coliforms. These maximum removal percentages are much higher than the mean ones (Table 5.1).

Single regression analyses on one variable indicate PFE_{30} as the parameter having the highest correlation with *BOD* and *COD* removal (Juanicó and Shelef 1991, 1994). PFE_{30} is also the dominant parameter when a second and a third variable are added in multivariate regression analysis. *MRT* ranks much below PFE_{30} in all the models. The effect of operational parameters overrode the effect of the environmental ones which were not detectable. It has been long known that sewage treatment reactors render effluents of better quality at longer *MRT*, and *MRT* is widely used for the design of all types of perfectly mixed steady-state flow treatment plants. Results from the study of several reservoirs in Israel indicate that in a nonsteady-state flow reactor, the reactor performance is determined by the relatively small fraction of fresh effluents (*PFE*) and not by the whole age distribution of the effluents (*MRT*). In a steady-state flow reactor the *PFE/MRT* ratio is constant, explaining why *MRT* can successfully substitute *PFE* in this particular case (Juanicó and Friedler 1994).

5.3.7
The Removal of Coliforms

The annual mean removal of total coliforms in the three reservoirs presented in Table 5.1 is only one order of magnitude, a low value for treatment units with so long a mean residence time. Similar results have been found in other continuous-flow reservoirs (Dor et al. 1987; Perle 1988; Liran et al. 1994; Goldberg 1994; and others). The low pathogen removal capacity of continuous-flow wastewater reservoirs obliges – in many cases – the performance of disinfection of the effluents (Rabkin and Eren 1982).

Fattal et al. (1993), Liran et al. (1994) and Goldberg (1994) performed independent studies on the parameters affecting the die-off of faecal coliforms in wastewater reservoirs, finding that die-off is closely related to photosynthetic activity. Die-off rate is higher in summer than in winter, due to the higher photosynthesis which is a function of solar radiation and water temperatures. These findings agree with those of Curtis et al. (1992), which identified the production of singlet oxygen associated with high *p*H values as main parameters in the die-off of pathogens in stabilization ponds.

Liran et al. (1994) found that the removal of faecal coliforms in the reservoir strongly changes throughout the year following the changes of the operational parameters of the reservoir (input/output rates and volume). There are also sharp differences in coliforms removal efficiency between the epilimnion and the hypolimnion (Fig. 5.10).

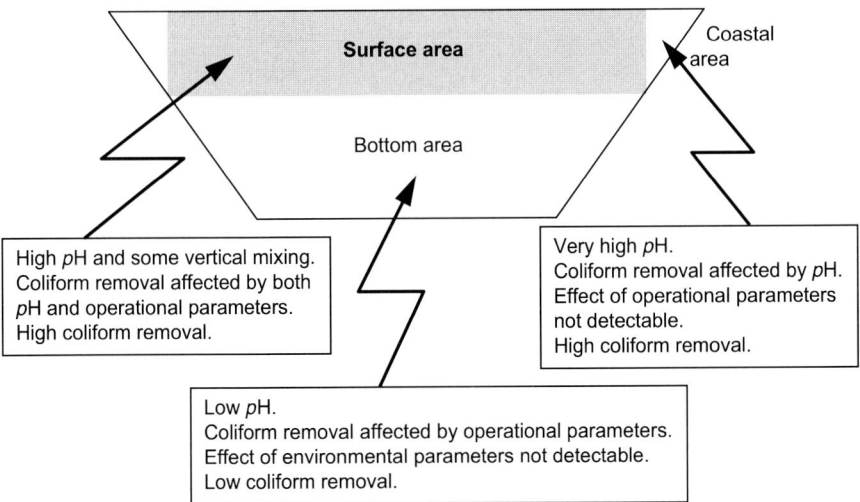

Fig. 5.10. The three different areas within a wastewater reservoir regarding the die-off of faecal coliforms, according to Liran et al. (1994)

Coliforms removal is high in the epilimnion of the reservoir characterized by high pH values generated by photosynthesis. When the reservoir is operated as a continuous-flow reactor, the coliforms removal in the epilimnion is related mainly to the percentage of fresh effluents with five days or less within the reservoir (PFE_1 and PFE_5). Coliforms removal is low in the hypolimnion of the reservoir without photosynthesis and low pH values. When the reservoir is operated as a continuous-flow reactor, the coliforms removal in the hypolimnion is related mainly to PFE_{20}–PFE_{30}.

The low *coliform removal/mean residence time* performance ratio of stabilization reservoirs operated as continuous-flow reactors is due to:

1. Their relatively large hypolimnion with no algae activity and low pH values where coliform removal is minimal. The negative effect of the large hypolimnion can be reduced by mixing the reservoir (stratification break-down).
2. The drastic increase in the percentage of fresh effluents within the reservoir by the end of the irrigation season. The operation of the reservoirs as sequential batch reactors solves this problem.

5.4
The New Sequential Batch Reservoirs

Sequential batch reservoirs (SBR) were developed as an answer to the quest for better quality effluents that could not be provided by the limited continuous-flow reservoirs. SBR are not considered merely storage units but an integral part of the sewage treatment system (Juanicó 1994d; Friedler and Juanicó 1996). In the SBR operational mode, the inflow to the reservoir is stopped before the reservoir starts to release effluents (Fig. 5.2).

The rate of *BOD* degradation or faecal coliforms die-off in the 'closed' batch reservoirs is the same than in continuous-flow ones, but the removal finally obtained is much higher because there are no new (fresh) effluents with high concentration of pollutants entering to the reservoir (*PFE* is zero). Several real-scale experiments performed by independent research teams in Israel (Kott et al. 1978; Juanicó 1996) indicate that the removal of pollutants with low degradation rates such as *BOD*, *COD* and detergents in batch reservoirs is up to one order of magnitude, and the removal of pollutants with high degradation (die-off) rate such as faecal coliforms is up to five orders of magnitude. Similar results have been obtained by different research teams in other parts of the world, e.g., Funderburg et al. (1978) in Texas, USA and Indelicato et al. (1995, 1996) in Southern Italy. The following two examples illustrate the effect of batch operation in *BOD* and coliform removal.

5.4.1
Example I. Batch Operation During Winter

The Sarid Reservoir is 7.3 m deep with an active volume of 405 000 m³. It receives effluents from the city of Afula after treatment in stabilization ponds (which are highly overloaded) with $BOD = 160$ mg l⁻¹ and faecal coliforms = 1 000 000/100 ml. Input of effluents to the reservoir was stopped on November 30 (early winter) and the reservoir was then operated in batch mode for several months. Composite samples from several points and depths were taken on 1st November (when effluents were still enter-

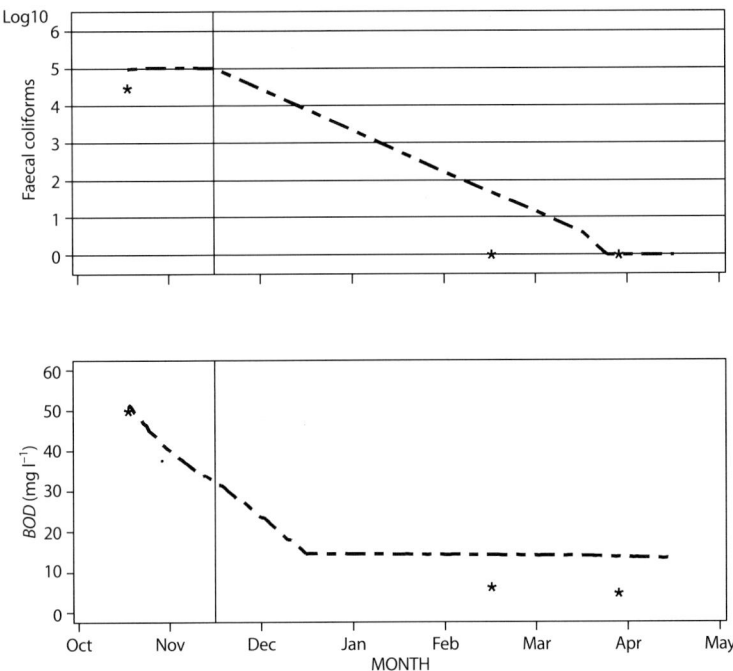

Fig. 5.11. Reduction of *BOD* and faecal coliforms in the Sarid Reservoir operated as a batch reactor in winter. Composite samples from various points and depths (*lines:* model forecasting; *stars:* measured values)

ing to the reservoir), 1st March (after three months of batch operation) and on 12th April (after more than four months of batch operation), and analysed for several physico-chemical and bacteriological parameters. The reduction of *BOD* and faecal coliforms in the reservoir were simulated by mathematical and statistical models. Figure 5.11 presents the values for *BOD* and faecal coliforms in the reservoir, comparing predicted with actual values. It can be observed than faecal coliforms numbers dropped by four and a half orders of magnitude in about 90 days. The latest sample taken in April also showed an almost zero concentration of faecal coliforms. *BOD* concentration was reduced till reaching a minimum of 10–20 mg l⁻¹. In both cases the predictions of the models were more conservative than the measured values.

5.4.2
Example II. Batch Operation During Spring

The reservoir 'III' is 6 m deep with an active volume of 400 000 m³. It receives effluents from another reservoir with a *BOD* of about 250 mg l⁻¹ and faecal coliforms of about 200 000. During the studied year, the 'III' reservoir had 50 000 m³ of effluents by the end of January, received 211 000 m³ of effluents during February, 105 000 m³ during March, and 21 000 m³ till 14 April. On 14 April all input of wastewater to the reservoir 'III' was stopped and it started to work as a batch reactor. Samples were taken at the east and west sides of the reservoir, at surface (0.2 m below surface) and bottom (0.5 m above

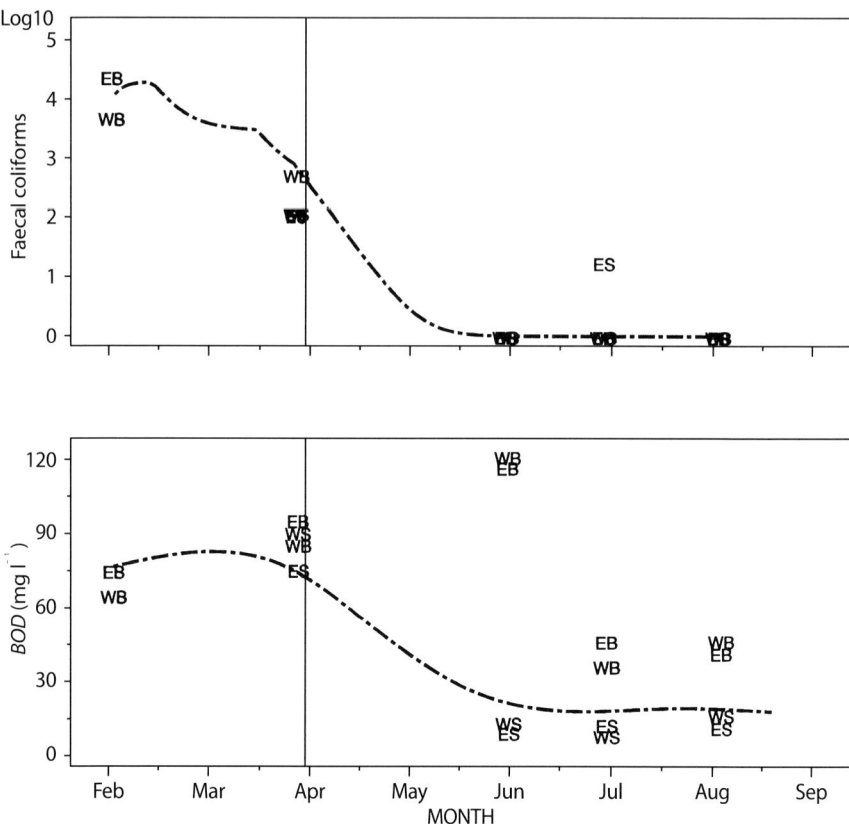

Fig. 5.12. Reservoir III. Actual field data and model predictions (*samples: EB* = East bottom, *ES* = East surface, *WB* = West bottom, *WS* = West surface; *dotted lines:* model predictions for surface quality; *vertical lines:* date when inflow to reservoir was cut off

bottom) and analysed for several parameters. Profiles of temperature, *p*H, dissolved oxygen, electrical conductivity and light penetration were also taken. Figure 5.12 presents the results of *BOD* and faecal coliforms analyses, compared with model predictions for surface data. The *BOD* model is a multiple regression one described in Juanicó and Shelef (1991, 1994). The faecal coliforms model is a simple kinetic one. Figure 5.12 shows that, after a batch operation of about 45 days, the 'III' reservoir produced effluents of unrestricted irrigation quality even for rigorous standards.

5.4.3
Several Reservoirs Working in Sequential Batch

When the input to the reservoir is closed, effluents must be stored in another reservoir. This requires more than one reservoir in each "treatment and storage unit". Optimal number of SBR reservoirs is 3–4, but 5–6 may be used in some cases. Figures 5.13 and 5.14 present the operation of several reservoirs in sequential batch working un-

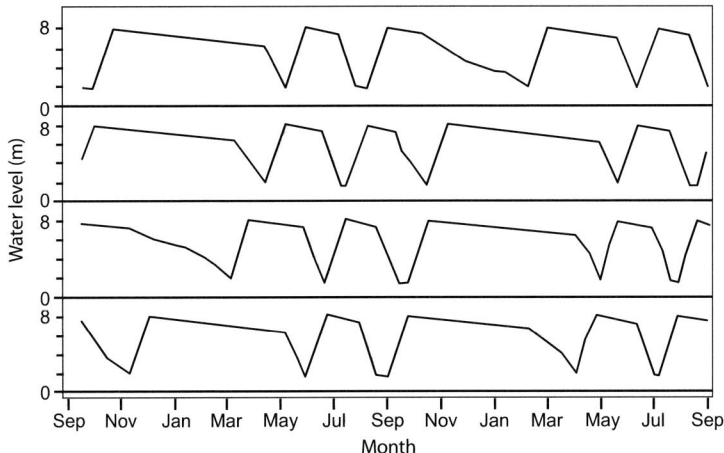

Fig. 5.13. Water level curves of four reservoirs of equal size, working sequential batch in parallel, supplying a water demand curve of "continuous discharge with intensive discharge season" (see Fig. 5.4). Reservoirs are emptied more than once a year

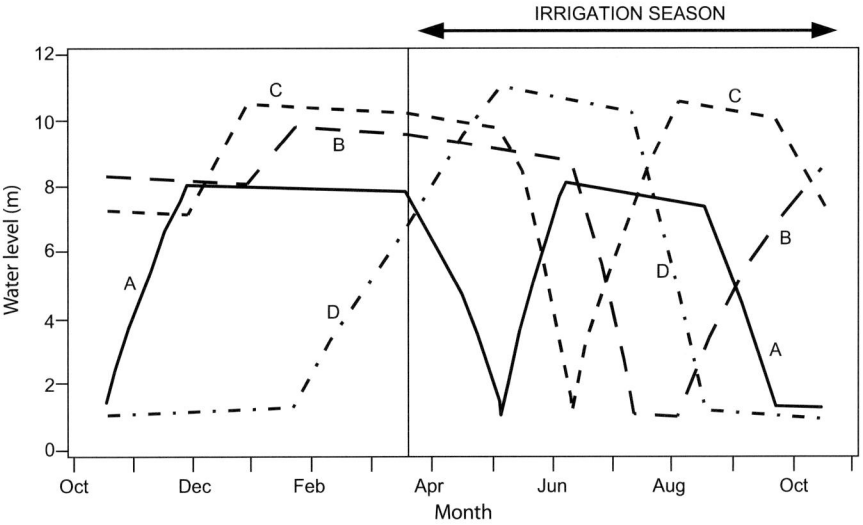

Fig. 5.14. Water level curves of four reservoirs of different sizes, working sequential batch in parallel, supplying a water demand curve of "discontinuous discharge – long season" (see Fig. 5.4). Reservoir A is emptied twice a year, C is emptied 1½ times a year, B and D are emptied only once a year

der different conditions. Note that at any given time effluents are taken from only one reservoir while the others are in filling, batch or idle stages. The fill-batch-emptying-idle cycles of the reservoirs are shorter during the high water demand periods (compare Figs 5.13 and 5.14 with Fig. 5.4) and some of the reservoirs may be emptied twice during the irrigation season.

SBR reservoirs require more storage capacity than continuous-flow reservoirs due to the period of no inflow of effluents to the reservoir which must be compensated by supplementary storage capacity in another reservoir. Continuous-flow reservoirs perform the poorest as treatment units, but are relatively small in relation to the amount of released effluents, and their operation is simple and elastic. SBR render effluents of very good quality, but are relatively large and their operation is more complicated. An optimization of the operation of these reservoirs shows that SBR in series perform better when the effluents are released over a period of a few months (Figs 5.3 and 5.4) while SBR in parallel perform better when the discharge season is longer or all the year round.

5.5
Maximum Organic Loading

5.5.1
Mean Surface Organic Loading

The relationship between the organic loading to the reservoir and its oxygen regime is more complicated than in conventional waste stabilization ponds, due to two main factors:

1. The cumulative effect of the wastewater entering a reactor of changing volume. For example, it has not the same effect to fill the reservoir with effluents with $BOD = 160$ mg l^{-1} during four winter months (low temperatures and solar radiation on a reservoir of increasing volume) and then with $BOD = 40$ mg l^{-1} during four summer months (high temperatures and solar radiation on a reservoir of decreasing volume), than the contrary ($BOD = 40$ mg l^{-1} during winter and 160 mg l^{-1} during summer).
2. The deeper the reservoir, the lower the surface organic loading that it can support without developing anaerobic conditions. Dissolved oxygen in reservoirs is provided by photosynthetic activity of algae and diffusion of atmospheric oxygen. Thus, deeper reservoirs with a small area/volume ratio have a poorer supply of oxygen than shallow water bodies. Besides, they have a large anaerobic hypolimnion with high oxygen demand. Deeper reservoirs require lower surface organic loading or the use of aerators and/or mixers to maintain aerobic conditions.

Another factor affecting the allowable loading to the reservoir is the quality of the input wastewater. Wastewater with $BOD = 100$ mg l^{-1} entering the reservoir from a stabilization pond imposes less oxygen demand than wastewater with $BOD = 100$ mg l^{-1} entering the reservoir from an activated sludge or aerated lagoons reactor, due to the oxygen produced by the algae present (or not) in the incoming wastewater. This is especially valid in the fall when the reservoir is almost empty and most of the effluents are 'fresh' in a period of low algae growth.

Thus, the proper determination of the maximum surface organic loading allowed in a specific reservoir is made through simulation of the behaviour of the whole reservoir (see Chapter 7). However, some 'rules of the dump' may be useful as a first approximation.

Planners in Israel have used an average surface organic loading of 50 kg BOD ha^{-1} d^{-1} as the maximum allowable loading. Some studies such as Dor and Raber (1990 and

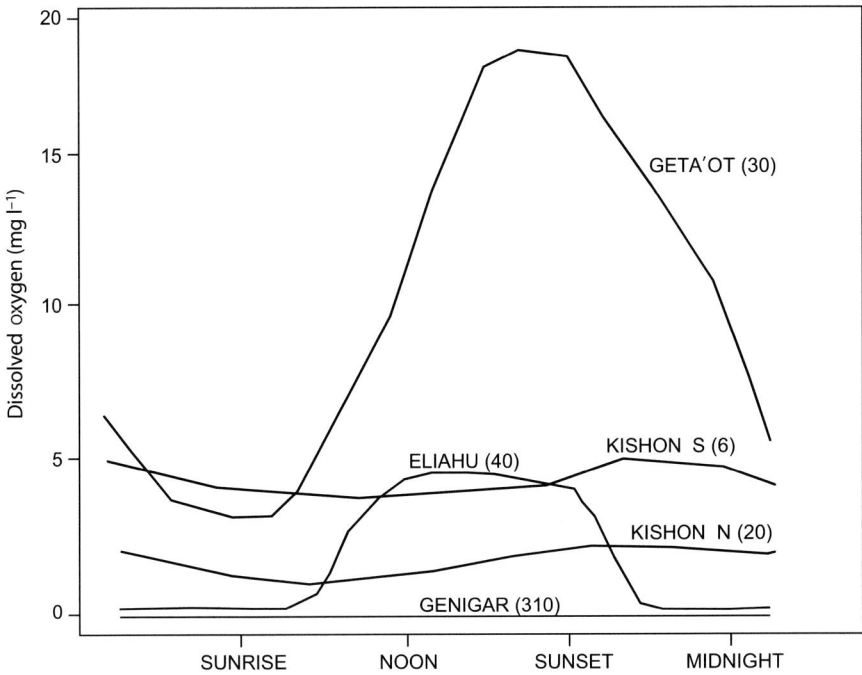

Fig. 5.15. The daily cycle of dissolved oxygen in five reservoirs in Israel, at 20 cm below the surface, in July (midsummer). The values in parentheses are the annual mean organic surface loading of the reservoirs in kg BOD ha^{-1} d^{-1}

in this book) have confirmed that most reservoirs with this loading present a good performance. Reservoirs receiving low surface organic loading (annual mean up to 30 kg BOD ha^{-1} d^{-1}) are fully aerobic or facultative (Table 5.1, Fig. 5.15): they present no offensive odours. However, the Eliahu Reservoir (a 9 m deep reservoir receiving 40 kg BOD ha^{-1} d^{-1}) is facultative, but dissolved oxygen at surface may drop to zero at night with sporadic emissions of offensive odours. Reservoirs receiving organic loading similar to those of facultative stabilization ponds (150 kg BOD ha^{-1} d^{-1}) or more are anaerobic most of the time with strong emissions of offensive odours. These findings point out that the value of 50 kg BOD ha^{-1} d^{-1} may be too high in some cases. A value between 30 and 40 kg BOD ha^{-1} d^{-1} would be a safer upper limit when the non-emission of offensive odours is imperative.

5.5.2
Increasing the Surface Organic Loading

It is possible to increase the surface organic loading to wastewater reservoirs beyond the limits recommended above by adding mixers and aerators to the reservoirs. Unfortunately, there is no 'rule of the dump' to calculate the relationship between maximum organic loading and the required mixing and aeration power; simulation by modelling in a case by case basis is still required.

5.5.3
Shocks of High Organic Loading

Wastewater reservoirs are able to receive sporadic shocks of high organic loading without any conspicuous change in their oxygen balance. This is due to the large volume of effluents stored in the reservoir receiving the shock. Thus, they act as an equalization tank. This capacity is important in the design of a sewage treatment and reuse system, because intensive sewage treatment systems, such as activated sludge and others, are known for suffering from periodic collapses of the process with release of low quality effluents. The reservoirs will 'absorb' these failures without problem. Most sewage treatment and reuse systems in Israel are made of a series of intensive or semi-intensive treatment units such as activated sludge or anaerobic and aerated lagoons, followed by a series of extensive units such as wastewater reservoirs (Juanicó 1994a, 1994b; Aharoni and Kanarek 1994).

5.6
The Tools for Design

As the reader could appreciate in the previous pages, the unsteady-state condition of wastewater reservoirs does not allow the use of simple kinetic models with analytical solutions for process design. Instead, simulation by means of mathematical modelling is required.

Three main models are necessary for the optimal design and operation of wastewater reservoirs:

1. *Hydraulic age distribution.* This model (see Chapter 7) describes the changes in the hydraulic age distribution under different operational regimes. It allows the quick elimination of design and operational alternatives which lead to high *PFE* values in the effluents discharged by the reservoirs.
2. *Ecological simulation.* This model (see chapter on it in this book) simulates the behaviour of the reservoir and the quality of the released effluents. It requires time and effort to run it and interpret the results, but it allows selection among potential design and operational alternatives previously selected by the hydraulic age distribution model, and enables fine tuning of the finally selected one.
3. *Multiple-reservoirs operation.* This model (not published yet but relatively easy to develop) simulates the simultaneous operation of several reservoirs in sequential batch mode and the matching of reservoirs discharge with the water demand curve. It allows one to decide between parallel and in series alternatives, determine the optimal number of sequential batch reservoirs, and select the optimal sequence of fill-batch-emptying-idle cycles.

5.7
Control and Monitoring

There are two aspects in control and monitoring of wastewater reservoirs which must be taken into consideration to avoid erroneous sampling and interpretation of data: sampling and data analysis.

5.7.1
Sampling

The quality of water in the epilimnion differs from that of hypolimnion. An evaluation of the quality of water in the whole reservoir requires sampling at different depths. However, the evaluation of the quality of the water released for irrigation can be based only on surface samples, because a properly designed outlet should take water only from the epilimnion.

Reservoirs have a large volume of deep water at a distance from the coast, surrounded by a very small volume of coastal shallow water (Fig. 5.10). The quality of shallow coastal water differs from that of deep water due to several factors: higher *p*H due to higher photosynthesis, resuspension of sediments by waves, long-shore and rip currents (Fig. 5.6), higher water temperature and evaporation rates, etc. Thus, it is not correct to sample the coastal water to evaluate the quality of the effluents in the epilimnion of the reservoir. Samples should be taken at least 4 m from the coast, with the help of a small boat or a fishing rod. Samples should not be taken on the leeward side of the reservoir (Fig. 5.6).

5.7.2
Data Analysis

The nonsteady-state characteristic of wastewater reservoirs imposes the use of weighted means to calculate the average quality of the reclaimed water released by the reservoir (Fig. 5.16). The concentrations of a given pollutant measured by sampling and analysis at different times during the irrigation season must be weighted by the amount of effluents released during the period which is represented by each sample.

Fig. 5.16. Analysis of data on the quality of the water released for irrigation from a continuous-flow wastewater reservoir. **a** Wrong procedure: calculate the average quality by making a simple arithmetic mean of the values obtained by sampling and analysis (*circles* 1 to 3). **b** Correct procedure: calculate the average quality by making a mean of the concentrations (*circles* 1 to 3) weighted by the volume of water released during the period related to each sample (*rectangles* I to III)

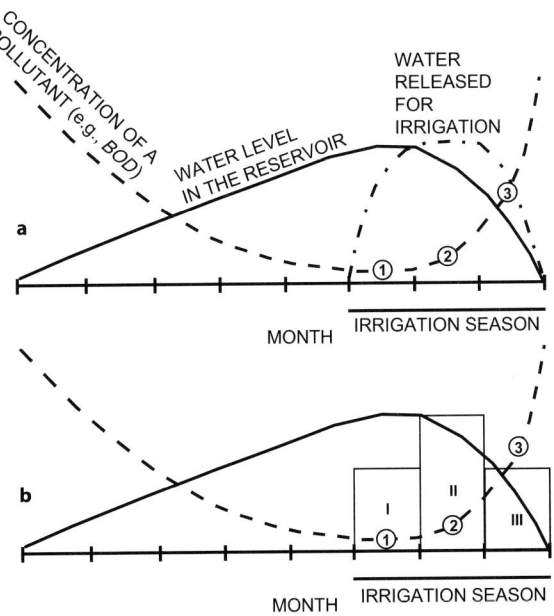

In the example presented in Fig. 5.16, the simple arithmetic mean of the concentration of a given pollutant in three samples (1, 2 and 3) would be:

Arithmetic mean = ([1] + [2] + [3]) / 3   ,

while the weighted mean will include the volumes of water that are represented by each sample (I, II and III):

Weighted mean = ([1] × I + [2] × II + [3] × III) / (I + II + III)   .

The use of non-weighted means introduces a systematic error which sometimes may be significant.

References

Aharoni A, Kanarek A(1994) The wastewater reclamation system of Natania. Performance of the Southern Reservoir during the 1993 year. Water and Irrigation 338:42–45 (in Hebrew)

Argaman Y, Redlich E, Juanicó M, Rom D (1988) Distribution of residence time in the reservoirs. In: The Kishon Complex Monitoring Program, Fifth Annual Rep., pp 111–141, Technion. (in Hebrew)

Avnimelech Y (1989) Modelling the accumulation of organic matter in the sediments of a newly constructed reservoir. Water Res 23(10):1327–1329

Avnimelech Y, Wodka M (1988) Accumulation of nutrients in the sediments of Ma'aleh HaKishon reclaimed effluents reservoir. Wat Res 22(11):1437–1442

Bar-Or Y, Keshet N (1996) Water quality in wastewater reservoirs. Wat Irrigat 358:29–37 (in Hebrew)

Berend J, Pano A (1976) Winter storage of wastewater effluents. Tahal Co. Report, Tel Aviv, 50 pp (in Hebrew)

Curtis T, Mara D, Silva S (1992) Influence of pH, oxygen, and humic substances on ability of sunlight to damage faecal coliforms in waste stabilization pond water. Applied Envir Microbiol 58(4):1335–1343

Dor I, Raber M (1990) Deep wastewater reservoirs in Israel: Empirical data for monitoring and control. Wat Res 24(9):1077–1084

Dor I, Schechter H, Bromley H (1987) Limnology of a hypertrophic reservoir storing wastewater effluent for agriculture at Kibbutz Na'an, Israel. Hydrobiologia 150:225–241

Eren J (1978) Succession of phyto- and zooplankton in a wastewater storage reservoir. Verh Internat Verein Limnol 20:1926–1929

Fattal B, Puyeski Y, Eitan G, Dor I (1993) Removal of indicator microorganisms in wastewater reservoir in relation to physico-chemical variables. Wat Sci Tech 27(7–8):321–329

Friedler E (1993) Mathematical modelling of stabilization reservoirs. DSc Thesis, Technion-Israel Institute

Friedler E, Juanicó M (1996) Treatment and storage of wastewater for agricultural irrigation. Wat Irrig Rev 16(4):26–30

Funderburg S, Moore B, Sorber C, Sagik B (1978) Survival of polivirus in model wastewater holding ponds. Progress Water Technol 10 (5–6):619–629

Goldberg T (1994) Model for the forecasting of die-off rate of E. coli in the wastewater reservoir of the kibbutz Na'an. MSc Thesis, The Hebrew University of Jerusalem, 152 pp (in Hebrew)

Indelicato S, Barbagallo S, Cirelli G (1995) Change in wastewater quality during seasonal storage. Procc. Conf. Natural and Constr. Wetlands WW Treatment and Reuse, Perugia, pp 195–204

Indelicato S, Barbagallo S, Cirelli G, Zimbone S (1996) Reuse of municipal wastewater for irrigation in Italy. Procc. 7th Internat. Conf. Wat Irrigat, Tel Aviv, pp 210–221

Juanicó M (1994a) Alternative schemes for municipal sewage treatment and disposal in industrialized countries: Israel as a case study. Ecol Eng 2:101–118

Juanicó M (1994b) The role of stabilization reservoirs on sewage treatment systems. Wat Irrig 335:19–21 (in Hebrew)

Juanicó M (1994c) Limnology of a warm hypertrophic wastewater reservoir in Israel. I. The physical environment. Int Revue ges Hydrobiol 79(3):423–436

Juanicó M (1994d) The role of stabilization reservoirs on sewage treatment systems. Wat Irrig 335:19–21 (in Hebrew)

Juanicó M (1995) Limnology of a warm hypertrophic wastewater reservoir in Israel. II. Changes in Water Quality. Int Revue ges Hydrobiol 80(3):415–428

Juanicó M (1996) The performance of batch stabilization reservoirs for wastewater treatment, storage and reuse in Israel. Water Res. 33(10-11):149-159

Juanicó M, Shelef G (1991) The performance of stabilization reservoirs as a function of design and operation parameters. Wat Sci Technol 23(7–9):1509–1516

Juanicó M, Friedler E (1994) Hydraulic age distribution in perfectly mixed non-steady-state reactors. ASCE J Environ Eng 120(6):1427–1445

Juanicó M, Shelef G (1994) Design, operation and performance of stabilization reservoirs for wastewater irrigation in Israel. Wat Res 28(1):175–186

Kott Y; Ben-Ari H, Betzer N (1978) Lagooned, secondary effluents as water source for extended agricultural purposes. Wat Res 12:1101–1106

Levenspiel O (1972) Chemical reaction engineering. John Wiley & Sons, 2nd Edition, 578 pp

Liran A, Juanicó M, Shelef G (1994) Bacteria removal in a stabilization reservoir for wastewater irrigation in Israel. Wat Res 28(6):1305–1314

Meron A, Eren J (1985) Effect of salinity on agricultural reclamation. Proc. Water Reuse Symp. III 1:543–553

Moreno M, Medina M, Moreno J, Soler A, Saez J (1988) Modelling the performance of deep waste stabilization ponds. Wat Resour Bull 24(2):377–87

Nature Reserves Agency (1997) Monitoring of streams, reservoirs and wastewater irrigation. Report on 1996 Activities, 90 pp, Israel (in Hebrew)

Pano A (1975) Storage of wastewater in Sarid and Mizra reservoirs. Tahal Report, 26 pp (in Hebrew)

Perle M (1988) The fate of indicator and pathogenic microorganisms in a wastewater renovation system. Master Thesis, Technion, 130 pp (in Hebrew)

Rabkin S, Eren I (1982) Disinfection of Haifa's effluents after residence in reservoirs. Period May–August 1982. Mekorot Water Supply Co, North Regional Report, 11 pp (in Hebrew)

Wachs A, Berend A (1968) Extra deep ponds. In: Glyna E, Eckenfelder W (eds) Advanc. water quality improv., Univ Texas Press, pp 450–456

Chapter 6

Hydraulic Age Distribution

Marcelo Juanicó · Eran Friedler

6.1
Introduction

The methodology to study the hydraulic age distribution in steady-state perfectly mixed reactors was established by Danckwerts (1953) in his classic work "Continuous-flow Systems." But the simple equations developed for the steady-state regime are not applicable to nonsteady-state regimes, which are characterized by irregular hydraulic age distribution. The Leslie Matrix Model was originally developed for the analysis of age structured biological population dynamics. This chapter deals with the application of the Leslie Matrix Model to the "age population" of effluents within stabilization reservoirs as representatives of nonsteady-state flow reactors.

Most of the perfectly mixed sewage treatment reactors in use (i.e., aerated lagoons, activated sludge, nonelongated stabilization ponds with long residence time, etc.) are designed as ideal steady-state flow systems with constant flow, volume and mean hydraulic residence time. These sewage treatment systems do not actually fulfil the ideal steady-state flow assumption, because of differences in flow between day and night, weekdays and weekend, summer and winter, impact of storm water discharges, etc. (Fig. 6.1). As the plants are designed with a constant volume, changes in flow rate determine concomitant changes in the hydraulic age distribution and mean residence time of the effluents within the reactor. Although this departure from the steady-state flow has been long known, it has not yet been introduced in the calculations for the process design of the systems. The available mathematical tools for process design are based on results from batch experiments applied to the steady-state flow regime. The analytical solutions to equations for steady-state flow regime are much easier to handle than those required by the actual changing flow regime of the treatment plants. It has been assumed that the deviations from steady-state flow are minimal (Levenspiel 1972) and that their effect can be diminished by means of small operational adjustments of the treatment plant. The present concern with the impact of storm water discharges on treatment plants is relatively new, and efforts to abandon the steady-state flow assumption are at an early stage of development (Rossman 1989; Beck et al. 1991).

Sequential batch reactors have an irregular flow with changing volume, following an operational regime of five steps: fill, batch react, batch settle, draw, idle (Irvine and Busch 1979). The filling rate is a very important operational parameter determining the performance of the reactor (Dennis and Irvine 1979). Although there is an increasing use of the sequential batch reactors for several purposes, no mathematical tools have been developed to analyse their hydraulic age distribution.

Fig. 6.1. Variation in sewage inflow into Sewage Treatment Plants. **a** Diurnal variations in the influent to the Greater Tel Aviv Treatment Plant (sequence selected at random from 1988 data). **b** Monthly variations in the influent to Haifa City Treatment Plant (averages from 1985 to 1991)

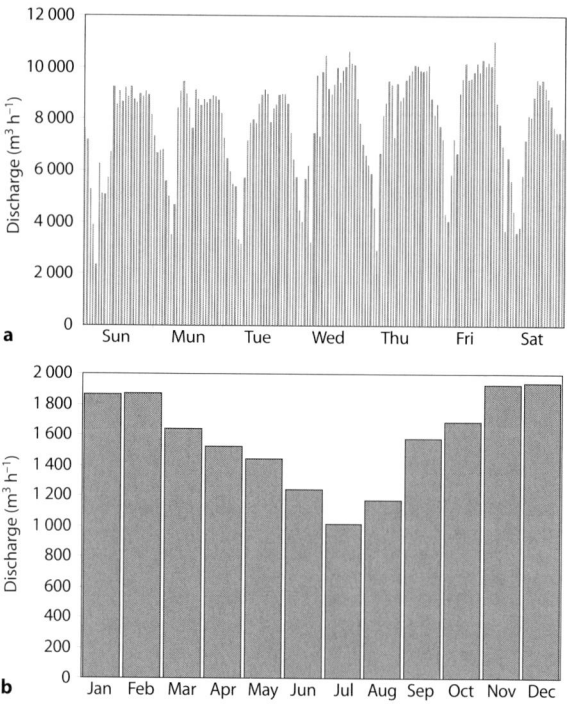

Stabilization reservoirs are employed in Israel for the seasonal storage of wastewater during the rainy winter, for irrigation use during the dry summer. They are similar to sequential batch reactors, and Juanicó and Shelef (1991, 1994) demonstrated that the filling/emptying regime determines most of the reservoir performance. These authors developed a new parameter (*PFE*: the percentage of fresh effluents within the reservoir) to study the performance of the reservoirs, and found very good correlations between *PFE* and the removal of *BOD* and faecal coliforms. They presented a computer algorithm to calculate *PFE*, but a mathematical tool for the analysis of the hydraulic age distribution is not yet available.

Steady-state flow reactors are only a particular situation of nonsteady-state ones, reached when inflow rate is constant and outflow rate equals inflow (leading to constant volume). Batch reactors are also a particular situation of nonsteady-state ones, reached when both inflow and outflow are zero and volume remains constant. A general solution for perfectly mixed, nonsteady-state flow reactors, will include the steady-state flow and batch reactors as particular solutions, as demonstrated below in this article.

The objectives of this chapter are:

1. To define a single model of the hydraulic age distribution in all perfectly mixed reactors, including steady-state flow and batch reactors as particular cases of the general model.
2. To provide the mathematical tools and computer algorithms for describing the hydraulic age distribution in formal terms, and to calculate the hydraulic mean resi-

dence time and the percentage of fresh effluents within the reactor (*PFE*). These tools will enable direct application of the model in the design and operation of perfectly mixed stabilization reservoir in particular and chemical reactors in general under any flow regime.

6.2
Concepts: The Hydraulic Age Distribution

6.2.1
PFE: Percentage of Fresh Effluents

In perfectly mixed reactors the hydraulic age distribution of effluent withdrawn from the reactor is identical to the hydraulic age distribution of the reactor contents. Some of these effluents have resided within the reactor for a long time, while some of them are "fresh" (i.e., resided within the reactor only for short time). For instance *PFE* of 10 days (e.g., PFE_{10}) is the volume of the effluents residing 1–10 days within the reactor, expressed as a percentage of the total reactor contents.

The difference in hydraulic age distribution between steady-state and nonsteady-state flow reactors, and the relationship between the hydraulic age distribution and reactor performance are illustrated herewith by two theoretical cases:

6.2.1.1
Case One: Two Similar Waste Stabilization Ponds with Different Flow Regimes

Waste stabilization ponds are reactors widely used in sanitary engineering for sewage treatment. Main design parameters are pond depth which varies between 0.8 to 2.5 m, the mean residence time ($\bar{t}$) which ranges between 15 to 30 days, and the surface organic loading which ranges from 80 to 150 kg *BOD* ha^{-1} d^{-1}. The values finally selected for the design depend on several parameters such as *BOD* of raw sewage, seasonal changes in local temperature and solar radiation, etc. Ponds are generally designed with fixed volume and since effluent leaves the pond *via* overflow device, outflow rates are a direct function of inflow rates. Perfectly mixed conditions have been assumed for design of stabilization ponds, although recent tracer experiments have demonstrated that a dispersed flow model better matches the actual hydraulic regime in most ponds. However, nonelongated ponds with long residence times approach perfectly mixed conditions very well.

Figure 6.2a shows the theoretical hydraulic age distribution of effluents in an ideal perfectly mixed steady-state flow pond (Q_{in} of 2 000 m^3 d^{-1}). Figure 6.2b depicts the age distribution in a pond with the same mean residence time as the first one (27 days), but with a non-steady-state flow regime (which is closer to the regime of actual ponds), and an *average* inflow of 2 000 m^3 d^{-1}. The shadowed area in the figure indicates the percentage of fresh effluents residing 10 days or less within the reactor (PFE_{10}). In the steady-state flow pond both *PFE* and $\bar{t}$ are constant and can be easily calculated as a relationship between flow and volume. As both are constant, the $PFE/\bar{t}$ ratio is also constant. In the nonsteady-state flow pond *PFE* and $\bar{t}$ are not constant varying according to the changes in the inflow rates, thus the whole hydraulic age distribution must be known to calculate them. The mean residence time depends on the whole hydraulic age distri-

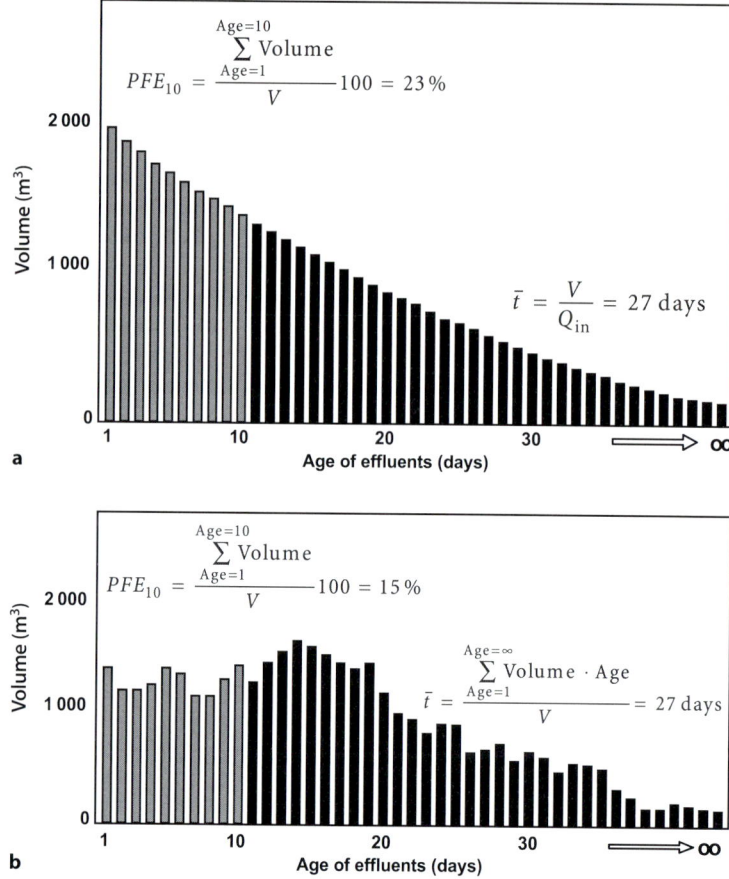

Fig. 6.2. Hydraulic age distribution of effluents in two waste stabilization ponds of constant volumes. $Q_{in\,avg} = 2\,000\ m^3\,d^{-1}$; $MRT(\bar{t}) = 27$ d. **a** Steady-state flow pond (constant inflow rate), $PFE_{10} = 23\%$. **b** Nonsteady-state flow pond (irregular inflow rate), $PFE_{10} = 15\%$

bution of the effluents. The PFE of n days, on the contrary, depends only on the changes in inflow rates that occurred during the last n days, and its variability will be higher than that of $\bar{t}$. The smaller the n of PFE_n, the higher its variability will be.

6.2.1.2
Case Two: The Relationship between Hydraulic Age Distribution and Reactor Performance

Figure 6.3a presents the hydraulic age distribution of effluents in an ideal reactor with constant inflow ($100\ m^3\,d^{-1}$), no outflow, and growing volume.

Figure 6.3b shows BOD concentration in each age class (influent BOD $100\ mg\ l^{-1}$) if a removal rate of 4% per day is assumed (a value matching actual BOD removal rate in waste stabilization ponds). Most of the BOD within the reactor belongs to the

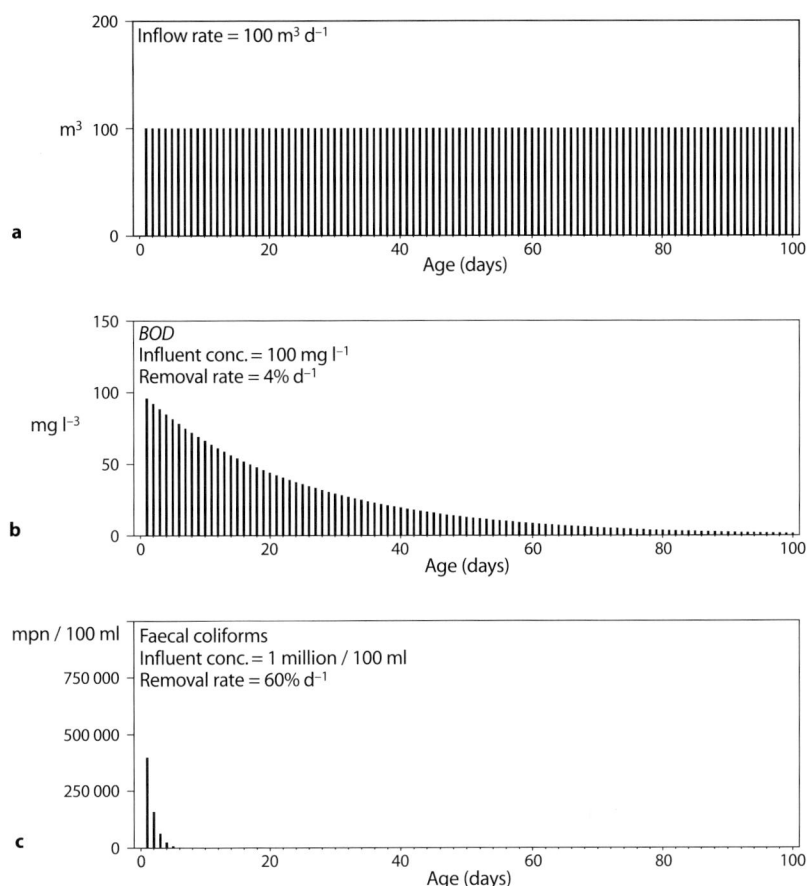

Fig. 6.3. Hydraulic age distribution and reactor performance. Age distribution in a reactor without out-flow, receiving 100 m³ d⁻¹ during 100 days. Effluents volume within the reactor is zero at the beginning and 10 000 m³ on day 100. Mean hydraulic residence time is 50 days. *BOD* concentration in each age class of effluents in the reactor. Faecal coliforms concentration in each age class of effluents in the reactor

0–30 days age classes (PFE_{30}). Juanicó and Shelef (1991, 1994) analysed *BOD* removal in stabilization reservoirs, which were operated as perfectly mixed non-steady-state flow reactors with a mean residence time of about 80–100 days. Multiple regression analyses indicated that *BOD* removal is more highly correlated related to *PFE* than to the mean residence time ($\bar{t}$), while PFE_{30} had the highest correlation coefficient than all other *PFEs* (their work is described in detail elsewhere in this book).

Figure 6.3c depicts faecal coliforms concentration in each age class if influent concentration is 1×10^6 MPN/100 ml and a removal rate of 60% per day is assumed (a value matching actual faecal coliforms removal rate in waste stabilization ponds). It can be appreciated that most of the faecal coliforms within the reactor belong to the 0–3 days age classes. Liran et al. (1994) analysed the removal of faecal coliforms in Geta'ot Reservoir, finding that PFE_1 and PFE_5 are much better descriptors of faecal coliforms removal than the mean residence time ($\bar{t}$). This example will be addressed again in the case study at the end of this chapter.

The close relationship between performance of perfectly mixed steady-state reactors and $\bar{t}$ has been long known, and $\bar{t}$ is largely used as one of the main reactor design parameters (Levenspiel 1972; Trambouze et al. 1988). But this relationship is only illusory, the performance of the reactor is actually related to *PFE* as shown in Fig. 6.3. The '*n*' of the *PFE* with the highest correlation with the performance of the reactor depends on the rate of the reaction considered. The apparent relationship between reactor performance and $\bar{t}$ occurs because under steady-state flow conditions, the *PFE*/$\bar{t}$ ratio is constant and thus $\bar{t}$ can substitute for *PFE* in the analysis of reactor performance.

6.3
The Leslie Matrix Model

6.3.1
Background

The Leslie Matrix Model describes age structured population dynamics. The first version of this model was introduced by Leslie (1945) and named after him. It is widely used in population dynamics (Cullen 1985; Yodzis 1989; Raimundo and Engel 1990). Population density is observed at regularly spaced times, separated by a fixed time interval, thus it can be partitioned into several age classes, one for each time interval. The following definitions are needed:

1. *Survival rate* p_j is the expected fraction of individuals in age class j that will survive to age class $j + 1$.
2. *Effective fertility* m_j of age class j is the expected number of offspring that are born to one parent in age class j and survive the upcoming census.
3. *The Leslie Matrix* is the matrix representation of the above transformation laws:

$$L = \begin{bmatrix} m_1 & m_2 & \cdot & \cdot & \cdot & m_n \\ p_1 & 0 & \cdot & \cdot & \cdot & 0 \\ 0 & p_2 & \cdot & \cdot & \cdot & 0 \\ \cdot & \cdot & \cdot & \cdot & \cdot & \cdot \\ 0 & \cdot & p_{n-2} & \cdot & \cdot & 0 \\ 0 & \cdot & \cdot & p_{n-1} & p_n & 0 \end{bmatrix} . \tag{6.1}$$

4. *The harvesting matrix* specifies the harvesting policy of the population. The harvesting policy can be specified differently for each age class (e.g., how many fish from each age the fishery industry can catch in a given fish stock).

Now the age structure of the population can be represented by a vector. Given the age structure vector at time zero, the population at any time can be predicted. For an autonomic population (i.e., a population that is not interfered by an external factor) it can be shown that for a long enough period the population will reach a stable age distribution (i.e., constant age structure vector), if the Leslie Matrix has a dominant eigenvalue. Lack of dominant eigenvalue will result in a cyclical behaviour.

6.3.2
The Leslie Matrix Model Applied to Age Class Structure of Effluents in a Reactor

Effluents in a reactor can be considered as a "population" divided into different age classes. Thus the Leslie Matrix Model method can be applied to represent the population dynamics and the age structure of the effluents. Unlike the classic case, effluents do not reproduce and thus the effective fertility equals zero.

Let the effluents be divided to n age classes. If V_j^i is the volume of effluents in age class j at time i, and F_j^i is the fractional volume of age class j at time i, then Eq. 6.2 defines the relationship between the two, where V^i is the total volume of effluents in the reactor at time i (Eq. 6.3).

$$F_j^i = \frac{V_j^i}{V^i} \quad , \tag{6.2}$$

$$V^i = \sum_{j=1}^{n} V_j^i \quad . \tag{6.3}$$

The combination of Eq. 6.2 and 6.3 leads to the following expression:

$$\sum_{j=1}^{n} F_j^i = 1 \quad . \tag{6.4}$$

If completely mixed conditions are assumed, effluents withdrawal from each age class is proportional to its fractional volume:

$$Qout_j^i = F_j^i Qout^i \quad , \tag{6.5}$$

where $Qout^i$ is the total withdrawal from the reactor. Influent enters only the youngest age class, and can be assumed to enter the reactor either at the beginning or at the end of a time interval. Herein, it is assumed that influent enters the reactor at the end of each time interval. Under this assumption no influent that entered the reactor at the time interval between time i and $i + 1$ leaves the reactor. This assumption is acceptable if the chosen time interval is considerably short in regard to the V/Q_{in} ratio.

Age classes can be grouped into three types:

1. *Youngest age class* ($j=1$), which contains effluents which entered the reactor at time interval i.
2. *Age classes* j, ($j = 2, ..., n - 1$).
3. *Oldest age class* ($j = n$), which contains the oldest effluents in the reactor. This type, unlike the first two, is a lump of all the age classes in the reactor which are n time intervals old or more.

6.3.2.1
Youngest Age Class (j = 1)

The volume of this class is given by

$$V_1^{i+1} = Qin^i \quad .$$

(6.6)

The fractional volume of this class can be derived by substitution of Eq. 6.2 into Eq. 6.6.:

$$F_1^{i+1} = \frac{Qin^i}{V^{i+1}} \quad .$$

(6.7)

6.3.2.2
Age Classes j (j = 2, ..., n − 1)

Let the mid-interval total volume be defined as the arithmetic mean of the volumes at the beginning and at the end of each time interval.

$$V^{i+(1/2)} = \frac{V^i + V^{i+1}}{2} \quad .$$

(6.8)

Now the volume of the jth class can be expressed as:

$$V_j^{i+1} \atop {}_{j=2,n-1} = V_{j-1}^i \left(1 - \frac{Qout^i}{V^{1+(1/2)}} \right) \quad .$$

(6.9)

Combining Eq. 6.2 and 6.9 yields an expression for the fractional volume of this class.

$$F_j^{i+1} \atop {}_{j=2,n-1} = \frac{V_{j-1}^i}{V^{i+1}} \left(1 - \frac{Qout^i}{V^{1+(1/2)}} \right) \quad .$$

(6.10)

Rearrangement of Eq. 6.10 leads to Eq. 6.11, which expresses the fractional volume of the jth class at time $i + 1$ as a function of the fractional volume of the $j − 1$ class at time i, the ratio of total volumes at these times, and the relative outflow.

$$F_j^{i+1} \atop {}_{j=2,n-1} = F_{j-1}^i \frac{V^i}{V^{i+1}} \left(1 - \frac{Qout^i}{V^{1+(1/2)}} \right) \quad .$$

(6.11)

6.3.2.3
Oldest Age Class (j = n)

As mentioned before, this class is a lump of all the "old effluents" in the reactor. Effluents from class n do not leave this class to an older class $n + 1$. Part of the effluents in this class is withdrawn from the reactor in proportion to the class's fractional volume. The volume of the nth class can be defined as:

$$V_n^{i+1} = \left(V_{n-1}^i + V_n^i\right)\left(1 - \frac{Qout^i}{V^{i+(1/2)}}\right) \quad . \tag{6.12}$$

An expression for the fractional volume is obtained by combining Eq. 6.12 and 6.2.

$$F_n^{i+1} = \left(F_{n-1}^i + F_n^i\right)\frac{V^i}{V^{i+1}}\left(1 - \frac{Qout^i}{V^{i+(1/2)}}\right) \quad . \tag{6.13}$$

6.3.2.4
Representation of Equations 6.2 to 6.13 in Matrix Form

At time step i the total volume at time step $i + 1$ is not yet known. α^i, β^i and γ^i can be defined as:

$$\gamma^i = V^{i+1} = V^i + Qin^i - Qout^i \quad , \tag{6.14}$$

$$\alpha^i = \frac{V^i}{V^{i+1}} = \frac{V^i}{\gamma^i} \quad , \tag{6.15}$$

$$\beta^i = 1 - \frac{Qout^i}{V^{i+(1/2)}} = 1 - \frac{Qout^i}{\left(\dfrac{V^i + \gamma^i}{2}\right)} \quad . \tag{6.16}$$

Substituting Eqs 6.14, 6.15 and 6.16 into Eqs 6.7, 6.11 and 6.13 yields:

$$F_1^{i+1} = \frac{Qin^i}{\gamma^i} \quad , \tag{6.17}$$

$$F_j^{i+1} = \alpha^i \beta^i F_{j-1}^i \qquad j = 2, \ldots n - 1 \quad , \tag{6.18}$$

$$F_n^{i+1} \quad \alpha^i \beta^i \left(F_{n-1}^i + F_n^i\right) \quad . \tag{6.19}$$

Combining Eqs 6.17–6.19 provides the matrix form of the age distribution structure of the effluents population in the reactor.

$$
\begin{bmatrix} F_1 \\ F_2 \\ \cdot \\ \cdot \\ \cdot \\ \cdot \\ \cdot \\ F_{n-1} \\ F_n \end{bmatrix}^{i+1}
= \alpha^i \beta^i
\begin{bmatrix} 0 & 0 & 0 & 0 & 0 & 0 & 0 & 0 \\ 1 & 0 & 0 & 0 & 0 & 0 & 0 & 0 \\ 0 & 1 & 0 & 0 & 0 & 0 & 0 & 0 \\ \cdot & & & & & & & \\ \cdot & & & & & & & \\ \cdot & & & & & & & \\ \cdot & & & & & & & \\ 0 & 0 & 0 & 0 & 0 & 1 & 0 & 0 \\ 0 & 0 & 0 & 0 & 0 & 0 & 1 & 1 \end{bmatrix}
\times
\begin{bmatrix} F_1 \\ F_2 \\ F_3 \\ \cdot \\ \cdot \\ \cdot \\ F_{n-2} \\ F_{n-1} \\ F_n \end{bmatrix}^i
+ \frac{Qin^i}{\gamma^i}
\begin{bmatrix} 1 \\ 0 \\ 0 \\ \cdot \\ \cdot \\ \cdot \\ \cdot \\ 0 \\ 0 \end{bmatrix}
\quad . \tag{6.20}
$$

Equation 6.20 can be expressed in a vector form (Eq. 6.23) where $\underline{F}$ is the age distribution vector at all times, $\underline{\underline{A}}$ multiplied by α^i and β^i is the transformation matrix from one time step to another, and $\underline{B}$ is a correction vector which is needed because the effluents population is not autonomic (i.e., influent enters the reactor from an external source). Vector $\underline{B}$ represents the source of fresh effluents from the inflow.

$$\underline{F}^{i+1} = \alpha^i \beta^i \underline{\underline{A}} \underline{F}^i + \frac{Qin^i}{\gamma^i} \underline{B} \quad . \tag{6.21}$$

Withdrawal of effluents from the reactor represents the harvesting policy – withdrawal from each age class is proportional to its fractional volume. Matrix $\underline{\underline{A}}$ multiplied by α^i and β^i is actually the Leslie Matrix combined with the harvesting matrix in the classic case (Cullen 1985). Therefore in a no inflow case, where the population is autonomic, the population will eventually concentrate in the oldest age class (n). When there is inflow to the reactor, the population is not autonomic, and the influent is added to the youngest age class.

6.3.2.5
Percentage of Fresh Effluents (PFE$_k$) and Mean Residence Time ($\bar{t}$)

The percentage of fresh effluents and the mean residence time can be defined as follows:

The Percentage of Fresh Effluents is the percent of effluents that have been in the reactor for k time intervals or less (note that k must always be less or equal to n).

$$PFE_k^i = 100\left(F_1^i + F_2^i + F_3^i + \ldots + F_k^i\right) = 100\sum_{j=1}^{k} F_j^i \quad k \le n \quad \forall k,n \quad , \tag{6.22}$$

where PFE_k equals 100 when k equals n, by definition.

The Mean Residence Time is the weighted mean of all the age classes, weighted by their own volume and age.

$$\bar{t}^i = 1F_1^i + 2F_2^i + 3F_3^i + \ldots + nF_n^i = \sum_{j=1}^{n}\left(F_j^i j\right) \quad . \tag{6.23}$$

The PFE$_k$/$\bar{t}$ Ratio. Dividing Eq. 6.22 by Eq. 6.23 and simplifying the expression yields:

$$\frac{PFE_k^i}{\bar{t}^i} = \frac{100}{1 + \dfrac{F_2^i + 2F_3^i + \ldots + (k-1)F_k^i + (k+1)F_{k+1}^i + \ldots + nF_n^i}{F_1^i + F_2^i + F_3^i + \ldots + F_k^i}} \quad . \tag{6.24}$$

This ratio is dependent on the various fractional volumes. Therefore, in non-steady-state conditions it will change from one time interval to another. In steady-state conditions the fractional volumes do not change and thus this ratio is constant.

6.3.2.6
Selection of the Oldest Significant Age Class

An open question is, how accurate should the computations be or, in other words, into how many age classes should the effluents population within the reactor be divided? There is no straightforward answer to this question. For *PFE* calculation, the number of needed age classes depends on the process. For mean residence time calculation, as a rule of thumb, the bigger the ratio V^i / Qin^i is the more age classes are needed.

Figure 6.4 depicts an example of the age classes needed for the analysis of a "steady-state continuous-flow" reactor. For a mean residence time of 5 time units, an oldest significant age class of about 30 time units will have a volume equals to 0.01 of the volume of the youngest class. If more accuracy is desired (e.g., an oldest age class with a volume of 0.001 of the volume of the youngest class) an age of about 40 time units will be necessary. The longer the mean residence time, the more age classes are needed for the same accuracy. Moreover, it is shown that as $\bar{t}$ is longer, the larger is the difference between the age of the oldest significant age class for different accuracies.

6.3.3
Age Distribution in Different Types of Reactors

The expressions developed above represent the age structure of the effluents in any perfectly mixed reactor. The analysis of some special types of reactors follows.

6.3.3.1
Batch Reactors

Type 1: "Inflow-Only" Reactor
This reactor fills up gradually and drains all at once, no outflow occurs during the filling period, therefore Eqs 6.14 and 6.16 are reduced to:

$$\beta^i = 1 \quad , \tag{6.25}$$

$$\gamma^i = V^i + Qin^i \quad . \tag{6.26}$$

Fig. 6.4. Oldest significant age class vs. mean residence time within a perfectly mixed steady-state flow reactor. Three different ratios between fractional volumes of oldest age class and of youngest age class (F_n / F_1)

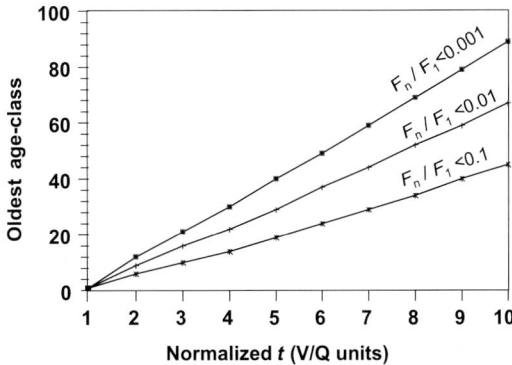

Substituting Eqs 6.25 and 6.26 into Eq. 6.21 yields a simplified form of the vector representation of age structure dynamics:

$$\underline{F}^{i+1} = \alpha^i \underline{\underline{A}} \underline{F}^i + \frac{Qin^i}{\gamma^i} \underline{B} \quad . \tag{6.27}$$

Type 2: "Outflow-Only" Reactor

This reactor fills instantly and drains over a relatively long period of time. The effluents will have an initial uniform age, the age will grow old but will stay uniform. If the initial age structure is not uniform, all age classes will grow older tending to concentrate at the oldest age class. Effluents withdrawal serves here as a harvest factor, which reduces all the population until the reactor is empty, while each age class is reduced in accordance with its fractional volume. Therefore, if the initial age structure vector is not uniform it will change from one time step to another. In this type of reactor Eq. 6.16 is reduced to:

$$\gamma^i = V^i - Qout^i \quad , \tag{6.28}$$

$$\underline{F}^{i+1} = \alpha^i \beta^i \underline{\underline{A}} \underline{F}^i \quad . \tag{6.29}$$

Substituting Eq. 6.28 into Eq. 6.21 and considering no inflow conditions, gives a reduced expression for the age structure vector.

Type 3: "Closed-Batch" Reactor

This reactor is filled at once, and emptied also immediately after some period of time. Here the effluents population is autonomic and is not harvested. The age structure vector does not remain constant, because all age classes grow older tending to concentrate in the oldest age class. In this case Eq. 6.21 is reduced to:

$$\underline{F}^{i+1} = \underline{\underline{A}} \underline{F}^i \quad . \tag{6.30}$$

Type 4: "Sequential-Batch" Reactor

This reactor is quite commonly used for waste and wastewater treatment (Irvine and Busch 1979). It is a combination of the three types mentioned above, including five different stages: filling, reaction, settling, discharge and idle.

Stage 1: Filling stage: The reactor operates as type 1.
Stage 2: Reaction stage: The reactor operates as type 3.
Stage 3: Settling stage: The reactor continuous to operate as type 3.
Stage 4: Discharge stage: The reactor operates as type 2.
Stage 5: Idle stage: If the reactor is not completely empty, it operates again as type 3.

6.3.3.2
Flow Reactors

Type 5: "Steady-State Continuous-Flow" Reactor

In this type of reactor the inflow equals the outflow and both are steady; the volume of effluents in the reactor also remains constant. Under these conditions Eqs 6.15 and 6.16 can be reduced to:

$$\alpha = \alpha^i = 1 \qquad \forall i \quad, \tag{6.31}$$

$$\beta = \beta^i = \left(1 - \frac{Q}{V}\right) \qquad \forall i \quad, \tag{6.32}$$

$$\underline{F}^{i+1} = \beta \underline{\underline{A}} \underline{F}^i + \frac{Q}{V} \underline{B} \quad . \tag{6.33}$$

Combining Eqs 6.31 and 6.32 into Eq. 6.21 gives a simplified expression for the vector representation of the age structure:

Starting with any initial age structure vector ($\underline{F}^0$) a constant age structure vector will be reached as time goes to infinity. The rate of approaching a constant vector depends on the initial age structure vector and the V/Q ratio. Considering Eq. 6.33 and the fact that in steady-state conditions $\underline{F}^{i+1} = \underline{F}^i$, the age structure of the effluents population can be represented by:

$$F_j = \frac{Q}{V} \beta^{**(j-1)} = \frac{Q}{V}\left(1 - \frac{Q}{V}\right)^{**(j-1)} \qquad j = 2, \ldots, n-1 \quad, \tag{6.34}$$

$$F_1 = \frac{Q}{V} \quad, \tag{6.35}$$

$$F_n = \beta^{**(n-1)} = \left(1 - \frac{Q}{V}\right)^{**(j-1)} \quad . \tag{3.36}$$

Figure 6.5 represents the age structure of effluents in a reactor under steady-state conditions. Note that effluents in the reactor moves from "young" age classes to "old"

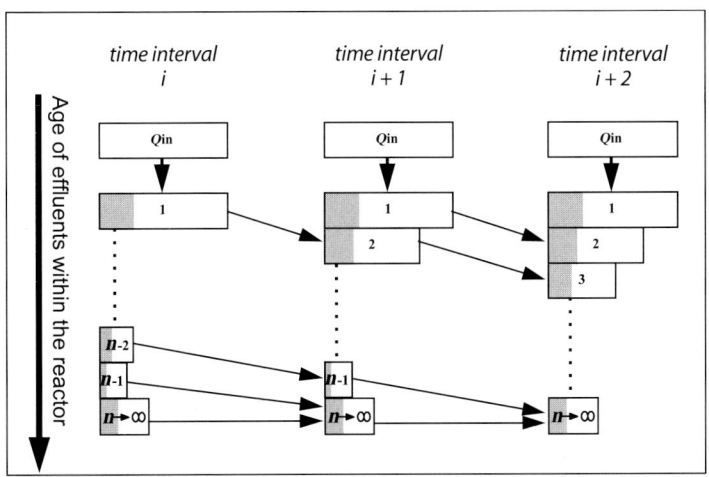

Fig. 6.5. Hydraulic age distribution in a perfectly mixed steady-state continuous-flow reactor. *Shadowed areas:* Volumes lost through outflow. *White areas:* Volumes remaining in the reactor

ones, but the fractional volume of each age class remains unchanged. Using infinite series analysis methods it can be shown that:

$$\sum_{\substack{j=1 \\ n\to\infty}}^{n} F_j = 1 \quad ,$$

(6.37)

and under steady state conditions the mean residence time is:

$$\bar{t} = \frac{V}{Q} \quad .$$

(3.38)

Expression 6.38 is well known from established chemical and hydrological engineering practice.

Type 6: "Nonsteady-State Fixed Volume Continuous-Flow" Reactor
In this type of reactor inflow varies with time, but as the total volume in the reactor remains constant, outflow always equals the inflow.

$$Q^i = Qin^i = Qout^i \qquad \forall i \quad .$$

(3.39)

This type of reactor is typical for many sewage treatment systems (e.g., activated sludge, aerated lagoons, stabilization ponds, etc.) where inflow varies with time (Fig. 6.1) and outflow leaves the reactor by overflow. Combining Eq. 6.39 into Eqs 6.14–6.16 leads to the following expressions:

$$\gamma^i = V = \text{constant} \qquad \forall i \quad ,$$

(6.40)

$$\alpha^i = 1 \qquad \forall i \quad ,$$

(6.41)

$$\beta^i = 1 - \frac{Q^i}{V} \quad .$$

(6.42)

Considering Eqs 6.40–6.42, Eq. 6.21 can be written as:

$$\underline{F}^{i+1} = \beta^i \underline{\underline{AF}}^i + \frac{Q^i}{V} \underline{B} \quad .$$

(6.43)

Type 7: "Nonsteady-State Variable Volume" Flow Reactor
In this type of reactor inflow and outflow vary independently with time and, as a result, the volume of effluents in the reactor also changes with time. Here the general representation of the age structure vector (Eq. 6.21) should be used.

6.3.4
Computer Algorithms

No analytical solution for computing both the percentage of fresh effluents (PFE) and the mean residence time ($\bar{t}$) can be found for nonsteady-state conditions. Therefore, computer algorithms are needed to make the calculations.

6.3.4.1
Algorithm for Computing PFE$_k^i$

PFE_k^i is a sum of all age classes from 1 to k (Eq. 6.22). Thus, age class k is the last age class needed for PFE_k^i calculation and the effluents population should be divided at least to $k + 1$ age classes. The $k + 1$ age class is the oldest age class as defined above. After choosing the needed number of age classes, the number of initial conditions to be set is determined by the number of age classes, one initial condition per one age class. In steady-state flow reactors initial conditions are easily determined, while in non-steady state flow reactors only approximated initial conditions can be set. When a cyclical behaviour of the system is observed (e.g., a reactor with a weekly or annual cycle), it is possible to make a first calculation for a whole cycle starting with approximated initial condition, and then use the end of this first run as the initial conditions for the calculation of a second cycle (the age structure vector at the end of each cycle equals, by definition, the one at the beginning of the next cycle).

6.3.4.2
Algorithm for Computing $\bar{t}$

There are no established criteria for choosing the oldest age class in this case. The experience gained by running the algorithm under several different conditions indicates, as a rule of thumb, that the oldest age class should have a volume of less than one percent of the youngest one (age classes with larger volumes have a significant

---

FORTRAN algorithm for the calculation of the percentage of fresh effluents (*PFE*) and the hydraulic mean residence time ($\bar{t}$) in perfectly mixed chemical reactor, under any flow regime. The algorithm is not optimized for execution in order to keep clearness.

```
        g = V(i) + Qin(i) – Qout(i)
        a = V(i) / g
        b = 1 – 2 * Qout(i) / (V(i) + g)

        F(i,1) = Qin(i) / g
        Do 10 j = 2, n-1
            F(i,j) = a * B *(F(i–1,j–1))
    10  Continue
        F(i,n) = a * b * (F(i–1,n–1) + F(i–1,n))

        MRT(i) = 0.0
        Do 20 j = 1, n
        MRT(i) = MRT(i) + j * F(i,j)
    20  Continue

        Do 30 j = 1, k
            PFE(i,j) = 0.0
    30  Continue

        Do 40 j = 1, k
            Do 40 m = 1, j
                PFE(i,j) = PFE(i,j) + 100 * F(i,m)
    40  Continue
```

FORTRAN notation: $a = \alpha^i$, $b = \beta^i$, $g = \gamma^i$, $MRT(i) = t^i$

effect on the calculation). For steady-state conditions the mean residence time is simply V/Q as demonstrated above. In nonsteady-state conditions the number of initial conditions should correspond to the total number of age classes. A trial and error process should also be practiced here.

To make a simultaneous calculation of both PFE and $\bar{t}$, it is necessary to select the oldest age class of the two.

6.4
A Case Study: The Geta'ot Reservoir in 1989

The Geta'ot Reservoir is an earthen reservoir, 5.5 m maximum depth, 50 000 m³ maximum volume, utilized for wastewater storage during winter to be used for cotton irrigation during the summer. Influent enters from a stabilization pond, with 2×10^6 to 7×10^6 faecal coliforms/100 ml. The outlet is made of a flexible pipe hanging from a raft which maintains the opening 1 m below water surface at all water levels.

The reservoir is operated as a nonsteady-state perfectly mixed reactor with variable volume (Fig. 6.6). The reservoir was full and operated as a batch reactor when sampling started (Batch 1 period in Fig. 6.6), then new effluents were introduced to

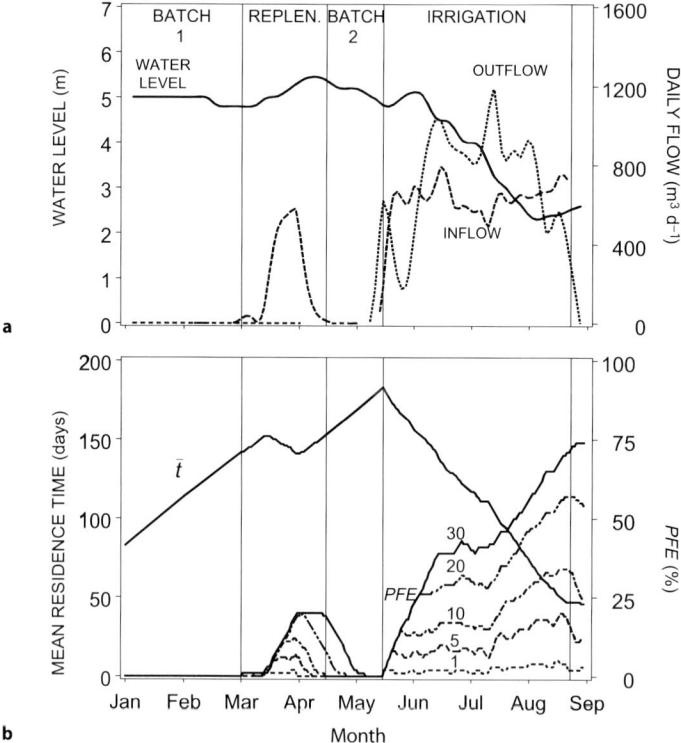

Fig. 6.6. Operational and hydraulic parameters of Geta'ot Reservoir during the studied period. Water level, inflow and outflow. Mean residence time ($\bar{t}$) and PFE (from 1 to 30 days)

Fig. 6.7. Faecal coliforms removal in the surface layer of Geta'ot Reservoir during the studied period. • • Observations; — Trend, obtained by means of a spline smoothing interpolation (SM35 of SAS 1989)

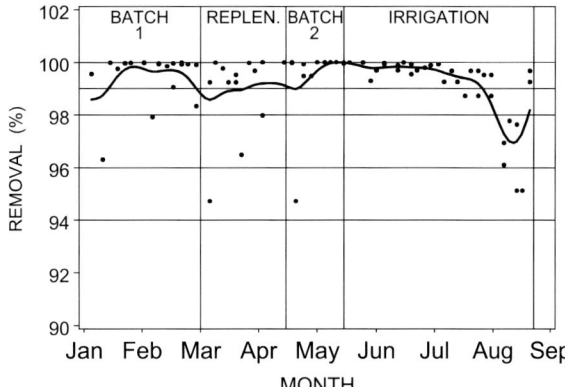

the reservoir in March–April to restore evaporation and seepage losses (Replenishment period), a second batch period followed (Batch 2 period), followed by the irrigation period which was characterized by strong inflow and outflow.

Sampling was performed in the reservoir almost weekly from January to September 1989. Faecal coliforms concentrations, several environmental parameters, as well as all the pertinent operational parameters were measured. The removal of faecal coliforms for each sampling time during the year was computed as the difference between inflow and outflow concentrations, expressed as a percentage of inflow concentration (Fig. 6.7).

The dependence of the removal of faecal coliforms on operational and environmental parameters was analysed by multiple regression analysis, working with one, two or three independent variables. Table 6.1 lists the five regression analysis models that yielded the highest coefficients of determination, for each number of independent variables. The mean residence time was not even the fifth most significant factor in any of the cases. PFE_{20} and PFE_{30} were of less importance than PFE_1 and PFE_5. This supports the theoretical model of Fig. 6.3c: most faecal coliforms in the reactor belong to effluents with ages of 5 or less days. This example, using actual data from a full scale reactor, shows the importance of the hydraulic age distribution for analyzing the performance of perfectly mixed nonsteady-state flow reactors. Further details on this particular study can be found in Liran et al. (1994).

Table 6.1. Regression models of faecal coliform removal in the surface layer of Geta'ot Reservoir during the irrigation period (*Source:* Liran et al. 1994)

1 Independent variable	R^2	2 Independent variables		R^2	3 Independent variables			R^2
pH	0.46	PFE_5	ABS	0.77	PFE_1	ABS	Chlorophyll	0.86
PFE_5	0.35	PFE_1	ABS	0.76	PFE_1	ABS	pH	0.84
PFE_1	0.32	pH	ABS	0.75	PFE_5	ABS	Chlorophyll	0.82
PFE_{20}	0.25	PFE_{20}	ABS	0.72	PFE_5	ABS	pH	0.81
PFE_{30}	0.23	PFE_{30}	ABS	0.68	PFE_1	ABS	COD	0.79

6.5
Conclusions

The hydraulic age distribution of the steady-state flow and batch reactors can be analysed as particular cases of the more general model on the nonsteady-state flow reactor.

The simple equations used for representation of hydraulic age distribution in the particular cases of steady-state flow and batch reactors are not applicable, even as an approximation, to the analysis of nonsteady-state flow reactors.

The percentage of fresh effluents within the reactor (*PFE*) is a better descriptor of the reactor performance than the mean residence time. While the mean residence time relies on the whole hydraulic age distribution, *PFE* relies only on the small portion of fresh effluents which actually determines the performance of the reactor.

High rate reactions (such as faecal coliforms removal) correlate better with *PFE* of few days (1–5 days in the discussed example) while low rate reactions correlate better with *PFE* of more days (30 days for *BOD* removal).

The hydraulic mean residence time, a parameter largely used for the design of sewage treatment plants, is of limited value because it can be used only when the flow regime is strictly steady-state or batch, and this rarely occurs in practice.

Even in the case of steady-state flow reactors, the well known relationship between the mean residence time and the performance of the reactor is only illusory. Under these conditions the ratio *PFE/MRT* is constant, and thus *PFE* can be substituted for the mean residence time.

The Leslie Matrix is an effective tool for the analysis of the hydraulic age distribution in perfectly mixed reactors, under any flow regime.

6.6
Notation

- $**$ = Power. To distinguish from superscript (Eqs 6.35, 6.36).
- $\underline{A}$ = "Normalized" transformation matrix for effluents population in the reactor.
- $\underline{B}$ = "Normalized" correction vector for non autonomic effluents population.
- F^i = Age structure vector of effluents in the reactor at the end of time interval i.
- F^i_j = Ratio between the volume of age class j and the total volume of effluents in the reactor (fractional volume) at the end of time interval i.
- PFE^i_k = Percentage of fresh effluents at the end of time interval i. Ratio between effluents that have been in the reactor k time intervals or less and the total volume of effluents within the reactor, expressed in percents.
- Qin^i = Inflow to the reactor during time interval i.
- $Qout^i$ = Total outflow of effluents from the reactor during time interval i.
- $Qout^i_j$ = Outflow of effluents of age class j during time interval i.
- $\bar{t}^i$ = Mean residence time of effluents in the reactor at the end of time interval i.
- V^i – Total volume of effluents in the reactor at the end of time interval i.
- V^i_j = Volume of age class j at the end of time interval i.
- α^i = Ratio between the total volume of effluents in the reactor at the end of time interval i, and the total volume of effluents at the end of time interval $i + 1$.

- β^i = Fraction of total effluents volume that does not leave the reactor between time intervals i and $i + 1$.
- γ^i = Total volume of effluents in the reactor at the end of time interval $i + 1$ calculated as the total volume at time interval i, plus inflow, minus withdrawal of effluents.

Acknowledgements

This chapter is based on a paper named "*Hydraulic age distribution in perfectly mixed non-steady-state reactors*" which was published in the *Journal of Environmental Engineering ASCE*, vol. 120(6), pp 1427–1445. The authors wish to thank the ASCE Manager of Journals for the permission to publish this chapter.

References

Beck MB, Adeloye A, Finney B, Lessard P (1991) Operational water quality management: Transient event and seasonal variability. Wat Sci Tech 24(6):257–265
Cullen MR (1985) Linear models in biology. John Wiley & Sons, N.Y., USA, 213 pp
Danckwerts PV (1953) Continuous-flow systems: Distribution of residence times. Cem Eng Sci 2(1):1–13
Dennis RW, Irvine RL (1979) Effect of fill : react ratio on sequencing batch biological reactors. Journal WPCF 51(2):255–263
Irvine RL, Busch AW (1979) Sequencing biological batch reactors – an overview. Journal WPCF 51(2):235–243
Juanicó M, Shelef G (1991) The performance of stabilization reservoirs as a function of design and operation parameters. Wat Sci Tech 23(7–9):1509–1516
Juanicó M, Shelef G (1994) Design, operation and performance of stabilization reservoirs for wastewater irrigation in Israel. Wat Res 28(1):175–186
Leslie PH (1945) On the use of matrices in certain population mathematics. Biometrika 33(3):183–214
Levenspiel O (1972) Chemical reaction engineering. John Wiley & Sons, N.Y., USA, 578 pp
Liran A, Juanicó M, Shelef G (1994) Coliform removal in a stabilization reservoir for wastewater irrigation in Israel. Wat Res 28(6):1305–1314
Raimundo S, Engel A (1990) Some considerations on a simplified version of Leslie's model. Ecol Model 52:41–52
Rossman L (1989) Risk equivalent seasonal waste load allocation. Wat Resourc Res 25(10):2083–2090
SAS (Statistical Analysis System) (1989) Version 6, SAS Inc., USA
Trambouze P, Van Landeghem H, Wauquier JP (1988) Chemical reactors: Design, engineering, operation. Institut Français du Petrole Publish. Paris, France, 608 pp
Yodis P (1989) Introduction to theoretical ecology. Harper & Row, N.Y., USA, 384 pp

Chapter 7

Modelling

Eran Friedler

7.1
Introduction

Models use available information to select the major processes defining a particular system and to enable the estimation of the system's response to changes. They can also assist in selecting the technology needed to solve a specific environmental problem, and in assessing the required level of treatment prior to discharge in order to restrict the damage to the environment to a tolerable level. The models' advantage is their ability to examine a large number of alternatives at a lower cost and in shorter time than full scale experiments. Using models enables one to screen alternatives and thus to submit only the selected ones to full-scale experiments. The main disadvantage of models is the numerous assumptions that must be made in order to simplify the system.

Preparing a water quality model in large water bodies in general, and an effluent quality model in stabilization reservoirs in particular, requires a combination of different disciplines: hydrology and hydrodynamics to describe the water balance, water movement and mixing mechanisms; meteorology to define the environmental conditions; chemistry, physics and biochemistry to describe the fate of dissolved and suspended matter; ecology and biology to describe the relationships between different populations, and between them and the environmental conditions.

Choosing a model's level of complexity is a difficult task. The tendency is to create over-complex models because it is easy to do so. Yet, simplistic models often describe the system better than complex ones, because complex models include many parameters requiring new assumptions, definitions and calibration leading to a raised level of uncertainty. Moreover, the greater the complexity the larger the number and types of data required. Data collection is a demanding process, in cost, time, and skilled manpower. It is obvious that there is no point in developing a highly sophisticated model if the data for the model's input are not in hand to make its calibration possible. Thus, the availability and quality of the data are often the limiting factor for the level of the model's complexity.

7.1.1
Milestones in the Development of Models of Aquatic Ecosystems

The need to control pollution of water resources and describe, in quantitative terms, their response to an increasing pollutant load, led to the development of ecological models of aquatic ecosystems. In the twenties, the Ohio River Commission in the USA conducted a comprehensive study of the influence of pollution sources on the river's water quality. This research led to the development of one of the first models of an aquatic environment the Streeter Phelps Model (Streeter and Phelps 1925). The model calculated the oxygen deficit resulting from two contrasting processes occurring simultaneously: oxygen consumption by bacteria which degrade organic matter discharged into the river (from a point-source), and dissolving of oxygen in water by atmospheric reaeration. These two processes were presented as first order reactions and expressed by a differential equation. The integration of the equation for a steady state condition, gave the oxygen deficit as a function of distance from the pollution source.

'Closed analytical models' (as the Streeter Phelps Model) failed to describe many of the complex interactions that exist in real aquatic ecosystems which usually function under non-steady-state conditions. Consequently, the need arose for a tool which may be adapted to the extensive variability that characterises real ecosystems. Following the development of computers, and especially in the 1950s and 1960s, numerical methods were developed, which enabled simulation of closer-to-reality situations. These trials contributed to the development of ecological models of natural aquatic ecosystems.

The first models of the behaviour of large water bodies were developed by Water Resources Engineering Inc. (1968) and by Huber et al. (1972) of MIT. These models were uni-dimensional and simulated thermal energy dispersion in water bodies. Based on these models, Markofsky and Harleman (1973) described the behaviour of dissolved oxygen and *BOD* in the water body. Later, Chen and Orlob (1975) used them as a basis for models predicting the concentrations of dissolved oxygen, *BOD*, and various nutrients.

7.2
Models of Stabilization Reservoirs

Stabilization reservoirs are a relatively new treatment unit and only few models for describing the processes occurring within the reservoirs have been developed. The first work was conducted by Shelef et al. (1977) Prediction model for maximum tolerable *BOD* concentration. This was followed by Avnimelech (1989) "Model of accumulation of organic matter in the sediment"; Juanicó and Shelef (1991) "Statistical models for prediction *BOD* and *COD* removal as a function of the operational regime"; Juanicó and Friedler (1994) "Model for the calculation of hydraulic age distribution in stabilization reservoirs". Friedler (1993) developed a mechanistic simulation model for the water body and the sediment of stabilization reservoirs.

7.2.1
Models for Predicting Maximal Permitted Dissolved *BOD* in Stabilization Reservoirs

Shelef et al. (1977) developed two simple models for the prediction of maximum permitted dissolved *BOD* concentration in a reservoir, not yet leading to anaerobic conditions. In one model the reservoir is assumed to be completely mixed, while in the other it is assumed to be thermally stratified. Both models have two state variables, namely: dissolved oxygen and *BOD*.

7.2.1.1
Model of Completely Mixed Reservoir

Oxygen balance in the water body is determined by oxygen input in the wastewater inflow entering to the reservoir, oxygen output in the effluent withdrawn for irrigation, atmospheric reaeration, photosynthetic production, and consumption for *BOD* degradation and microbial respiration.

Assumptions

- Oxygen concentration in inflow is zero.
- Inflow and outflow flows are small relative to the volume of effluent in the reservoir.
- Constant rates of atmospheric reaeration, microbial respiration, and oxygen consumption for *BOD* degradation.
- The rate of photosynthetic oxygen production depends only on solar radiation intensity and is constant throughout the water column.
- Initial conditions – at 06:00 a.m. the reservoir is anaerobic.

Results
Simulation was performed for a single 24 hour period, for reservoirs having various dissolved BOD_{final} concentrations and different depths. The model predicts that a reservoir in which water level exceeds 2 m and dissolved BOD_5 of 50–80 mg l^{-1} will be totally anaerobic. When dissolved BOD_5 decreases to 10–20 mg l^{-1}, the same reservoir will be aerobic (at least during daylight hours).

Limitations
By simulating only a single day and ignoring inflow and outflow flows, the model describes the reservoir as though it were functioning under hydraulic batch steady-state conditions. A description far from reality, in which the reservoir is functioning under nonsteady-state conditions.

With the exception of solar radiation (which is modelled to change during the day and thus to cause diurnal variation in oxygen production rate), all other rates were considered as constant, whereas they change with the environmental conditions. Furthermore, solar radiation intensity declines with depth leading to a decreasing oxygen production rate (even under completely mixed conditions). Oxygen production rate depends not only on solar radiation intensity, but also on other parameters such as temperature, availability of nutrients and, above all, algae concentration.

7.2.1.2
Model of a Stratified Reservoir

This model was developed assuming that oxygen produced in the upper layers does not immediately reach the lower layers but is advected by the vertical currents in the reservoir. Dissolved oxygen balance is modelled by the same manner as in the mixed reservoir model. In this model, in contrast to the mixed reservoir model, the water body is divided into horizontal water layers, for each one the oxygen balance equation is solved. Oxygen transfer between layers occurs via molecular diffusion and vertical advection. Atmospheric reaeration influences only the top water layer, while benthic oxygen demand influences the oxygen balance of the lowest water layer.

Assumptions
The same assumptions as in the completely mixed reservoir model were made, with the exception of photosynthetic oxygen production which is limited to the top 25 cm, and benthal oxygen demand which is constant.

Results
Simulation was carried out for a single 24 hour period, for reservoirs with dissolved BOD_{final} concentration varying between 10–220 (mg l⁻¹), water level between 0 and 10 m, and currents between 0 and 70 (m h⁻¹).

- When quiescent conditions prevail, the top layer (the photosynthetic layer) is aerobic around the clock, independent of the water level within the reservoir, even at high *BOD* concentrations.
- The aerobic zone expands during the day and contracts during the night.
- At a given water level and *BOD* concentration, mixing causes the expansion of the aerobic layer up to a certain point, above which further increase of mixing intensity will cause a contraction of the aerobic layer until it will disappear. This point drops as dissolved *BOD* concentration rises.
- At a constant *BOD* concentration and mixing rate, increasing the reservoir's water depth will result in the expansion of the aerobic layer. However, when the depth is raised beyond a certain optimal value the aerobic layer contracts. The water depth which will cause the aerobic layer to disappear is inversely proportional to the dissolved *BOD* concentration and mixing intensity.

Limitations
The limitations of this model are the same as to those of the mixed reservoir model, with the exception of the oxygen production which is assumed to occur only in the

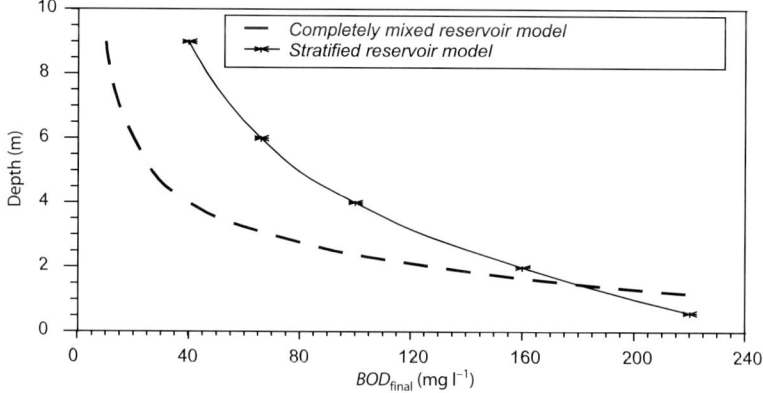

Fig. 7.1. Maximum allowed BOD_{final} as a function of reservoir depth (*Source:* Shelef et al. 1977)

top 25 cm. These two models are not dynamic models as might seem at first sight. The ecosystem representation in these models is very simplified, both from the point of view of processes considered and the operational regime. This simplification allows the models to be used only as a tool for estimating the permitted BOD_{final} concentration in the reservoir (which is the purpose for which they were designed).

Comparison of the results of the two models with measurements of dissolved oxygen concentration distribution in the water column of several reservoirs showed that within a depths range of 1–4 m, the observations match the completely mixed model, whereas at depths in the range of 5–7 m, the observations match the stratified model with low mixing velocities.

The models allow the prediction of the maximum dissolved BOD_{final} concentration which will still avoid malodour emissions from the reservoirs (Fig. 7.1). The curve for a mixed reservoir describes a case in which, for at least 10% of the day, the water column will be oxic in order to prevent anaerobic conditions. The curve for a stratified reservoir describes a case in which at least during daylight the aerobic zone will be 30–50 cm thick, in order to prevent sulfides emission.

7.2.2
Model for the Prediction of the Accumulation of Organic Matter in the Sediment of Stabilization Reservoirs

Avnimelech (1989) developed this model in order to predict the accumulation of organic matter in the sediment of a newly constructed stabilization reservoir as a result of a change in the sediment ecosystem: from alluvial ecosystem (before the reservoirs were filled for the first time) to aquatic ecosystem. Ma'aleh HaKishon Reservoirs served as a case study in this case.

Having only limited goals, the model is rather simplistic, containing one state variable (organic matter in the sediment), the concentration of which is determined exclusively by sedimentation and degradation rates. Thus, the material balance equation is expressed as follows:

$$\frac{dC_{OG_s}}{dt} = J_s - k_s\, C_{OG_s} \quad , \tag{7.1}$$

where:

- C_{OG_s} = Concentration of organic matter in the sediment $[M/L^2]$
- J_s = Sedimentation flux of suspended organic material $[(M/L^2)/T]$
- k_s = Rate constant of degradation of organic matter in the sediment $[(M/M)/T]$

Assumptions

- Constant flux of organic matter degradation.
- Sediment degradation is considered to be first order reaction dependent exclusively on organic matter concentration in the sediment.

These two assumptions actually turn the model into a steady-state model, where the above Eq. 7.1 can be solved analytically.

Results

From the analytical solution of the material balance equation, accumulation of organic matter in the sediment was predicted. From the integral equation, it was appreciated that only a finite concentration of organic matter can accumulate in the sediment. This maximum concentration is equal to sedimentation flux divided by the degradation rate (J_s/k_s). Further, the time that would elapse between the transition from alluvial ecosystem to aquatic one, and the instant when certain proportion of steady state concentration of organic matter in the sediment is reached could be estimated (e.g., the time until 70%, 80%, or 90% of steady state concentration is reached).

For the case study reservoirs the initial concentration of organic matter in the sediment was determined as 0.31 kg C m^{-2} (from soil samples), sedimentation flux of organic matter as 0.31 kg C m^{-2} yr^{-1} (using data gained from input-output mass balances and from sediment traps), and degradation rate of organic matter in the sediment as 0.42 y^{-1} (from literature data). Under these conditions the model predicts that steady state concentration of organic matter in the sediment of Ma'aleh HaKishon Reservoirs would be 0.75 kg C m^{-2}, while it would take about four years until the concentration in the sediment will reach 90% of the steady-state concentration.

Limitations

The assumption of constant sedimentation flux of organic matter is the main limitation of the model. This assumption actually determines steady state conditions, which do not correspond to the operational regime of the reservoirs. Another limitation is the assumption of a constant degradation rate, since it is well established that this rate is highly dependent on temperature, dissolved oxygen concentration, and other factors. In light of these limitations, the model could not serve as a tool for prediction, but as an aid for estimation of the anticipated maximum concentration of organic matter in the sediments of reservoirs (a task for which it was developed).

7.2.3
Statistical models for Predicting *BOD* and *COD* Removal as a Function of the Operational Regime

Juanicó and Shelef (1991, 1994) developed a series of empirical statistical models using a multi-variable regression technique to correlate between the efficiency of *BOD* and *COD* removal and the operational parameters of the reservoirs. These models, although being black-box models which do not describe processes which occur in the water body, contribute significantly to the identification of the main parameters influencing the quality of effluent within the reservoirs.

Methodology

The models were developed on the basis of measurements in four stabilization reservoirs. Table 7.1 lists the ranges of all data used in the development of these models. The dependent variables were *BOD* removal and *COD* removal. All possible correlations between the dependent and the independent variables were examined. In order to find non-linear correlations the dependence of the dependent variables was also checked regarding base 10 logarithm and the second power of all the in-

Table 7.1. Ranges of available data used to develop the multiple regression models (*Source:* Juanicó and Shelef 1991, 1994)

Water quality parameters				Operational parameters			
Variable	Units	Min. value	Max. value	Variable	Units	Min. value	Max. value
BOD removal	%	0	97	Residence time	days	26	163
COD removal	%	0	94	PFE_{10}[a]	%	0	45
Inflow *BOD*	mg l^{-1}	37	1 000	PFE_{30}[b]	%	3	100
Inflow *COD*	mg l^{-1}	133	2 600	Water level	m	1.5	10.4
BOD loading	kg/ha/d	0	490	Volume	m^3	9 000	6.0×10^6
COD loading	kg/ha/d	0	1 280	Area	m^2	7 000	7.3×10^5
Temperature	°C	12	33	Area/volume	m^2/m^3	0.122	0.800
pH	–	6.4	8.4				

[a] PFE_{10}: Percentage of effluent residing in the reservoir 10 days or less;
[b] PFE_{30}: Percentage of effluent residing in the reservoir 30 days or less.

dependent variables. From all possible models, three models were chosen. In the chosen models the determination coefficient (R^2) was the highest, the intercept values and the regression coefficients were statistically significant (99% probability), they relied on a large number of points, and the residuals were homogeneous (not biased).

Results
Agreement between the models and the observations is good. The relatively low correlation coefficients (ca. 70%) stem from a small number of major deviations from the model at the end of summer and in autumn, when water level and residence time are minimal and PFE_{30} is maximal. In this period the assumption of perfect mixing when calculating the hydraulic parameters does not apply, since development of hydraulic short-circuiting and dead volumes that cause fluctuations in water quality. The main findings of the models are listed herewith:

- All models indicated that PFE_{30} comprises the main parameter determining the efficiency of *BOD* and *COD* removal. There is a negative correlation between the removal and the PFE_{30} (Table 7.2).
- The mean residence time was not found to be a significant factor in determining the removal efficiency. This is explained by the fact that since the residence time averages all age groups it is therefore less sensitive to the operational changes occurring in the reservoirs. A high dependency of the processes in the reservoir on PFE_{30}, and not on the residence time, apparently occurs in all wastewater treatment systems that operate in a nonsteady state.
- *BOD* removal efficiency was not found to be dependent on inflow *BOD* concentration, whereas *COD* removal was found to be dependent on inflow *COD* concentration.

Table 7.2. The selected multiple regression models to predict *BOD* and *COD* removal in stabilization reservoirs (*Source:* Juanicó and Shelef 1991, 1994)

Model	N^a	R^2	Dependent variable	Independent variable			Partial correlation (Type II[b]), (%)
1	65	72	*BOD* removal	+85.16		(intercept)	
				−0.421	*	PFE_{30}	45
				−6.49	*	Water level	35
				−75.16	*	LOG_{10} (area/volume)	29
2	85	72	*COD* removal	+75.75		(intercept)	
				−0.0074	*	PFE_{30}^2	55
				+0.0099	*	Inflow *COD*	40
				−0.232	*	Water level^2	15
3	85	71	*COD* removal	84.5		(intercept)	
				−0.0081	*	PFE_{30}^2	43
				+0.01	*	Inflow *COD*	40
				+33.82	*	LOG_{10} (area/volume)	9

[a] Number of observations used to compute the model;
[b] Correlation between dependent variable and each independent variable, when the effect of all other independent variables was excluded from the regression equation.

- Negative correlation was revealed between removal efficiency and the water level in the reservoir. The explanation lies in the oxygen regime: the higher the reservoir water level the less oxygen reaches the lower layers, and the degradation rate declines.

Limitations

The major limitations of these models is precisely their being empirical models. Since they are, their prediction ability is good only for interpolation but not for extrapolation (i.e., they can provide correct predictions only for conditions similar to those for which they were developed). In different conditions they are liable to provide erroneous results. This is demonstrated by their inability to predict effluent quality when the reservoir water level is low and the flow regime is changing. Furthermore, these models, being empirical models, in fact relate to the reservoir as a black box and do not describe the processes taking place in the water body itself.

7.2.4
Model of Hydraulic Age Distribution in Stabilization Reservoirs

Juanicó and Friedler (1994) developed this model in order to analyse the hydraulic age distribution in non-steady-state perfectly mixed reactors in general and in stabilization reservoirs in particular. The rationale behind the development of this model was that the mean residence time was found to be an inadequate descriptor for non-steady-state systems, and there was a need to mathematically develop a better descriptor. This descriptor being PFE_n – Percentage of Fresh Effluent (i.e., effluent residing within the treatment unit n units of time or less). This model is actually a tool for calculating various hydraulic descriptors (PFE_n, mean residence time, etc.), which then can be used in the design and operation of stabilization reservoirs in particular and any perfectly mixed reactor in general. This work is described elsewhere in this book.

7.3
Simulation Model of Stabilization Reservoirs

7.3.1
Introduction

Friedler (1993) developed the bases for this model. The model is a semi-empirical mechanistic model for describing hydraulic, hydrological, physical, physical-chemical and biological processes that occur in the reservoirs. The model enables the prediction of the quality of effluent withdrawn from the reservoirs as a function of the operational regime, the quality of incoming wastewater, the reservoir physical design and the weather conditions.

Stabilization reservoirs are hypertrophic systems functioning under nonsteady-state conditions. On one hand they are similar to other extensive treatment units, while on the other hand they are large enough to allow for limnological phenomena to become significant. Thus, the model integrates knowledge both from the field of wastewater treatment and from limnology of surface water bodies.

The model developed is actually a meta-model combining a number of sub-models each describing specific aspects of the effluent quality in stabilization reservoirs: hydraulics and hydrology; diurnal variation of solar radiation; solar radiation penetration through the water column; algae community; zooplankton community; organic matter accumulation, sedimentation and degradation both in the water column and in the sediment; pathogen bacteria die-off both in the water column and in the sediment; and sediment-water column interaction. This integration of sub-models into a single meta-model makes it possible to quantify the relative influence of each factor on the quality of the effluent in the reservoir, to find the mutual influences and links between the different factors, and to obtain an overall picture of the reservoir's behaviour in various conditions.

7.3.2
Assumptions

7.3.2.1
Uni-Dimensional Assumption and Division into Layers

The model assumes that the reservoir can be represented as a vertical series of horizontal layers. That is, each quantitative value is averaged in the horizontal layer and variation is considered only in the vertical axis (Fig. 7.2). This assumption is consistent with acquired field data and theoretical data on various water bodies (stabilization ponds, reservoirs and lakes), while taking into consideration the reservoir's size, shape (roughly rectangular or circular), and the long mean residence time, ca. 100 days (Krenkel and French 1982; Orlob 1983; Losordo and Piedrahita 1991; and others).

The water body is divided by the model into six layers of differing thickness and volume. This number of layers was found to be most appropriate while considering minimum and maximum water depths in different reservoirs (ranging between 1 and 1.5 m at the end of the irrigation season, and in the 6 to 14 m range, when the reservoirs are full; respectively). The volume of each layer varies in respect to its water

Fig. 7.2. Layers of water (between *continuous lines*) and sediment (between *dashed lines*) in the model

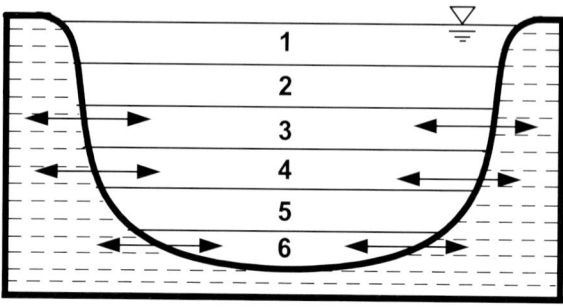

balance. The differing thickness of the water layers enables the calculation of exact mass balance, inclusive of the period when the reservoir volumes fluctuate sharply (irrigation season).

The sediment of the reservoir is divided into layers of fixed height, each of ca. 5 cm (Fig. 7.2). This thickness was selected due to numerical considerations for matching the differences between the water layers (of variable thickness) and the sediment layer (of fixed thickness). The structure of sediment layers makes it possible for matter to pass between all water layers and the sediment enveloping its bottom part.

7.3.2.2
Mass Conservation Principle

The model assumes it to be possible to describe the dynamics of each physical, chemical and biological component by the mass conservation principle in every layer of water and sediment. These dynamics are expressed by differential equations for each horizontal water and sediment layer. The solution of the mass balance equations gives the concentration of matter as a function of time and depth. The model relates to the following factors:

1. Mass increase due to inflow (input of effluent into the reservoir, rain)
2. Mass decrease due to outflow (effluent withdrawal from the reservoir for irrigation, evaporation and seepage)
3. Internal changes in mass due to growth, respiration, mortality, decay, grazing, etc. These changes may be positive (production, growth) or negative (consumption, decay)
4. Settling of mass from upper layers to lower layers (removal from the top layer, addition to the lower layer) and to the appropriate sediment layers (removal from the water body, addition to the sediment)
5. Mass transport between layers in the water body, as a result of mixing of the water body or part of it, and due to diffusion between the layers
6. Oxygen transport in the water-air interface

Diffusion in the mass-balance equations expresses mass transfer as a result of mixing flows caused by wind or thermal mixing, and molecular mass transfer (Brownian movement). Mass transfer due to mixing is greater, by at least one order of magnitude, than mass transfer by molecular diffusion. Therefore, in most cases the model neglects the latter and makes allowances only for mass transfer resulting from mixing. The exception is dissolved oxygen for which the model constantly calculates mass transfer by molecular diffusion, since molecular diffusion is the major source of oxygen for the hypolimnion during the thermal stratification period.

7.3.2.3
Simplified Biological Presentation

Biological links and processes must be simplified to allow them to be presented in the model. In stabilization reservoirs there are dozens, even hundreds of different species of organisms. However, experience has shown that in order to describe the major processes occurring in the reservoir it is unnecessary to present each species separately. Thus, the species were grouped in cells, one for each functional group, i.e., one cell for algae, one for zooplankton, and one for pathogen bacteria (or indicator); there is also a cell for organic matter.

7.3.2.4
The Thermal Regime of the Reservoir is Entered as a Forcing Function

Presenting the thermal regime in the model as a forcing function makes it possible to study the influences (positive or negative) exerted by thermal stratification in the reservoir on the quality of the effluent. The model's objective is not to simulate the thermal regime, but to test it's impact on the effluent quality. Furthermore, weather conditions and reservoir operation regimes in Israel have resulted in permanent summer thermal stratification in some reservoirs and completely mixed conditions (or ephemeral stratification) in others. Two similar reservoirs, located near each other, may differ in existence or non-existence of permanent summer thermal stratification.

The formation and breakdown of the 'permanent' thermal stratification are determined on the basis of field data. However, stratification is broken if the water depth is low or the thickness of the hypolimnion is small (both disturb the thermal stability of the water body). In this case ephemeral thermal stratification occurs (stratification prevails during the daylight hours and is broken in the night). Thus, the reservoir may be in one of three states:

1. A single homogeneous water body with uniform vertical distribution of all water quality variables (typical winter conditions in all reservoirs; summer conditions in some reservoirs)
2. Two water bodies with low rate of water transfer between them: epi- and hypolimnion (summer conditions in part of the reservoirs)
3. Two water bodies with limited mass transfer (water) between them during the day, a homogeneous mixed water body during the night (summer conditions in part of the reservoirs)

The epilimnion (which presents high variability in depth when thermal stratification prevails) was divided into five horizontal water layers, whereas the hypolimnion, in which the vertical variability is much smaller, was conceptualised as a single horizontal water layer. Table 7.3 summarizes the mixing of the water body in the model as a result of the thermal regime of the reservoir.

Table 7.3. Water body mixing by the model in three possible thermal regimes

Season	Thermal stratification Permanent	Thermal stratification Ephemeral	Mixed layers	Timing of mixing	Remarks
Winter	No	No	1–6	Every timestep	Winter conditions in all reservoirs; Summer conditions in some reservoirs
Summer I	No	Yes	1–6	Every night	Summer conditions in some reservoirs
Summer II	Yes	Yes	1–5	Every night	Summer conditions in some reservoirs

7.3.3
Forcing Functions

Forcing functions are the external variables which are not an integral part of the system being examined, but influence it. Some of the forcing functions can be controlled while others cannot, some are known and others are not and are to be determined by the model calibration process.

7.3.3.1
Rain

Rain's main impact is on the water balance. Rainfall joins the upper water layer. Since the concentration of most matter in the rain is negligible relative to effluent, the rain actually dilutes the effluent in the top layer. An exception could be dissolved oxygen, whose concentration in the rain water may be assumed to be saturated. Rain water therefore usually contributes oxygen to the upper water layer. However, the impact of rain dissolved oxygen on the oxygen concentration in the reservoir was examined in the framework of the sensitivity analyses and was found to be negligible.

7.3.3.2
Evaporation

Evaporation also affects the water balance of the reservoir. Only water (H_2O) evaporates and thus it increases the concentration of various material in the upper layer of effluent. The evaporation rate from the water surface of each reservoir is estimated at about 80% of the evaporation rate in 'Evaporation Tub Class-A' (Lokiec 1983; Romem 1991; Rozenzvieg 1986; and others) in the nearest meteorological station.

7.3.3.3
Solar Radiation

Solar radiation is the energy source for algae which are the primary producers in the reservoir. Solar radiation also influences the die-off rate of pathogens.

The average daily radiation and the duration of sunshine during each day is determined by the specific meteorological conditions of each reservoir. The simulation of the following phenomena is performed on the basis of these data:

- Diurnal variation of solar radiation
- Changes in the angle of the direct solar radiation on the water surface, as a function of the geographical latitude, date, and time of day. The resulting changes in the penetration of solar radiation into the water body (the fraction that is not reflected).
- Light attenuation in water column as a function of algae and organic matter concentration

7.3.3.4
Operational Regime – Effluent Inflow to the Reservoir and Its Withdrawal: Timing, Quantity and Spatial Distribution (i.e., to Which Layer/s and from Which Layer/s)

Nine different flow regimes were established. These regimes (Table 7.4) make it possible to choose the layers into which, and from which, effluent is pumped. Every water layer receives or releases effluent in proportion to its thickness, relative to the total thickness of the zone into which effluent is entered or withdrawn. Quality of effluent pumped for irrigation is determined by means of weighted average of each state variable, weighted by concentration in each layer (from which the effluent is withdrawn), and the withdrawal from each layer.

7.3.3.5
The Physical Structure of the Reservoir

The physical structure of the reservoir influences the system on two levels:

- The physical dimensions of the reservoir determines links between the height of water surface above the bottom, volume of water, area of the water surface and area of the wetted cross section. Functions defining these links are called 'reservoir func-

Table 7.4. Vertical allocation of inflows and outflows to/from the reservoir in the model

Regime	Inflow Epilimnion	Hypolimnion	Outflow Epilimnion	Hypolimnion
1	Yes	Yes	Yes	Yes
2	Yes		Yes	
3	Yes			Yes
4		Yes	Yes	
5		Yes		Yes
6	Yes		Yes	Yes
7		Yes	Yes	Yes
8	Yes	Yes	Yes	
9	Yes	Yes		Yes

tions' and are specific for each reservoir. The physical structure influences the extent of mixing, the depth of the hypolimnion, the extent of solar radiation penetration into the lower water layers, and other parameters influencing the effluent quality.

- The type characteristics of the soil and the sealing layer determine the hydraulic conductivity, and thus, influence the seepage rate. However, the hydraulic conductivity and the thickness of the sealing layer are usually not known, but must be determined by calibration. Dissolved matter in the effluent leaves the reservoir with the seeping effluent, while suspended matter does not exit with this effluent.

7.3.3.6
Quality of Incoming Effluent

The quality of the incoming effluent should either be taken from field data or assumed.

7.3.3.7
Water Body Temperature

Temperature is determined externally using field data from different reservoirs in Israel. The temperature of each layer relates to the season of the year, depth of the reservoir and the presence, or absence of thermal stratification in the reservoir (Juanicó 1993).

7.3.4
State Variables

Choosing the state variables to be entered or not into the model is one of the more difficult tasks in proper modelling (Jørgensen and Mejer 1979; Beck 1991; Orlob 1992). The choice of state variables has a direct influence on the complexity level of the model and thus must consider both theoretical and practical aspects. In this model atten-

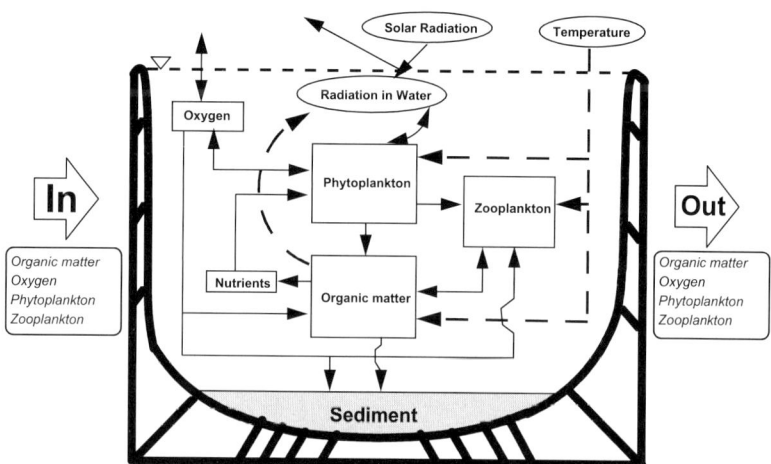

Fig. 7.3. Interaction between state variables of the model

tion was given to the system characteristics and the availability of data and parameters needed for running the model. Only principal processes occurring in reservoirs were included, while secondary processes were deliberately neglected. As a result, the model is not a sophisticated and complex representation of the reservoir as
an aquatic eco-system, but a quantitative tool for predicting the effluent quality in the
reservoir. Five 'quality' state variables were chosen for the water column (algae, dissolved oxygen, indicator bacteria, organic matter, and zooplankton), while two 'quality' state variables were chosen for the sediment (organic matter and indicator bacteria). The reciprocal relationships between the water quality variables as conceived by
the model are described in Fig. 7.3. The main equations governing the biological processes occurring within the water column and the sediment are presented in Table 7.5.

7.3.4.1
State Variables in the Water Column

Algae

- *Growth*
  - Dependence on light intensity is described by inhibition model – the Steele equa-
    tion (Steele 1962). This matches observations made in various stabilization res-
    ervoirs.
  - Dependence on water body temperature is described according to the Van't Hoff
    Arrhenious equation. However, when the temperature of the effluent in the res-
    ervoir departs from the tolerable temperature range the algae's activity is halted.
  - Dependence on nutrient concentration (nitrogen and phosphorus) is represented
    indirectly by means of Michaelis-Menton equation.
- *Respiration.* Algae respiration is dependent only on the temperature and is described
  according to the Van't Hoff Arrhenious equation. However, when the reservoir efflu-
  ent departs from the tolerable temperature range algae respiration stops.
- *Mortality* (i.e., natural mortality and not predation)
  - Voinov and Akhremenkov (1990) described algal mortality dependency on the wa-
    ter body temperature in three equations for three temperature ranges. However, due
    to lack of data it was impossible to quantify the constants of this equation. Thus,
    mortality rate as a function of temperature is determined as a step function.
  - No quantitative description was found for the dependence of algal mortality on
    dissolved oxygen concentration. Thus a three-step function was formulated.
- *Settling.* Algal settling is related to stress. However, this phenomenon can not be
  quantitatively expressed, so the model assumes that living algae do not settle at all,
  whereas the dead algae join the suspended organic matter pool and then settle (as
  organic matter).

Dissolved Oxygen
The oxygen balance equation is solved for each layer while considering the following
processes:

- Total oxygen production and consumption in the water body and the consumption
  in the sediment layers enveloping it.

- Oxygen transfer by atmospheric reaeration at the water-air interface at the surface of the top water layer. The rate of transfer is a function of wind velocity, temperature, area to volume ratio, and the difference between actual oxygen concentration and its saturation concentration (Eq. 7.2).
- Oxygen addition to the top layer by oxygen in rainwater.

Table 7.5. Equations describing the biological processes occurring in the reservoirs

Variable	Process	Governing equation
Algae	Growth	$Gr_A = Gr_{A(20)}^{max} \theta_{Ag}^{(T-20)} \left[\dfrac{I}{I_{opt}} \exp\left(1 - \dfrac{I}{I_{opt}} \right) \right] \dfrac{C_{OG}}{K_{AOG} + C_{OG}} F_{GA(T)}$ $F_{GA(T)} = \begin{cases} 1 & T_{min} < T < T_{max} \\ 0 & T < T_{min} \quad or \quad T > T_{max} \end{cases}$
	Respiration	$R_A = R_{A(20)} \theta_{Ar}^{(T-20)} F_{RA(T)}$ $F_{RA(T)} = \begin{cases} 1 & T_{min} < T < T_{max} \\ 0 & T < T_{min} \quad or \quad T > T_{max} \end{cases}$
	Mortality	$M_A = M_A F_{MA(T)} F_{MA(O2)}$ $F_{MA(T)} = \begin{cases} 1 & T_{min} < T < T_{max} \\ 2 & T < T_{min} \quad or \quad T > T_{max} \end{cases}$ $F_{MA(O2)} = \begin{cases} 1 & C_{O2} > 0 \quad (mgl^{-1}) \\ 2 & C_{O2} = 0 \quad (mgl^{-1}) \\ 2.5 & \left(\begin{array}{l} C_{O2} = 0 \quad (mgl^{-1}) \\ more\,than\,24\,hours \end{array} \right) \end{cases}$
Indicator bacteria	Mortality	$M_{PT} = M_{PTd(20)} \theta_{PTM}^{(T-20)} + \alpha_{PTM} I F_{PT(O2)}$ $F_{PT(O2)} = \begin{cases} 1 & C_{O2} > 0 \quad (mgl^{-1}) \\ 0.67 & \left(\begin{array}{l} C_{O2} = 0 \quad (mgl^{-1}) \\ more\,than\,24\,hours \end{array} \right) \end{cases}$
Organic matter	Degradation	$D_{OG} = D_{OG(20)}^{max} \theta_{OG}^{(T-20)} \left(\dfrac{C_{OG}}{K_{OG} + C_{OG}} \right) F_{D(O2)}$ $F_{D(O2)} = \begin{cases} \dfrac{\left[1 + \left[\cos\left(a \dfrac{\pi}{2} \right) \right] \right]^2}{2} & C_{O2} \geq 0 \quad (mgl^{-1}) \\ 0.4 & \left(\begin{array}{l} C_{O2} = 0 \quad (mgl^{-1}) \\ more\,than\,24\,hours \end{array} \right) \end{cases}$ $a = max\left\{ \dfrac{C_{O2}^* - C_{O2}}{C_{O2}^*} ; \; 0 \right\}$

Table 7.5. *Cont.*

Variable	Process	Governing equation
Zooplankton	Grazing	$Gz = Gz_{(20)}^{max}\, \theta_{Zg}^{(T-20)}\, \dfrac{\omega_A C_A + \omega_{OG} C_{OG}}{K_{GZ} + (\omega_A C_A + \omega_{OG} C_{OG})} F_{GZ(O2)}\, F_{GZ(T)}$
		$F_{GZ(O2)} = \dfrac{1}{1 + \exp\left[1.885\left(2.438 - C_{O2}\right)\right]}$
		$F_{GZ(T)} = \begin{cases} 1 & T_{min} < T < T_{max} \\ 0 & T < T_{min} \quad \text{or} \quad T > T_{max} \end{cases}$
	Respiration	$R_Z = R_{Z(20)}\, \theta_{Zr}^{(T-20)}\, F_{ZR(O2)}$
		$F_{ZR(O2)} = \begin{cases} 0 & C_{O2} < 0.13 & (mgl^{-1}) \\ 0.595 + 0.673\log(C_{O2}) & 0.13 \le C_{O2} \le 4 & (mgl^{-1}) \\ 1 & C_{O2} > 4 & (mgl^{-1}) \end{cases}$
	Mortality	$M_Z = \dfrac{10^{[1.121-0.261(T-9)]} + 10^{[-2.978+0.145(T-2)]}}{100} F_{ZM(O2)}$
		$F_{ZM(O2)} = \begin{cases} 1 & C_{O2} > 1 \quad (mgl^{-1}) \\ \dfrac{10}{1 + 9 C_{O2}} & 0 \le C_{O2} \le 1 \quad (mgl^{-1}) \\ \infty & \begin{pmatrix} C_{O2} = 0 \quad (mgl^{-1}) \\ \text{more than 24 hours} \end{pmatrix} \end{cases}$
Sediment organic matter	Degradation	$D_{SOG} = D_{SOG(20)}^{max}\, \theta_{SOG}^{(T-20)} \left(\dfrac{C_{SOG}}{K_{SOG} + C_{SOG}} \right) F_{SD(O2)}\, F_{SD(water)}$
		$F_{D(O2)} = \begin{cases} \dfrac{\left[1 + \left[\cos\left(a\dfrac{\pi}{2}\right)\right]\right]^2}{2} & C_{O2} \ge 0 \quad (mgl^{-1}) \\ 0.4 & \begin{pmatrix} C_{O2} = 0 \quad (mgl^{-1}) \\ \text{more than 24 hours} \end{pmatrix} \end{cases}$
		$a = MAX \left\{ \dfrac{C_{SO2}^* - C_{O2}}{C_{SO2}^*} ; \quad 0 \right\}$
		$F_{SD(water)} = \begin{cases} 1 & \text{Sediment under water} \\ 2 & \text{Sediment above water} \end{cases}$
Sediment indicator bacteria	Mortality	$M_{SPT} = M_{SPTd(20)}\, \theta_{SPTM}^{(T-20)}\, F_{SPT(O2)}$
		$F_{SPT(O2)} = \begin{cases} 1 & C_{O2} = 0 \quad (mgl^{-1}) \\ 2 & C_{O2} > 0 \quad (mgl^{-1}) \\ 5 & \text{Sediment above water} \end{cases}$

- Oxygen from upper layers reaches lower layers through mixing and diffusion. Molecular diffusion of oxygen is negligible in the epilimnion since oxygen passage through mixing is several orders of magnitude greater than by diffusion. In the epilimnion-hypolimnion interface diffusion is significant since, in effect, it serves as the main source of oxygen for the hypolimnion.

$$S_{O_2} = k_{O_2(20)} \; \theta_{O_2(T)}^{(T-20)} \; \frac{A_W}{V} \left(C_{O_2\text{sat}} - C_{O_2} \right)$$

$$k_{O_2(20)} = \begin{cases} 1.5084 \times 10^{-2} \sqrt{U} & 0 \le U \le 5.5 \\ 1.1520 \times 10^{-3} U^2 & U > 5.5 \end{cases} \quad (U \text{ in m s}^{-1}) \; . \tag{7.2}$$

Indicator Bacteria (Coliforms or Faecal Coliforms)

Mortality rate of indicator bacteria depends on solar radiation intensity and temperature. Moreover, when anaerobic conditions develop in a certain water layer, mortality rate declines to 0.67 of its value in aerobic conditions since then light intensity is lower, no oxygen is present, and smaller changes in the pH occur.

Indicator bacteria settle from each layer to the enveloping sediment layer and to the water layer beneath it at a constant velocity.

Organic Matter

Organic matter is degraded mainly by bacteria. Bacterial population in the water body will be aerobic, facultative, or anaerobic according to the concentration of dissolved oxygen.

- Dependence on water body temperature is described according to the Van't Hoff Arrhenious equation.
- Dependence on organic matter concentration is described by the Michaelis-Menton equation. Dependence of aerobic decay on oxygen concentration is generally described by the Michaelis-Menton equation. However, bacteria present in hypertrophic water bodies are resistant to low oxygen concentrations (Zison et al. 1978; Voinov and Akhremenkov 1990), so in effect from oxygen concentrations of ca. 4 mg l^{-1} and above, the organic matter degradation behaves as a zero order reaction in regard to dissolved oxygen concentration. When the dissolved oxygen concentration falls below 4 mg l^{-1} the aerobic decay rate decreases. However, 'pockets' of facultative and/or anaerobic bacteria starts to degrade the organic matter when oxygen concentration falls to low/anoxic. The importance of anaerobic degradation rises as oxygen concentration further declines and oxygen concentration approaches zero.
- The two reasons set out above indicate that from an overall viewpoint (i.e., total degradation rate of organic matter) there is a need to perform superposition of the aerobic processes and anaerobic ones. The function constructed for describing this superposition matches similar equations developed by Svirezhev et al. (1984) and by Voinov and Akhremenkov (1990).

Suspended organic matter settles from each layer to the enveloping sediment layer and to the water layer beneath it at a constant velocity.

Zooplankton

Zooplankton feed on algae and suspended organic matter and thereby help reduce turbidity of the water body and improve effluent quality. The vertical migration of zooplankton was not entered into the model since it is very difficult to quantitatively describe this vertical movement. This approach has been adopted in many models.

- *Grazing*
  - Dependence on water body temperature is described according to the Van't Hoff Arrhenious equation. However, when the temperature of the effluent in the reservoir digresses from the tolerable temperature range grazing is halted.
  - Dependence on dissolved oxygen concentration is described by an equation based on the one presented by Svirezhev et al. (1984). The original equation was modified in order to express the more resistant nature to low oxygen concentrations of zooplankton of stabilization reservoirs. In the amended equation, when the oxygen concentration is 3 mg l^{-1} the grazing rate is ca. 75% of its maximum rate, and when the oxygen concentration is 5 mg l^{-1} the rate is 99% of the maximum, while in the original equation for these concentrations the rates are ca. 50% and 88% respectively.
- *Respiration*
  - Dependence on water body temperature is described according to the Van't Hoff Arrhenious equation.
  - Dependence on dissolved oxygen concentration is described by an equation based on Lampert (1984) and Kobayashi and Hoshi (1984). These equations were modified in order to express the specific conditions of the reservoirs and findings reported in the literature. The modified equation shows that when oxygen concentration falls below 0.13 mg l^{-1} zooplankton's respiration stops, and when oxygen concentration exceeds 4 mg l^{-1} respiration rate is at its maximum.
- *Mortality*
  - Dependence on water temperature is presented by a modified equation presented by Leidy and Ploskey (1980). The original equation is appropriate for zooplankton acclimatised to climates colder than the climate in this country. Therefore it had to be adapted to the climate in Israel.
  - Dependence on dissolved oxygen concentration is expressed by an equation based on Svirezhev et al. (1984) and Vionov and Akhemenkov (1990), and on studies performed in hypertrophic water bodies (Kobayashi and Hoshi 1984; Weider and Lampert 1985; and others). Accordingly, the developed equation raises the basic mortality rate as oxygen concentration falls below 1 mg l^{-1}, and the mortality rate is multiplied tenfold when the oxygen concentration reachs zero. In addition, when anaerobic conditions develop the whole zooplankton population is wiped out.

Operational State Variables

- Elevation of the lower and upper water surfaces of the layer above the reservoir bottom
- Volume of the layer
- Area of the upper and lower water surfaces
- Area of the adjacent/enveloping sediment (i.e., area of the wetted cross section of the whole layer)
- Thickness of the layer

It should be noted that a one to one relationship exists between the elevation of the lower and upper water surfaces of each layer and the other operational state variables. These relationships are defined specifically for each reservoir using the 'reservoir functions' mentioned above.

7.3.4.2
State Variables in Sediment

The height and area of each sediment layer are constants. The height of each sediment layer is 5 cm, while its area is determined by the 'reservoir functions'. Nevertheless, since the thickness and position of each water layer in the reservoir change with time, the corresponding sediment layers should be allocated each water layer in every time frame.

The state variables in the sediment are the organic matter and the indicator bacteria settling from each water layer to the sediment layers that envelop it. In lakes a developed benthos population is present in the sediment. In stabilization reservoirs high organic loading bring about anoxic conditions in the sediment which inhibit the development of the benthos and thus most of the degradation of organic matter will be carried out by facultative or anaerobic bacteria.

Each sediment layer may be covered by water, or exposed to the atmosphere.

- When a sediment layer is covered by water it will be anaerobic throughout. This assumption matches the field data of sediment cores taken from different lakes showing that even when the water body covering the sediment is totally aerobic, only a negligible part of the sediment will be aerobic too.
- Sediment layers may be exposed to the atmosphere as a result of a declining water level. In this case organic matter will be degraded by intensive aerobic degradation processes and indicator bacteria die-off will be greatly expedited.

Organic Matter in Sediment
Accumulation results from organic matter settling from the appropriate water layer to the respective sediment layer.

Organic matter degradation in the sediment will be carried out by anaerobic decay. Though the sediment is anaerobic, it will always have a demand for oxygen which will be expressed in the appropriate water layer.

- Dependence on water body temperature is described according to the Van't Hoff Arrhenious equation.
- Dependence on organic matter concentration is described by the Michaelis-Menton equation.
- Dependence on dissolved oxygen present in the water layer above/adjacent to the sediment layer is expressed by the same equation developed for degradation in the water body but with a lower concentration threshold for the decline of the decay rate because the bacteria in the sediment are more resistant to low oxygen concentrations. The oxygen threshold concentration below which the organic matter degradation in the sediment declines in the sediment is 1.5–2 mg l^{-1} (Environmental Laboratory 1982, Leidy and Ploskey 1980, Bryant and Bauer 1987).

- When a sediment layer is exposed to the atmosphere due to the decline of the water level, the degradation rate rises to double the maximum rate in the particular temperature.

Indicator Bacteria in the Sediment

Accumulation of indicator bacteria in the sediment occurs as a result of sinking from the appropriate water layer to the particular sediment layer.

Die-off of indicator bacteria in the sediment is a function of temperature and dissolved oxygen concentration only. Solar radiation does not generally reach the sediment. If it does, its intensity is low, and in addition the suspended matter which sank into the sediment shields it from the radiation that does reach the sediment. As a result, the influence of solar radiation on the rate of decay of indicator bacteria (pathogens) in the sediment is negligible. Mortality rate in the sediment is lower than that in the water body. Liran et al. (1994) showed that the die-off rate of indicator bacteria in sediment is 0.03 No No^{-1} d^{-1}. However, when aerobic conditions prevail in the water layer above the sediment the decay rate doubles. And, when a sediment layer is exposed to the atmosphere, the decay rate is multiplied five fold due to the hostile environmental conditions.

7.3.4.3
Major Potential State Variables not Entered into the Model

Nitrogen and Phosphorous

In natural water bodies N and P often limit biological activity in general and algal activity in particular (the primary producers). The kinetics of algal growth as a function of N and P concentration is usually described by the Michaelis-Menton equation. Half-saturation constants reported in the literature, both for lakes and stabilization ponds, lie in the range of 0.01–0.4 mg l^{-1}, and 0.002–0.06 mg l^{-1} for N and P respectively. The concentration of N and P found in stabilization reservoirs is in the range of tens of mg l^{-1}, and of several mg l^{-1} respectively. Therefore, it may be assumed that in stabilization reservoirs they will not limit algal growth. This assumption falls in line with statements of various researchers. The significance of this is that in hypertrophic systems (including stabilization reservoirs) algal growth rate as a function of these nutrients is actually a zero order reaction (the plateau region in the Michaelis-Menton equation). As a result of the above and the problem of data availability, it was decided not to enter N and P into the model as state variables, but to represent them indirectly as a fraction of the total concentration of organic matter in the reservoir. Working with existing data, relationships were established between their concentrations and that of organic matter in the water body. From these relationships their half-saturation constants were expressed on the basis of organic matter.

Active Biomass

In most studies of stabilization reservoirs and lakes, in contrast to intensive wastewater treatment systems (activated sludge), it is not usual to present the bacteriological population as an independent entity, but to append it as a part of the organic matter. This approach was adopted in the model developed for the stabilization reservoirs.

7.3.5
The Logical Structure of the Model and the Computer Programme

The model has a hierarchic structure with three levels of subprogrammes which assures conceptual clarity. At the top of the pyramid (level I) is the main programme whose main function is the provision of a frame for the model as a whole. The main programme calls the programmes in level II which are the model's main subprogrammes divided into subjects. They are more detailed, carry out numerous calculations and call the programmes in level III which generally carry out series of calculations of a single category only.

The model can operate with different time steps, from a few minutes to several days. Experience shows that for the hydraulic part of the model a daily time step is short enough to describe the changes occurring in the reservoirs. However, the Environmental Laboratory (1982) and James (1993) hold that working with a 24 hour time step, and averaging solar radiation in this time, will result in an overestimation of ca. 100% of the photosynthetic production. Solar radiation, and the biological components in the model, which have a fast change rate, impose a time step of 2–3 hours. This is sufficiently short to describe the changing conditions, but still long enough not to get into overlong computer running time.

7.3.6
Results

Rate Constants
A comprehensive comparative survey of the literature was carried out to quantify the process rate constants, stressing a number of aquatic eco-systems similar to stabilization reservoirs: lakes, rivers, stabilization ponds, and fish ponds. A comparison was made of the constants values in these ecosystems in line with the degree of their similarity to conditions prevailing in stabilization reservoirs. An attempt was made to find constants from models and studies conducted in hot regions matching the Israeli climate, but since most of the existing studies were from moderate climates the constants had to be adapted to the Israeli one. A statistical analysis was performed for each constant to find the value that best suites the model. This was done while considering results obtained by statistical analysis that was performed on an existing data on stabilization reservoirs in Israel. Most constants reached by the analysis suited the model very well. After calibration and verification a small number of the constants were changed, but the new values too fell within the range of values reported in the literature.

Initial Conditions – State Variables
Algae, organic matter, and zooplankton concentrations are expressed differently (i.e., chlorophyll, numbers, BOD, COD, TOC, etc.) but in order to set initial conditions to the model it was necessary to express their concentration on a uniform basis. The basis chosen was total organic carbon (TOC). The methods employed to achieve this goal are described herewith:

- *Algae concentration.* Algae concentration is usually expressed as chlorophyll concentration or on a numerical basis. The correlation between chlorophyll and algae biomass is better than the correlation between the number of algae and their biomass. Therefore, as far as was possible, chlorophyll data were taken as the basis for the model initialisation. In establishing the link between chlorophyll concentration and algae organic carbon concentration, it should be remembered that the ratio of organic carbon to chlorophyll varies in different species. The values of this ratio which are quoted in the literature range in the 0.05–0.003 mg C/µg chl_a span, while the average value which is employed in many models is 0.025 mg C/µg chl_a (Gaillard 1981; Environmental Laboratory 1982; Jørgensen et al. 1991; and others) This average value was taken as the conversion factor for this model.

- *Organic matter.* Organic matter concentration in wastewater is expressed by various means (*BOD, COD, DOC, TOC*, etc.), while most data in Israel are measured by *BOD* or *COD*. As the above measures represent various fractions of the total organic matter, their congruence is not complete, and the ratios between them vary according to the type of wastewater, and its degree of treatment. Therefore, links between the above measures have to be established for each type of wastewater (Waite 1978; Medy and Idelovitch 1980; Ebise and Inoue 1991; and others). It is accepted that correlation between *COD* and *TOC* is better than the one between *BOD* and *TOC*, thus, when possible, *COD* data were taken as the basis for the model. The transformation functions used in the model were developed by Medy and Idelovitch (1980) for municipal sewage in Israel. The organic matter was separated into its components: organic matter in algae biomass, in the zooplankton biomass, and "net" organic matter. Statistical analysis of existing data from stabilization reservoirs in Israel revealed that about 50% of the organic matter in reservoirs is floating matter and about 50% dissolved matter, which agree with reported data in the literature.

- *Zooplankton.* Zooplankton concentration is usually expressed as animals per unit of volume. The dominant species in the zooplankton population in stabilization reservoirs fluctuate greatly with the season of the year, and the size of the individuals is not constant. This makes it hard to describe the zooplankton population in quantitative-weight terms, a familiar problem in limnological models whose solution is complicated. The major species were examined and a characteristic weight was found for each species. The specific weight of a "general" individual was worked out by means of weighted average of the characteristic weights of the dominant species, weighted by the relative prevalence of each species. By these calculations, it was established that the characteristic weight of a "general" individual was 1.67×10^{-3} mg C/animal.

7.3.6.1
Model Calibration and Verification

The model was calibrated for the Geta'ot Reservoir (1986–1987 hydrological year) and verified on the Genigar Reservoir (1986–1987 hydrological year) and the northern Ma'aleh HaKishon Reservoir.

Geta'ot Reservoir – General Information
The reservoir is located in the northern coastal strip of Israel, between Acre and Nahariya about 2 km east of the seashore. The maximum water level (above the bottom) is 5.5 m, the maximum volume is 50 000 m³, and the maximal area is 11 500 m².

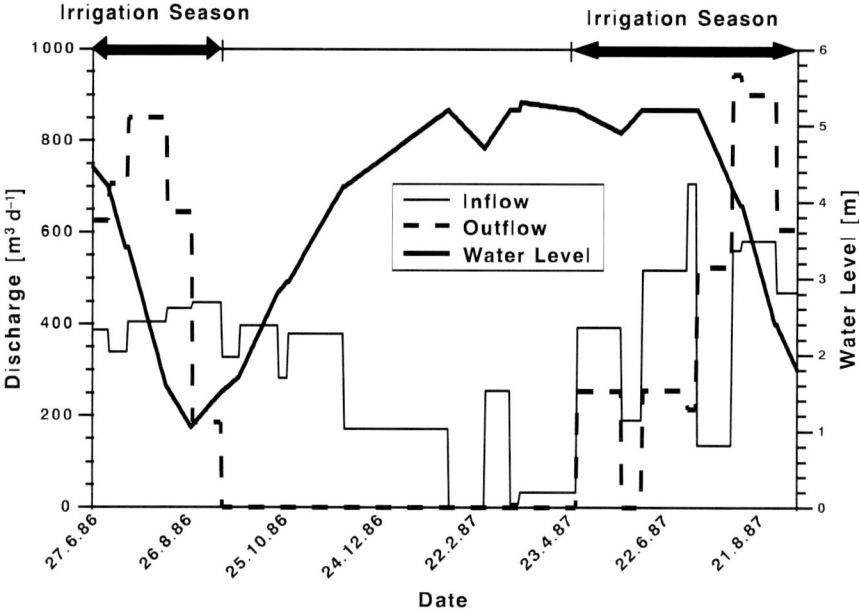

Fig. 7.4. Operational data of Geta'ot Reservoir (June '86–Sept. '87)

Wastewater flows into the reservoir throughout the year (at changing flow rates), while effluent withdrawal is performed during summer for irrigation of cotton and other crops. Data of the hydrological year of 1986–1987 served as the basis for model calibration (Fig. 7.4).

- *Wastewater inflow.* Raw wastewater of the nearby kibbutz flows into two parallel anaerobic ponds (alterating operation), and effluent from these ponds flows into a facultative pond from which effluent is pumped into the reservoir. The average concentrations of COD_{total}, chlorophyll and faecal coliforms are 260 mg l^{-1}, 940 µg l^{-1}, and 4×10^6 MPM/100 ml respectively.
- *Effluent withdrawal.* Effluent is withdrawn from the reservoir from an opening located about one meter below the water surface. The effluent is filtered and the backwash of the filters is returned to the reservoir.
- *Oxygen.* The reservoir has a medium organic loading and it can be defined as a facultative reservoir. High algae concentrations in summer cause high oxygen concentrations in the top water layer, while the bottom one is anaerobic.
- *Operational and thermal regime.* Wastewater enters the hypolimnion and the effluent is withdrawn from the epilimnion (regime 4; Table 7.4). When water level in the reservoir falls below 4 m above the bottom, due to the vertical location of the inlet wastewater enters the whole water column (regime 5; Table 7.4). Permanent thermal stratification started on 15.4.1986, while on 11th August it was broken for one day and the whole water body was mixed (apparently due to particularly strong winds). The thermal stratification resumed the next day and ended finally on 15th November.

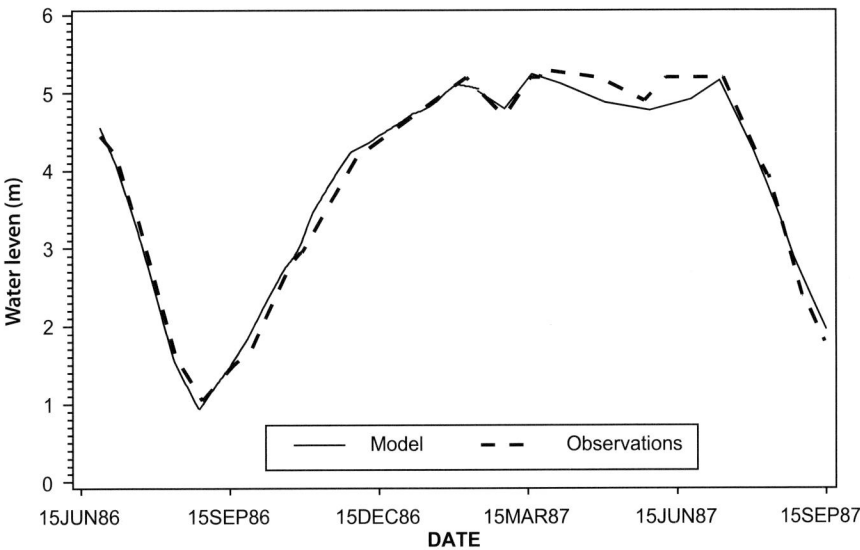

Fig. 7.5. Water level in Geta'ot Reservoir – model vs. observations

Comparison of Predicted and Observed Data

Figure 7.5 depicts the water level in the reservoir throughout the simulated hydrological year. It can be seen that the agreement between the simulated and real data is excellent, while the relative average monthly deviation between the balance calculated by the model and the balance calculated from the measurements was ±7%.

Algae. The illustration focuses on the seasonal variation of algae concentration in the epilimnion (Fig. 7.6a). The lines in the figure show average algae concentration in water layers 2, 3, and 4 (belonging to the epilimnion) as predicted by the model, while the asterisks show measured algae concentration at different depths (belonging to the epiliminion). When thermal stratification begins (early May) a massive algal bloom starts as a result of the rise of the water temperature and intensity of solar radiation. During the thermal stratification period the reservoir is not mixed and epilimnion algae do not leave it. High solar radiation intensity in the epilimnion leads to a high photosynthesis rate which allows the algae population to develop. Thermal stratification actually separates the reservoir into two water bodies with limited connection between them: the epilimnion in which both oxygen producing and oxygen consuming processes take place; and the hypolimnion in which the major occurring processes are oxygen consuming ones turning the hypolimnion anaerobic.

Algal bloom breaks down in early June due to zooplankton grazing. Algae concentration in the epilimnion remains low during June and July as a result of grazing and the decline of the nutrient concentration in the epilimnion. This phenomenon of a clear water phase coming immediately after the massive algae bloom occurs in many reservoirs and water bodies. In August, when the zooplankton population weakens due to a rise in the *PFE* (Percentage of Fresh Effluent – effluent with

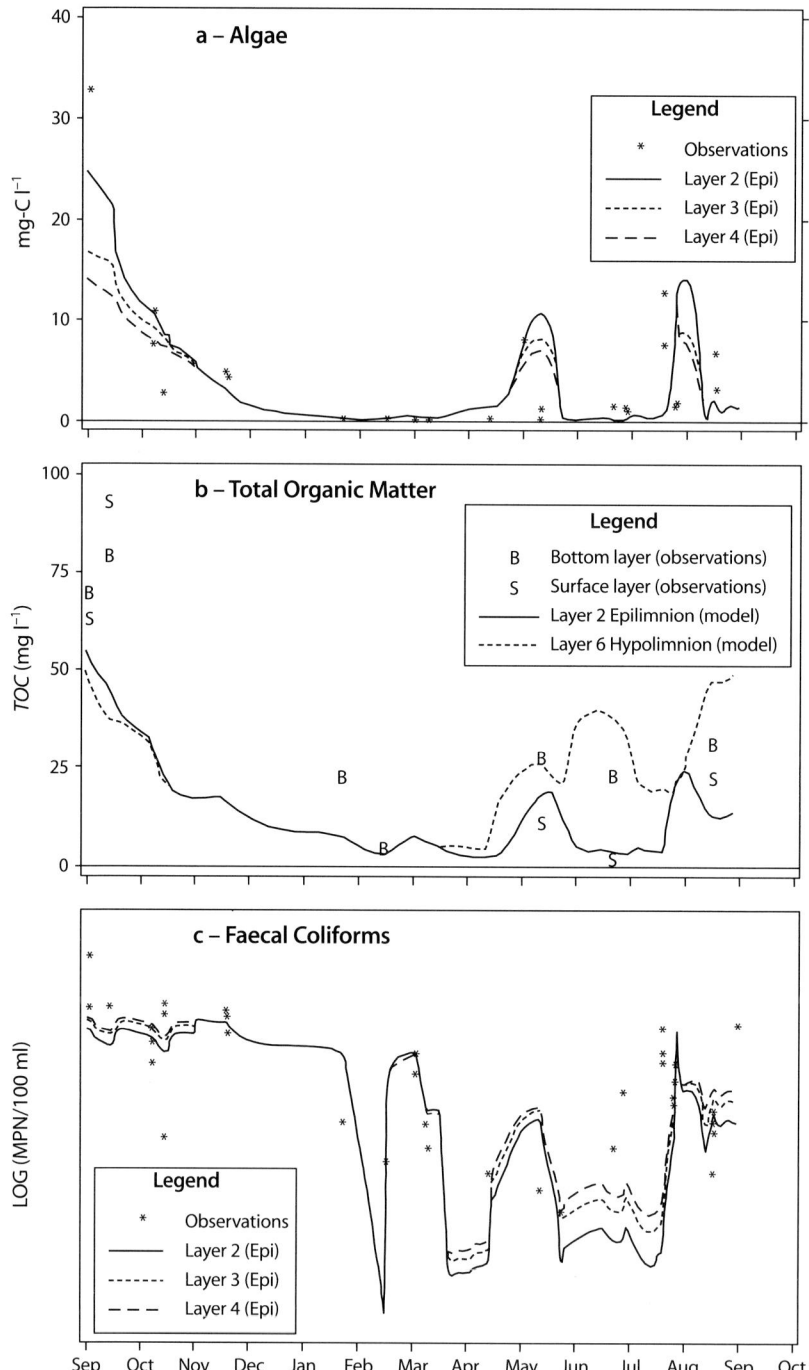

Fig. 7.6. Geta'ot Reservoir: **a** Algae, **b** total organic matter and **c** faecal coliform – model vs. observations

relatively short residence time within the reservoir), causing a decline of effluent quality and depleted oxygen concentrations, another bloom occurs. Since the reservoir is shallow in this period, the bloom takes place throughout the whole water column.

Organic matter (as expressed by *TOC*). The existing database had no "net" organic carbon measurements, only COD_{total} measurements, thus these values were transformed to *TOC* values. Total organic carbon concentrations as predicted by the model were obtained by adding up the organic carbon in algae, zooplankton, and "net" organic matter. Good agreement was found between model predictions and the observed data (Fig. 7.6b).

In winter, when the reservoir is mixed, uniform organic concentration in all water layers prevailed. *TOC* concentration is relatively high in October–November, when the reservoir is almost empty and holds a high proportion of *PFE*. *TOC* concentration declines as the effluent volume rises and the fresh effluent percentage drops. The moderate rise of *TOC* concentration occurring in mid-March stems from inflow of fresh effluent to compensate for losses due to seepage and evaporation.

During the thermal stratification period in summer, layer 2 and 6 represent the average concentrations in the upper part of the epilimnion and in the hypolimnion respectively. In this case too (as with algae), it can be seen that as thermal stratification commenced, *TOC* concentration in the epilimnion rose, mainly due to algae bloom. During June–July, *TOC* concentration in the epilimnion was low, while in the hypolimnion it rose, due to the rising *PFE* in the reservoir. In August, thermal stratification broke for few days which was expressed by a uniform *TOC* concentration throughout the water body. When thermal stratification resumed, *TOC* concentration in the hypolimnion rose, while the concentration in the epilimnion dropped, due to settling of organic matter from upper layers and inflow of wastewater into the hypolimnion. During this period *TOC* concentration generally rose due to a steep rise of the *PFE* in the reservoir.

Zooplankton. A comparison between the predicted and observed zooplankton showed that the model does well in predicting the biomass of the zooplankton, but it was unable to predict their numbers. This stems from the greatly varying compo-sition of zooplankton types, thus the characteristic weight of the "general" individual was not constant leading to deficient transformation from weight basis to numerical basis.

Faecal coliforms. Faecal coliform concentration was found to be highly influenced by the effluent inflow regime. In February no effluent entered the reservoir, causing a sharp decline of faecal coliforms concentration (3–4 orders of magnitude; Fig. 7.6), while in March coliform concentration rose sharply due to high inflow of effluent. In April, low effluent inflow causes a further decline in the coliform concentration. During June–July, thermal stratification and effluent inflow into the hypolimnion brings about low coliform concentration in the epilimnion. In August, due to the significantly higher *PFE*, and the decline of the water level to below 4 m (which induces the effluent to enter the whole of the water body), coliform concentration in the epilimnion rises considerably.

Dissolved oxygen. Figure 7.7a presents a dissolved oxygen profile at midday (12–15 p.m.) on a typical summer day. The intense solar radiation boosted photosynthesis causing a very high oxygen concentration in the uppermost layers of the epilimnion (far above saturation). Thermal stratification prevents the mixing of epilimnion and hypolimnion

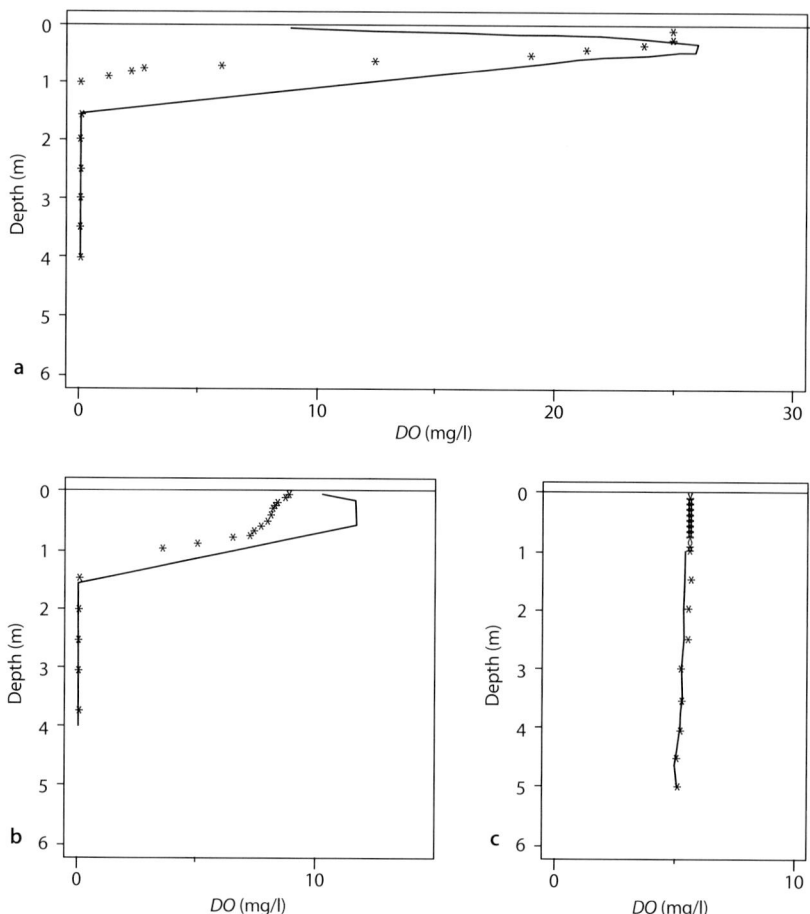

Fig. 7.7. Geta'ot Reservoir – Dissolved oxygen profiles on typical summer and winter days – model vs. observations. **a** summer day: noon; **b** summer day: midnight; **c** winter day: noon

so that the latter remains anaerobic at midday too. Oxygen concentration of the same summer day in the upper layers was shown to decline during the night because there was no oxygen production by photosynthesis; enhanced oxygen demand for algae respiration (present at high concentrations in the epilimnion); oxygen demand for organic matter decomposition; epilimnion mixing which results in more uniform oxygen distribution in the epilimnion, and transfer passage at the air-water interface. The hypolimnion remains anaerobic at night too. In winter, the whole of the reservoir was mixed and the oxygen profile therefore uniform in the whole water column. Algae activity is restricted in winter, both by low temperatures and weak solar radiation, so that the main source of oxygen is atmospheric reaeration.

The model was further verified on Genigar Reservoir (850 000 m³, 8 m depth) an anoxic-anaerobic reservoir which receives mostly raw sewage (85–90% of the total inflow) and some marginal flood water, and on Northern Ma'aleh HaKishon Reservoir (6 400 000 m³, 10.5 m deep) a facultative reservoir which receives effluent from an intensive wastewater treatment plant. For both reservoirs the model predictions showed good agreement with observed data.

7.3.6.2
Sensitivity Analyses and Examination of Alternatives

Two theoretical reservoirs – R7 and R10 – were examined in sensitivity analysis. The reservoirs were of a truncated cone shape differing mainly in their maximum depth. Table 7.6 summarises some physical data of the reservoir.

In order to display the behaviour of the reservoirs in various design and operation conditions, 16 distinct alternatives were studied. The alternatives studied comprised different combinations of regimes: filling, vertical distribution, thermal, and the extent of organic load.

Operational data. Solar radiation, evaporation, and rainfall data were determined from the weather summaries of the Israel Meteorological Service for the Northern Coastal Plain. Wind velocities were taken from the averages in the Climate Atlas for the Physical and Environmental Planning in Israel (Bitan and Rubin 1991). Irrigation season started on 30 April and ended on 30 September. The season's duration, timing, and distribution of effluent withdrawal were determined in line with the accepted regime for cotton irrigation in the Jeesrael Valley, Israel. Wastewater inflow varied in accordance with the inflow regime that was tested (Table 7.8).

Quality data. The influent was designated as activated sludge effluent which was conveyed in pipelines leading from the treatment plant to the reservoir, while the residence time in the pipelines was in the order of several of hours. From this, the quality of influent was determined, while three quality levels were tested (Table 7.7). Since the influent was defined as activated sludge effluent, neither algae nor zooplankton were present. Thus, the COD_{total} presented in the table actually presents the concentration of organic matter. Since the influent was conveyed in a pipe for several hours before entering the reservoir, no dissolved oxygen was present.

Table 7.6. Some physical characteristics of R10 Reservoir

Parameter	Units	R7	R10
Maximum depth	m	7.00	10.00
Minimum operational depth	m	1.00	1.48
Maximum volume	m³	5 000 000	5 000 000
Minimum operational volume	m³	679 000	679 000
Maximum surface area	m²	757 400	551 500
Area of reservoir bottom	m²	673 500	451 300
Slopes of embankments	m/m	1 : 4	1 : 4

Table 7.7. Quality characteristics of influent used for sensitivity analyses and examination of alternatives

Quality level	COD (mg l^{-1})	TOC[a] (mg l^{-1})	Algae (mg C l^{-1})	Zooplankton (mg C l^{-1})	Dissolved oxygen (mg l^{-1})	Faecal coliform (MPN/100 ml)
High	140	50	0	0	0	3.16×10^4
Medium	240	85	0	0	0	1.58×10^5
Low	290	100	0	0	0	3.16×10^5

[a] Conversion from COD to TOC was performed in accordance with the function set by Medy and Idelovitich (1980).

Thermal regime and vertical distribution of inflow and outflow. The thermal regime and the vertical distribution regime of inflowing and withdrawn effluent was set in accordance with the regime that was tested. Water temperatures were determined based on the work of Juanicó (1993).

On the basis of reservoirs R10 and R7, 16 different alternatives of design and operation were studied. R10 with alternative 1 served as the baseline case. Table 7.8 lists the design and operational regimes that were tested by the model.

Sensitivity Analyses
In the framework of testing the sensitivity of the "process" part of the model the impacts of some 50 different constants were examined. In order to obtain a uniform measure of the model's sensitivity to each constant, all constants were changed in the same proportion, i.e., every constant was enlarged or reduced by 10%. The model was run with each change of every constant (a total of ca. 100 runs), and an estimation was made of the degree of the model's sensitivity to the constants.

The model's stepwise, "uneven," form of the results for the winter season stems from the use of monthly averages to define water temperature, wind force and average daily solar radiation (which was due to the dearth of data that dictated the use of monthly averages). In summer, when biological processes rates are high and the variations in reservoir volume are sharp, entering monthly averages is not felt.

Initial conditions of effluent quality variables in the water body affect the model's results for a period of up to four months. The duration of this period is determined by the level of precision in laying down the initial conditions, and by the initial volume of effluent in the reservoir. Initial conditions of the sediment quality variables have a long-term effect on the results, for a period longer than one hydrological year. Particular importance should therefore be assigned to finding the real initial conditions that prevail in the sediment of a reservoir to be modelled.

The degradation rate of organic matter in the sediment, and the subsequent rate of oxygen consumption, are influenced by two factors: temperature and oxygen concentration in the water layer adjacent to the sediment.

Changes of ±10% of the self-shading factor and absorption of light by organic matter increased or decreased by 2 times the average radiation intensity in the hypolimnion in spring, when the reservoir is full. The effect is significant in spring, and not in summer, because in summer light extinction due to the presence of algae plays

Table 7.8. Design and operation alternatives studied by using SRES model

Alternative	Reservoir	Inflow regime	Vertical regime	Inflow layer	Outflow layer	Thermal stratification	Quality level
1 (baseline)	R10	Steady	4	Hypo	Epi	Ephemeral	H
2	R10	Steady	4	Hypo	Epi	Permanent	H
3	R10	Steady	4	Hypo	All	Permanent	H
4	R10	Steady	7	Hypo	Epi	Permanent	H
5	R10	Steady	4	Hypo	All	Permanent	M
6	R10	Steady	7	Hypo	All	Permanent	M
7	R10	Steady	1	All	Hypo	Permanent	M
8	R10	Steady	5	Hypo	Epi	Permanent	M
9	R10	Steady	2	Epi	Epi	Permanent	M
10	R10	Steady	7	All	Epi	Permanent	M
11	R10	Steady	4	Hypo	Epi	Ephemeral	L
12	R10	Steady	4	Hypo	Epi	None	L
13	R10	Steady	4	Hypo	Epi	Permanent	L
14	R10	SBR1	4	Hypo	Epi	Ephemeral	L
15	R7	Steady	4	Hypo	Epi	Ephemeral	L
16	R7	Steady	4	Hypo	Epi	Ephemeral	H

Inflow Regime: timing of inflow into the reservoir, *Steady:* steady inflow throughout the year, *SBR1:* inflow is cease one day before the start of the irrigation season and renewed after the end of the season, *Vertical Regime:* vertical distribution regime of inflow and outflow (see Table 7.4), *Epi:* Epilimnion, *Hypo:* Hypolimnion, *All:* the whole water column, *Thermal Stratification:* general thermal stratification regime (specific regime can be set automatically by the model when certain conditions exist), *Ephemeral:* thermal stratification during daytime, mixing during night, *None:* no stratification occurs (mixing day and night), *Permanent:* permanent thermal stratification (day and night), *H:* high, *M:* medium, *L:* low (see Table 7.7).

a dominant role. This factor was not widely reported in the literature and thus must be researched to define it exactly.

Changing the values of the temperature factor in the decomposition of organic matter, brought about only small changes (ca. 1%) in the concentration of organic matter.

Variations of ±10% of the maximum algae growth rate at 20° C caused a change of ±15% in the summertime algae concentration; ±20–30% in oxygen concentration at noon, and ±25% in the zooplankton concentration. Variations of this factor caused minor changes in the timing of the algae's appearance, their bloom and their disappearance from the reservoir.

Stoichiometric factor linking algae biomass production and oxygen production:

- Lowering the factor value caused a small (ca. 5%) rise of algae concentration. The oxygen concentration at the peak of the algae bloom declined at night by ca. 30%, and during the day by ca. 15–20%. As a result, the zooplankton population was unable to develop and its maximal concentration was one order of magnitude less than in the baseline case. Zooplankton are sensitive to low dissolved oxygen concentrations. Being the consumers, their inability to develop may result in higher concentrations of algae and organic matter. It should be noted that, nevertheless, the changes in the concentration of the *TOC* which, by definition, represents all the organic matter in the reservoir, were not as sharp as the changes noted in the concentrations of "net" organic matter and zooplankton.

- Raising the value of this factor caused smaller changes than its reduction. A certain decline occurred in algae concentration (ca. 5%), a rise of ca. 15% in oxygen concentrations at noon, and a small rise in the zooplankton concentration.

Changes in the optimal radiation intensity required for algae photosynthesis caused various reactions in the water layers which were related to the average radiation intensity in each layer:

- When the factor was lowered by 10% a negative reaction was obtained in the upper layer of the epilimnion, i.e., at noon the algae produced less oxygen (ca. 10%) due to the higher inhibition of photosynthesis. In the layer below it, algae produced more oxygen at noon (ca. 6%) as a result of less difference between the radiation intensity in this layer and the optimal radiation intensity. In deeper water layers changes were negligible.
- When the factor was raised by 10% the reaction was the opposite. In the top layer algae produced more oxygen at noon (ca. 6%), while in the layer below it they produced less oxygen (ca. 10%). In deeper water layers changes were negligible.

Temperature factor for zooplankton respiration and grazing. Since these two factors present the dependence of the metabolic rate of one organism on temperature, changes in them were made simultaneously. Reducing their values resulted in a reduction of zooplankton peak concentration in summer by ca. 15%, and a fortnight long delay in its appearance. On the other hand, zooplankton survived, albeit at a very low concentration, until the middle instead of the beginning of April. Increasing these factors caused the opposite effect: Zooplankton peak concentration was greater and earlier and they survived only to end of March instead of to early April.

Zooplankton assimilation efficiency. Lowering this constant caused a decline of ca. 50% in the zooplankton peak concentration, a delay of about a fortnight in its appearance, and shorter survival in winter. As a result, the decline in organic matter and algae concentrations caused by zooplankton grazing in September was more moderate. Raising this constant caused a rise of ca. 20% in the peak concentration, earlier appearance, and longer survival in the reservoir.

Changes in the maximum growth rate of zooplankton at 20° C caused strong variations in zooplankton concentration and survival in the reservoir. Lowering it caused a decline of ca. 50% in the peak concentration, a delay in its appearance, and shorter survival. Raising it caused a rise of ca. 20% in the peak concentration, earlier appearance and longer survival.

Effluent Quality as a Function of Various Alternatives for Design and Operation
The large number of studied alternatives makes it impracticable to extensively discuss the results obtained for each alternative in the framework of this book, thus only the main findings will be given. The quality of effluent in all alternatives except alternative 14 (Table 7.8) was relatively good at the start of the irrigation season, and dete-

Table 7.9. Maximum and minimum concentrations in effluent withdrawn for irrigation in the studied alternatives

Alternative	Algae Min (mg C l⁻¹)	Algae Max[a] (mg C l⁻¹)	Dissolved oxygen Min (mg l⁻¹)	Dissolved oxygen Max[a] (mg l⁻¹)	TOC Min (mg l⁻¹)	TOC %[b] removal	TOC Max[a] (mg l⁻¹)	TOC %[b] removal	Faecal coliform Min (MPN/100 ml)	Faecal coliform %[b] removal	Faecal coliform Max (MPN/100 ml)	Faecal coliform %[b] removal
High quality influent (see Table 7.7)												
1	0	13.5	0	15	6	88	22	55	1.6E2	99.50	1.3E3	96.02
2	0	11.5	0	11.5	6	88	21	57	2.5E2	99.21	1.3E3	96.02
3	0	6	0	17	2	96	10	80	1.6E1	99.95	1.3E3	96.02
		10.5		8			21	57				
4	0	3.5	0	12	5	90	9.5	81	2.5E2	99.21	1.3E3	96.02
		9		10			19.5	61				
16	0	7	1.2	14	5	90	14.5	71	1.0E2	99.68	6.3E2	98.00
		2		6			6	88				
Medium quality influent (see Table 7.7)												
5	0	6	0	16	4	95	13	85	1.0E2	99.94	6.3E3	96.01
		9.5		6			24	71				
6	0	3	0	10	7	92	14	83	1.6E3	99.00	6.3E3	96.01
		10		5.5			25	70				
7	0	3	0	7	9	89	14	83	1.6E3	99.00	6.3E3	96.01
		10		5.5			25	70				
8	0	9.5	0	2	4	95	25	70	1.6E3	99.00	1.3E4	92.03
9	0	11	0	14	11	87	26	69	1.6E3	99.00	6.3E3	96.01
		9.5		6			25	70				
10		11	0	22	8	90	22	74	7.9E2	99.49	6.3E3	96.01
		21		40			31	63				
		10		6			33	61				
Low quality influent (see Table 7.7)												
11	0	6.5	0	4.5	12	88	24	76	2.0E3	99.37	1.3E4	96.02
12	0	4	0.1	3	13	87	22	78	4.0E3	98.74	1.0E4	96.84
13	0	6	0	15	5	95	13	87	1.6E2	99.95	1.0E4	96.84
		7		5			24	76				
14[c]	0	5	(0) 2	6.5	0	100	15	8 570	0	100	6.3E3	98.00
15	0	17	0.5	10.5	10	90	30		1.0E3	99.68	1.0E4	96.84

[a] When more than one value appear, each one relates to a different algal bloom which occurred in the reservoir;
[b] Maximum removal efficiency corresponds to minimum concentration in the withdrawn effluent, while minimum efficiency corresponds to maximum concentration;
[c] In this alternative (in contrast to the others), maximum quality was achieved towards the end of the irrigation season.

riorated toward the end of the season. This was expressed by greater concentrations of *TOC*, algae and faecal coliforms. In all alternatives maximum oxygen and *TOC* concentrations appeared together with the algae bloom.

Table 7.9 summarises the minimum and maximum concentrations in effluent pumped for irrigation as predicted by the model. In those alternatives where permanent thermal stratification prevailed, two algae blooms appeared – in July and in September. In alternatives with no permanent thermal stratification, only one algae bloom appeared in September when the water level in the reservoir was low.

From the alternatives studied it may be said that R7 reservoir leads better with high organic load than R10 reservoir because it is shallower. The effluent withdrawn from this reservoir for irrigation contains less faecal coliform, its oxygen concentration is higher, and the *TOC* concentration is also lower at the beginning of the season. However, during the algae bloom the *TOC* concentration is ca. 25% higher than in the effluent in R10 reservoir which was operated with the same regime.

Alternative 14 deserves special attention: pumping of influent into the reservoir is halted throughout the irrigation season and renewed only after its end. This practice resulted in constant improvement of effluent quality along the irrigation season, while the best effluent quality was obtained toward the end of the season. About a month after the inflow of influent was stopped, faecal coliforms disappeared from the water body, and the concentration of organic matter declined to half its value at the start of the season. The algae bloom was relatively weak as a result of a shortage of nutrients. Thus, toward the end of the season *TOC* concentration also declined to zero. Figure 7.8 presents *TOC* and faecal coliform in the effluent pumped for irrigation. The curves in the figure actually stand for eight independent curves, each demonstrating withdrawn effluent quality in a three-hour time interval. It can be noticed that during the algal bloom period *TOC* concentrations in the effluent were somewhat higher during the day than during the night. This was due to algae growth rate dependency on solar radiation. Diurnal variation in faecal coliforms concentrations was much less noticeable.

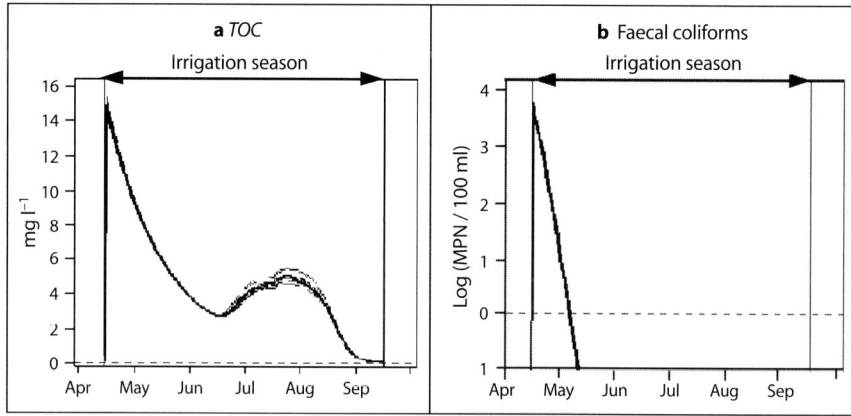

Fig. 7.8. Alternative 14 – *TOC* (**a**) and faecal coliform (**b**) in effluent withdrawn for irrigation

7.4
Summary and Conclusions

The Prediction Capability of the Model

The verification of the model, which was carried out on differing reservoirs under different organic load, demonstrated that the model makes a good prediction of the processes occurring in the stabilization reservoirs and of the quality of effluents obtained from them. The good prediction ability turns the model into a useful research and design tool for the study of the influence of various alternatives on the design and operation of stabilization reservoirs, their efficiency as wastewater treatment units, and the quality of the effluent obtained from them.

The model predicts the oxygen concentration in the water column very well, with the exception of its concentration on the water surface, and that only during daytime thermal stratification, for two reasons:

- It is unable to quantify the partial mixing occurring in the epilimnion in the day during the thermal stratification period.
- The description of oxygen transfer in the water-air interface, according to the Henry Law, is inadequate for the process occurring in the range of super-saturation concentrations.

The model makes a good prediction of zooplankton concentration on a weight basis. The prediction is good enough for describing their role as the consumers in the reservoirs. However, since the composition of the zooplankton population varies greatly and different species of zooplankton have different typical body weights differing in orders of magnitudes, the model is unable to predict their concentration on a numerical basis (i.e., number of animals per unit volume).

Conclusions Obtained from the Model Regarding Stabilization Reservoirs

The regime of influent pumping into the reservoirs is the most important parameter determining the quality of the obtained effluent. The timing of wastewater inflow and the vertical distribution regime of wastewater inflow and effluent outflow influence both the effluent and sediment quality in the reservoir, and the quality of effluents pumped out for irrigation.

The model indicated that halting the inflow of wastewater into the reservoir at the start of the irrigation season resulted in a quick improvement of the withdrawn effluent leading to high quality effluent. It is therefore recommended to change over from seasonal with continuous all-year inflow storage (the most common practice at present), to perennial storage using several reservoirs operated as batch reactors.

The presence or absence of thermal stratification in the reservoir was found to be most significant in determining the appearance of algae bloom and the oxygen regime. The quality of the effluent obtained from the reservoir with permanent thermal stratification is less uniform than the effluent quality in a mixed or ephemerally stratified reservoir.

- The model indicates that there are two algae blooms in a reservoir with permanent thermal stratification:
  1. The spring algae bloom starts immediately upon the onset of thermal stratification. In this period the reservoir is not mixed and therefore algae in the epilimnion

do not leave it. The strong solar radiation in the epilimnion, and the availability of nutrients enhance intense photosynthesis allowing the algae population to develop. This bloom is ended due to zooplankton grazing. Algae concentration remains low as long as the zooplankton are able to exist in the reservoir. The thermal stratification actually separates the water body into a layer in which oxygen is produced and consumed (epilimnion), and a layer in which it is only consumed (hypolimnion) which usually turns anaerobic.

2. The second algae bloom occurs when the zooplankton population is weakened by the increase in the percentage of fresh effluent in the reservoir which reduces the oxygen concentration. In this period the reservoir is shallow, the solar radiation penetrates all the water column to the bottom and the algae bloom therefore occurs in the whole water column.

- Only one algae bloom occurs in a mixed or ephemerally stratified reservoir. In such a thermal regime, the mixing of the water body disperses the algae in the whole of the water column, i.e., algae are dispersed into depths in which the solar radiation intensity is low and does not enable significant photosynthesis to take place. As a result, algae bloom appears only in the later period of the irrigation season, when the water level in the reservoir declines and the radiation reaching the lower layers is sufficiently intense to maintain significant photosynthesis.

By comparing two reservoirs with the same volume but different depths, it was demonstrated that a shallow reservoir (R7) has a better oxygen regime than a deeper reservoir (R10). For this reason a shallow reservoir can handle higher organic loading. A shallow reservoir has a greater area-volume ratio which improves the oxygen regime of the water body.

- In winter, when the main oxygen source is atmospheric aeration, raising the area-volume ratio enhances oxygen transfer into the water body.
- In summer, when the main oxygen source is algal photosynthesis, raising this ratio results in an expansion of the volume of the stratum in which the photosynthesis takes place.

The predator-prey relationships of zooplankton and algae and suspended organic matter, which produce the clear water phase in the epilimnion, contribute significantly to the improvement of effluent quality.

Organic matter decomposition rate in the sediment and subsequently the oxygen consumption rate of the sediment was found to be influenced by two factors: temperature and oxygen concentration in the water layer adjacent to the sediment (although the sediment as a whole is anaerobic, it was found to present oxygen demand supplied from the sediment-water interface). In winter, low temperature is the main limiting factor, while in summer, when the temperature is high, the main limiting factor is the oxygen concentration in the water layer adjoining the sediment.

The model indicates that organic matter concentration in the sediment reaches dynamic equilibrium after several years of operation. In winter, organic matter usually accumulates, due to the higher settling flux and lower decomposition rate. In summer, as the temperature rises, the decomposition rate accelerates and the concentration of organic matter in the sediment declines. The organic matter concentration in the sediment when the dynamic equilibrium sets in depends on the influent quality and its operational regime. This concentration is higher, by several orders of magnitude, than the organic matter concentration in the water body and may reach as much as several thousands g C m^{-2}.

The model indicates that faecal coliforms concentration in the sediment reaches dynamic equilibrium after several years of operation. In winter faecal coliforms usually accumulate due to the higher settling flux and lower die-off rate. In summer the die-off rate rises resulting in the reduction of faecal coliform concentration in the sediment declines. The concentration of faecal coliforms in the sediment when the dynamic equilibrium sets in depends on influent quality and the operational regime. This concentration is higher, by several orders of magnitude, than faecal coliforms concentration in the water body.

From the alternatives studied it could be concluded that the determination of the maximum organic load which will still ensure the desired effluent quality should be performed specifically for each reservoir. This maximum load is a function of the physical structure of the reservoir, the operational regime, the thermal regime, and the vertical distribution of the inflow and outflow.

7.5
Notation

Dimensions: L = length, L^2 = area, L^3 = volume, M = mass (kg, mg, etc.), T = time (day, hours, second, etc.)

- α_{PTM} = Proportion factor for light effect of indicator bacteria mortality
- θ_{Ag} = Temperature coefficient for algal growth
- θ_{Ar} = Temperature coefficient for algal respiration
- $\theta_{O2(T)}$ = Temperature coefficient for oxygen transfer
- θ_{OG} = Temperature coefficient for organic matter degradation
- θ_{PTM} = Temperature coefficient for indicator bacteria die-off
- θ_{SOG} = Temperature coefficient for sediment organic matter degradation
- θ_{SPTM} = Temperature coefficient for sediment indicator bacteria die-off
- θ_{Zg} = Temperature coefficient for zooplankton grazing on algae and suspended organic matter
- θ_{Zr} = Temperature coefficient for zooplankton respiration
- ω_A = Algae preference factor for zooplankton grazing
- ω_{OG} = Organic matter preference factor for zooplankton grazing
- a = Coefficient
- A_W = Area of water surface
- C_A = Algae concentration
- C_{O2} = Dissolved oxygen concentration
- C_{O2}^* = Dissolved oxygen concentration below which oxygen starts to limit aerobic degradation of organic matter
- C_{O2sat} = Dissolved oxygen saturation concentration
- C_{OG} = Organic matter concentration
- C_{SO2}^* = Dissolved oxygen concentration below which oxygen starts to limit aerobic degradation of organic matter in the sediment

- D_{OG} = Organic matter degradation rate
- $D_{OG(20)}^{max}$ = Maximum degradation rate of organic matter at 20° C
- D_{SOG} = Sediment organic matter degradation rate
- $D_{SOG(20)}^{max}$ = Maximum degradation rate of sediment organic matter at 20° C
- $F_{D(O2)}$ = Oxygen factor for organic matter degradation
- $F_{GA(T)}$ = Temperature factor for algae growth
- $F_{GZ(O2)}$ = Oxygen factor for zooplankton grazing
- $F_{MA(O2)}$ = Oxygen factor for algae mortality
- $F_{MA(T)}$ = Temperature factor for algae mortality
- $F_{PT(O2)}$ = Oxygen factor for indicator bacteria mortality
- $F_{RA(T)}$ = Temperature factor for algae respiration
- $F_{SD(O2)}$ = Oxygen factor for sediment organic matter degradation
- $F_{SD(water)}$ = Water/atmosphere factor for sediment organic matter degradation
- $F_{SPT(O2)}$ = Oxygen factor for indicator bacteria mortality in the sediment
- $F_{ZM(O2)}$ = Oxygen factor for zooplankton mortality
- $F_{ZR(O2)}$ = Oxygen factor for zooplankton respiration
- Gr_A = Algae growth rate
- $Gr_{A(20)}^{max}$ = Maximum algae growth rate at 20° C
- Gz = Zooplankton grazing rate
- $Gz_{A(20)}^{max}$ = Maximum zooplankton grazing rate at 20° C
- I = Solar radiation intensity
- I_{opt} = Solar radiation intensity for which algae growth is at its maximum rate
- $k_{O2(20)}$ = Oxygen transfer rate coefficient at 20° C
- K_{AOG} = Half saturation constant for algae growth in relation to organic matter (as a pool nitrogen of and phosphorous)
- K_{OG} = Half saturation constant for organic matter degradation in relation to organic matter
- K_{SOG} = Half saturation constant for sediment organic matter degradation in relation to sediment organic matter
- M_A = Algae mortality rate
- M_{PT} = Indicator bacteria mortality rate
- $M_{PTd(20)}$ = Indicator bacteria mortality rate at 20° C
- M_{SPT} = Sediment indicator bacteria mortality rate
- $M_{SPTd(20)}$ = Sediment indicator bacteria mortality rate at 20° C
- M_Z = Zooplankton mortality rate
- PFE = Percentage of Fresh Effluent
- R_A = Algae respiration rate
- $R_{A(20)}$ = Algae respiration rate at 20° C
- R_Z = Zooplankton respiration rate
- $R_{Z(20)}$ = Zooplankton respiration rate at 20° C
- S_{O2} = Atmospheric reaeration rate
- T = Water temperature
- T_{min} = Minimum tolerable water temperature
- T_{max} – Maximum tolerable water temperature
- U = Wind velocity at the water surface height
- V = Volume

References

Avnimelech Y (1989) Modelling the accumulation of organic matter in the sediment of a newly constructed reservoir. Wat Res 23(10):1327–1329

Beck MB (1991) Principles of modelling. Wat Sci Tech 24(6):1–8

Bitan A, Rubin S (1991) Climatic atlas of Israel for physical and environmental planning and design. Dept. of Geography, Tel-Aviv Uni. Isr. Meteor. Service, Minst. of Transport (in Hebrew), Tel-Aviv, Israel

Bryant CW, Bauer EC (1987) A simulation of benthal stabilization. Wat Sci Tech 19(2):161–168

Chen CW, Orlob GT (1975) Ecologic simulation for aquatic environments. In: Patten BC (ed) System analysis and simulation in ecology, vol III, Academic Press, N.Y., USA, pp 475–528

Ebise S, Inoue T (1991) Change in the C:N:P ratios during passage of water areas from rivers to a lake. Wat Res 25(1):95–100

Environmental Laboratory (1982) CE-QUAL-R1: A numerical one dimensional model of reservoir water quality; User's manual. US Army engineer waterways experiment station, Vicksburg, Miss., USA, pp 1–623

Friedler E (1993) Mathematical model of stabilization reservoirs. DSc Research Thesis, 188 pp (in Hebrew). Contains a separate volume of appendices, 110 pp (in English). Technion-Israel Institute of Technology, Haifa, Israel

Gaillard J (1981) A predictive model for water quality in reservoirs and its application to selective withdrawal. Hydrowater Program, Col. State Univ., Fort Collins, Colorado, USA, 232 pp

Huber WC, Harleman DRF, Ryan PJ (1972) Temperature prediction in stratified reservoirs. Jour Hydr Div ASCE, 98(4):649–666

James A (1993) Modelling water quality in lakes and reservoirs. In: James A (ed) An introduction to water quality modelling. John Wiley & Sons, Chichester, U.K., 2nd edn, pp 233–261

Jørgensen SE, Mejer H (1979) A holistic approach to ecological modelling. Ecol Model 7:169–189

Jørgensen SE, Nielsen SN, Jørgensen LA (1991) Handbook of ecological parameters and ecotoxicology. Elsevier Science Pub, Amsterdam, The Netherlands, pp 1–1263

Juanicó M (1993) Stabilization reservoirs in Israel – Typical temperatures for process design. Unpublished data

Juanicó M, Shelef G (1991) The performance of stabilization reservoirs as a function of the design and operation parameters. Wat Sci Tech 23(7–9):1509–1516

Juanicó M, Friedler E (1994) Hydraulic age distribution in perfectly mixed nonsteady-state reactors. ASCE Journal of Environmental Engineering 120(6):1427–1445

Juanicó M, Shelef G (1994) Design, operation and performance of stabilization reservoirs for wastewater irrigation in Israel. Wat Res 28(1):175–186

Kobayashi M, Hoshi T (1984) Analysis of respiratory role of haemoglobin in *Daphnia magna*. Zool Sci 1:523–532

Krenkel PA, French RH (1982) State-of-the-art of modelling surface water impoundments. Wat Sci Tech 14(1–2):241–261

Lampert W (1984) The measurement of respiration. In: Downing JA, Rigler FH (eds) A manual on methods for the assessment of secondary productivity in fresh waters. IBP Handbook 17, Blackwell Scientific Publ., London, England, pp 413–468

Leidy GR, Ploskey GR (1980) Simulation modelling of zooplankton and benthos in reservoirs: Documentation and development of model constructs. USDI Fish and Wildlife service. National Reservoir Research Program, Fayetteville, Arkansas, USA, 221 pp

Liran A, Juanicó M, Shelef G (1994) Coliform removal in a stabilization reservoir for wastewater irrigation in Israel. Wat Res 28(6):1305–1314

Lokiec EF (1983) Behaviour of a reservoir for seasonal effluent storage during the filling and storage period. MSc Research Thesis (in Hebrew), Technion-I.I.T., Haifa, Israel, 137 pp

Losordo TM, Piedrahita RH (1991) Modelling temperature variation and thermal stratification in shallow aquaculture ponds. Ecol Model 54:189–226

Markofsky J, Harleman DRF (1973) Prediction of water quality in stratified reservoirs. Jour Hydr Div ASCE 99(5):729–745

Medy M, Idelovitch E (1980) Monitoring of gross organics in water and wastewater. Tahal water planning for Israel LTD. Sewage reclamation department, Tel-Aviv, Israel, 45 pp

Orlob GT (1983) Mathematical modelling of environmental quality: Streams, lakes, and reservoirs. John Wiley & Sons, N.Y., USA, 518 pp

Orlob GT (1992) Water-quality modelling for decision making. Jour of Wat Res Plan Manag ASCE 3:295–307

Romem M (1991) Manual of reservoir design. Industry and Building Res. Inst., Assoc. of Engineers and Architects in Israel. (in Hebrew), Tel-Aviv, Israel, 436 pp

Rosenzvieg D (1986) Calculation of evaporation losses and direct rain on reservoirs. Wat Irrig 239: 35–36 (in Hebrew)

Shelef G, Rubin E, Riter A, Goldshmit Y, Eren J, Oron G (1977) Seasonal storage of wastewater. Fac. of Civ. Eng., Fac. of Cem. Eng., Technion-I.I.T. (in Hebrew), Haifa, Israel, 60 pp

Steele JH (1962) Environmental control of photosynthesis in the sea. Limnol Oceanog 7:137–150

Streeter HW, Phelps EB (1925) A study of the pollution and natural purification of the Ohio River, III. Factors concerned in the phenomena of oxidation and reaeration. US Public Health Service, Bulletin 145, Washington, DC, USA, pp 1–75

Svirezhev YM, Krysanova VP, Voinov AA (1984) Mathematical modelling of fish pond ecosystem. Ecol Model 21:315–337

Voinov AA, Akhremenkov AA (1990) Simulation modelling system for aquatic bodies. Ecol Model 52:181–205

Waite TD (1978) An evaluation of non-specific measurements on organic pollutants in waters and wastewaters. Australian water resources council (Technical Paper N° 34). Aust Gov Pub Service, Canberra, Australia, 162 pp

Water Resources Engineering Inc. (1968) Prediction of thermal energy distribution in streams and reservoirs. Water Resources Eng. Inc. Report to California Department of Fish and Game, USA, 90 pp

Weider LJ, Lampert W (1985) Differential response of *Daphnia* genotypes to oxygen stress: respiration rates, haemoglobin content and low-oxygen tolerance Oecologia 65:487–491

Zison SW, Mills WB, Deimer D, Chen CW (1978) Rates, constants, and kinetics formulations in surface water quality modelling. Environmental Research Laboratory, Office of research and Development, USEPA, Athens, Georgia USA 316 pp

Chapter 8

Sediment-Water Interrelationship

Yoram Avnimelech

8.1
Introduction

The bottom soil of water bodies such as reservoirs, ponds, and shallow lakes acts as the storage, regulation and buffer organ of the system. As an illustration, the quantities of several components in the bottom soil can be compared to the equivalent terms in the water. The range of concentrations in the water of organic carbon, nitrogen and phosphorus of shallow lakes reservoirs and impoundments are in the order of $10-10^2$, $0.1-10^1$ and $10^{-2}-10^1$ ppm (parts per million) respectively. The equivalent concentrations in bottom soils are in the order of 10^4-10^5 ppm organic carbon, 10^3 ppm total nitrogen and 10^3 to 10^4 ppm of total phosphorus. Thus, the concentrations in the bottom soil are about 3 orders of magnitude higher than those in the water. The concentration of phosphorus in the soil may be up to 5 orders of magnitude higher than in the water. Total phosphorus concentration in the water of the Sea of Galilee, Israel, is in the order of 10^{-2} ppm, while that in the sediment is in the order of 10^3 ppm (about 0.3%). Putting this differently, the amount of C or N in 1 cm deep bottom soil layer is equivalent to that found in a water column of about 10 m or, in the case of P, in a water column 1000 m deep! Very similar data are relevant for the distribution of other components, such as heavy metals.

Moreover, the concentrations of many components in the water is relatively constant and stable over time and space, due to control mechanisms such as solubility limitations, light limited algal density, particles sedimentation, etc. The concentrations of most chemical components in the bottom soil are not limited and may differ by orders of magnitude among different impoundments. The bottom soil may be in a chemical equilibrium with the water column and thus, the soil composition may be reflected in the quality of the water. However, the existence of an actual equilibrium between the bottom soil and the water is questionable in most cases, due to kinetic considerations.

Processes of nutrient binding to the sediment or release to the water are always present. One interesting case of such interaction occurs in new reservoirs or in impoundments in which the load increased due to intentional or unintentional input of nutrient rich effluents. In such cases a massive binding of nutrients to the bottom soil is expected. The opposite process is also expected when a loaded (eutrophic) impoundment is cleaned, i.e., the nutrient input is restricted.

8.2
Accumulation of Nutrients in Bottom Soils

Build-up of nutrients in the bottom soil takes place in newly constructed (or newly loaded) impoundments.

A very clear build up of ammonium takes place in newly flooded reservoirs (Gunnison et al. 1980). Avnimelech and Wodka (1988) found that ammonium in the bottom soil of Ma'aleh HaKishon Reservoir (a reservoir holding treated effluents described by Eren, this volume), increased from the original level of 0.005 mg g^{-1} to 0.06 and 0.6 mg g^{-1} following one and two years of operation, respectively. Ammonium concentrations in bottom soils of fish or shrimp ponds (systems loaded with nitrogen rich organic matter), reach levels of several hundred ppm. Ammonium accumulation is typical in systems where organic matter is metabolized under anaerobic conditions such as bottom soils of eutrophic impoundments (Waring and Bremner 1964; Reddy et al. 1980). The energetic efficiency of anaerobic metabolism is lower than that of aerobic metabolism (Reddy et al. 1986). The result is that less protein is produced per unit carbon metabolized under anaerobic conditions. The nitrogen requirement associated with the decomposition of organic matter under anaerobic conditions is about one third of that required under aerobic conditions (Williams et al. 1968). This, in addition to the absence of nitrification, leads to a marked build up of ammonium.

Avnimelech and McHenry (1984) compared the concentrations of organic carbon, total nitrogen, and phosphorus in the watershed soils (the soils from which the earth works were made for the construction of the reservoir) with that in the sediments of reservoirs. A general enrichment of the reservoir sediments was found. Organic carbon and nitrogen were enriched up to a limiting value of 24 and 1.7 g kg^{-1}, respectively. In watershed soils containing levels higher than these limiting values, nutrients were released from the soil. This process is probably due to the fact that the microbially mediated release of organic components from the solid phase is related to their concentrations. No such limiting values were found in respect to phosphorus or clay contents. These components were accumulating in the sediments with no clear limiting value.

Bottoms of impoundments are constantly enriched through the sedimentation of particles, mostly organic, from the water body. Daily sedimentation rates of ca. 20 g per m^2 were found in the Ma'aleh HaKishon Reservoir (Avnimelech and Wodka 1988). Organic carbon contents of the sedimented material collected in traps was about 10%, total nitrogen and total phosphorus amounted to about 1% of the collected material.

8.3
Release of Nutrients from the Bottom

The bottom soil can serve both as a source of chemical components released to the water as well as a sink for components from the water. Different mechanisms control the release of the organic components stored in the soil and of the inorganically bound components. The degradation of the organic materials in the bottom and the concomitant release of soluble components to the water column is the major mechanism that liberates nitrogen or soluble organic components back to the water.

It is usually considered that the degradation of organic matter in the sediment is controlled by a first order kinetics:

$$\frac{dC}{dT} = -KC \quad .$$

(8.1)

Berner (1980) showed that the diagenesis of organic matter in marine bottom soils is controlled by a series of first order rate constants, when the more readily available components degrade first at a relatively high rate and the residual fraction degrades subsequently with lower rate constants.

Avnimelech et al. (1984) evaluated the reaction rate constant by analyzing the concentration gradient of organic carbon, nitrogen, and phosphorus in cores sampled in 64 shallow lakes and reservoirs across the USA. The depth along the sediment core was assumed to be proportional to the time passed since the deposition of the sedimented material. The age of each layer was evaluated through the location of Cs^{137} peaks. The decrease in the concentration of C, N, and P was in a good agreement with a first order kinetic model. The frequency distribution curves of the degradation constants in the different impoundments were somewhat skewed, probably due to the fact that the degradation is much faster in reservoirs that were dry during some periods. The modes of the degradation constant (seen as the better estimate of the true degradation constant of flooded sediments) were found to be 4, 5 and $4 \times 10^{-3}\,yr^{-1}$ for organic carbon, nitrogen, and phosphorus, respectively. Similar degradation constants were found in lakes by other authors (Farrington et al. 1977; Johnson et al. 1982; Murray et al. 1978; Serruya 1971). It should be emphasized that the core sampling and analysis was done on layers 1–2 inches thick, equivalent to a deposition period of a few years. Faster processes cannot be evaluated with this setup.

A different set of decomposition rate constants were found in more reactive systems, such as fish pond sediments and soil bottom of the Ma'aleh HaKishon Reservoir. The decomposition constant of organic matter in the bottom soil of fish ponds was found by measuring the decrease in organic matter in sediment that was collected in a trench and disconnected from further accumulation for a known time (Avnimelech 1984). The decomposition rate constant was found to be $4 \times 10^{-1}\,yr^{-1}$, i.e., about 10 times higher than the one reported above. The same decomposition rate constant was found in Ma'aleh HaKishon Reservoir. The method to evaluate the rate constant in the Ma'aleh HaKishon bottom soil will be described further in this chapter.

Phosphorus and other components (mostly inorganic, heavy metals included) are held in the sediment as insoluble salts or as adsorbed species. Thermodynamic equilibria constants govern such reactions. Such equilibria lead to a release of soluble ions whenever the solution is under-saturated in respect to the solubility or the adsorption isotherms. The significance of such processes depends on the rate of release and the rate of diffusion of the soluble species in the water column. Eren et al. (1977) found that phosphorus solubility equilibrium seems to hold over time in earthen fish ponds, concomitant with a series of changes (P fertilization, diel changes in pH etc.). These ponds are rather shallow (about 1 m deep) and thoroughly mixed by the strong prevailing daily winds. The probability of such equilibrium in deeper impoundments is low, especially in stratified impoundments.

An important release mechanism is the resuspension of bottom sediment particles. Reddy et al. (1986), in their study of Lake Appopka, estimated that one resuspension event per day leads to a nutrient flux three times higher than the equivalent diffusion fluxes. Top layer of bottom sediments, especially those rich in respect to organic matter are flocculant and may be easily detached. Both currents and bioturbation (especially that induced by fish) may resuspend bottom particles and thus physically lead to an upward flux of materials associated with the resuspended particles.

The evaluation of resuspended materials is not simple and quite uncommon.

Sediment traps are the basic tools to evaluate material fluxes at or near the bottom. Yet, the interpretation of sediment traps data is quite difficult. The material trapped may originate from two different fluxes: the flux of materials sedimenting from the water body and a flux of resuspended material, i.e., bottom material raised and resettled. The separate evaluation of the two fluxes is neither easy nor straightforward.

Avnimelech and Wodka (1988) were able to evaluate the two fluxes separately in the Ma'aleh HaKishon Reservoir by using three independent sets of measurements related to accumulation in the bottom soil. The first one (the nutrient balance) was based upon the monitoring of the quantities of water getting into the reservoir and getting out of it, together with the monitoring of concentrations in the water body. The difference between the amount of a given material introduced and that found in the water and in the outgoing flow represent losses from the water column. In the case of phosphorus, an element that is not lost to the atmosphere, this deficit represents sedimentation. The second method to evaluate sedimentation was sampling and analysis of the bottom soil before flooding as well as approximately 1 and 2 years later. The amount of phosphorus accumulating in the bottom soil was found to be practically the same as that calculated by the balance approach. The third method (the evaluation of sedimented phosphorus through the measurement of sedimentation by sediment traps) gave a value approximately twice as high as compared to the two previous methods. This implied that resuspension in Ma'aleh HaKishon Reservoir (depth of ca. 12 m) was about equal to sedimentation (Sediment collected = sedimentation + resuspension).

Serruya (1978) and later Avnimelech and Kochba (in press) used a different approach, based upon the fact that the composition of sedimented and resuspended materials are different. The sedimented materials are mostly organic in nature, made mostly out of algae in lakes and out of feed materials and their products in fish ponds. In fish ponds, if no allochtoneous material is brought in (a condition that is usually fulfiled), one should not expect to find soil forming elements and minerals in the sediment flux. The presence of components such as Al, Si or Fe in sediment traps is indicative to a resuspension flux getting into the trap. Through the determination of the concentrations of these elements in the bottom soil and in the material collected in the sediment traps, and by application of a dilution analysis, it was possible to evaluate separately the sedimentation and resuspension fluxes. In fish ponds, shallow water bodies affected by both wind-induced and fish-induced bottom soil disturbance, resuspension was found to be about 10 and more times higher than sedimentation. It was estimated that a layer of about 3 mm of the bottom soil is resuspended daily (Avnimelech and Kochba, in press). The resuspension mechanism may be a major transfer vehicle of chemicals and nutrients between the soil and the water and vice versa.

8.4
Nutrient Balance of Reservoirs

Nutrient balance in reservoirs includes the following elements: input, accumulation and output. A nutrient balance for Ma'aleh HaKishon was constructed by Avnimelech and Wodka (1988) and presented in Fig. 8.1.

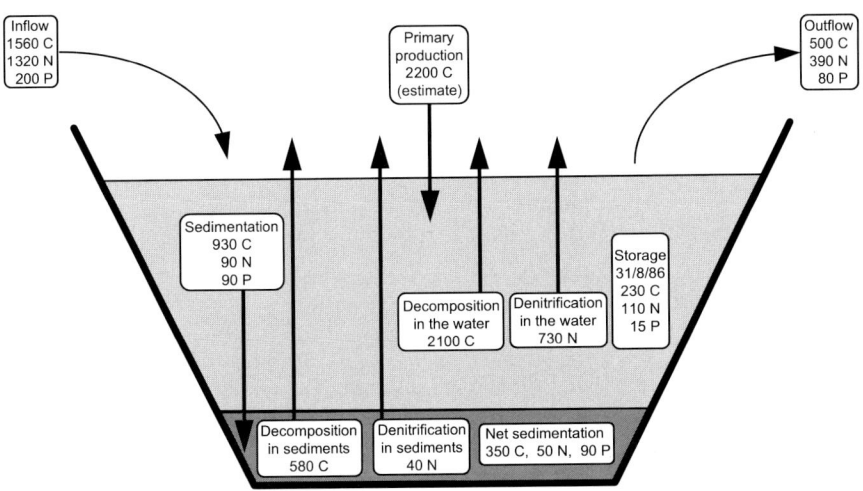

Fig. 8.1. Nutrient balances (tons) and flow chart of Ma'aleh HaKishon Reservoir. April 1984–August 1986

Input was calculated from the volume of effluents pumped into the reservoir and the monitored concentrations of nutrients in the water. These data were collected and furnished by the Mekorot Water Company, which operates this reservoir. An additional input of organic carbon was primary productivity which was monitored as a part of the reservoir study (Dor et al. 1987). Nutrient output was computed in a similar manner, using the volume and composition data for the effluents released for irrigation.

The storage of nutrients in the water body was obtained through the analyses of water composition in the final balance day (August 1986) and multiplication of appropriate concentrations by the volume of water in the reservoir system. Losses of nutrients were defined as the difference between input and (output + storage in the water):

$$Losses = Input - (output + storage\ in\ the\ water)\ . \tag{8.2}$$

The major nutrient loss mechanism is sedimentation, evaluated from sediment trap data, corrected for resuspension as mentioned before. The difference between the computed loss of a given nutrient and its sedimentation is due to its degradation in the water column (respiration, denitrification). An additional process derived from the collected data is the degradation in the bottom layer. The accumulation of phosphorus in the sediment was practically equal to the calculated sedimentation. Yet, the accumulation of organic carbon and nitrogen in the sediment was lower than the computed sedimentation. This expected deviation is due to respiration (aerobic or anaerobic) and denitrification taking place in the bottom layer.

Nutrient balances and flow charts computed from the above mentioned data (Fig. 8.1) represent a comprehensive description of that reservoir, and one of the most detailed nutrient flow charts published.

8.5
Modelling of Soil Bottom Processes

The accumulation of biologically active components (organic carbon, nitrogen, ammonium, *BOD* etc.) is a function of the sedimentation of these components on one hand and their biodegradation and release on the other hand (Avnimelech 1989).

This process can be formulated as the sum of constant sedimentation flux (*B*) and a first order degradation process, both as related to a given component, C_i:

$$\frac{dC_i}{dt} = B - K_i C_i \quad .$$

(8.3)

Equation 8.3 is a simplified formulation, assuming that both the addition of a given component as well as its degradation rate are constant. This is only an approximation of course, yet it may hold if time units are large enough (e.g., years rather than days). In addition, the term C_i needs clarification. C_i is considered here as a two-dimensional concentration term (e.g., mg cm^{-2}), for a given soil depth, assumed to be the active depth. In addition, it is assumed that no significant accretion occurs, and the organic fraction is not significantly diluted by sedimenting inorganic particles.

Upon integration of Eq. 8.3 we get:

$$C_i = \frac{B - e^{-Kt}(B - KC_{i0})}{K_i} \quad .$$

(8.4)

One interesting feature of Eq. 8.4 is that for a long time range (actually, when $t \times K_i$ are high), the concentration term, C_i approaches a limit:

$$C_i = \frac{B}{K_i} \quad .$$

(8.5)

The steady state defined by Eq. 8.5 is obtained at a point when the release of C_i from the bottom soil is equal to its rate of addition. The rate of approach to the steady state is a function of K_i. Due to the high K_i for the Ma'aleh HaKishon Reservoir, it was predicted that the reservoir will be close to a steady state situation following a period of 2–3 years. A very good agreement between predicted values and measured was found when sediments were sampled following a period of about 4 years of operation, during a drainage of the reservoir.

References

Avnimelech Y (1984) Reactions in fish pond sediments as inferred from sediment cores data. In: Rosenthal H, Sarig S (eds) Research on aquaculture. European Mariculture Soc Apec Publ N° 8, Bredene, Belgium, pp 41–54

Avnimelech Y (1989) Modelling the accumulation of organic matter in the sediments of a newly constructed reservoir. Wat Res 23:1327–1329

Avnimelech Y, McHenry JR (1984) Enrichment of transported sediments with organic carbon, nutrients and clay. Soil Sci Soc Am J 48:259–266

Avnimelech Y, Wodka M (1988) Accumulation of nutrients in the sediments of Ma'aleh HaKishon reclaimed effluents reservoir. Water Res 22:1437–1442

Avnimelech Y, McHenry JR, Ross JD (1984) Decomposition of organic matter in lake sediments. Envir Sci Technol 18:5–11

Berner RA (1980) A rate model for organic matter decomposition during bacterial sulfate reduction in marine sediments. In: Biogeochemistry of organic matter at the sediment water interface, pp 35–44, CNRS Int. Colloq

Dor I, Kalinsky I, Eren J, Dimentman C (1987) Deep wastewater reservoirs in Israel. I. Limnological changes following self-purification. Water Sci Technol 19(12):317–322

Eren Y, Tsur R, Avnimelech Y (1977) Phosphorus fertilization of fish ponds in the Upper Galilee. Bamidgeh, Israel J Aquaculture 31:3–8

Farington JW, Henrichs SM, Anderson R (1977) Fatty acids and lead-210 geochronology of sediment core from the Buzzard Bay, Massachusetts. Geochim Cosmochim Acta 41:289–296

Gunnison D, Brannon JM, Smith I, Burton GA (1980) Changes in respiration and anaerobic nutrient regeneration during the transition phase of reservoir development. In: Barica J, Mur LR (eds) Hypertrophic ecosystems. Dev Hydrobiol 2:151–158

Johnson TC, Evans JE, Eisenreich SJ (1982) Total organic carbon in Lake Superior sediments: Comparison with hemipelagic and pelagic marine environments. Limnol Oceanogr 27:481–491

Murray JW Grundmanis V, Smethie WM (1978) Interstitial water chemistry in the sediments of Saanich inlet. Geochim Cosmochim Acta 42:1011–1026

Reddy KR, Khaleel R, Overcash MR (1980) Nitrogen, phosphorus and carbon transformations in a coastal plain soil treated with animal wastes. Agric Waste Intern J 2:225–238

Reddy RC, Feijtel TC, Patrick WH (1986) Effects of redox conditions on microbial oxidation of organic matter. In: Chen Y, Avnimelech Y (eds) The role of organic matter in modern agriculture. Martinus Nijhoff Publ

Serruya C (1971) Lake Kinneret. Nutrient chemistry of the sediments. Limnol Oceanogr 16:510–521

Serruya C (1978) Sediment chemistry. In: Serruya C (ed) Lake Kinneret W. Junk bv Publ. The Hague, pp 205–215

Waring SA, Bremner JM (1964) Ammonium production in soil under waterlogged conditions as an index of nitrogen availability. Nature 201:951–952

Williams WA, Mikkelsen DS, Mueller KE, Ruckman JE (1968) Nitrogen immobilization by rice straw incorporated in lowland rice production. Plant Soil 28:49–60

Chapter 9

Specific Construction Details

Meir Romem

Reservoirs for wastewater storage are similar to other reservoirs, with a few differences herein described.

9.1
Earthen Reservoirs

Reservoirs (Fig. 9.1) are earth impoundments made to temporarily store wastewater. The reservoir is excavated with embankments built with the excavated soil. The engineer will try to equilibrate the volume of excavated soil and the volume necessary to build the embankments, while taking into consideration the excess of excavated soil necessary for the consolidation of the embankments. The volume of excavated soil will be bigger than the filled soil, with a relationship of 15–25% according to the soil type.

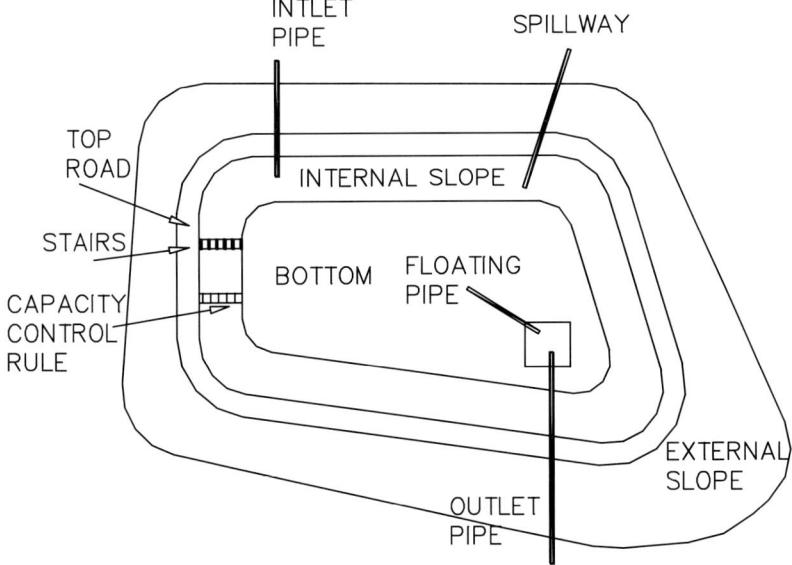

Fig. 9.1. Typical lay-out of a reservoir

9.2
The Spillway

The spillway will be constructed only for emergencies. Wastewater release is forbidden in order to prevent the pollution of streams and the environment. On the other side, the freeboard will be bigger to prevent overflow. A large storage capacity will help to prevent overflow.

9.3
The Floating Outlet Pipe

In wastewater reservoirs it is necessary to pump the water out from the upper layers in order to receive effluents of better quality. Surface layers are more oxygenated, with lower *BOD* and pathogens. If the water is pumped from the deep anaerobic layer with low *p*H, it will stink, producing nuisances to the workers and the environment. Besides, surface pumping avoids the drag off of sediments from the bottom of the reservoir that may contain heavy metals, refractory materials and pathogens.

9.3.1
Elements of the Outlet Pipe (Fig. 9.2)

The unit includes:

1. The main iron pipe
2. Floaters to maintain the pipe near the water surface
3. The inlet to the main pipe protected with an iron netting that filters out gross objects
4. Rotation axle of the main pipe
5. Concrete block to anchor the outlet pipe (*a* and *c*)
6. Cables to prevent movement of the floating pipe produced by the wind

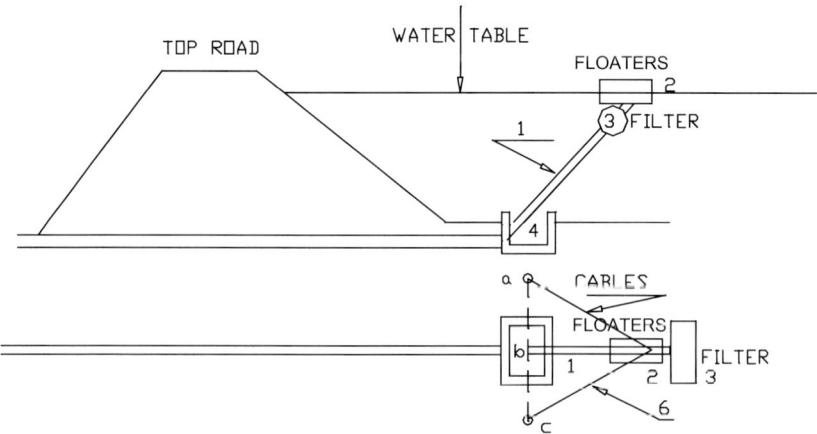

Fig. 9.2. Floating pipe – general view

9.3.2
Details of the Design (Fig. 9.2)

1. The floaters (2) are joined to the pipe (1) with a chain or with another system that permits changing the depth of the opening.
2. The maximum angle between the floating pipe and the bottom when the reservoir is full will be no more than 70–75°.
3. Two points at the concrete boxes to anchor the cables (a–c) will be in a straight line with the axle of rotation of the floating pipe (b). The three points a–b–c will be in straight line (see Fig. 9.2).
4. Iron netting will protect the entrance of the water to the main pipe. Around the netting it is necessary to install a vertical-wall ring to prevent floating objects blocking the entrance to the main pipe.
5. Special attention in the joints between the cables and the concrete box.
6. The dimension of the floaters will be calculated according to the weight of the floating system.
7. The floaters are generally made of plastic materials but it is possible to make them of stainless steel.
8. The rotation axle is constructed by one pipe into another pipe. The internal diameter of the external pipe is the same as the external diameter of the internal pipe.
9. Near the floaters in the floating pipe there is a stand to prevent the main pipe from touching the bottom of the reservoir.
10. The stand should not touch the isolating sheet at the bottom. It is necessary to install a concrete box in this place.

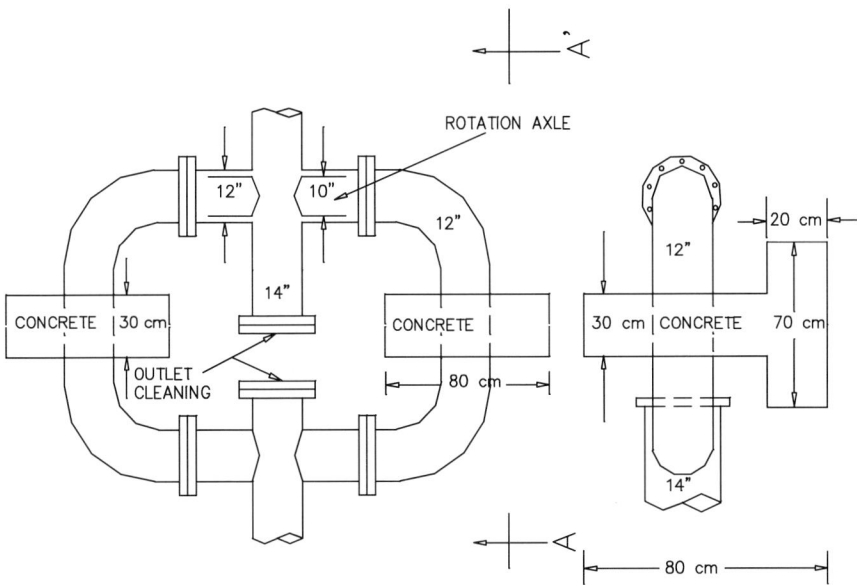

Fig. 9.3. Down part of the apparatus

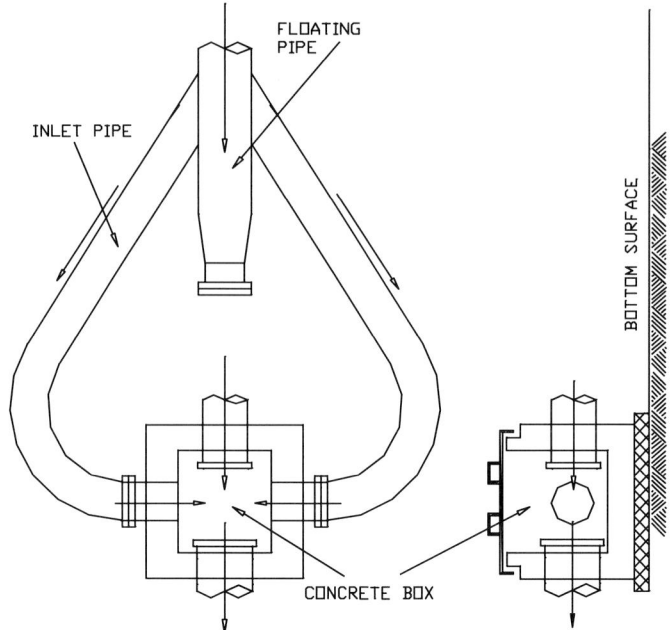

Fig. 9.4. Floating pipe

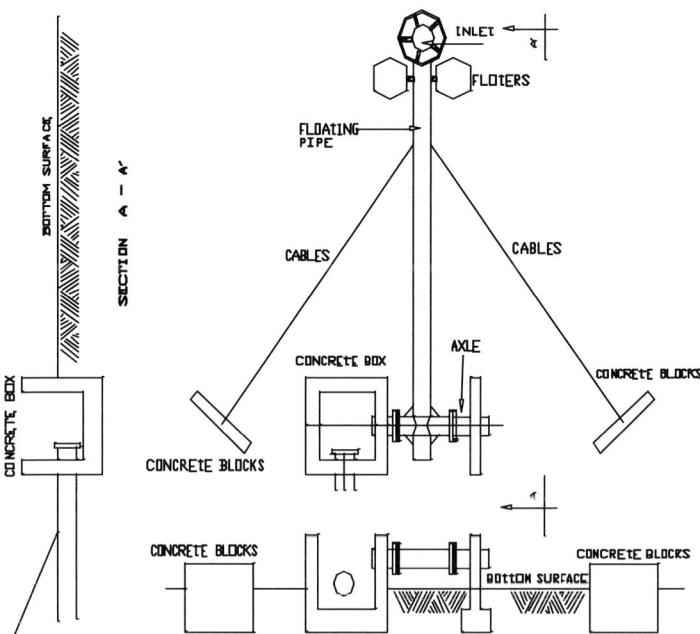

Fig. 9.5. Floating pipe with eccentric concrete box

9.3.3
Details of the Joints

Figures 9.3–9.6 present variations in the design of the joints. Diagrams represent the general design of floating pipes. There are practically no differences between different systems, except for small differences in the joint between the floating pipe and the outlet pipe.

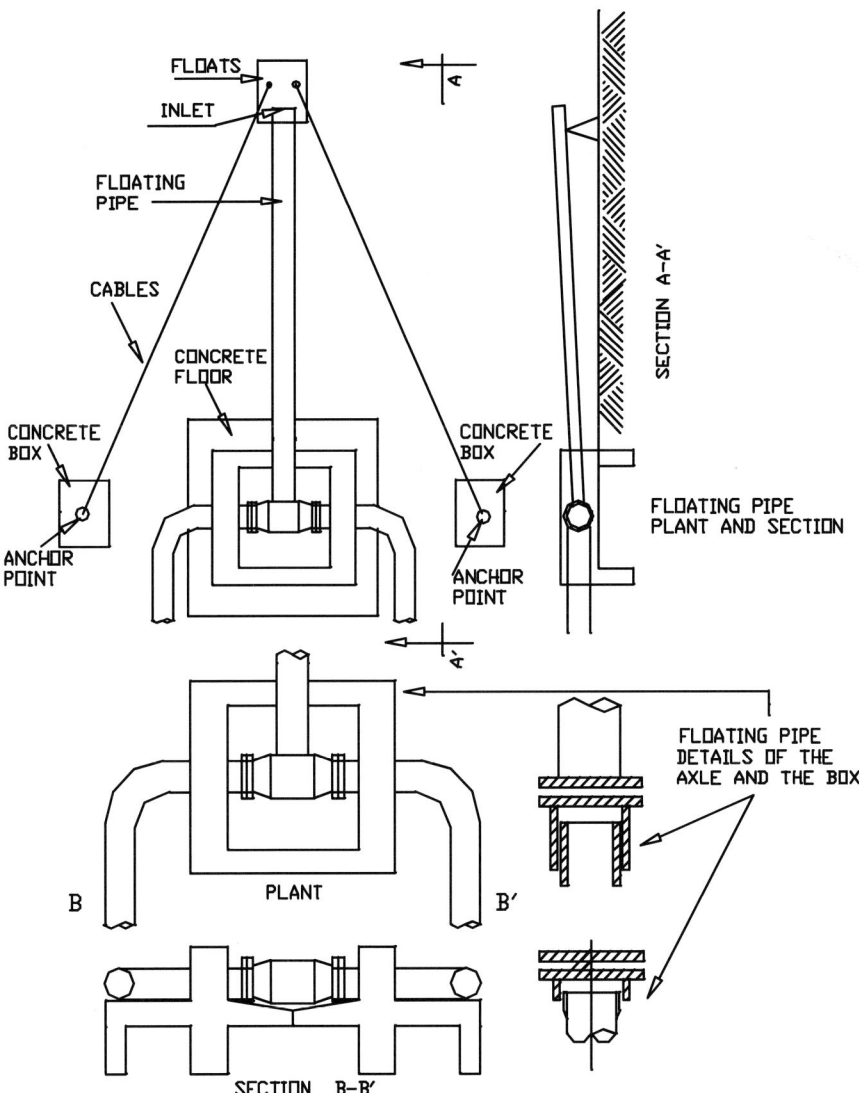

Fig. 9.6. Floating pipe with two pipes to the concrete box

9.4
The Road at the Top of the Embankment

The rains produce runoff over the external slope of the embankment. The erosion produced by water can destroy this slope. If the reservoir is lined with exposed membranes it is preferable to design the road with a small slope to the water (Fig. 9.7). The runoff will flow over the sheet without damaging the embankment.

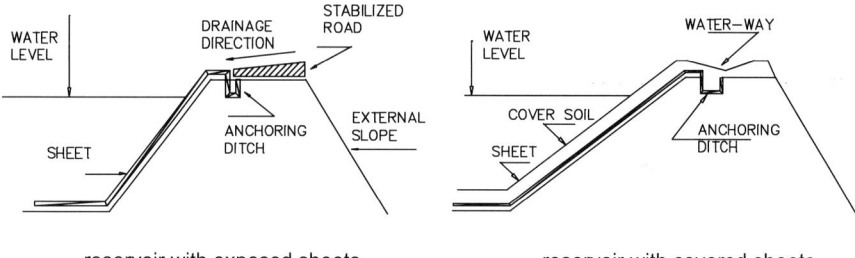

reservoir with exposed sheets   reservoir with covered sheets

Fig. 9.7. The road at the top of the embankment

Chapter 10

Nitrogen and Nitrification

Aharon Abeliovich · Drora Kaplan

10.1
Introduction

Although nitrifying bacteria usually receive little attention, primarily because they are very difficult to work with, they play a major role in structuring our environment (Abeliovich 1992), directly by affecting the rate of nitrogen recycling, and indirectly through their role in global climatic processes. The greenhouse gases NO and N_2O, both byproducts of the microbial nitrification process, also affect the atmospheric ozone layer by being involved in the catalytic destruction of the ozone in the atmosphere.

We will describe here the central position of nitrifying bacteria in affecting one highly specialized ecosystem, that of the wastewater reservoirs. Denitrification and denitrifying bacteria are not discussed here as these reactions do not play any unique role in this ecosystem. Nitrate and nitrite can be present and can accumulate in the reservoir only if very high quality effluents are stored in it, thus maintaining transparent and aerobic conditions throughout all the water column. Otherwise, any nitrate or nitrite generated will be rapidly consumed by respiration when oxygen is depleted.

Due to their size and depth, the physical behaviour of wastewater reservoirs is very similar to that of small lakes. During the warm season, when the water turns turbid, as is the general case due to massive development of algal population, a stable thermocline builds up, separating and sealing an upper warm layer of water from a cooler bottom water layer. When a stable thermocline is thus established, the chemistry and biology of the top and bottom water layers are very different. Only if very high quality effluents are introduced into the reservoir can it be expected that the entire water profile will be oxygenated at all times. Otherwise, during the summer, the organic load that arrives with the effluents results in an aerobic epilimnion and an anaerobic hypolimnion. In extreme cases overloading a reservoir will turn its epilimnion anaerobic as well (Abeliovich 1982). During the winter the situation is very different, as low night temperatures induce total mixing of top and bottom water layers, as is the case with holomictic lakes. At the same time the low winter temperatures also reduce the rates of all biological activities, and in particular the rate of nitrification, which, as will be described later, slows at temperatures below 16–17° C.

10.2
Ammonia in Wastewater Reservoirs

10.2.1
General Aspects

For various reasons, domestic effluents that are intended for irrigation contain high concentrations of ammonia. Therefore, the effluents stored in the reservoirs can contain between 20–80 mg l^{-1} of ammonia (Abeliovich 1982). This is advantageous for the farmer as he can save nitrogen fertilization, but it also has far reaching effects on the ecophysiology of the reservoir, and this in more than one way. Ammonia influences the reservoir through its inhibitory effect on photosynthesis, and through its frequently incomplete nitrification. This occasionally results in accumulation of nitrite which at times may reach very high concentrations. The latter effect has practical implications, as the nitrite interferes with chlorination of the effluents and can thus interfere with efficient disinfection of the water before its discharge for irrigation.

The sources of ammonia in the stored effluents are urea, proteins, nucleic acids, and amino sugars present in the raw wastewater. Ammonia is the end product of all deamination reactions that take place during the biodegradation of nitrogen containing organic matter. Oxidation of ammonia is carried by a specialized group of nitrifying bacteria, and the process requires a stable set of specific environmental conditions. As nitrification is a strictly aerobic process, oxygen is a major limiting factor. This is because in eutrophic environments nitrifyers grow much slower than aerobic heterotrophic bacteria. Nitrifyers are therefore at a disadvantage when they have to compete with the rest of the bacteria for oxygen in this biotope, although their affinity to oxygen is very high (Abeliovich 1985, 1987). Another limiting factor for nitrifying bacteria in the reservoir is excess light at the surface of the reservoir, which, as will be described later, has a strong inhibitory effect on nitrifying bacteria. As availability of photosynthetic oxygen is just the same associated with illumination of surface water, these restrictions make nitrification of ammonia in wastewater reservoirs a very slow process. The outcome of the inhibitory effect of ammonia on photosynthesis and the inhibitory effect of light on nitrification is that ammonia is very persistent in the stored effluents. As a result of the interplay between these factors, oxygenation of the water in the reservoir is frequently suboptimal even when *BOD* is low.

Although nitrifyers are found in the reservoir in the entire water column down to the anaerobic bottom water layers throughout the year (Abeliovich 1987), nitrification can take place only where and when oxygen is present, and when the water is within the right temperature range and light regime. Nitrification is therefore restricted to a narrow upper water layer of the reservoir, and to only those daylight hours when photosynthesis provides sufficient oxygen without light being inhibitory to the activity of nitrifying bacteria.

10.2.2
Inhibition by Ammonia of Algal Photosynthesis and Growth

Un-ionized ammonia penetrates freely through cell and organelle membranes; it is known as an uncoupler of photosynthesis in isolated chloroplasts (Avron and

Shavit 1965) and as an inhibitor of photosynthesis in algae (Abeliovich and Azov 1976). Since the un-ionized ammonia fraction in the water is pH dependent, its inhibitory effect on photosynthesis, oxygen evolution and algal growth depends both upon the water pH and the ammonia concentration (Abeliovich and Azov 1976). It is also toxic to various photosynthetic organisms: *Prymnesium parvum* (Shilo and Shilo 1953) among several groups of algae, marine diatoms (Natarajan 1970) *Anabaena* sp. (Belkin and Boussiba 1991) and *Scenedesmus obliqus* (Abeliovich and Azov 1976), and the toxic and inhibitory effect of ammonia seems to be very common among microorganisms carrying oxygenic photosynthesis, although there is at least one exception to this rule, the relatively resistant alkalophilic *Spirulina platensis* (Belkin and Boussiba 1991). These observations also defined the boundaries of toxicity of ammonia: the combination of the presence of 2.0–3.0 mM ammonia and pH values of >8.0 in the water prevents photosynthesis and growth of algae (Abeliovich and Azov 1976).

The major factors that determine water pH in the reservoir are alkalinity, respiration and photosynthesis, through their effect on the concentrations of CO_2, bicarbonates and carbonates, which are the major buffer system in these waters. When searching for the manifestations of this phenomenon in the field (Abeliovich 1982, 1983), it was possible to show that in the reservoir there exists a negative correlation between the concentration of ammonia and the ratio of photosynthesis to respiration (Fig. 10.1). The reason for this is that in contrast to oxygenic photosynthesis, respiration is not affected by ammonia at the concentrations prevailing in the reservoir. Therefore, the increase in ammonia concentration inhibits the rate of photosynthesis but not that of

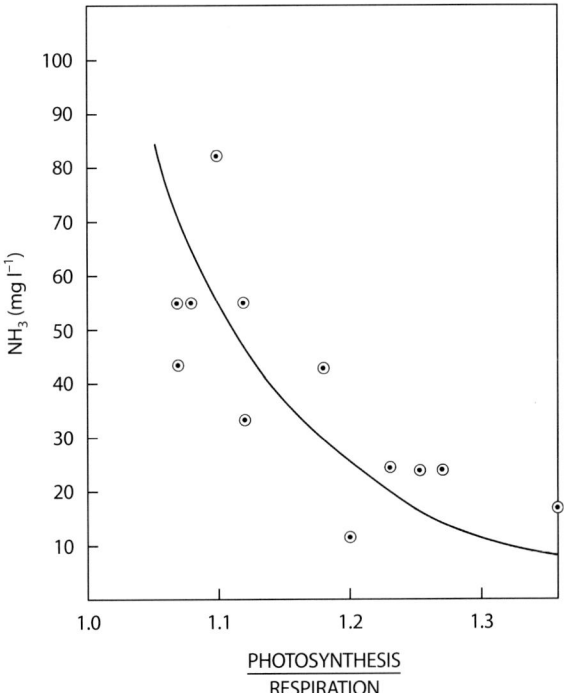

Fig. 10.1. Correlation between the concentration of ammonia and the ratio of photosynthesis (gross) to respiration. The rate of photosynthesis was measured at the reservoir's water surface (Ram Reservoir, Beer Sheba, Israel) in August between 6:00–9:00 a.m. and the rate of respiration was determined in the laboratory (after Abeliovich 1983)

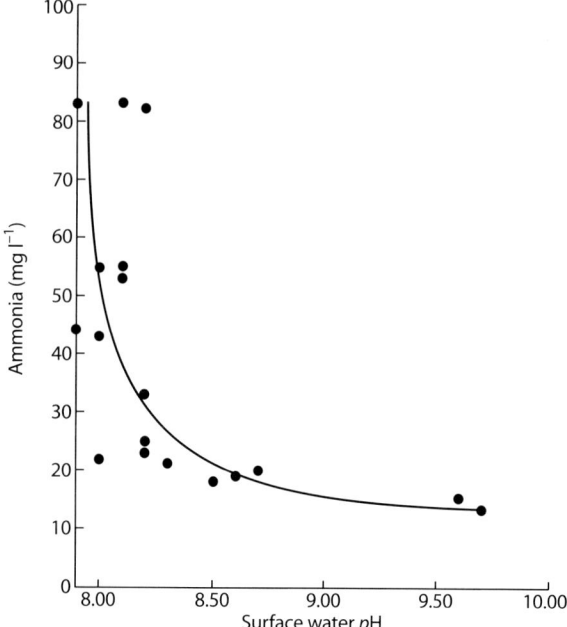

Fig. 10.2. Correlation between concentration of ammonia and pH at the surface of the water of the Ram Reservoir (Beer Sheba, Israel) (after Abeliovich 1983)

respiration, resulting in a decrease in the ratio of photosynthesis to respiration. Similarly, and for the same reasons, there is a strong negative correlation between the concentration of ammonia and the pH at the surface of the water (Fig. 10.2).

Obviously, the key to nuisance-free operation of a wastewater reservoir is in keeping its upper water layer oxygenated throughout most or all of the diurnal cycle. This is possible only if the rate of photosynthesis is kept maximal and the rate of respiration kept minimal. There are, however, built-in obstacles: *PAR* (photosynthetically active radiation) is available only 50% of the time (average), and this only to the upper water layer, which is frequently only a fraction of the water volume. With a compensation point (the depth at which, due to light attenuation, respiration and photosynthesis rates are equal) usually at 0.1–1.0 m (1–10% of the depth of the reservoir), this means that not more than 10% of the water volume on the average is photosynthetically oxygenated for some time during each diurnal cycle, and turbulence by wind can deliver the oxygenated water into deeper layers. In this case a stable thermocline at a depth of 2.5–3.0 m can significantly decrease the volume of water that has to be oxygenated, and wind indeed can help oxygenate all the epilimnion.

Under favorable conditions, in pure cultures of various algae isolated from oxidation ponds or reservoirs (*Scenedesmus* sp., *Chlorella* sp., or *Spirulina* sp.), the ratio of oxygenic photosynthesis to respiratory oxygen consumption is within the range of 5–10. This means that whether a stable thermocline is established or not, the potential for oxygenating the surface water layers of the reservoir is always there, provided that nothing prevents the algae from carrying maximal rates of photosynthesis.

Because CO_2-carbonate is the major buffer system in the reservoir, photosynthesis is a process that rapidly elevates the surface water pH, which at times can reach values of 10.5 (Abeliovich 1969). This elevation of the pH by photosynthesis can be counterbalanced, if enough organic carbon is available for bacterial and algal respiration. Although respiration is a much slower process, it can compensate for the rapid surface CO_2 uptake because it takes place in the entire water volume all the time, as long as oxygen is available.

Of all the factors that affect the delicate balance we wish to establish and maintain in the reservoir so that it will function without generating any environmental nuisances while storing domestic wastewater, among them oxygen, pH, algae and bacteria, carbon sources and ammonia, it is the latter which plays the key role and might present a major problem for reservoir operators. This is because already concentrations of 2–3 mM ammonia, which are well within the normal range of concentrations present in secondary effluents, affect photosynthesis and therefore affect oxygenation of the surface water. In the presence of ammonia photosynthesis can proceed only until it elevates the water pH to that value at which the equilibrium between ionized and unionized ammonia will establish an inhibitory concentration of un-ionized ammonia for oxygenic photosynthesis (Abeliovich and Azov 1976). Thus, it is possible to have an anoxic and odorous reservoir even as it holds effluents of good quality in terms of BOD if the concentration of ammonia is high enough to arrest photosynthesis. Therefore, for optimal performance and environmentally safe operation, not only the BOD should be kept low in the stored effluents, but the concentration of ammonia in the effluents should be kept low as well, below those concentrations that affect photosynthesis at the pH values prevailing in the reservoirs.

10.3
Nitrifying Bacteria in Reservoirs

10.3.1
Abundance of Nitrifying Bacteria

Nitrifying bacteria are abundant in wastewater stabilization ponds and reservoirs at the whole range of oxygen concentrations, from strict anaerobiosis to saturation (Abeliovich 1983, 1985). Oxygen saturation for nitrification in reservoir water is already achieved at a very low concentration of 0.05 mg l^{-1} (Fig. 10.3) which probably represents that oxygen concentration which provides just maintenance energy, as oxygen saturation for growth is achieved only at higher concentration of 0.1–1.0 mg l^{-1} (Fig. 10.4).

10.3.2
Detection of Nitrifying Bacteria in Wastewater Reservoirs

Because most nitrifying bacteria cannot usually be cultivated on organic solid media (Gerhardt et al. 1994), alternative ways for quantifying them must be used. They can be detected in water by several methods; each has its advantages and disadvantages, and none can therefore be recommended as a single method for providing quantitatively reliable figures. The most commonly used is cultivation of serial dilutions of a water sample in test tubes with ammonia or nitrite. The presence of nitrifyers can then be detected by following the appearance of nitrite and/or nitrate. This is a very slow

Fig. 10.3. Combined data of nitrification vs. O_2 concentration in a chemostat fed by oxidation pond effluents and maintained at variable retention times (50–150 h). When the chemostat was operated at low O_2 (0.05–0.2 mg l^{-1}) concentrations, these values were maintained for only 6 h during the day, with O_2 concentrations being below detection level (<0.01 mg l^{-1}). Each point represents the result of one assay done after steady state in the operation of the chemostat was achieved (3–5 days)

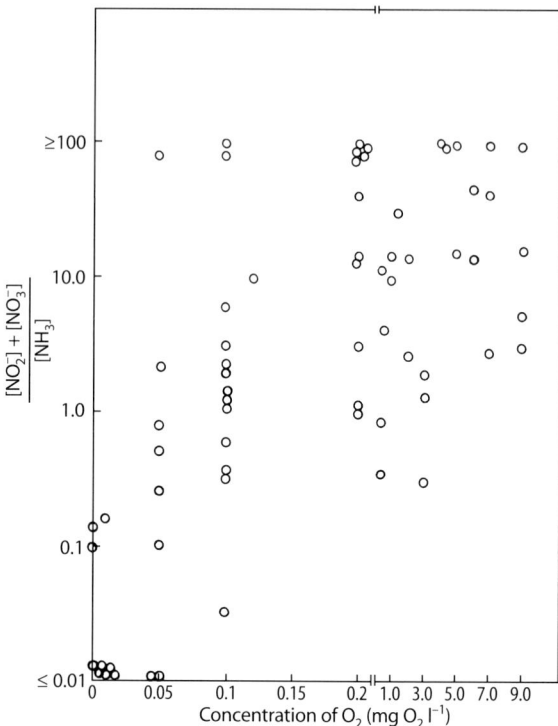

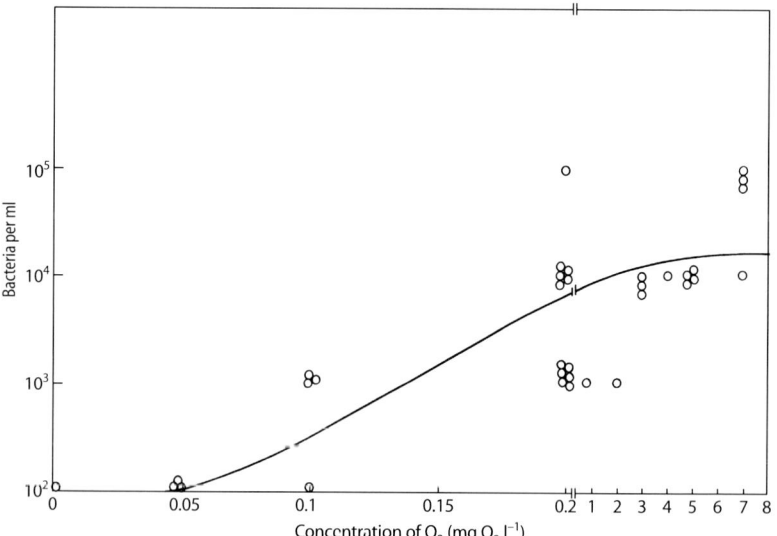

Fig. 10.4. Oxygen concentration vs. number of *Nitrosomonas* sp. cells (as determined by serial dilutions) in a chemostat fed by water from an overloaded anaerobic oxidation pond. Retention time of the water was 100 hours (after Abeliovich 1987)

assay because of the slow growth rate of the bacteria. The number of nitrifyers is usually determined after 21 days of incubation of the dilutions, but we have found that positive dilution tubes keep appearing even after 60 days. Another possibility is to count nitrifying bacteria using fluorescent labeled antibodies (e.g., FITC) prepared against whole cells of axenic isolates (Abeliovich 1987). This is a very convenient method, but it always leaves open the possibility of the presence of a population of nitrifyers not detected by the antibodies. Finally, it was recently shown that it is possible to detect very small numbers of nitrifyers by amplifying the nitrifyers' DNA using the polymerase chain reaction with specific primers prepared for each group of bacteria (Nejidat and Abeviolich 1994). Other methods based upon identification of specific DNA sequences can also be used, but they all are in danger of being too specific, causing a significant number of nitrifying bacteria to be overlooked because they lack the appropriate nucleotide sequences. Methods based upon immunochemical reactions or DNA specificity do not identify specifically viable and culturable bacteria, unless based on detection of specific mRNA sequences, and can introduce another error due to counting dead bacteria. Therefore, any attempt to estimate the number of nitrifyers in the wastewater reservoir should rely on the parallel use of several independent methods.

Nitrification of ammonia in wastewater reservoirs is a very slow process (Abeliovich 1987), for reasons mentioned above. For example, in one study carried in Ma'aleh HaKishon Reservoir (Israel) the concentration of ammonia dropped by just 50% in 120 days, while the reservoir was oxygenated and nitrifying bacteria were present at all times throughout the water column (Fig. 10.5).

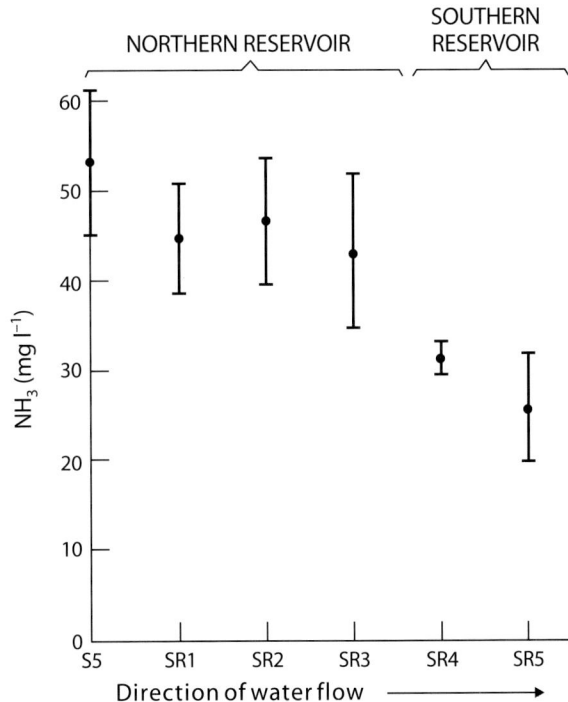

Fig. 10.5. Change in ammonia concentration during effluent movement from northern inlet to southern outlet in the Ma'aleh HaKishon Reservoir. Values are averages for samples from February to August 1985. Locations: S5 Inlet to the northern reservoir; SR1 50 m from the inlet; SR2 centre of northern reservoir; SR3 connection to southern reservoir; SR4 centre of southern reservoir; SR5 outlet from southern reservoir. Data were kindly provided by Y. Eren, Mekoroth Ltd. (after Abeliovich 1987)

10.3.3
Nitrifying Bacteria in Anaerobic Environments

Nitrifying bacteria have been occasionally observed in anaerobic biotopes such as marine sediments (Blackburn 1983) and wastewater treatment plants (Van der Graaf et al. 1990). Several physiological activities under anaerobic conditions have also been described: Ritchie and Nicholas (1972) described the generation of N_2O and NO from nitrite by *Nitrosomonas europaea* under strict anaerobic conditions, while Poth (1986) described the reduction of nitrite to dinitrogen under microaerophilic conditions, and Bock et al. (1995) showed later that *Nitrosomonas europaea* can carry a denitrification reaction under anoxic conditions using nitrite as an electron acceptor and hydrogen as an electron donor.

We have consistently found, in field studies in wastewater reservoirs, both ammonia and nitrite oxidizing bacteria in the anaerobic hypolimnion when these are overloaded and stratified for 7–8 months (Abeliovich 1985, 1987; Abeliovich and Vonshak 1993). During the summer months their numbers in the anoxic zone increased by 3–4 orders of magnitude, while at the same time no such increase occurred in the epilimnion of the same reservoir.

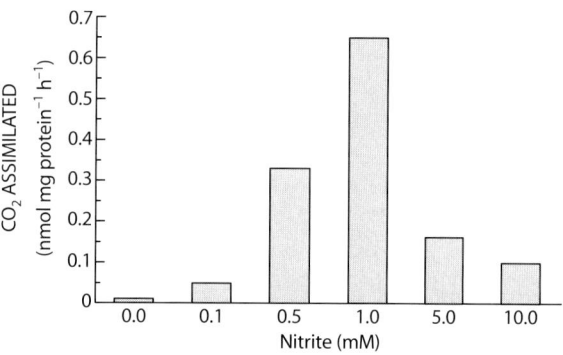

Fig. 10.6. Incorporation of $^{14}CO_2$ into TCA precipitable material by *Nitrosomonas europaea* under anaerobic conditions. 2×10^8 cells ml^{-1} were incubated in a buffered medium (7) supplemented with Na-pyruvate (0.05 mM), Na$^{14}CO_3$ (0.1 mM mM^{-1}, 5.9 mM), and different concentrations of nitrite

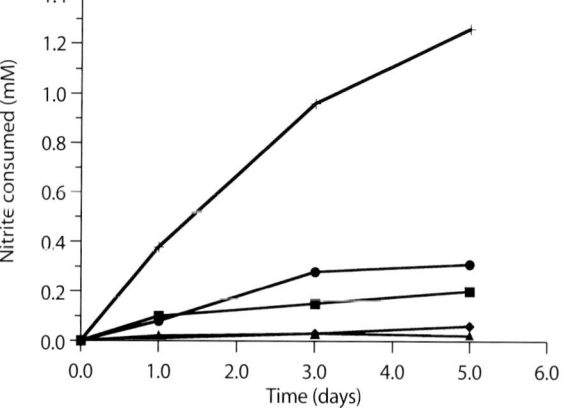

Fig. 10.7. Consumption of nitrite by *Nitrosomonas europaea* under anaerobic conditions with and without ammonia. Cells (10^8 ml^{-1}) were incubated in medium supplemented when necessary with nitrite (brought every 48 h to 0.4 mM), pyruvate (5 mM) and ammonia (11.3 mM). Incubation mixtures contained: + ammonia and pyruvate; ● ammonia; ■ pyruvate; ◆ no ammonia and no pyruvate; ▲ heat inactivated cells (10 min at 100° C) with ammonia and pyruvate (after Abeliovich and Vonshak 1992)

In laboratory experiments, *Nitrosomonas europaea* was capable of anaerobically assimilating CO_2 (Fig. 10.6) into cellular matter (see also Abeliovich and Vonshak 1992 for details). Nitrite was consumed in this reaction, and pyruvate and ammonia were required (Fig. 10.7). However, neither nitrate nor nitrite are ever present in any detectable concentrations in the hypolimnion of overloaded reservoirs, and therefore cannot account for the massive proliferation of nitrifies observed there (Abeliovich 1985). Therefore, although the potential for survival of nitrifyers in anoxic environments could be established in laboratory experiments, the actual metabolic activities that enables a population of nitrifyers to proliferate in the anoxic hypolimnion of the wastewater reservoir are still obscure.

10.4
Unbalanced Nitrification of Ammonia

10.4.1
Nitrification of Ammonia in Wastewater Reservoirs

Nitrification of ammonia in wastewater reservoirs is often an incomplete process, frequently leading to the accumulation of high concentrations of nitrite. Routine water quality monitoring one large reservoir (capable of storing $\approx 13 \times 10^6$ m³ water), Ma'aleh HaKishon, during the years 1984–1993 (Shelef et al. 1984–1993) revealed that nitrite accumulated almost every year in spring-summer, reaching concentrations at times of 30–50 mg l^{-1} nitrite in the stored water. In a field study carried out in the Ma'aleh HaKishon Reservoir during 1994–1996 we also observed accumulation of high concentrations of nitrite for extended periods of time, up to 10 mg l^{-1} nitrite N (<30 mg l^{-1} nitrite). High concentration of nitrite was particularly persistent during 1996 (Fig. 10.8). Accumulation of nitrite usually starts in the spring, when the water temperature rises high enough to allow onset of activity of ammonia oxidizers (18–20° C). Unfortunately,

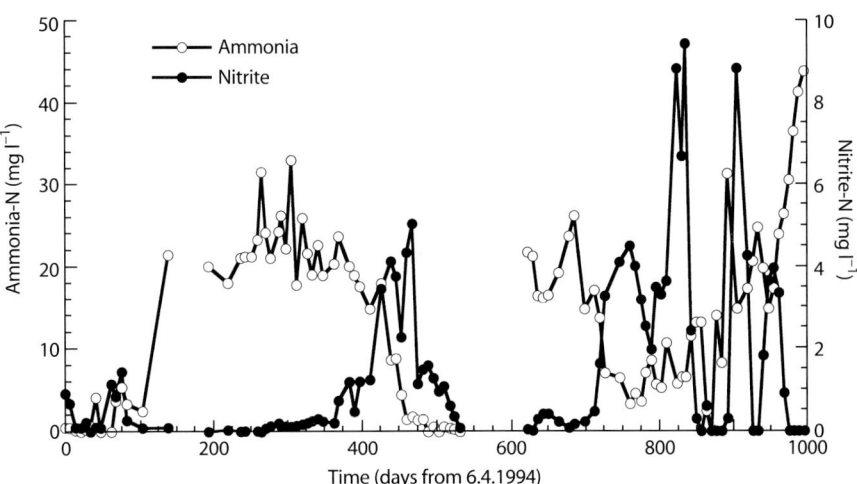

Fig. 10.8. Changes in the concentrations of ammonia-N and nitrite-N in Ma'aleh HaKishon northern reservoir during the years 1994–1996

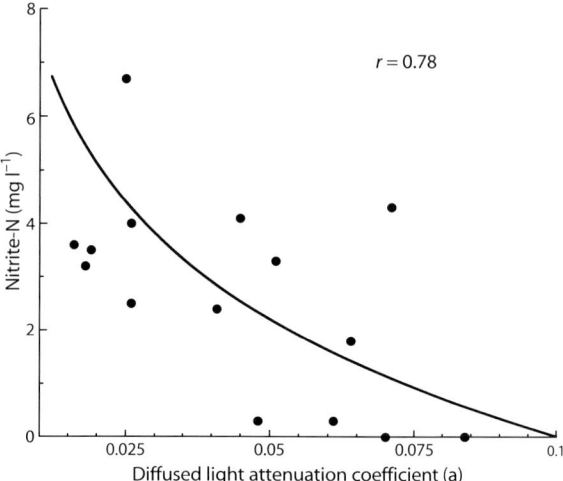

Fig. 10.9. Correlation between water transparency (presented as light attenuation coefficient) in northern Ma'aleh HaKishon Reservoir and the accumulation of nitrite during 1996. Light attenuation was measured at noon with a Licor radiation sensor (Li-190SA) connected to a data logger (Li-1000). Diffused light attenuation a was calculated as $a = -1 / X \ln I / I_0$, I is the measured light intensity at depth X, and I_0 is the light intensity just below the water surface

this phenomenon coincides with the onset of the irrigation season, and the presence of high concentrations of nitrite in the water interferes with the disinfection of the effluents by chlorination which is mandatory before the water is released to the farmers.

Nitrite can accumulate in the water not only as the end product of aerobic ammonia oxidation, but also due to incomplete denitrification under anaerobic conditions. However, this possibility, that the source of the nitrite is denitrification should be excluded, as this reservoir is always aerobic throughout all the water column due to the high quality of the stores effluents.

This phenomenon of unbalanced nitrification in the reservoir can be attributed to increased light penetration due to low water turbidity, and might be a result of the nitrite oxidizing bacteria being more photosensitive than the ammonia oxidizing bacteria, as in this field study we found a strong correlation between transparency of the water and accumulation of nitrite (Fig. 10.9). Also, it can be seen from the data presented in Fig. 10.10 that there is a general correlation between the appearance and of a "bloom" of ammonia oxidizers and the accumulation of nitrite, and that as their number dropped there was also a decrease in the concentration of nitrite (e.g., days 300–500 and 700–900). However, although one would expect the concentrations of nitrite, which is supposed to be the energy source for nitrite oxidizers, to drop when nitrite oxidizers proliferate in the reservoir, this was not what actually happened in the field. Nitrite concentrations remained high for long periods of time in spite of the presence of a massive population of nitrite oxidizers, as if nitrite were not the major energy source for these bacteria under the conditions prevailing in the Ma'aleh HaKishon Reservoir during the summer (Fig. 10.11, days 300–500 and 700–900). In an earlier study (Abeliovich and Vonshak 1993) it was shown in laboratory experiments that light has a strong inhibitory effect on oxidation of ammonia in reservoir water containing organic matter, and that this inhibitory effect was unique to reservoir water and could not be demonstrated when defined mineral medium prepared with distilled water was used. It is likely, therefore, that the reservoir water presents alternative energy sources to the nitrite oxidizing bacteria when nitrite oxidation is arrested by light.

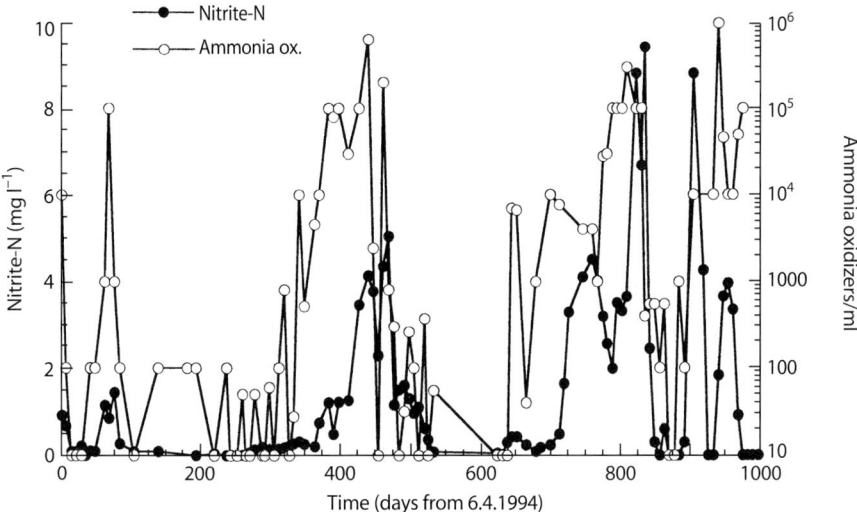

Fig. 10.10. Concentrations of nitrite and numbers of ammonia oxidizers in northern Ma'aleh HaKishon Reservoir during 1994–1996

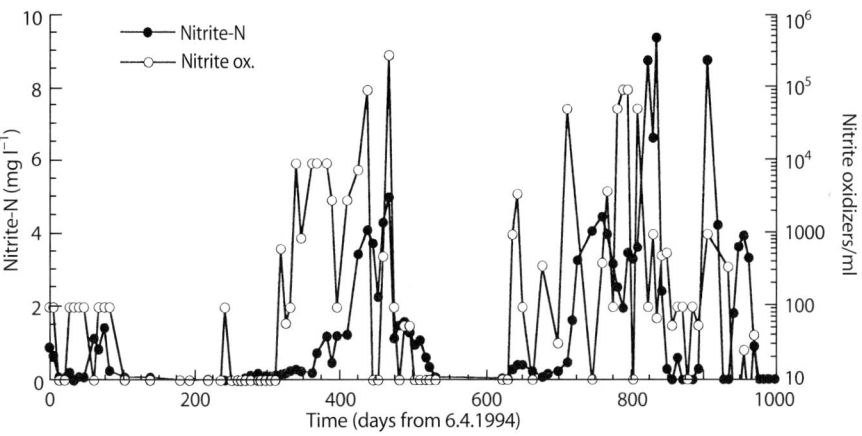

Fig. 10.11. Concentrations of nitrite and numbers of nitrite oxidizers in northern Ma'aleh HaKishon Reservoir during 1994–1996

10.5
Summary

The observed rate of nitrification of ammonia in any particular wastewater reservoir is a result of a complex interplay between several interrelated key factors, of which four were discussed here in detail: water temperature, light penetration, organic load and oxygen availability.

Water temperature affects nitrification both directly and indirectly. In the Israeli temperate climate, the nitrification rate slows down gradually as the temperature drops below 20° C, until it is unobservable when the water temperature drops to 14–15° C. Therefore ammonia concentrations peak during the winter season. However, temperature can affect nitrification in the reservoir in more than one way, particularly if it stores nutrient rich effluents: directly through its effect on the nitrification rate, and indirectly through its effect on the stratification of the reservoir. If the reservoir stores nutrient-rich water, this causes algal blooms to develop in the early spring as the temperature rises, resulting in increased turbidity. The compensation point can then be found at just 5–10 cm below the water surface and this results in the establishment in early spring of a stable thermocline at a depth of 200–300 cm. The thermocline seals off most of the stored water in the anaerobic hypolimnion where no nitrification can take place. Under these conditions, only a fraction of the ammonia in the reservoir is available for nitrification. Also, within the epilimnion of the turbid reservoir, oxygen for nitrification is available within a very narrow water layer, and for all practical purposes, nitrification is insignificant. With regard to nitrification the situation is not much better if the reservoir receives high quality effluents and a food chain based on algae grazers is established, (as is the case with the Ma'aleh HaKishon Reservoir). Here, light attenuation is minimal, deep water layers are heated, and no thermocline is established, as there is a constant mixing and oxygenation of all the water column. However, because microbial nitrification is sensitive to light, minimal light attenuation results in total or partial arrest of nitrification, which is manifested at times in the form of unbalanced nitrification (as described above). At the same time if, as a result of these different inhibitory ecophysiological mechanisms, ammonia concentrations in the water remain high enough, oxygenic photosynthesis might be arrested, reducing oxygen availability, either to all the water column in the case of a well mixed reservoir, or to the epilimnion in the case of a stratified reservoir. On the other hand, high turbidity enables nitrifyers to survive, but then no nitrification can take place in most of the water volume, as it is anaerobic throughout the warm season, while in the shallow epilimnion the interplay between oxygen and light enables only a limited rate of nitrification. Diffusion of oxygen from the atmosphere into the reservoir's water is negligible when compared to the oxygen demand even when high quality effluents are stored. Light, through photosynthesis, remains the predominant source of oxygen for all aerobic metabolic activities of the microbial flora, including nitrification. Therefore, everything that affects photosynthesis, affects nitrification too. Thus, the same light which is essential for the oxygenation of the water, inhibits nitrification. It is this contradiction between the requirement of light for photosynthesis and its inhibitory effect on nitrification that leads to the erratic and at times very slow nitrification rates we observe in the wastewater reservoir. Still, it must be remembered that if the water is used exclusively for agricultural irrigation purposes, the inhibition of nitrification can be seen as an advantage, as it prevents the loss of nitrogen through denitrification, and saves the farmer expensive nitrogen fertilization.

Acknowledgements

This review is based in part on studies that were funded by the German-Israeli Water Technology Fund (MOS & BMBF), and the Water Commissioner, Israeli Ministry of Agriculture.

References

Abeliovich A (1969) Waterblooms of blue-green algae and oxygen regime in fish ponds. Verb Internat Verein Limnol 17:594–601

Abeliovich A (1982) Biological equilibrium in deep wastewater reservoirs. Wat Res 16:1135–1138

Abeliovich A (1983) The effects of unbalanced BOD and ammonia in oxidation ponds. Wat Res 17:299–301

Abeliovich A (1985) Nitrification of ammonia in wastewater: Field observations and laboratory studies. Wat Res 19:1097–1099

Abeliovich A (1987) Nitrifying bacteria in wastewater reservoirs. Appl Environ Microbiol 53(4):754–760

Abeliovich A (1992) Transformations of ammonia and the environmental impact of nitrifying bacteria. Biodegradation 3:255–264

Abeliovich A, Azov Y (1976) Toxicity of ammonia to algae in sewage oxidation ponds. Environ Appl Microbiol 31(6):801–806

Abeliovich A, Vonshak A (1992) Anaerobic metabolism of nitrifying bacteria. Arch Microbiol 158:267–270

Abeliovich A, Vonshak A (1993) Factors inhibiting nitrification of ammonia in deep wastewater reservoirs. Wat Res 27:1585–1590

Avron M, Shavit N (1965) Inhibitors and uncouplers of photophosphorylation. Biochim Biophys Acta 109:317–331

Belkin S, Boussiba S (1991) Resistance of Spirulina platensis to ammonia at high pH values. Plant Cell Physiol 32:953–958

Blackburn TH (1983) The microbial nitrogen cycle. In: Krumbein WE (ed) Microbial geochemistry. Blackwell Scientific Publications Ltd., Oxford

Bock E, Schmidt I, Stueven R, Zart D (1995) Nitrogen loss caused by denitrifying Nitrosomonas cells using ammonium or hydrogen as electron donors and nitrite as electron acceptor. Arch Microbiol 163:16–20

Gerhardt P, Murray RGE, Wood WA, Kreig NR (1994) Methods for general and molecular bacteriology. American Society for Microbiology, Washington, DC

Nejidat A, Abeliovich A (1994) Detection of Nitrosomonas europaea by polymerase chain reaction. FEMS Microbiol Ecol 120:191–194

Natarajan KW (1970) Toxicity of ammonia to marine diatoms. Jour Wat Pollut Control Fed 42:R184–190

Poth M (1986) Dinitrogen production from nitrite by Nitrosomonas isolate. Appl Environ Microbiol 52:957–959

Ritchie GAF, Nicholas DJD (1972) Identification of the sources of nitrous oxide produced by oxidative and reductive processes in Nitrosomonas europaea. Biochem Jour 126: 1181–1191

Shelef G et al. (1984–1993) Monitoring Kishon water reclamation project. Annual reports (in Hebrew), Sherman center for research in environmental and water resources engineering. Technion-Israel Institute of Technology, Haifa

Shilo M, Shilo M (1953) Conditions which determine the efficiency of ammonium sulfate in the control of Prymnesium parvum in fish breeding ponds. Appl Microbiol 1:330–333

Van der Graaf AA, Mulder A, Slijkhuis H, Robertson LA, Kuenen JG (1990) Anoxic ammonium oxidation. In: Christiansen C, Munk L, Viladsen J (eds) Proceedings of the 5th European Congress on Biothechnology. Munksgaard, Copenhagen, pp 388–391

Chapter 11

Phytoplankton

Inka Dor

11.1
Introduction

This chapter on the phytoplankton of wastewater reservoirs is based on sixteen reservoirs which were investigated in detail in Israel (Dor et al. 1987a,b; Dor and Raber 1990; Abeliovich this volume; Dor unpublished results). The studied reservoirs are distributed all over the country and they differ considerably in the origin of their effluents which may be a mixture of domestic wastewater and floods, while others also receive manure or industrial wastes. Their volumes vary from 50 000 to 12 000 000 m^3 and the depths from 5 to 14 m. Organic contents of the wastewater stored in the reservoirs (as BOD_5 filtrate) and total suspended solids (TSS) range widely: 4–160 mg l^{-1} and 6–150 mg l^{-1} respectively. The most common type of reservoirs have a medium loading with BOD_{5f} = 30–60 mg l^{-1}. Ammonia (as N) and total nitrogen contents are around 15–20 and 30–40 mg l^{-1} respectively. Total phosphate is usually around 5–7 mg l^{-1}. However, at the extremely high loads of dairy effluents, having in certain cases BOD_5 up to 10 000 mg l^{-1}, ammonia content in the reservoir may reach 200 mg l^{-1} and total phosphate 300–500 mg l^{-1} (see Abeliovich this volume).

11.2
Wastewater Reservoir as an Algal Habitat

In hypertrophic ecosystems of the type here described, highly enriched with mineral and organic matter, a prominent growth of algal biomass occurs. Like in stabilization ponds, algae have a powerful influence on the biology, chemistry and performance of the wastewater reservoirs. An increase in biomass affects light conditions in the water column. In reservoirs having low to medium content of chlorophyll, the compensation depth in the summer is usually around 150–400 cm. Consequently, the upper water layer up to this depth is oxygenated mostly by the photosynthetic activity of algae under the influence of solar radiation. Higher concentrations of chlorophyll, reaching up to 1 000 µg l^{-1}, result in high turbidity and shallow compensation depth, down to 20–30 cm only. Due to absorption of solar radiation, bioturbidity induces warming of the upper water mass with a resulting thermal stratification. A sharp thermal stratification lasting from the early spring to the autumn has been described recently also in the similar deep wastewater stabilization reservoirs in Spain (Llorens et al. 1992). It should be also mentioned that compensation depth remains very limited during the winter wind-mixing of the water column, which is connected with a high mineral turbidity. Accordingly, photosynthetic oxygen production in this season remains confined to the uppermost few tens of centimetres only (Gloyna 1971) and it is mainly wind action which supplies oxygen to the totally mixed water column. In the Na'an Reservoir Dor et al. (1987b) recorded dissolved oxygen of only 0.5–1.5 mg l^{-1} during the winter mixing. During the early spring thermal stratification, connected with high chlorophyll concentrations and an intense photosynthetic activity in the euphotic zone, dissolved oxygen accumulated and its concentrations reached more than 20 mg l^{-1} at noon i.e., up to 300% of saturation. However only uppermost water was at such supersaturation while deeper layers of the water column were severely light limited and had a very low contents of oxygen. The limitation of light exerts an enormous selective pressure. The mixotrophic algae have an advantage under these conditions, utilizing organic substrate for their nutrition. Indeed, in the heavily loaded high rate oxidation ponds, up to 25–50% of algal carbon requirement could be derived directly from organic matter and not via photoassimilation (Abeliovich 1980). That evidently favours growth of facultative heterotrophs such as *Chlorella vulgaris*, one of the most successful species in a variety of open sewage treatment systems (Azov et al. 1980).

Other environmental factors such as pH are also affected by algal biomass and exhibit seasonal fluctuations. In Na'an Reservoir during the winter, when bacterial respiration prevailed, pH values were around 7. During the spring and summer, pH reached 9 in the illuminated epilimnion, as a consequence of CO_2 take up by photosynthesis. Phosphorus precipitates at high pH values, sinking to the bottom while ammonia stripping occurs through pond surface. Besides, nitrogen removal takes place via algal assimilation of ammonia into biomass. High pH also causes die-off of pathogens, yielding a microbiologically improved effluent.

The presence of phytoplankton and oxygen during the spring and summer stimulates the development of herbivorous zooplankton, which reproduce rapidly. Grazing by zooplankton appears to be an important factor affecting phytoplankton population (Thompson and Rhee 1994). The activity of these filter feeders decreases the algal concentrations with concomitant improvement of transparency and deepening of the photic zone. At this stage the high-oxygen euphotic zone expands considerably in various reservoirs, depending on the reservoir type.

11.3
Composition and Seasonality of Phytoplankton

The following description and discussion is mostly confined to algae recorded in the medium-loaded reservoir Na'an, which is a typical representative of most wastewater reservoirs in Israel. The results are related to the same period described in the chapter "The Na'an Reservoir" – see case studies in this volume. Algae recorded in the reservoir during the above research are listed in Table 11.1 and illustrated in Plates 11.1–11.3. During this period the corresponding mean organic matter concentrations were 56–60 mg l^{-1} as BOD_5 filtrate. Of the 18 genera recorded (represented by 20 species), 17 are assigned among the pollution tolerant genera (Palmer 1969). Palmer (1980) based on 165 authorities, ranks the algae according to its tolerance to pollution as follows (in order of decreasing tolerance): *Euglena, Chlamydomonas, Scenedesmus, Chlorella, Ankistrodesmus, Cyclotella, Closterium, Micractinium, Pediastrum, Arthrospira, Oocystis, Actinastrum, Coelastrum, Spondylomorum, Golenkinia, Selenastrum* and *Dictyosphaerium*. Additional genus not mentioned in the above early list of Palmer is *Closteridium*, which however appeared in the later work of Palmer (1980), among "algae most abundant and widespread in sewage ponds."

Table 11.1. Wastewater reservoir algae (Na'an Reservoir as an example)

Cyanophyceae

Arthrospira jenneri (Hassal) Kutz., Plate 11.1d

Chlorophyceae

Actinastrum hentzschi Lagerheim, Plate 11.3f
Ankistrodesmus convolutus var. *minutus* (Nag.) Rabh., Plate 11.1f
Chlamydomonas sp., (not illustrated)
Chlorella vulgaris Beijernick, Plate 11.1b
Closteridium lunula Reinsch., Plate 11.3d
Closterium braunii Reinsch., Plate 11.2f
Coelastrum microporum Nageli, Plate 11.1e
Dictyosphaerium echrenbergianum Nageli, Plate 11.2a
Golenkinia radiata (Chod.) Wille, Plate 11.2d
Micractinium pusillum Fresenius, Plate 11.3f
Oocystis pusilla Hansgirg, Plate 11.2e
Pediastrum duplex var. *clathratum* (A. Braun) Lagerheim, Plate 11.3c
Pediastrum sp., Plate 11.2c
Scenedesmus quadricauda (Turp.) de Brebisson, Plate 11.3b
Scenedesmus obliquus (Turp.) Kutzing, Plate 11.2b
Selenastrum minutum (Nag.) Collins, (not illustrated)
Spondylomorum quaternarium Ehr., (not illustrated)

Euglenophyceae

Euglena minuta Prescott, Plate 11.1a

Diatomeae

Cyclotella meneghiniana Kutz., Plate 11.3a

According to Gloyna (1971) species of *Euglena* show a high degree of adaptability to various pond conditions and may be present during all seasons and under most climate conditions. The next in adaptability as indicated by this author are *Chlamydomonas, Micractinium, Ankistrodesmus, Scenedesmus* and *Chlorella*.

Similar phytoplankton composition, with less genera recorded, was reported by Azov et al. (1980) from high rate oxidation ponds. Under a very high BOD_5 of up to 250 mg l^{-1}, *Scenedesmus dimorphus* prevailed, while at the lower BOD_5 of 60 mg l^{-1} *Micractinium pusillum* was dominant. Only slightly higher diversity was recorded in Ma'aleh HaKishon Reservoir, operated under much lower BOD_5 of 5–15 mg l^{-1}. The additional genera in the above reservoir included *Nannochloris, Planktosphaeria* and *Stichococcus* (Kalinsky 1987). In a recent paper Soler et al. (1991) provide a list of predominant algae in a deep wastewater self-depuration lagoon in Spain. Among the predominant genera mentioned are again *Chlamydomonas, Euglena, Golenkinia, Chlorella, Micractinium* and *Scenedesmus*.

According to above, we can state safely that the algae listed in Table 11.1, while representing a wide range of local wastewater reservoirs, fit well the general pollution algae literature.

Concentrations and seasonal distribution of the four dominant algae in the Na'an Reservoir are presented in Figs 11.1 and 11.2. As compared to other algae, *Chlorella* is

Fig. 11.1. Concentrations
and seasonal distribution of
Chlorella and *Chlamydomonas*
in Na'an Reservoir

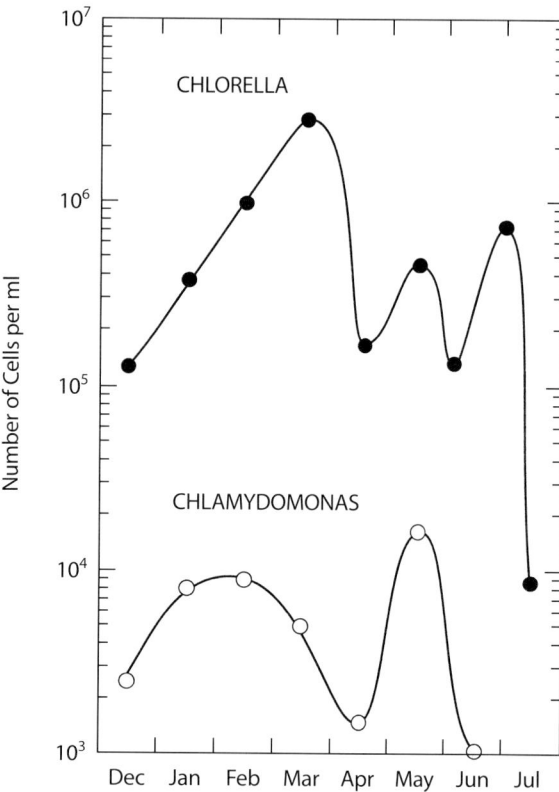

by far the most abundant and seasonally stable. It declines during the winter till 10^5 cells ml^{-1}, then increases rapidly to the spring peak of 3×10^6 cells ml^{-1}. Later in the spring the numbers drop and show fluctuations around 10^4–10^5 cells ml^{-1} as a result from the late spring multiplication of zooplankton filter feeders decimating algal population. *Chlorella* constituted 90–97% of the total cell count most of the year, with a drop to only 76% during the summer decline in the cell number. Similarly, the Dan Region (Tel Aviv) oxidation ponds exhibit an overall remarkable seasonal stability, with *Chlorella* as the predominant alga, while seasonal variations occur mainly with the non-predominant species (Shelef and Kanarek 1995). Almost identical seasonal fluctuation is displayed by *Chlamydomonas*, which however always shows much lower concentrations of 10^2–10^3 cells ml^{-1}. Also *Ankistrodesmus* and *Micractinium* exhibited the same seasonal pattern with an early spring peak reaching 10^3 cells ml^{-1}, followed by the later drop in numbers. Among other forms *Euglena* and *Selenastrum* were quite stable all over the year and represented by low concentrations not exceeding 10^3 cells ml^{-1}. Remaining algae listed in Table 11.1 appeared irregularly and in still lower concentrations, prevailing mostly in the summer.

The concomitant change of algal biomass (represented by chlorophyll a) was from 840 µg l^{-1} (March peak) to only 34 µg l^{-1} (April drop) (see Dor et al. 1987b, reproduced in this volume).

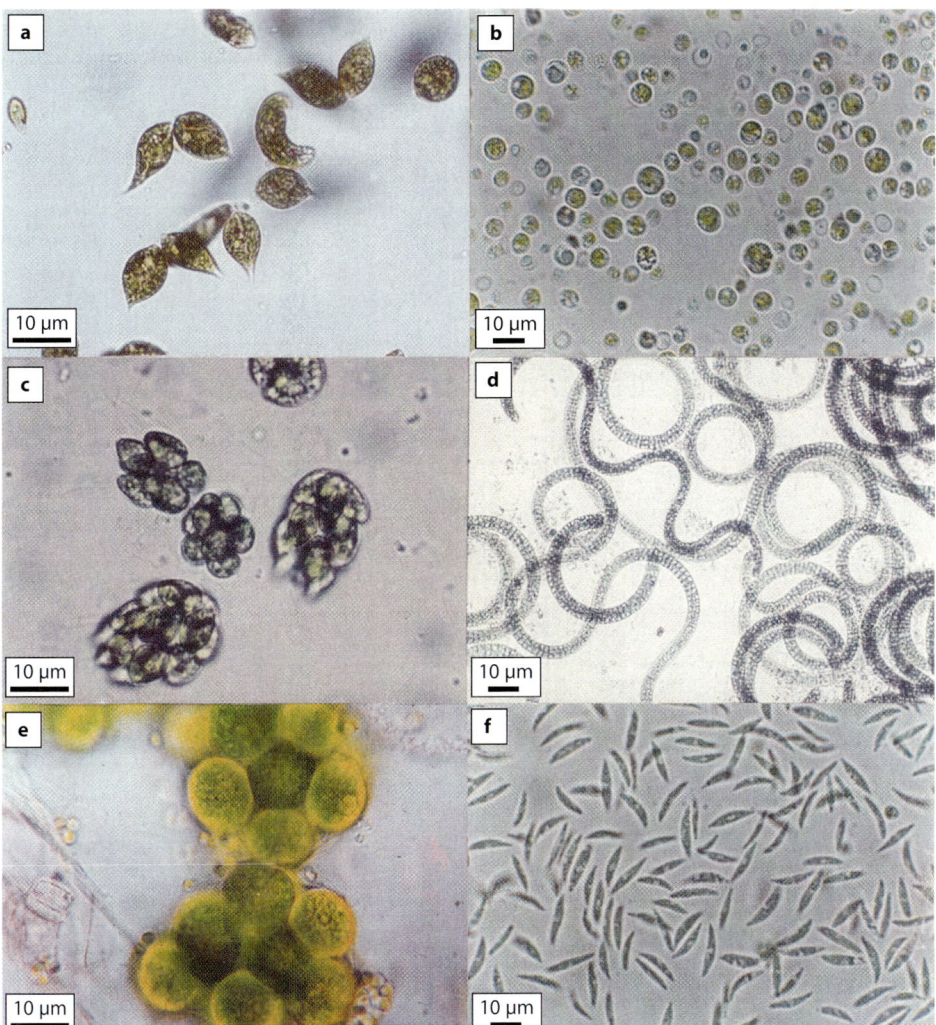

Plate 11.1. Wastewater reservoir algae: **a** *Euglena minuta* Prescott; **b** *Chlorella vulgaris* Beijernick; **c** *Spondylomorum quaternarium* Ehr.; **d** *Arthrospira jenneri* = *Spirulina jenneri* (Hassal) Kutz.; **e** *Coelastrum microporum* Nageli; **f** *Ankistrodesmus convolutus* var. *minutus* (Nag.) Rabh.

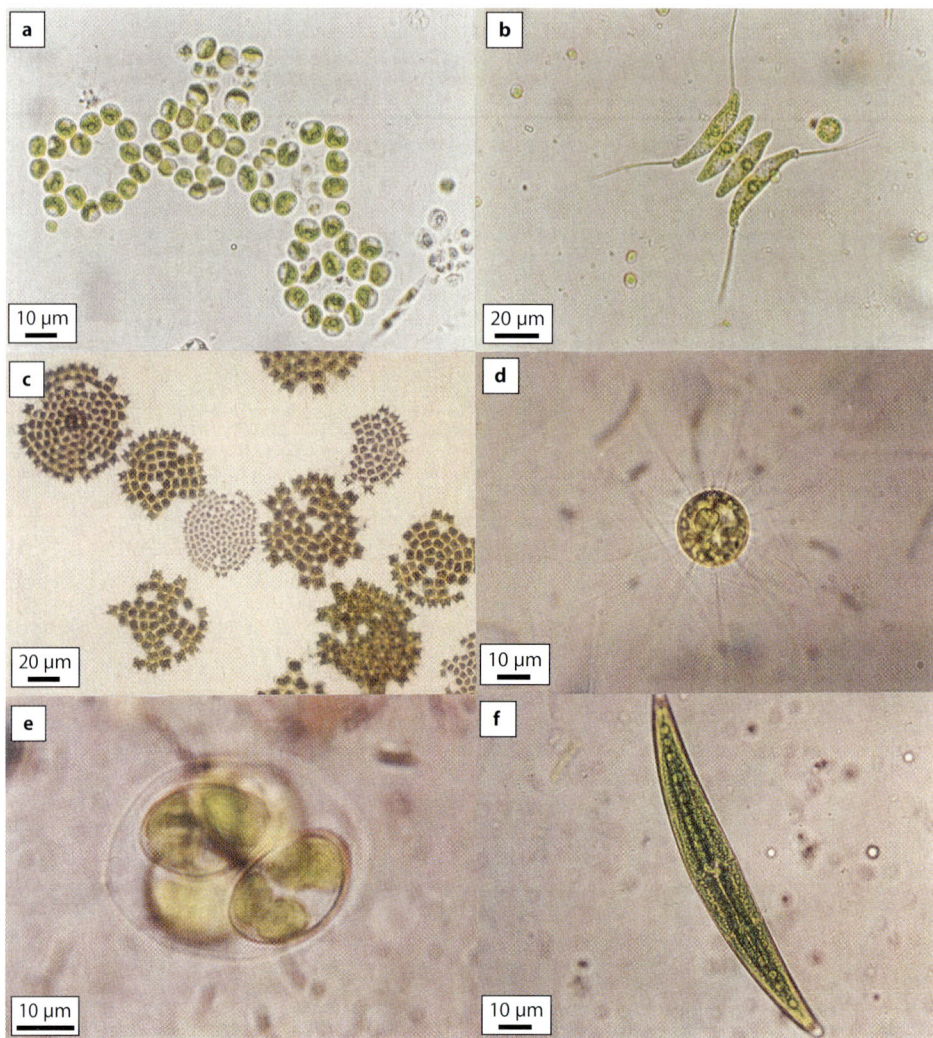

Plate 11.2. Wastewater reservoir algae: **a** *Dictyosphaerium ehrenbergianum* Nageli; **b** *Scenedesmus* sp.; **c** *Pediastrum* sp.; **d** *Golenkinia radiata* (Chod.) Wille; **e** *Oocystis pusilla* Hansgirg; **f** *Closterium braunii* Reinsch.

Plate 11.3. Wastewater reservoir algae: **a** *Cyclotella meneghiniana* Kutzing; **b** *Scenedesmus quadricauda* (Turp.) de Brebisson; **c** *Pediastrum duplex* var. *clathratum* (A. Braun) Lagerheim.; **d** *Closteridium lunula* Reinsch.; **e** *Actinastrum hantzschii* Lagerheim; **f** *Micractinium pusillum* Fresenius

Fig. 11.2. Concentrations and seasonal distribution of *Ankistrodesmus* and *Micractinium* in Na'an Reservoir

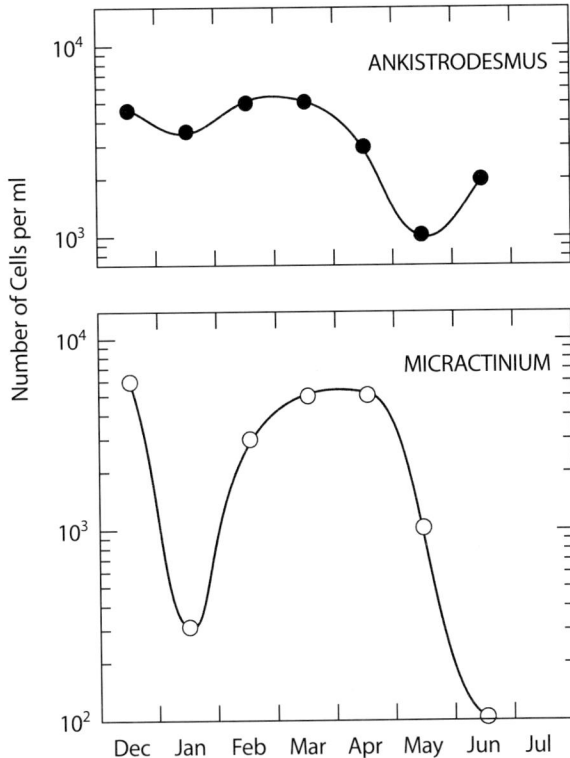

11.4
Algae-Bacteria Relationships

Like in any type of open sewage treatment systems, algae and bacteria populating wastewater reservoirs grow in large numbers and in a close coexistence. The mutual dependence of both groups of microorganisms on their metabolic by-products was recognized early in this century. Obviously, heterotrophic bacteria utilize the oxygen released by photosynthetic algae while algae absorb CO_2 and other nutrients produced by the aerobic degradation of organic matter by bacteria. These reciprocal relationships were repeatedly described as "symbiosis" in the early works (Ludwig et al. 1951; Lange 1967; El-Sharkawi and Moawald 1970; Kuentzel 1970; Wiedeman 1970; Humenik and Hanna 1971) and up to recently (Tchobanologus and Burton 1991; Laliberte et al. 1994; Hammouda et al. 1995). However, the above obvious synergism constitutes only part of the algal-bacterial relationships which are much more complex.

Although the algae are considered to be photosynthetic, they can accomplish this process only during daylight hours and – due to the high turbidity of the medium in the wastewater – only in the upper layer of the water column. During the darkness hours and in the deep aphotic layer, algae are unable to perform photosynthesis. Indeed, 11 dominant algal strains isolated from a stabilization pond by Wiedeman (1970), exhibited mixotrophic nutrition, utilizing a large variety of organic molecules, which evidently allows their survival in the aphotic zone of the wastewater body. This fact places wastewater algae as competitors of heterotrophic bacteria, which are more numerous and smaller, having therefore an advantage in absorbing organic nutrients. The competition may lead to antagonism. Indeed, a rich evidence is found in the literature for existing antagonistic effects between algae and bacteria inhabiting wastewater bodies.

Production of bacterial growth inhibitor by *Chlorella* was first reported by Pratt (1942) and the inhibitory agent excreted by the alga was named "chlorelline". Substances having a pronounced antibacterial activity were later obtained from filtrates of *Chlorella pyrenoidosa*, and have been shown to be oxidation products of unsaturated fatty acids (Spoehr et al. 1949). Steeman-Nielsen (1955) noted indications suggesting that in the light *Chlorella* sp. and certain marine algae produce substances inhibiting bacterial respiration. Fogg (1962) stated that pathogenic and coliform bacteria die out rather quickly in sewage oxidation ponds, in which there is abundant algal growth, which could be a consequence of the production of antibacterial agents. Accorinti (1964) isolated bacterial inhibitors from another wastewater algae – *Scenedesmus obliqus*. Hellebust (1974) cited a number of works proving that various planktonic algae produce extracellular antibacterial substances, which are most probably oxidized products of fatty acids. Chrost (1975a, 1975b) reported on the existence of antibacterial effects of the substances produced by algae during the blooms and discussed the resulting dependence between phytoplankton and bacteria development. Dor (1978) tested two algae which are eminently successful in the presence of organic pollution: *Chlorella vulgaris* and *Scenedesmus obliqus*. The aim of this study was to test the effect of these algae – isolated from a high rate oxidation pond – on the natural bacterial flora of the sewage entering this pond. Strains of bacteria were isolated from the inflowing raw sewage and identified as *Citrobacter* sp., *Pseudomonas* sp., *Klebsiella* sp. and *Aeromonas* sp. An additional culture of *Staphylococcus* sp. was also tested. The influence of algae was investigated according to three independent methods:

a *Zones of inhibition:*
   Algal filtrates were extracted and tested by means of cup test (Spoehr et al. 1949); the transparent zones of inhibition on the plates were measured. According to the zone of inhibition size, all five bacterial strains were variously sensitive to the substances extracted from the culture filtrates of both algae tested.

b *Effect on the rate of respiration:*
   The raw domestic sewage (populated by the above strains of bacteria) was incubated 1) with algal inoculum, 2) with algal cultures filtrates and 3) with tap water as a control. In addition the sewage was also incubated inside a dialysis bag suspended in the freshly inoculated algal culture on sewage. Rate of respiration was measured in the sub-samples. The curves of respiration rates (as rates of oxygen consumption) showed a sharp decline either in the presence of algal inoculum or algal filtrates. Sewage suspended in a dialysis bag inside algal culture of *Chlorella* or *Scenedesmus* on sewage exhibited a prominent inhibition of the respiration as compared to the control.

c *Bacterial counts:*
   Numbers of bacteria in the sewage dropped prominently either in the presence of *Chlorella* or *Scenedesmus*.

These experiments demonstrated unequivocally not only that sewage algae excrete bacterial inhibitors but also that these substances are active in natural concentrations.

In a later work (Dor and Benzion 1980) the effect of heterotrophic bacteria on the algae was evaluated. It was demonstrated that *Chlorella vulgaris*, incubated in the pasteurized domestic sewage supplemented with antibiotics, grew much better than in the presence of a natural bacterial flora of the same sewage. When incubated similarly in an autoclaved sewage but in the presence of single bacterial strains isolated from the raw sewage, *Chlorella* always exhibited a decreased rate of divisions and lower yields as compared to the axenic control. Among 20 bacterial strains tested, not one was stimulatory – all of them interfered with the growth of algae. The extent of inhibition varied according to the strain added.

Benzion and Dor (1981) also tested a possible selective repression in the pond, in favour of a stimulatory bacterial flora. Among ten bacterial strains isolated from the pond none was stimulatory toward *Chlorella* and *Scenedesmus*. Also in this case various bacterial strains inhibited growth of both algae to different degrees. Moreover, growth of *Chlorella* and *Scenedesmus* was also severely inhibited in the autoclaved and re-enriched organic medium, used previously for bacterial growth.

An overview of the complex interactions between algae and bacteria in aquatic ecosystems was presented by Cole (1982). According to this author inhibition of algal growth by bacteria can be induced by bacterial metabolism modifying environment: a) anaerobic conditions may evolve when supply of organic matter is sufficient, and b) bacterial metabolism may lower the pH, both, through the production of organic acids and through the oxidation of variety of compounds. An additional damaging effect may be exerted by lytic bacteria. In the competition for nutrients, the small, numerous, and rapidly growing bacteria can dominate the larger, less numerous, and slowly growing algae. Other inhibitory phenomena can be caused by soluble substances released by certain bacteria; however these remain as yet unidentified. Stimulation of algal growth by bacteria may result through nutrient regeneration by heterotrophic remineralization of various organic substrates as well as through vitamin production by certain bacteria. Considering inhibition of bacterial growth by a number of algae, Cole quotes the production of antibiotic substances by a variety of species. It is also known that high pH values, due to the photosynthetic activity, are lethal to many bacterial strains. In addition to these inhibitory actions, bacteria can be also stimulated by the algae: besides the photosynthetic oxygen stimulating bacterial respiration, some of the organic matter from dead algal cells will be metabolized by bacteria while living phytoplankton releases soluble organic compounds which are utilized by bacterial decomposers.

Such complexity of interactions suggests that mutual selective pressures exerted by algae and bacteria may regulate the composition and concentration of phyto and bacterioplankton in the waters rich in organic matter.

References

Abeliovich A (1980) Factors limiting algal growth in high rate oxidation ponds. In: Shelef G, Soeder CJ (eds) Algae Biomass. Elsvier, pp 205–215

Accorinti J (1964) Attempts to isolate the inhibitors from *Scenedesmus obliquus*. Phyton 21(1):103–107

Azov Y, Shelef G, Moraine R, Levi A (1980) Controlling algal genera in high rate wastewater oxidation ponds. In: Shelef G, Soeder CJ (eds) Algae Biomass. Elsvier, pp 245–253

Benzion S, Dor I (1981) Bacterial inhibition of algal growth in photosynthetic sewage treatment systems. In: Shuval H (ed) Developments in arid zone ecology and environmental quality. Balaban Publishers, pp 249–286

Chrost RJ (1975a) Inhibitors produced by algae as an ecological factor affecting bacteria in water ecosystems. I. Dependence between phytoplankton and bacteria development. Acta Microb Polonica Ser B7/24:125–133

Chrost RJ (1975b) Inhibitors produced by algae as an ecological factor affecting bacteria in water ecosystems. II. Antibacterial activity of algae during blooms. Acta Microb Polonica Ser B3:167 –176

Cole JJ (1982) Interactions between bacteria and algae in aquatic ecosystems Ann Rev Ecol Syst 13:291–314

Dor I (1978) Effect of the green algae *Scenedesmus obliquus* and *Chlorella vulgaris* on heterotrophic bacteria in polluted waters. Verh Internat Verein Limnol 20:1930–1933

Dor I, Benzion S (1980) Effect of heterotrophic bacteria on the green algae growing in wastewater. In: Shelef G, Soeder CJ (eds) Algal Biomass, pp 421–429

Dor I, Raber M (1990) Deep wastewater reservoirs in Israel: Empirical data for monitoring and control. Wat Res 24(9):1077–1084

Dor I Kalinsky I, Eren J, Dimentman C (1987a) Deep wastewater reservoirs in Israel: Limnological changes following self-purification. Wat Sci Tech 19(12):317–322

Dor I, Schechter H, Bromley HJ (1987b) Limnology of a hypertrophic reservoir storing wastewater effluent for agriculture at Kibbutz Na'an, Israel. Hydrobiologia 150:225–241

El-Sharkawi FM, Moawald SK (1970) Stabilization of dairy wastes by algal-bacterial symbiosis in oxidation ponds. J W P C F 42:115–125

Fogg GE (1962) Extracellular products. In: Lewin RA (ed) Biochemistry of algae. Academic Press, pp 475–489

Gloyna EF (1971) Waste stabilization ponds. World Health Organization, Geneva, 175 pp

Hammouda O, Gaber A, Abdel-Raouf N (1995) Microalgae in wastewater treatment. Ecotoxicology and Environmental Safety 31:205–210

Hellebust JA (1974) Extracellular products. In: Stewart WDP (ed) Algal physiology and biochemistry. Blackwell, London, pp 838–863

Humenik FJ, Hanna GP (1971) Algal-bacterial symbiosis for removal and conservation of wastewater nutrients. J W P C F 43(4):580–594

Kalinski I (1987) Limnology and self-purification in Ma'aleh HaKishon – a deep wastewater reservoir. Unpublished Msc. Thesis, The Hebrew University, 136 pp (in Hebrew)

Kuentzel LE (1970) Bacteria-algae symbiosis – a cause of algal blooms. Proc. Nat. Symp. on Hydrobiol., Bioresources of shallow water environments, pp 321–331

Laliberte G, Proulx D, De Pauw N, De La Noue J (1994) Algal technology in wastewater treatment. In: Rai LC, Gaur JP, Soeder CJ (eds). Algae and water pollution. Heft 42, Archiv für Hydrobiologie, pp 283–302

Lange W (1967) Effect of carbohydrates on the symbiotic growth of planktonic blue-green algae with bacteria. Nature 215:1277–1278

Llorens M, Saez J, Soler A (1992) Influence of thermal stratification on the behaviour of a deep wastewater stabilization pond. Wat Res 26(5):569–577

Ludwig HF, Oswald WJ, Gotaas HB, Lynch V (1951) Algae symbiosis in oxidation ponds. Sewage and Ind Wastes 23:1337–1355

Palmer CM (1969) A composite rating of algae tolerating organic pollution. J Phycol 5:78–82

Palmer CM (1980) Algae and water pollution. Castle House Publ. Ltd., 123 pp

Pratt R (1942) Studies on *Chlorella vulgaris*. V Some properties of the growth inhibitor formed by *Chlorella* cells. Amer J Bot 29:142–148

Shelef G, Kanarek A (1995) Stabilization ponds with recirculation. Wat Sci Tech 31(12):389–397

Soler A, Saez J, Llorens M, Martinez I, Torella F, Berna LM (1991) Changes in physico-chemical parameters and photosynthetic microorganisms in a deep wastewater self-depuration lagoon. Wat Res (6):689–695

Spoehr HA, Smith JHC, Strain HH, Milner HW, Hardin GJ (1949) Fatty acid antibacterials from plants. Carnegie Inst Washington, Publ 586, 67 pp

Steeman-Nielsen, E (1955) The production of antibiotics by plankton algae and its effect upon bacterial activities in the sea. Papers in Marine Biol and Oceanogr (London), pp 281–286

Ichobanoglous G, Burton FL (1991) Wastewater engineering, treatment, disposal and reuse. Mc Graw-Hill Inc., pp 436–437

Thompson PA, Rhee GY (1994) Phytoplankton responses to eutrophication. In: Kausch H, Lampert W (eds) Algae and water pollution. Archiv für Hydrobiologie, Heft 42:124–166

Wiedeman VE (1970) Heterotrophic nutrition of waste-stabilization pond algae. In: Zajc JE (ed) Properties and products of algae. Plenum Press, pp 107–114

Chapter 12

Fauna

Chanan Dimentman · Francis Por

12.1
Introduction

The first wastewater reservoir in Israel was built in 1969 in Yoqne'am and had a capacity of 330 000 m³. Today there are more than 200 wastewater reservoirs and they are – besides Lake Kinneret – the most significant type of inland water environment in the country. Therefore, wastewater storage reservoirs represent an environmental opportunity for a variety of faunal elements. Half of the reservoirs are fed exclusively by raw or treated sewage: these constitute the subject of the present article. Other reservoirs receive also freshwater from flood run-off, the National Water Carrier as well as springs and streams and from underground sources. These waters are mixed in various proportions with treated sewage as well as with water from hypertrophic fish ponds. A few reservoirs also receive a mixture of brackish waters. Other types of hypertrophic aquatic ecosystems in the region are the polluted streams and the waters of the Hula Nature Reserve.

12.2
Limnological Parameters

Maximum water depth in the wastewater reservoirs is between 5 to 14 m, depending of on the construction of each of them. Deep reservoirs tend to be stratified during most of the year. Most of the reservoirs are practically emptied by the end of the irrigation season after 3–4 summer months of intensive irrigation. The irrigation season is longer in the arid areas of the country.

Minimum temperature recorded in the hypertrophic reservoirs of Israel is about 12° C and the maximum is about 29° C. The most frequent annual range is between 14° C and 28° C. The maximal vertical temperature gradient recorded was between 15° C in the hypolimnion and 28.5° C in the epilimnion. The thermal regime of the hypertrophic reservoirs is practically identical with that of the other standing water bodies of Israel. The inflow of effluents influences the temperature of water near the inlet of the reservoirs.

High turbidity is characteristic for the hypertrophic water bodies, with Secchi disk measurements reaching a minimum of 0.1 m. However, several periods of very clear water, with Secchi readings up to almost 4 m have been recorded.

The presence of dissolved oxygen in the reservoirs is affected mainly by the quality of the inflow, by the depth of the reservoir, and by the residence time. Reservoirs fed exclusively by raw and poorly treated sewage are often anoxic during the night and in the early morning hours. Other reservoirs are as a rule aerobic, at least near the surface, even during the night. The amount of dissolved oxygen as well as its distribution in the water column are fluctuating in function of the magnitude of the diel vertical mixing and of the relation between photosynthetic oxygen production and its consumption. As a consequence, the hypertrophic reservoirs present many varied patterns of diel and seasonal regimes of dissolved oxygen.

12.3
Faunal Limits and Definitions

We shall not discuss the fauna of the natural water bodies that became secondarily hypertrophic, like several hypertrophic basins and ponds in Israel. These are as a rule connected to peripheral clean waters, such as springs, streams, etc. In periods of stress, these clean water bodies serve as refugia for the primary aquatic fauna and enable re-expansion into the main water body when conditions improve. The man-made wastewater hypertrophic reservoirs discussed here present no such refugia. They present maximum water depth toward the end of the spring and then follows a rapid fall in the water level during peak utilization and evaporative losses in the late summer, followed by a gradual refilling through sewage inflow. The turbidity threshold for the metazoan organisms is situated between 10 cm to 20 cm Secchi values. The lowest oxygen concentrations tolerated by metazoans are between 0.2 to 0.5 mg l^{-1} However, in several cases this limit was found to be higher, i.e., between 1.0 and 2.0 mg l^{-1}. The sediment in the reservoirs is probably permanently anaerobic because of the very high organic content and therefore there are no periods of oxygenation which enable a benthic infauna to develop and resting stages to survive. Water temperature is not a stressing factor, but it directly influences the taxonomic composition and abundance of the fauna in the hypertrophic reservoirs, as well as their seasonal and diel dynamics. The different stressing factors define the faunistic parameters of the hypertrophic reservoirs discussed here.

The fauna of these reservoirs is composed mainly of secondary aquatic species, i.e., species which have the capacity to recolonize from the surroundings the temporary or stressed water bodies, either actively or by passive transport of resting stages. Animals which need permanent presence of water within ecologically tolerable conditions, i.e. the primary aquatic species such as fish, most of the mollusks, and the higher crustaceans, are not present. The effluents which feed the reservoirs have a limnological quality which is generally well beyond the survival values for metazoans. Perhaps the faunistically most important feature of the hypertrophic reservoirs is the lack of natural viable populations of fish and of other macro-predators in their aquatic fauna.

12.4
Faunal Composition and Frequency

Only a small part of the inventory of the aquatic fauna of the wastewater reservoirs is based on published papers (Eren 1978; Bromley 1981; Dor et al. 1987a, 1987b). Most of the information is spread in different research and survey reports which circulated since 1976, many of them in Hebrew. These sources are: Eren (1976), Bromley (1982), Dor (1982), Por (1982), Eren et al. (1986, 1987), Dimentman and Eitan (1988), Shelef and

Table 12.1. An inventory of the aquatic fauna of hypertrophic reservoirs in Israel

Protozoa
Sarcodina
   Amoeba sp.
Ciliophora
   Astylozoon sp.
   Coleps sp.
   Cyclidium sp.
   Didinium nasutum (O.F. Müller)
   Euplotes sp.
   Halteria sp.
   Paramecium sp.

Platyhelminthes gen. sp.

Rotifera
Monogononta
   Ascomorpha sp.
   Asplanchna brightwelli Gosse
   Asplanchna sieboldi (Leydig)
   Brachionus angularis Gosse
   Brachionus bidentata Anderson
   Brachionus budpestinensis Daday
   Brachionus calyciflorus Pallas
   Brachionus plicatilis O.F. Müller
   Brachionus rubens Ehrenberg
   Cephalodella sp.
   Epiphanes senta (O.F. Müller)
   Filinia longiseta (Ehrenberg)
   Hexarthra sp.
   Keratella sp.
Bdelloidea gen. sp.

Nematoda
   Panagrolaimus ruffoi Andrássy

Annelida
Oligochaeta
   Dero sp.

Arthropoda
Crustacea
Cladocera
   Daphnia longispina O.F. Müller
   Daphnia magna Straus
   Moina micrura Kurz

Copepoda
   Arctodiaptomus similis (Baird)
   Cyclops scutifer Sars
   Mesocyclops ogunnus Onabamiro
   Metacyclops minutus (Claus)
Ostracoda
   Cypris pubera O.F. Müller
   Herpetocypris sp.
   Heterocypris incongruens (Ramdohr)
   Heterocypris salina (Brady)
   Potamocypris mastigophora (Methuen)
Arachnoidea
Hydracari
   Hydrachna erythraea (Brullé)
Insecta
Odonata
   Anax parthenope (Selys)
   Crocothemis sp.
Hemiptera
Corixidae
   Sigara lateralis (Leach)
   Micronecta sp.
Notonectidae
   Anisops sardea Herrich-Schaeffer
Coleoptera
Dytiscidae
   Eretes sticticus (L.)
   Herophydrus musicus (Klug)
   Hydroglyphus pusillus (Fabricius)
Hydrophilidae
   Berosus bispina Reiche & Saulcy
   Enochrus ater Kuwert
   Enochrus bicolor Fabricius
   Sternolophus solieri (Castelnau)
Diptera
Chironomidae
   Chironomus spp.
   Polypedilum sp.
Culicidae
   Culex pipiens L
Syrphidae
   Eristalis tenax (L.)

CHORDATA
Reptilia
   Mauremys caspica rivulata Valenciennes

Juanicó (1989), Katznelson and Dimentman (1991), and Freund et al. (1992). Table 12.1 lists the 57 aquatic animal taxa that have been reported from the hypertrophic reservoirs in Israel till today. Of these, 37 could be identified to the species level, and 18 to the generic level only. Zooplankton dominates the fauna of these ecosystems in number of species, number of individuals and biomass. The most common taxa of zooplankton are the *Ciliophora* with at least 8 species and the *Rotifera* with at least 14 species. The *Cladocera* are represented by three species and the *Copepoda* by four.

The zoobenthos of the hypertrophic reservoirs is restricted to the shallow littoral and to some elevated substrates, whereas the very bottom is practically devoid of metazoans. The benthic organisms found are restricted to the *Rotifera, Bdelloidea, Nematoda* and *Oligochaeta. Ostracoda* appeared only in low loading wastewater res-

Table 12.2. An inventory of the waterfowl of hypertrophic reservoirs in Israel

AVES	*Larus cachinnans* Pallas	*Calidris ferruginea* (Pontoppidan)
Alcedinidae	*Larus canus* L.	*Calidris melanotus* (Vieillot)
Alcedo atthis (L.)	*Larus fuscus* L.	*Calidris minuta* (Leisler)
Ceryle rudis (L.)	*Larus heuglini* Bree	*Calidris temminckii* (Leisler)
Halcyon smyrensis (L.) (B)	*Larus melanocephalus* Temminck	*Calidris tenuirostris* (Horsfield)
	Larus minutus Pallas	*Gallinago gallinago* (L.)
Anatidae	*Larus ridibundus* L.	*Limicola falcinellus* (Pontoppidan)
Anas acuta L.		*Limnodromus scolopaceus* (Say)
Anas clypeata L.	Motacillidae	*Limosa lapponica* (L.)
Anas crecca L.	*Anthus spinoletta* (L.)	*Limosa limosa* (L.)
Anas penelope L.	*Motacilla alba* L.	*Lymnocryptes minimus* (Brünnich)
Anas platyrhynchos L. (B)	*Motacilla cinerea* Tunstall	*Neumenius arquata* (L.)
Anas querquedula L.	*Motacilla citreola* Pallas	*Neumenius phaeopus* (L.)
Anas strepera L.	*Motacilla flava* L.	*Phalaropus fulicarius* (L.)
Aythya ferina (L.)		*Phalaropus lobatus* (L.)
Aythya fuligula (L.) (B)	Phoenicopteridae	*Philomachus pugnax* (L.)
Aythya marila (L.)	*Phoenicopterus ruber* L.	*Tringa erythropus* (Pallas)
Aythya nyroca (Güldenstädt) (B)		*Tringa glareola* L.
Bucephala clangula (L.)	Podicipedidae	*Tringa nebularia* (Gunnerus)
Cygnus olor (Gmelin)	*Podiceps auritus* (L.)	*Tringa ochropus* L.
Netta rufina (Pallas)	*Podiceps nigricollis* C.L. Brehm	*Tringa stagnatilis* (Bechstein)
Oxyura leucocephala (Scopoli)	*Tachybaptus ruficollis* (Pallas) (B)	
Tadorna ferruginea (Pallas)		*Tringa totanus* (L.)
Tadorna tadorna (L.)	Rallidae	*Xenus cinereus* (Güldenstädt)
	Fulica atra L. (B)	
Ardeidae	*Gallinula chloropus* (L.) (B)	Sternidae
Butorides striatus (L.)	*Porphyrio porphyrio* (L.)	*Chlidonias hybrida* (Pallas)
Ixobrychus minutus (L.) (B)	*Porzana parva* (Scopoli)	*Chlidonias leucoptera* (Temminck)
	Porzana porzana (L.)	*Chlidonias nigra* (L.)
Charadriidae	*Porzana pusilla* (Pallas)	*Gelochelidon nilotica* (Gmelin)
Charadrius alexandrinus L.	*Rallus aquaticus* L.	*Sterna hirundo* L. (B)
Charadrius asiaticus Pallas		
Charadrius dubius Scopoli (B)	Recurvirostridae	Sylviidae
Charadrius hiaticula L.	*Himantopus himantopus* (L.) (B)	*Acrocephalus arundinaceus* (L.)
Charadrius pecuarius Temminck	*Recurvirostra avosetta* L.	*Acrocephalus melanopogon* (Temminck)
Chettusia leucura (Lichtenstein)		*Acrocephalus schoenobaenus* (L.)
Hoplopterus spinosus (L.) (B)	Remizidae	*Acrocephalus scirpaceus* (Hermann) (B)
Pluvialis fulva (Gmelin)	*Remiz pendulinus* (L.)	*Acrocephalus stentoreus* (Hemprich
Pluvialis squatarola (L.)		& Ehrenberg)
	Rostratulidae	*Cettia cetti* (Temminck)
Emberizidae	*Rostratula benghalensis* (L.)	*Locustella luscinioides* (Savi)
Emberiza schoeniclus (L.)		
	Scolopacidae	Threskiornithidae
Haematopodidae	*Actitis hypoleucos* (L.)	*Platalea leucorodia* L.
Haematopus ostralegus L.	*Arenaria interpres* (L.)	*Plegadis falcinellus* (L.)
	Calidris alba (Pallas)	
Laridae	*Calidris alpina* (L.)	Turdidae
Larus armenicus Buturlin	*Calidris canutus* (L.)	*Luscinia svecica* (L.)

(B) Recorded breeding in these ecosystems.

ervoirs. Among the demersal aquatic insect fauna of the littoral, the *Hemiptera* and the *Coleoptera* are dominant. *Chironomidae* were occasionally abundant. In some cases larvae of *Culex* and of *Eristalis* were also reported. Fish do not colonize the hypertrophic reservoirs naturally. However, fish have been introduced for manipulative purposes in several cases (see below).

The only naturally occurring aquatic vertebrate which inhabits the hypertophic reservoirs is the turtle Caspian terrapin *Mauremys caspica rivulata* (Gasith and Sidis 1983; Sidis 1983). Under special conditions of some reservoirs, elements of the wetland and even terrestrial invertebrate fauna can be very abundant. Such is the case of the terrestrial oniscoid isopod *Porcellinoides pruinosus*, which takes advantage of the humidity beneath stones and slabs above the waterline of certain reservoirs. Several species of spiders, mostly of the *Tetragnathidae* and the *Araneidae* are common around reservoirs with rich chironomid populations (G. Levy personal communication).

12.5
Aquatic Birds

Many species of water birds are attracted by the novel environment of the hypertrophic wastewater reservoirs. They use these water bodies for food and as roosting and stopover stations, and even as breeding places. This function of the reservoirs is especially important in Israel, which is a central hub for bird migration in the Old World. A total of 104 species of water birds have been reported in association with these ecosystems (Zuaretz 1970–1976, 1977; Paz and Zuaretz 1976; Yom-Tov and Mendelssohn 1988; Shirihai 1996; and our own sightings between 1970–1997) (Table 12.2). The most frequently encountered birds were surface-feeding ducks. They sieve zoo- and phytoplankton as well as suspended organic matter from the water surface of the reservoirs. All the 7 species of *Anas* reported in the reservoirs are known filter feeders.

12.6
Spatial and Temporal Distribution

The spatial distribution of the aquatic animals in the hypertrophic reservoirs is influenced:

1. By the gradient of conditions from the entrance point to the most remote corner of the reservoir, usually its exit;
2. By the winds which drive the plankton toward the windward shore. This happens especially at night, when the zooplankton concentrates at the surface of the water.

The factor which most affects the vertical distribution is light intensity and penetration. The vertical profiles of temperature, oxygen and other physico-chemical factors also play an important role. Paramount to the existence and distribution of the zoobenthos is the presence and concentration of dissolved oxygen, as well as of the toxic compounds. Indeed, in most of the hypertrophic reservoirs, oxygen is absent on the bottom and no metazoans are found there.

The various taxa of demersal insects exhibit different patterns of distribution. The *Chironomidae* larvae can be found all over the reservoir, whereas the *Culex* and *Eristalis*

Table 12.3. Vertical distribution of *Daphnia magna* and *Mesocyclops ogunnus* during day and night (26-27.6.1986) in the Ma'ale HaKishon reservoir

	Daphnia magna		*Mesocyclops ogunnus*	
Depth (m)	day (%)	night (%)	day (%)	night (%)
0.5	1	9	21	44
1	3	42	15	15
2	50	39	15	25
3	20	7	17	11
4	19	2	23	5
5	6	1	9	0

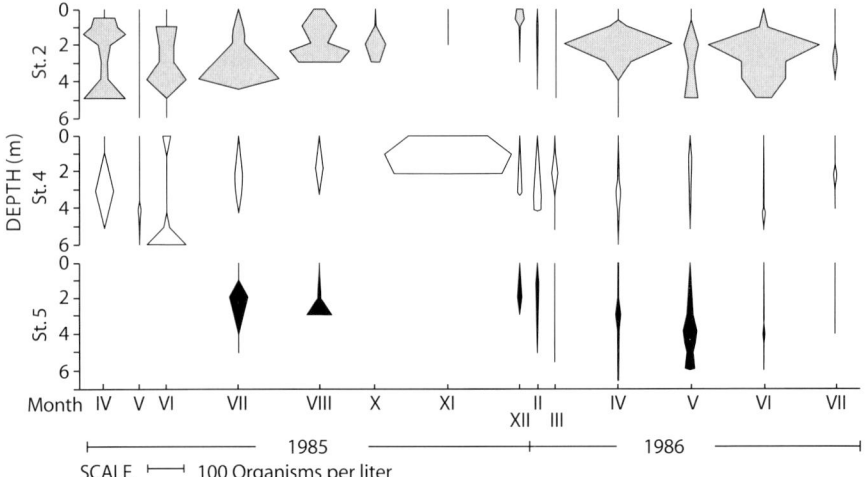

Fig. 12.1. Vertical distribution and density of Cladocera (*Daphnia magna*) in the Ma'aleh HaKishon Reservoir (April 1985–July 1986)

larvae and pupae are found near the banks. Among the *Hemiptera*, the *Corixidae* are found throughout the reservoir while the *Notonectidae* concentrate near the shore.

There are two types of cycles in the distribution of the aquatic fauna of the reservoirs, the diel and the seasonal cycles. The zooplankton accumulates in the surface layer during the night and migrates to deeper layers during the day (Table 12.3). On cloudy days the zooplankton stays at the surface also during day hours. The extent of the vertical migration differs in the different taxa, age groups and among sexes. *Cladocera* tend to migrate more than *Copepoda*; young specimens of both taxa migrate more than their adults; females of *Copepoda* migrate more than males. Vertical migration naturally occurs also in the absence of predation by fish. Comparative data on the vertical distribution of *Cladocera*, *Copepoda* and *Rotifera* during daytime from the Ma'aleh HaKishon Reservoir are presented in Figs 12.1–12.3. In the case of *Daphnia magna*, representing the *Cladocera* in the system (Fig. 12.1), maximal densities were

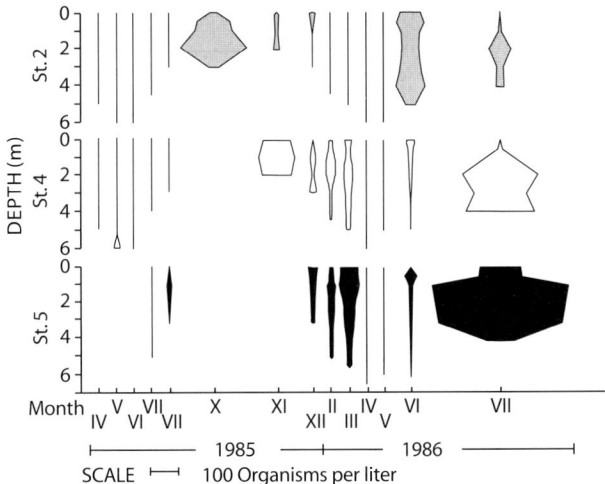

Fig. 12.2. Vertical distribution and density of Cyclopoida in the Ma'aleh HaKishon Reservoir (April 1985–July 1986)

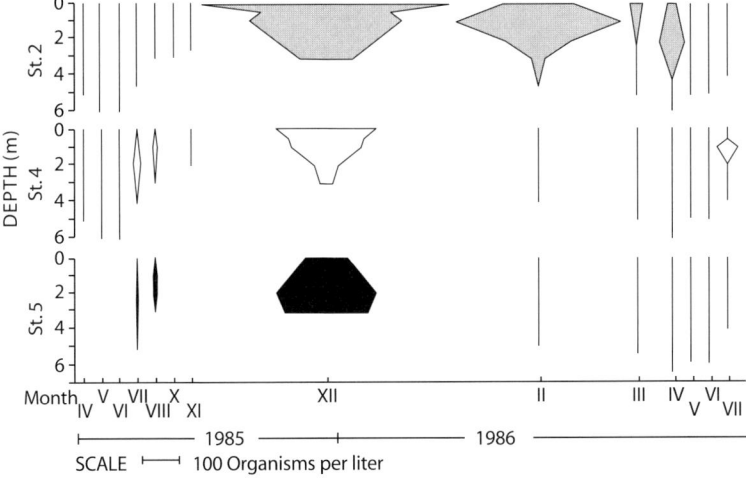

Fig. 12.3. Vertical distribution and density of Rotifera in the Ma'aleh HaKishon Reservoir (April 1985–July 1986)

found between 2–4 m, with decreasing values at shallower and deeper depths. As known for many other zooplankton species, *Daphnia* probably avoids the maximal illumination near the water surface. On cloudy days, the maximal concentration of *Daphnia* moved up to the upper 0.5 m of the water column. The maximal densities of the *Cyclopoida* (Fig. 12.2) and of the *Rotifera* (Fig. 12.3) were found in most of the cases in the upper two meters of the water column.

Seasonal dynamism is a function of natural environmental factors such as tempera-ture, as well as of the management of the reservoirs. The latter defines the water level and the water quality of the reservoir. Observations in the Ma'aleh HaKishon Reser-voir during 1985–1986 showed that *Daphnia magna* was present at low densities dur-ing the cold months with water temperatures between 12° C and 17° C. The density of this species increased only when the temperature rose to 18° C in the month of April. Similarly, the cyclopoid *Mesocyclops oqunnus*, an African species, was absent during the winter and early spring and appeared only at a temperature of 16° C and above. On the contrary, *Cyclops scutifer*, a European species, was present only at temperatures below 22° C and absent in the summer.

12.7
Food Chains

The herbivorous and the detritivorous food chains are much more interwoven than in unpolluted waters. In most of the hypertrophic reservoirs there is practically no benthic primary production. All the primary consumers in the hypertrophic reser-voirs are suspension feeders, feeding on organic particles, bacteria, and phytoplank-ton. The suspended organic particles of the sewage can be consumed either directly or after being settled by a bacterial coating. The primary consumers are mainly Pro-tozoa and Rotifera, as well as Cladocera and larvae of Culicidae. The predators (sec-ondary consumers) are the Ciliophoran *Didinium*, the Rotifera *Asplanchna*, all the Cyclopoida and most of the demersal Insects (Odonata, Hemiptera and Coleoptera) as well as the mosquito fish *Gambusia* where it has been introduced.

The detritivorous animals in these ecosystems feed on bacteria. Among them some Protozoa such as *Amoeba* and *Coleps*, the Nematoda *Panagrolaimus*, the Oligochaeta *Dero*, the different species of Ostracoda, and the dipteran larvae of the genera *Chironomus*, *Polypedilum* and *Eristalis*, all living mainly in the shallow benthos and on submerged hard substrates.

It is not possible to separate clearly between herbivory and detritivory in hyper-trophic ecosystems, and animal species behave to a large extent as opportunistic om-nivores. Since species diversity is often severely limited in this stressed environments, it is more adequate to speak of linear food chains rather than of complex food webs.

The three or four classic trophic levels are represented only in the longest food chains. The short food chains are an expression of the increasing severity of the environment fac-tors, as has been also reported in hypersaline aquatic ecosystems (Por 1981). A certain stability of the ecosystem is found in the extremes: reservoirs with good water qual-ity which present long and complex food chains, or in extremely loaded reservoirs where the biotic inventory is limited to photosynthetic bacteria and/or *Euglenidae*.

12.8
The Effect of Grazing

The influence of the benthic and demersal insects on the reservoirs is limited. On the contrary, the zooplankton exerts a considerable influence. Grazing activity of abun-dant populations of filter feeders such as the cladocerans *Daphnia* and *Moina*, and of *Brachionus*, *Filinia* and other herbivorous rotiferans, may considerably reduce the density of live and inert particles in the reservoir. The potential seston removal by the large cladoceran filtrators is higher than that of the smaller rotiferans. This is the result not only of the higher filtration rate of the cladocerans, but also of the wider size range of the particles they can strain and ingest. The dominance of the large filter feeders over the small ones corresponds to the "size efficiency hypothesis" (Brooks and Dodson 1965). The result of the strong grazing activity is an increase in water transparency and the consequent expansion of the euphotic zone. A deepening of the euphotic zone from 0.4 m to 1.0–1.4 m in the reservoir of Na'an resulted from an explosive development of rotiferans which in turn led to a steep decline of the algal biomass from 10^6 cells per ml in March 1980 to 10^4 cells per ml in April. At the same time, the bacterial cell count dropped respectively from 10^6 per ml to 10^5 per ml (Dor et al. 1987a). In the reservoir of Ma'aleh HaKishon, (Eren et al. 1987) the depth of the compensation level fluctuated between 1 m and 5 m, as a function of the grazing activity of *Daphnia magna*. When grazing by the large planktonic filtrators exceeds phytoplankton reproduction, the rate of photosynthetic oxygen production decreases, and concomitantly the respiratory depletion of the oxygen increases. In extreme cases, drastic reduction in dissolved oxygen concentration leads to the collapse of the filter feeders themselves (Eren et al. 1987). Intensive respiration by the zooplankton and by the heterotrophic bacteria combined with reduced CO_2 fixation by the phytoplankton may result in an increase in dissolved inorganic Carbon (HCO_3^-). In consequence the peak values of pH decrease during the daylight hours. The thermal profile is also influenced by deeper light penetration in the water column which results from increased seston consumption: light is being absorbed by a thicker layer of water, and as a consequence the surface waters are cooler. Finally, one more effect of the clearing of the reservoir waters by the filtrators is the growth of filamentous green algae on submerged hard substrates. Similar effects of overgrazing have been observed in hypersaline water bodies (Dimentman 1982). In both these types of extreme aquatic environments, overgrazing occurs owing to the absence of predation by planktivorous fish. While competition for food may occur only during short periods of limited food supply, a change in the composition of the zooplankton community, owing to abiotic seasonal factors, may have a noticeable influence on the phytoplankton. For instance, when a dense population of *Daphnia magna* was replaced by *Brachionus calyciflorus* in the reservoir of Ma'aleh HaKishon, the percentage of the large-sized green algae *Coelastrum* and *Lepocinclis* in the total algal population increased from 26% to 58%, whereas that of the smaller algae *Selenastrum* and *Chlorella*, dropped from 72% to 30% (Eren et al. 1987).

12.9
The Reservoirs as an Environmental Opportunity

The reservoirs are used by a variety of terrestrial and semi-aquatic species for several reasons: they provide humidity, drinking water, food and a roosting place. Spiders are found in dense populations along the shores of reservoirs where they can feed on winged Chironomids and other insects. The humidity below the boulders and pebbles on the embankments of the reservoirs attracts large number of the terrestrial oniscoid isopod *Porcellinoides pruinosus* who also finds abundant detritic food there. Seed eating birds in semi-arid and arid regions are attracted to these new aquatic environments. For instance, large numbers of the sand grouse *Pterocles* gather to drink around the reservoirs with gentle slopes in the Negev desert. The males of this species also gather water in the plumage and deliver it at considerable distances to their nestling (Bodenheimer 1935; Cramp 1985). Four species of sand grouses (*P. alchata*, *P. coronatus*, *P. orientalis* and *P. senegallus*) are frequently found around reservoirs and oxidation ponds (Ashkenazi pers. comm. and our own observations). A variety of birds feed on the biota of the reservoirs. The Anatidae ducks feed chiefly on phyto- and zooplankton, whereas the diving ducks *Aythya* feed on benthic food items. Wading birds, such as plovers (Charadriidae), avocets (Recurvirostridae), sandpipers (Scolopacidae) and others, feed on littoral benthos. Unlike most of the natural water bodies, the banks of the reservoirs are usually steep and devoid of vegetation. Therefore the attractiveness of this environment for waders is relatively limited. Adult Chironomids and other insects provide food for swallows (Hirudinidae) swifts (Apodidae) warblers (Sylviidae) and other insectivorous birds.

12.10
Unwanted Effects of the Fauna of the Hypertrophic Reservoirs

Clogging of filters and emitters of irrigation systems by algae and fauna is a serious technical problem with economical consequences (Katznelson and Neev 1987; Singer-Zacks 1987; Sagi 1989, 1990a, 1991; Feigin et al. 1991; Teltsch this volume). The intake must therefore be placed far from the downwind side of the reservoirs where a large biomass tends to accumulate. Another deleterious effect which needs appropriate investigation is the accumulation and long distance spread of chemical and viral pollutants by the water fowl visiting the hypertrophic reservoirs. Mosquito larvae are rarely found in the reservoirs, and since they invariably belong to *Culex pipiens* their presence results in mere molestation.

12.11
Fish Introduction for Manipulative Purposes

Several species of fish have been introduced in the reservoirs in Israel. Sagi (1990b) reported on the introduction of silver carp (*Hypophthalmichthys molitrix*) of big head carp (*Aristichthys nobilis*) and of the hybrids of these two species into reservoirs with various degrees of sewage and clear water mixture. By feeding on larger organisms of the plankton, such as *Microcystis, Oscillatoria, Daphnia* and various cyclopoid copepods, the introduced fish counteracted clogging of the filters and irrigation intakes. The efficiency of this size-selective filtration was estimated by Sagi (1990b) to be around 100 m^3 of water in 24 h, per kg of fish. While the density of the above-mentioned plankters decreased, that of the smaller algae increased. However, an overall decrease was clearly expressed in the total chlorophyll values. This practice is limited by the fact that it can be applied only to reservoirs with relatively good water quality. Also it is noteworthy that toward the end of the irrigation season, when the amount of dissolved oxygen decreases and concentration of ammonia increases, fish death occurs. The situation with another introduced fish, the mosquito fish (*Gambusia affinis*) is similar. This fish is an efficient predator of mosquito larvae but it is also sensitive to low oxygen and high ammonia values. Therefore, the introduction of the fish has to be repeated each year in spring.

12.12
Faunistic Succession and Typology of the Hypertrophic Reservoirs

From the more polluted to the less polluted reservoirs, several stages of hypertrophy can be distinguished. The most polluted ones do not have fauna at all. The pioneer faunistic communities are composed only of Sarcodina and Ciliophora (Protozoa). In the second stage Rotifera are dominant, accompanied by Ciliophora. The further stages differ among themselves according to the existence or not of stratification within the water column. The stratified reservoirs are inhabited only by zooplankton, composed of a variety of Rotifera and several species of Cladocera and Copepoda (Crustacea). The non-stratified reservoirs are inhabited also by zoobenthos: Nematoda, Oligochaeta, Chironomida and air breathing insects such as beetles (Coleoptera) water bugs (Hemiptera) and *Eristalis* larvae these latter restricted to the very shallow littoral. In the low loaded reservoirs, Ostracoda, water mites (Hydracari) and more insect species are ecountered. At this stage, introduction of various fishes becomes possible. Much has been written about the typology of the polluted aquatic ecosystems according to different criteria such as Dissolved Oxygen, Biological Oxygen Demand (*BOD*) and biological indicator species (see Caspers 1966; Sladecek 1973; Helawell 1989). Since there are different types of polluted waters, ranging from slightly polluted streams to hypertrophic fish ponds, followed by wastewater storage reservoirs and to chemically polluted reservoirs, there are many saprobity systems mentioned in literature (see Sladecek 1973). There is also a wide variety of regional and climatic differences which render a general classification very difficult, if not impossible (Bromley and Por 1975).

The only attempt to classify the hypertrophic reservoirs of Israel has been made by Dor (1987). This author used a combination of physico-chemical and biological criteria and came up with two types of categories: one based on the presence/absence of stratification and the second based on the bacterial and algal floras. Three floral categories were described by Dor (1987): reservoirs dominated by photosynthetic red bacteria, by *Euglena*, and by *Chlorella*. This floristic characterization of the different types of hypertrophic reservoirs in Israel could be implemented and enlarged by the superposition of the above mentioned faunistic indicators.

12.13
Conclusions

Wastewater reservoirs are another type of water body with extreme environmental parameters, and therefore comparable with naturally stressed water bodies such as the continental hypersaline waters. Moreover, they are another category of the complex hypertrophic waters to which belong the fish ponds, sewage loaded streams, lakes in the amusement parks, and other excessively nutrient rich waters like those of the Hula Nature Reserve. The hypertrophic wastewater reservoirs should not be written off by ecologists and protectors of nature. They can be considered as one of the modalities by which used waters can be returned to the benefit of aquatic biota. As mentioned above, hypertrophic reservoirs are important feeding and roosting places for migrating birds and a source of water, humidity, and food for other animals, especially in arid areas. This role is as yet far from being fully exploited. The ecological and even public advantages of the reservoirs can be maximized in the future through adequate planning and management.

Acknowledgements

The first author acknowledges the assistance in the field work given by M. Barry, I. Dor, G. Eitan, J. Eren, R. Katznelson and G. Sagi. Identifications of various taxa have been thankfully supplied by J. Andrassy (Budapest), S. Ashkenazi (Israel), C. Bader (Basel), H. J. Bromley (Israel), F. Hebauer (Deggendorf), G. Levi (Israel), C. Manicastri (Rome), K. Martens (Bruxelles), H. Pener (Israel) and W. Schneider (Mainz). The authors wish also to thank Drs I. Dor and M. Juanicó – editors of this volume – for the incentive and opportunity they gave us in publishing this article.

References

Bodenheimer F (1935) Animal life in Palestine. L. Mayer, Jerusalem, 507 pp
Bromley HJ (1981) Seasonal succession of zooplankton in a sewage reservoir. In: Shuval H (ed) Developments in arid zone ecology and environmental quality. Balaban ISS, Philadelphia
Bromley HJ (1982) Seasonal succession of zooplankton in a sewage reservoir. In: Dor I (ed) Na'an Reservoir study, Part 2. Limnological and sanitary characteristics. Report submitted to the Israel Water Commissioner, pp 61–67, 72
Bromley HJ, Por FD (1975) The metazoan fauna of a sewage-carrying wadi, Nahal Soreq (Judean Hills, Israel). Freshwater Biology 5:121–133
Brooks J, Dodson I (1965) Predation, body size, and composition of plankton. Science 150:28–30
Caspers H (1966) Stoffwechseldynamische Gesichtspunkte zur Definition der Saprobitatsstufen. Verhandlungen internationale Vereinigung für theoretische und angewandte Limnologie 16:801–808
Cramp S (ed) (1985) The birds of the western palearctic, vol IV, Oxford University Press, Oxford, 960 pp
Dimentman Ch (1982) Limnology and ecology of the Navit pools. Israel Journal of Zoology 31:68
Dimentman Ch, Eitan G (1988) Development of small animals in wastewater reservoirs, as a factor which causes clogging problems in irrigation systems. Report 1, 3 pp; Report 2, 14 pp Submitted to the Israel Water Commissioner (in Hebrew)
Dor I (1982) Na'an Reservoir study. Part 2 Limnological and sanitary conditions. Report submitted to the Israel Water Commissioner, 84 + V pp (in Hebrew)
Dor I (1987) Comparative study of wastewater reservoirs. Report submitted to the Israel Water Commissioner, 49 pp (in Hebrew)
Dor I, Schechter H, Bromley HJ (1987a) Limnology of hypertrophic reservoir storing wastewater effluent for agriculture at Kibbutz Na'an, Israel. Hydrobiologia 150:225–241
Dor I, Kalinsky I, Eren J, Dimentman Ch (1987b) Deep wastewater reservoirs in Israel. I. Limnological changes following selfpurification. Wat SciTech 9:317–322
Eren J (1976) Storage of the Haifa wastewater in Kfar Hasidim Reservoir. Report for the period March–September 1976. Mkorot Water Co, 19 pp (in Hebrew)
Eren J (1978) Succession of phyto and zooplankton in a wastewater storage reservoir. Verh Intern Verein Limnol 20:1926–1929
Eren J, Dor I, Dimentman Ch (1986) Investigation of Ma'aleh HaKishon – a deep wastewater reservoir. Report submitted to BMFT-Germany and NCRD-Israel, 25 pp
Eren J, Dor I, Dimentman Ch (1987) Investigation of Ma'aleh HaKishon – a deep wastewater reservoir. Final Report submitted to BMFT-Germany, and NCRD-Israel, 62 pp
Feigin A, Ravina I, Shalhevet J (1991) Irrigation with treated sewage effluents. Springer-Verlag, Berlin, 224 pp
Freund C, Dimentman Ch, Post AF (1992) Survey of the planktonic communities in the Gush Dan oxidation and maturation ponds, during 1991. Report submitted to "Mekorot" Water Company, 38 pp
Gasith A, Sidis I (1983) The distribution and nature of the habitat of the Caspian terrapin *Mauremys caspica rivulata* (Testudinea, Emydidae) in Israel. Israel Jour Zool 32.91–102
Hellawell JM (1989) Biological indicators of freshwater pollution and environmental management. Elsevier Applied Science, London, 546 pp
Katznelson R, Dimentman C (1991) A study of the maturation and recharge basins of the Gush Dan Sewage Plan. Report submitted to "Mekorot" Water Company, 59 pp (in Hebrew)
Katznelson R, Neev A (1987) A sieve series for analyzing particle size distribution Wat Irrig 247:33–37 (in Hebrew)
Paz U, Zuaretz S (1976) Annual Report of bird watching in Israel, for the year 1973. Israel Nature Reserves Authority, 132 pp (in Hebrew)

Por FD (1981) A classification of hypersaline waters based on trophic criteria. PSZN Journal of Marine Ecology (Naples) 1:121–131

Por FD (1982) Zooplankton studies in the Na'an sewage reservoir. In: Dor I (ed) Na'an Reservoir Study. Part 2. Limnological and sanitary characteristics. Report submitted to the Israel Water Commissioner, pp 58–60

Sagi G (1989) Water quality in the year 1988. Wat Irrig 269:37–40 (in Hebrew)

Sagi G (1990a) Water quality and clogging of filtration and irrigation systems. Wat Irrig 280:51–55 (in Hebrew)

Sagi G (1990b) Fish for improvement of the water quality in irrigation reservoirs. Wat Irrig 286:43–47 (in Hebrew)

Sagi G (1991) Water quality for irrigation in the past season and recommendations for the forthcoming season. Wat Irrig 295:17–20 (in Hebrew)

Shelef G, Juanicó M (1989) Criteria for the design and operation of stabilization reservoirs. Final Report submitted to BMFT-Germany and NCRD-Israel, vol 1, 156 pp, vol 2 (Data Base), 182 pp

Shirihai H (1996) The birds of Israel. Academic Press, London, 692 pp

Sidis I (1983) Ecology of the Caspian terrapin (*Mauremys caspica rivulata*) in polluted and unpolluted habitats in Israel. MSc thesis Tel Aviv University, Tel Aviv, 108 pp (in Hebrew)

Singer-Zacks M (1987) Water quality and emitter clogging relationship in wastewater irrigation. MSc thesis The Hebrew University of Jerusalem, 152 pp (in Hebrew)

Sladecek V (1973) System of water quality from the biological point of view. Archiv für Hydrobiologie. Beiheft Ergebnisse der Limnologie 7:1–218

Teltsch B (in this volume) The screen filter clogging capacity. In: Dor I, Juanicó M (eds) Hypertrophic reservoirs for wastewater storage and reuse. Ecology, Performance and Engineering Design. Springer-Verlag, Berlin

Yom-Tov Y, Mendelssohn H (1988) Changes in the distribution and abundance of vertebrates in Israel during the 20th century. In: Yom-Tov Y, Tchernov E (eds) The zoogeography of Israel. The distribution and abundance at a zoogeographical crossroad. Monographiae Biologicae 62. Dr. W. Junk Publications, Dordrecht, pp 515–547

Zuaretz S (1970–1976) Tazpit (Summaries of observations) (in Hebrew)

Zuaretz S (1977) A ten-years survey of water bird breeding in Israel (1967–1976). In: Paz U (ed) Nature preservati on in Israel, researches and surveys. Report 2. Nature Reserves Authority (in Hebrew), pp 63–126

Chapter 13

Odorous Compound

Boris Ginzburg · Jenia Gun · Inka Dor · Ovadia Lev

13.1
Introduction

Deep wastewater reservoirs are simple-to-operate, small-scale treatment systems, which benefit from the long retention time of the wastewater in the ponds. The fact that deep wastewater reservoirs can be used for the treatment of sewage led to a wide distribution of these facilities throughout Israel.

The increased popularity of deep wastewater reservoirs, their wide distribution in Israel, and the fact that they are area-extensive, exposed them to the public eye. The fast-developing urbanization process in Israel drives residences to the close vicinity of existing wastewater treatment systems. Thus, the aesthetic characteristics, i.e., the properties that laymen can sense directly, without mediating instrumentation, receive more attention. The intuitive association of malodours with health risks and the sensitivity of the olfactory system to traces of odorous compounds boost the concern over odour nuisance.

Despite the importance of odour attributes there has been no prior research on odour emissions from deep water reservoirs. This is probably due to the fact that the necessary tools for identification of trace odorous compounds became available only recently and because the methodology for the examination of odours in wastewater is still in its infancy. Our knowledge on the chemical nature of odorous emissions from water has dramatically improved during the last few years with the proliferation of hyphenated GC apparatus, the rapid progress in advanced preconcentration techniques (Grob and Zurcher 1976) and the elaboration of odour characterization by descriptive sensory techniques (Khiari et al. 1992; An-Kuo Meng et al. 1992). Still, most attention has been directed to the investigation of odorous compounds in fresh and drinking water and to the elucidation of their biological sources. Relatively little scientific effort was devoted to the characterization of odorous materials in water treatment effluents. From the 39 papers presented in the 1994 proceeding of "Off-flavours in the aquatic environment" (Hrudey et al. 1995), only one relates to odorous compounds in water effluents (Brownlee et al. 1994) – all others deal with odorous compounds and their sources in fresh water.

The account of specific organic compounds in wastewater effluents is also rather dull. Notable exceptions are the detailed works of Schroder (1991, 1993a,b), Paxeus (1996), Paxeus et al. (1992), and Paxeus and Schroder (1996) which concentrate on polar, odourless compounds in wastewater.

There are several factors that complicate the research of odorous compounds in wastewater. a) The human olfactory sense is highly sensitive. Human beings can sense some compounds down to 0.0001 ppb level, i.e., 0.1 ng l^{-1} (e.g., geosmine, methanethiol), which is much below the minimum detection limit of the most sensitive chromatographic detectors. A number of efficient preconcentration methods that were developed during the last several years, and in particular the CLSA (Close Loop Stripping Apparatus) (Grob and Zurcher 1976) help overcome this obstacle. b) The enormous number of odorous compounds complicates the identification process. Intensive chromatographic separation is required before identification process can be attempted. c) Unlike visual (lumens), audio (decibels) and even the taste senses (which is comprised of only four fundamental ingredients, bitter, salty, sweet and sour) the olfactory sense is less amenable to quantification since it is comprised of a non-linear combination of a large number of yet undefined basic ingredients. The science of defining these different odours is still in its infancy (Bartels et al. 1986). Recently, a taste and odour classification wheel for drinking water has been developed to standardize the nomenclature in the olfactory sensing of drinking water (Khiari et al. 1997). Odours in fresh and drinking water are classified into 8 groups:

1. earthy/musty/moldy
2. grassy/hay/straw/wood
3. marshy/swampy/septic/sulfurous
4. fragrant: vegetable/fruity/flowery
5. fishy
6. medicinal/phenolic
7. chemical/hydrocarbon/miscellaneous
8. chlorinous, ozonous

We have used this classification wheel to categorize the odorous compounds in the deep wastewater reservoirs. It was clear from the very first tests that this wheel is not optimized for wastewater effluents but, due to the lack of an alternative objective method of classification, this method was still used for the current study. It is clear that a new odour classification method will have to be developed for further wastewater research.

13.2
Materials and Methods

13.2.1
Sampling at the Na'an Reservoir

The Na'an Reservoir was selected as a test case. The Na'an Reservoir served as a model reservoir for a number of previous studies on domestic wastewater reservoirs (Dor et al. 1987a,b; Dor and Raber 1990) and it is a subject of several articles in this book. The Na'an Reservoir is a 12 m deep, 0.7 million m³ capacity reservoir, receiving effluents from two oxidation ponds working in parallel with an accumulated hydraulic retention time of 14 days. The treatment system receives the domestic sewage of the Ramla city and supplies water for restricted irrigation in nearby sunflower and cotton fields.

Three field sampling trips were conducted. The first was conducted in June 1996, just before the irrigation period. The second took place in October 1996 at the end of the summer irrigation period, before the winter rains. The third sampling was conducted in March 1997, immediately after the rainy season. Water samples were taken from ca. 20 cm below water level, stored in dark 1.5 litre bottles equipped with Teflon stoppers and kept in a refrigerator for analysis. Addition of 0.1 g l^{-1} $HgCl_2$ was used in order to prevent biodegradation of the wastewater stored in the sample bottles. Chromatographic analysis took place within 24 hours of sampling.

13.2.2
Materials

The following compounds were used for comparison with those found in wastewater: dimethyldisulfide, indole, skatole, octanal, nonanal, decanal and limonene were obtained from Aldrich. Methylisoborneol and geosmine were purchased from Salford Chemicals. Dimethyltrisufide was synthesized according to a recently published procedure (Korchevin and Turchaninova 1989; Ginzburg et al. 1997). Dimethyltetrasulfide was prepared by a similar procedure and could be used for qualitative confirmation of its presence in water but it was not sufficiently pure for quantitation. All solvents used were of chromatographic grade or equivalent.

13.2.3
Analytical Methods

Semi-volatile compounds were concentrated by a lab-built closed loop stripping apparatus (CLSA) prior to GC analysis. The stripping apparatus used in our laboratory consisted of a 1 litre thermostated (60° C) stripping bottle, a Teflon lined diaphragm pump, (KNF, Neuberger) an active charcoal trap (5 mg charcoal traps, ASSY Tekmar) and an external sleeve heater to prevent water condensation in the charcoal trap. Auxiliary tubes and fittings were made of Teflon or stainless steel. These materials can tolerate the heat treatment at 200° C for 2 hours that is required for cleanup of the apparatus between experiments. Air stream was circulated through the stripping bottle, heating sleeve, charcoal trap and the diaphragm pump for 2 hours. The trap was subsequently extracted by several microlitres of methylene chloride and injected into the GC. The closed loop apparatus exhibited 92% and 84% recovery of dimethyldisulfide and dimethyltrisulfide, and only 35% and 40% recovery of indole and skatole respectively.

The CLSA preconcentration was inefficient for the determination of the more polar compounds such as amines and alcohols because of their low vapour pressure, and therefore conventional extraction technique was used. Three aliquots each of 20 ml methyltretabutylether (MTBE) solvent and sodium chloride salting-out agent was used for the extraction of the organic compounds from the 1 000 ml sample. The organic solvent was then slowly concentrated to 0.1 ml final volume by gentle evaporation in a nitrogen stream. This procedure was used for the analysis of the April and October samples.

An HP5890 gas chromatograph equipped with a temperature programmable on-column injector and an EI (electron ionization) mass selective detector (HP5971 MSD) and NBS49K mass spectral library was used for quantification and identification of the odorous compounds.

Indole, skatole, MIB and geosmine were detected by GC/MS operated in the more sensitive Selected Ion Monitoring (SIM) method.

Characterization of the odorous fraction was carried out by GC with olfactory sensing. The olfactory characterization was performed on HP5890 GC equipped with a FID (flame ionization detector) by sniffing the outlet port of the FID with the flame extinguished. The experienced sniffer could then characterize both the intensity and the type of the odorous peaks, producing a so-called "aromagram." A subjective five level intensity scale was used in our studies.

A 30 m 0.32 mm ID RTX-1 capillary column with 1 mm film thickness was used in all experiments reported in this chapter. The following temperature programme was used: 40° C initial temp. at 1 min, first ramp at 5° C min^{-1} up to 185° C, second ramp 30° C min^{-1} up to 250° C and 5 min at final temp. 250° C. The on-column injector was set to 3 ° C above oven temperature.

BOD_5 measurements and bacterial standard Plate Count were performed using the conventional, standard methods (Eaton et al. 1995). Turbidity measurements were conducted using HACH Ratio/XR turbidimeter.

Evaluation of the Threshold Odour Number, TON, the dilution factor (with distilled water) required to eliminate the odour from the water samples, was conducted by a 7 member trained student panel according to a reported standard method (Eaton et al. 1995)

13.3
Results and Discussion

Table 13.1 summarizes the BOD_5, NTU and TON in different locations along the treatment system. The inlet pipe, the two parallel oxidation ponds and the wastewater reservoir were sampled and analysed. The TON of the influent (inlet pipe) was not re-

Table 13.1. Threshold odour number values, BOD_5 and turbidity at the Na'an system in summer

	Raw sewage	Oxidation pond 1	Oxidation pond 2	Main reservoir
TON value	–	140	160	64
BOD_5 (mg l^{-1})	260	180	200	40
Turbidity (NTU)	160	185	170	71

Table 13.2. Threshold odour number values, BOD_5 and turbidity at Na'an Reservoir by season

	June 1996	October 1996	March 1997
TON value	64	75	50
BOD_5, mg l^{-1}	40	60	40
Turbidity (NTU)	71	75	68

corded for safety reasons. The reduction in BOD_5 values along the treatment system is reflected in a similar reduction in the *TON* level following the order: raw sewage > oxidation ponds > reservoir. The turbidity was found to be, as expected, a less significant parameter and was affected by the concentration of algae at the location of sampling rather than by the suspended solids of wastewater origin. This explains the larger turbidity in the reservoir as compared to water samples taken from the inlet pipe.

A comparison of relevant water quality parameters and the corresponding *TON* are depicted in Table 13.2. June values represent the water quality of an almost full reservoir with the longest possible mean retention time. Best water quality was therefore expected during this period. Indeed, the BOD_5 values were relatively low (40 mg l^{-1}). October values represent water effluents that were collected after the summer irrigation period, so the reservoir contained wastewater that was fed during the summer period and hence fresh wastewater with minimal average retention time should be present. Indeed, the BOD_5 level was the maximal and so were the *TON* and turbidity levels (see Table 13.2). The third sample was taken at the end of March, immediately after the last winter rain. The reservoir was almost full, but it was functioning as a water catchment and the sewage was diluted by rain and flood-water. Thus, the water quality, manifested in the degradable organic matter (*BOD*) and malodours, was better as compared to other seasons. In fact, the *TON* level was found to be a rather misleading parameter. The reservoir water in June and October had strong, disgusting, septic odour, while the odour intensity in March was mild and less repelling. However, upon dilution (for determination of *TON*) the odour threshold was found to be almost identical to that of the summer values. The odour assessing panel decided to classify the odour in all cases as "septic," because our dull vocabulary (of odour attributes of wastewater) did not contain a better classification. Thus, the *TON*, which is a very useful measure for the characterization of water with only one dominant odour attribute is misleading when used for the classification of multi-odour wastewater. This inadequacy reflects the inherent nonlinearity of the olfactory system. It seems that the odorous compound which has the lowest ratio of odour threshold to concentration will determine the *TON* level, while the most repelling odour compounds which become effective only at high concentration determine the nuisance intensity. The synergistic effect of odorous compounds is also responsible for the nonlinearity and inherent limitations of the *TON* measure.

Figures 13.1 and 13.2 present the total ion chromatograms and aromagrams of three samples collected during different seasons in the Na'an Reservoir. The relative abundance was normalized by the response of 100 ng l^{-1} dodecanone-2 internal standard. The different patterns of the peaks in the chromatogram were selected to reflect the eight primary classes of odours that were presented in the introduction. A secondary classification is indicated above the aromogram peaks. The increase in *BOD* during the summer period is reflected in the much larger number and intensity (relative abundance) of the aromagram (Fig. 13.2) and total ion chromatograph peaks (Fig. 13.2) in the summer as compared to the end of winter. The resemblance between the two summer samples and the difference between the summer and winter samples can be observed, both in the aromagrams as well as in the total ion chromatograms.

Table 13.3 presents a list of compounds that were tentatively identified in the three sampling trips. In most cases the identification is based on library matching. Less than

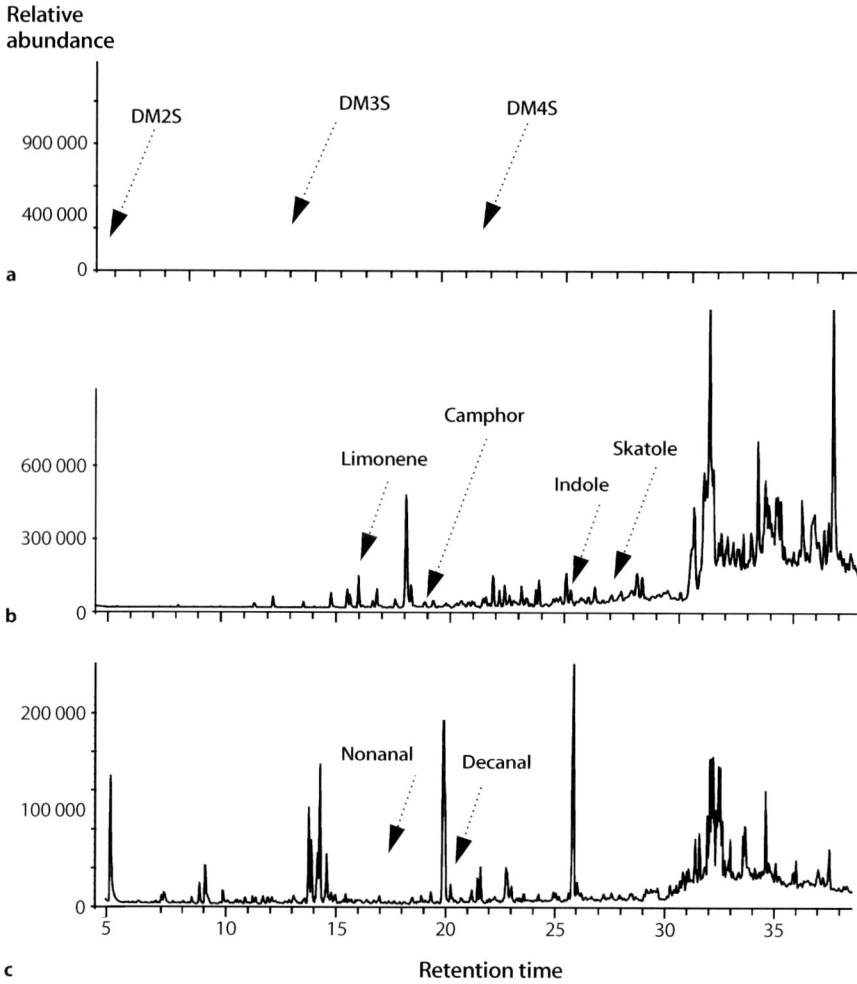

Fig. 13.1. GC/MSD total ion chromatograms of volatile organic compounds in the Na'an Reservoir. Samples collected in **a** June 1996; **b** October 1996; and **c** April 1997

50% of the large peaks of volatile compounds were identified and only 30% of the odorous peaks appearing in the aromagram were identified.

The odorous compounds that could be identified can be classified into the following four groups:

i. *Decomposition products of microorganisms.* Important groups are sulfur containing compounds such as dialkyloligosulfides (dimethyldisulfide, dimethyltrisulfide, dimethyltetrasulfide) and nitrogen containing substances such as indole, skatole, and benzothiazole. Other aromatic and aldehyde compounds (such as decanal, nonanal) are known to exert repelling obnoxious odours as well but their level was not high

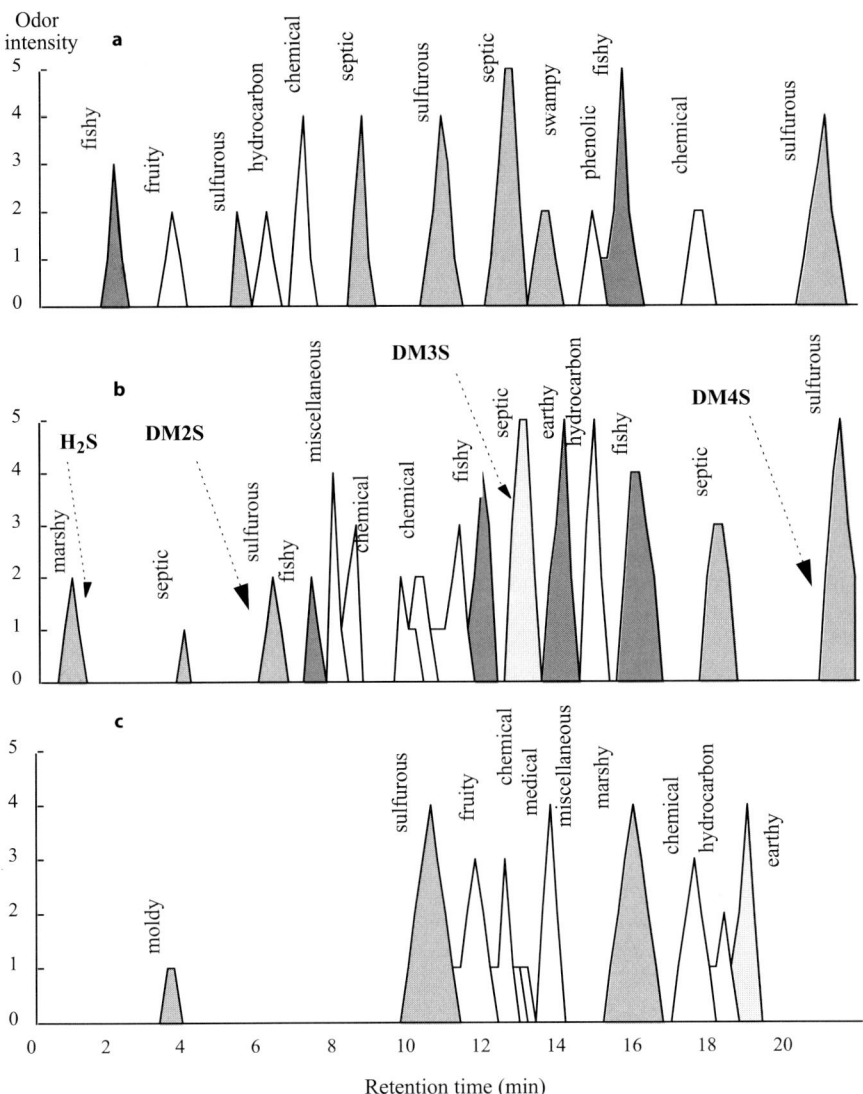

Fig. 13.2. Aromagram of volatile organic compounds in the Na'an Reservoir. Samples collected in **a** June 1996; **b** October 1996; and **c** April 1997

enough in this study, and peaks in the aromagrams could not be associated with aldehydes.

ii. *Decomposition products of organic compounds.* These were originally present in the sewage or odorous compounds and were not biodegraded by the biological treatment. Substituted phenols and benzenes such as biphenyl and hydroquinone and the carboxylic acid compounds are examples of this category.

Table 13.3. Volatile compounds at the Na'an Reservoir classified according to retention time

Retention time	Compound	Relative abundance in relation to the peak of internal standard decanon-2 at 100 mg l^{-1}			Group (classified a–d as in the text)
		1996 June	1996 October	1997 March	
1.60	Hydrogen sulfide	–	–		d
4.69	Methane, oxybis(chloro)			600	c
6.00	Pyridine			20	c
6.10	Disulfide, dimethyl	150	100		a
7.08	Toluene	1 220	550	210	c
7.18	Ethene, tetrachloro-	160	50	120	c
9.10	Benzene, 1,4-dimethyl-	22	10	250	c
9.90	Benzene, 1,3-dimethyl-	10		80	c
10.32	Benzene, 1,2-dimethyl-	550	250	120	c
12.11	Formamide, N,N-diethyl-			45	c
12.14	Ethanol, 2-butoxy-	20	75		c
12.81	2-Hexanol, 2-methyl-	290	100		c
13.46	Trisulfide, dimethyl	600	310		a
13.61	Octanal			40	c
13.75	Benzene, 1,2,3-trimethyl-	240	85	50	c
13.80	2-Propanol, 1-(2-methoxy-1-methylethoxy)	130	250	375	c
14.68	Phenol		120		c
14.82	3-Octanol, 3-methyl-	3 700	120		c
15.05	Benzene, 1,3-dichloro-	260			c
15.84	Nonanal	340	245	55	c
16.06	Limonene	690	200		b
16.85	Benzene, methyl(1-methylethyl)-	165	90		c
17.75	Benzene, 1,2,3,5-tetramethyl-	250	45		c
17.90	Phenol, 4-methyl-		1 180		c
18.11	1,6-Octadien-3-ol, 3,7-dimethyl-	590	170		b
18.50	Phenol, 2,4-dimethyl-		110	50	c
19.07	Camphor	170	40		b
19.63	Cyclohexanone, 2-methyl-5-(1-methylethyl)	660	100		b
20.25	Decanal	25	35	110	c
21.26	Benzothiazole	10	25	60	c
21.30	Tetrasulfide, dimethyl	10	10		a
21.38	1,2-Benzenediol		155		c
23.61	Formamide, N,N-dibutyl-			45	c
24.27	Benzenamine, 2,3-dichloro-		140	40	c
24.81	Tridecane	390	90	120	c
25.08	Decanoic acid			35	c
25.26	1(2H)-Naphthalenone, 3,4-dihydro-	175	110	45	b
26.74	1,1'-Biphenyl	450			c
27.26	Phenol, 2,6-bis(1,1-dimethylethyl)-		20	60	c
27.68	Tetradecane	160	320	140	c
28.60	Geranylacetone	240	145	50	b
29.75	Phenol, 2-(1,1-dimethylethyl)-	340		100	c
30.27	Dodecanoic acid			110	c
30.96	Hexadecane	410	710	570	c
31.16	Phenol, 2,4-bis(1-methylethyl)-	200	320	1 810	c
31.40	Diethyl phthalate	270	430	402	c
32.79	Heptadecane	940	450	720	c
32.92	Phenol, 4-nonyl-	540	210	360	c
33.00	Tetradecanoic acid	160	310	750	c
38.28	Dioctyl phthalate	380	6 870	420	c

iii. *Excretory products of living microorganisms.* Terpenes, terpenoides and other metabolites of algae, cyanobacteria, fungi, bacteria and actinomycetes such as MIB, geosmine, limonene, and camphor.

iv. *Hydrogen sulfide.* This compound belongs to classes i-iii above but since it is widespread and important, it deserves a class of its own. We assume that the small amounts of hydrogen sulfide that appeared in the summer sampling (Fig. 2b) is produced in the sampling beaker rather than in the reservoir since the upper layer of the reservoir was found to be oxygen rich (Dor et al. 1987a – reproduced in this volume).

13.4
Conclusions

Deep wastewater reservoirs treating domestic sewage are a potential source of odour nuisance. The odour intensity and the number and abundance of the odorous compounds, as reflected in the aromagrams and GC/MSD total ion chromatograms, are much larger in summertime than in the rainy season.

Only a fraction of the volatile odorous compounds were identified in this study, of which the sulfur (such as hydrogen sulfide, dimethylsulfides) and nitrogen containing species (such as indole and skatole) were positively identified as sources of the odour nuisances.

Acknowledgements

We thankfully acknowledge the help of the Ministry of Science in sponsoring this research. We thankfully acknowledge the assistance of Ms. T. Hariff and Dr. V. Glezer.

References

An-Kuo Meng, Brenner L, Suffet HI (1992) Correlation of chemical and sensory data by principal component factor analysis. Wat Sci Tech 25/2:49–56

Bartels JHM, Burlingame GA, Suffet IH (1986) Flavor profile analysis: Taste and odour control of the future. Journal AWWA 5:51–58

Brownlee BG, Kenefick SL, MacInnis GA, Hrudey SE (1994) Characterization of odorous compounds from bleached Kraft pulp mill effluent. Wat Sci Tech 31(11):35–42

Dor I, Raber M (1990) Deep wastewater reservoirs in Israel: Empirical data for monitoring and control. Wat Res 24(9):1077–1084

Dor I, Schechter H, Bromley H (1987a) Limnology of a hypertrophic reservoir storing wastewater effluent for agriculture at Kibbutz Na'an. Hydrobiologia 150:225–241

Dor I, Kalinsky J, Eren Y, Dimentman C (1987b) Deep wastewater reservoirs in Israel: Limnological changes following self-purification. Wat Sci Tech 19(12):317–322

Eaton AD, Clesceri LS, Greenberg AE (eds) (1995) Standard methods for examination of water and wastewater. APHA, AWWA and WEF, Washington, DC

Grob K, Zurcher F (1976) Stripping of trace organic substances from water. Equipment and procedure. Journal of Chromatography 117:285–294

Ginzburg B, Chalifa I, Zohari T, Hadas O, Dor I, Lev O (1997) Identification of oligosulfide odorous compounds and their source in the lake of Galilee. Submitted, Wat Res

Hrudey SE, McGuire MJ, Whitefield FB, (eds) (1995) Off-flavours in the aquatic environment 1994, Wat Sci Tech 31/11:R9–R10 (the whole volume)

Khiari D, Brenner L, Burlingame GA, Suffet IH (1992) Sensory gas chromatography for evaluation of taste and odor events in drinking water. Wat Sci Tech 25/2:97–104

Khiari D, Barrett SE, Suffet IH (1997) Sensory GC analysis of decaying vegetation and septic odors. J AWWA, April

Korchevin NA, Turchaninova LP (1989) New method for synthesis of diorganyl polisulfides. Zh Obshch
    Khim 59:1785–1790
Paxeus N (1996) Vehicle washing as a source of organic pollutants in municipal wastewater. Wat Sci Tech
    33/6:1–8
Paxeus N, Schroder HF (1996) Screening for non-regulated organic compounds in municipal wastewater
    in Goteborg, Sweden. Wat Sci Tech 33/6:9–15
Paxeus N, Robinson P, Balmer P (1992) Study of organic pollutants in municipal wastewater in Goteborg,
    Sweden. Wat Sci Tech 25/11:249–258
Schroder HF (1991) Polar hydrophilic compounds in drinking water produced from surface water. J of
    Chromatography 554:251–266
Schroder HF (1993a) Pollutants in drinking water and wastewater. J of Chromatography 643:145–161
Schroder HF (1993b) Surfactants: non-biodegradable, significant pollutants in sewage treatment plant
    effluents. J of Chromatography 647:219–234

Chapter 14

Degradation of Organosynthetic Pollutants

Lea Muszkat

14.1
Introduction

The pollution of water resources by pesticides, detergents, solvents, and a variety of industrial organics is a pressing worldwide problem. It is especially acute in regions of intensive industrial and agricultural activity. There, a severe burden on the environment results from pesticide production and application, chemical industry, and from chemical waste. Many of the organic chemicals of anthropogenic origin are persistent, tend to accumulate in living organisms, and are capable of penetration into groundwater (Cohen et al. 1986; Pionke and Glotfelty 1989; Muszkat et al. 1993a,b). Some of them exhibit long-term toxicity. Therefore, the removal of organic pollutants from water and from effluents prior to their disposal or reuse is of special significance. However, wastewater treatment by intensive systems like activated sludge fails in remediation of the organic pollutants (Juanicó 1993).

In Israel the storage of wastewater in deep seasonal reservoirs has been a common practice since the early seventies. The reservoirs receive wastewater and runoff water during the rainy winter for irrigation during the dry summer. The storage of wastewater in seasonal reservoirs proved to have several advantages: it prevents the free flow of effluents and the contamination of waterways, increases the availability of water for irrigation and, finally, the reservoirs proved to have a significant self purification capacity. Low requirements for maintenance and operation are by themselves a significant advantage. The seasonal storage reservoirs have been the subject of many recent studies which focused on their performance and treatment capacity (Juanicó and Shelef 1994; Juanicó et al. 1995; Aharoni and Kanarek 1994), limnological aspects (Eren and Dor 1985, 1986; Dor et al. 1987), and on microbial and physico-chemical variables (Abeliovich 1985; Abeliovich and Vonshak 1993; Fattal et al. 1993) and others. The fate of organic pollutants in wastewater storage reservoirs has not been reported till now. Nor has it been correlated in the field with possible implications of irrigation with these effluents, and with its potential impact on soil pollution. In the reported studies, organic pollutants were mentioned only with regard to generic parameters like *COD, BOD* and *DOC*, which are of common use for evaluation of organic load. No speciation is available and no information on the toxicity and the environmental behaviour of the organic chemicals is provided by these data. The need for detailed characterization of influents and especially of effluents with relation to organic contaminants imposes the use of newly established analytical methods capable of separation, like gas chromatography (GC) or high performance liquid chromatography (HPLC) in combination with mass spectrometry (MS), or with tandem mass spectrometry (MS/MS), using soft ionization techniques. These techniques are today growing in use for the analysis of water and wastewater.

Our studies in the Na'an Reservoir were performed after our work on agricultural soils with a history of long-term irrigation with effluents from this reservoir. In these soils, no accumulation of micropollutants was apparent, while a significant accumulation of micropollutants was found in other agricultural soils irrigated with effluents from an intensive wastewater treatment plant (Muszkat et al. 1993b). These results stressed the need for a deeper insight into the behaviour of micropollutants in storage reservoirs.

The present work was aimed at examining the fate of organic pollutants during their storage in the reservoirs and furthermore, at assessing the impact of possible soil pollution in agricultural areas due to irrigation by effluents from the corresponding reservoirs. Using electron impact, EI-GC/MS measurements, the study included the wastewater treatment systems of Givat Brenner and Na'an, as well as the reservoirs of the Judean Hills (Revadim a and b, Nachshon, Harel, Anot and Zova).

14.2
Results and Discussion

14.2.1
Organic Pollutants Identified

The Tzora operational reservoir serves as a short-term hydraulic regulator of the wastewater collection system rather than for wastewater treatment. The organic compounds identified in the Tzora operational reservoir include chemicals from domestic and industrial origin. A partial list of the chemicals detected at the inlet and the outlet of the reservoir is presented in Table 14.1.

The following groups of pollutants were detected at the examined reservoirs: detergents, fatty acids and esters, and industrial chemicals.

Table 14.1. Removal of organic chemicals in the Tzora Operational Reservoir after a residence time of about 48 h (partial list)

Compound	Molecular weight	Inlet ($\mu g\,l^{-1}$)	Outlet ($\mu g\,l^{-1}$)
Dimethyldisulfide	94	4	nd
2-propanol, 1-(2-methoxy-1-methylethoxy	148	22	8.3
2-propanol, 1-(2-methoxypropoxy)	148	21	11.5
Tetrahydrothiophen	120	3.5	1.4
1-(1-methyl-2-(2-propenyloxy)ethoxy) 2-propanol	174	4	1.5
Cresol	108	11	7
Indole derivatives	–	5	3
Tri(2-butoxyethyl) phosphate	398	9	5
Nonyl phenols (9 isomers)	220	94	82
Octyl phenols (7 isomers)	206	66	51

14.2.1.1
Detergents

A high concentration of detergents and related compounds was observed in the sampled waters, including long chain acids, alcohols, and esters from soaps, (i.e., heptyl ester of 2-propanoic acid, originating from non-ionic surfactants). Also included are alkyl benzenes, originating from the anionic detergents alkylbenzene sulfonates, and alkyl phenols, from the nonionic detergents alkyl phenol ethoxylates.

14.2.1.2
Fatty Acids and Esters

Long chain fatty acids of $C_{12-15, 18}$, their methyl and ethyl esters, originate from domestic waste including food oils, soaps and human excreta. The abundance of these compounds (29–43 ppb) reflects the high rate of domestic contribution to the total waste discharge. Of domestic origin are also sterols, vaniline and indole derivatives, detected in the wastewater. In addition, antioxidants (i.e., BHT) and preservatives have been observed which are constituents used by the food industry.

14.2.1.3
Industrial Chemicals

The abundance of domestic origin chemicals, and on the other hand the relatively low levels of industrial chemicals like plasticizers (15 ppb of all phthalates isomers), phthalates and adipates, reflects the low extent of chemical industrial waste in the total discharge. Phosphate esters (tri(butoxyethanol)-phosphate and tributyl-phosphate) have also been observed, which are used in industry as plasticizers, flame retardant hydraulic fluid sand defoamers, and to a lesser extent also in metal working industries.

Solvents
The solvents detected in the reservoirs are based on alcohols, ethers and ketones which are less harmful than chlorinated solvents. They are widely used for cleaning and degreasing in industry and also in car washing and are discharged into the sewers. During the treatment process, most of these solvents disappear, many of them by volatilization. In the Tzora operational reservoir anaerobic conditions prevail, which is reflected by unoxidised sulfur compounds (dimethyldisulfide, tetrahydrothiophene and sulfur, S_8). These disappear when transferred to the storage reservoirs.

Alkyl Phenols
At the operational Tzora reservoir more than 200 ppb of alkyl phenols have been detected. These include 18 different substances, mainly nonylphenols and octylphenols which are degradation products of the ethoxylated and propoxylated alkyl phenol detergents. Alkyl phenols of the higher alkyl groups proved to be relatively nondegradable (Gaffney 1976). Some of them, in particular nonylphenol ethoxylates, are harmful to aquatic organisms (Lewis 1991) and have been recently restricted in some European countries. Hence, the relative abundance of nonylphenols at the outlet of the storage reservoirs (Table 14.1) is problematic due to their ecotoxicological char-

acteristics. Studies of photooxidation of alkyl phenols in natural water through sensitization by fulvic acids (Faust and Hoigne 1987) has shown that their reactivity decreases with increasing numbers of carbon atoms. Thus, the estimated half life of nonylphenol is about 80 days in well-mixed water for a sunny day, compared with $t_{1/2}$ of 5 days for 2,4,6-trimethylphenol. It seems that the increase in the hydrophobic interactions of the alkyl group with dissolved organic matter decreases the availability of the phenol to attack by more polar species like peroxyradicals. This behaviour is to be contrasted with that of the antioxidants di- and trimethylphenol, where the reaction with peroxyradicals occurs readily.

14.2.1.4
Wastewater Analysis

The data given in Tables 14.1–14.5 present only a partial pattern of the micropollutants, since GC/MS is suitable only for analysis of nonpolar and volatile organic chemicals. Its application for wastewater analysis is therefore limited, due to the requirement of volatility and thermostability. Normally, GC/MS is able to elucidate only about 10–20% of all the organic substances in the effluents (Watts et al. 1981), while the polar substances, high molecular weight chemicals and also the thermolabile, would not be identified. High performance liquid chromatography, HPLC, may be suitable for separation of these compounds and HPLC combined with tandem mass spectrometry (MS/MS) analysis proved to be most suitable for hydrophilic organic pollutants. Using MS/MS after loop injection allows bypassing the analytical column and direct analysis of a mixture without the need of separation (Schroder 1991, 1992). It has been shown that during the treatment processes the nonpolar pollutants tend to be transformed to more hydrophilic compounds.

Many of the non-biodegradable compounds in the wastewater effluents may be surfactants. The elimination mechanisms in soil filtration processes are insufficient to eliminate the nonbiodegradable nonionic surfactants. Among these Schroder detected the nonbiodegradable poly(ethyleneoxide) and poly(propyleneoxide) chains of non-ionic surfactants, a mixture of C-5, C-6 and C-7 alkanol-polypropylene glycol ethers and the corresponding oxidation products. Clusters of ions have been observed in industrial wastewater using thermospray TSP-LC/MS.

14.2.2
Reservoirs in the Judean Hills

The operational reservoir of Tzora has been studied together with the seasonal storage reservoirs of Revadim a, Revadim b, Nachshon, Harel, Anot and Zova, all of which receive waters from the operational reservoir. The Tzora Reservoir receives wastewater from Jerusalem after flowing in the Soreq river for a distance of 45 km. Wastewater stays for about 48 h in the Tzora operational reservoir before being directed to the storage reservoirs in the area of Judean Hills. Sampling was carried out at the inlet and outlet of the Tzora Reservoir.

Figure 14.1a shows the TIC (total ion chromatogram) of effluent in the inlet to the Tzora Reservoir. This may be compared with the TIC of the water sampled at the final stage of the treatment, e.g., at the outlet of Harel Reservoir, after six months storage

Fig. 14.1. a TIC (Total Ion Chromatogram) of effluents at the beginning of treatment, the inlet to the Tzora operational reservoir; **b** TIC of reclaimed water sampled at the outlet of Harel Reservoir, after storage of six months

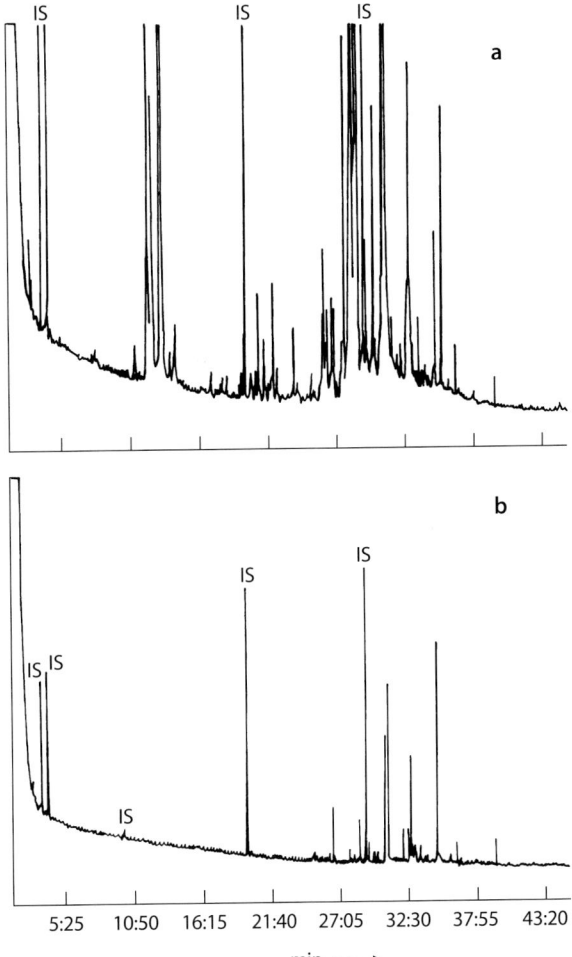

(Fig. 14.1b). A significant reduction of the pollution by organic toxicants can be noticed. Comparison with effluents after storage of 14 days only (outlet of Revadim Reservoir, Fig. 14.2) also reveals significant removal of micropollutants, with disappearance of signals due to major contaminants like alkyl phenols (signals of retention time 28–30 min), alkyl benzenes (retention time 30–30.5 min) and polymer degradation products (signals at 15–16 min). Apparently, the improvement in water quality is much greater in the case of the Harel Reservoir with a long storage period.

14.2.2.1
Seasonal Variation in Water Quality

The composition of wastewater in the reservoirs is subjected to continuous changes (Dor and Raber 1990; Juanicó 1993). The period of storage is one of the factors affect-

Fig. 14.2. TIC of reclaimed water at the outlet of Revadim Reservoir, after storage of 14 days

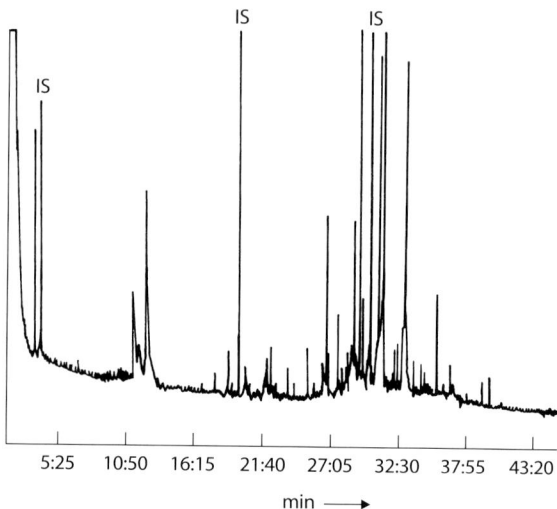

Table 14.2. Nachshon Reservoir, November 22, 1993

Compound	($\mu g\, l^{-1}$)
Tributylamine	2
BHT	0.5
Tributyl phosphate	0.5
Benzothiazolone	2.4
Nonyl phenyl and alkyl phenyl ethers	nd

Table 14.3. Nachshon Reservoir, August 22, 1994[a]

Compound	($\mu g\, l^{-1}$)
2-propanol, 1-(2-methoxy-1-methylethoxy)	4.5
Ethanol, 2-(2- butoxyethoxy)	1
2-propanol, 1-(1-methylethoxy)	3
2-propanol, 1-(2- methoxypropoxy)	12
1,1-dioxide, tetrahydrothiophene	1
Benzeneacetonitrile	1
Benzamide, N,N-diethyl 3-methyl (Deet)	4.5
4-(1,1,3,3-tetramethylbutyl) phenol	14
Dimethyl phenol	2
Diisopropylphenol (non linear)	6.2
Nonyl phenol (7 isomers)	26
Purine-2,6- dione, 1,3,7-trimethyl	2
Dodecylphenol	2
Tri, (ehanol,2-butoxy) phosphate	4

[a] Also detected: long chain aliphatic acids, alcohols and esters.

Fig. 14.3. TIC of reclaimed water at the Nachshon Reservoir; **a** Sampling at winter (end of November), after storage of five months; **b** Sampling on 28 August, the end of the irrigation season

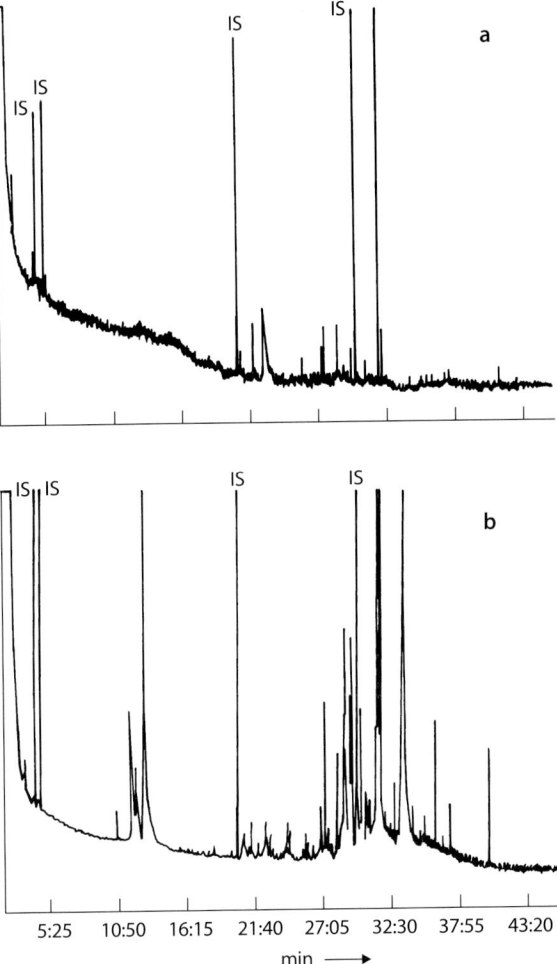

ing the quality of treated effluents. Long storage time would obviously result in better degradation of organic pollutants. The storage period may differ according to the rate of filling and pumping during the season. In wintertime, the percentage of water from rain and flood may predominate, while during summer the reservoirs collect only wastewater. The quality of treated effluents shows severe deterioration toward the end of the irrigation season, because of heavy loading with fresh incoming effluents. This temporal variation in water quality, mentioned in the case of Revadim and Harel ponds (Figs 14.1 and 14.2) is exemplified by the TIC of Nachshon Reservoir (Fig. 14.3). TIC of Fig. 14.3a was obtained from the sampling at winter (end of November) after storage of five months, while Fig. 14.3b shows the TIC from the end of the irrigation season (August 28). In the winter sampling only few micropollutants were observed at low levels (Table 14.2) while water from the end of August (Table 14.3) contains significant levels of domestic and industrial pollutants.

Table 14.4. Removal of organic pollutants in the Na'an Reservoir

Compound	Units	Inlet	Outlet
Pthalates	$\mu g\, l^{-1}$	67	17
Alkyl phenols	$\mu g\, l^{-1}$	31	13
Alkyl benzenes	$\mu g\, l^{-1}$	3	1
Hydrocarbons	$\mu g\, l^{-1}$	15	6
Anionic detergents	$mg\, l^{-1}$	5	0.5

14.2.3
Na'an Reservoir

The Na'an Reservoir receives effluents from the city of Ramla and kibbutz Na'an. Prior to their flow into the reservoir they undergo primary treatment and short residence time in two oxidation ponds. Data on typical micropollutants in the inlet and outlet of the reservoir are shown in Table 14.4. Significant removal of typical pollutants and of anionic detergents can be noticed. It seems that in the seasonal wastewater reservoirs an intensive microbial activity is maintained which is reflected by the high levels of *BOD* (biochemical oxygen demand), resulting in an increased biodegradation of organosynthetic pollutants.

These results match those of our previous study where the penetration of micropollutants into the soil has been examined. The studies were conducted in agricultural areas irrigated with effluents treated in deep reservoirs. No accumulation of organic micropollutants in the soils was noticed in these areas (Muszkat et al. 1993b). The contrary occurred in another case studied by us, where soil samples were made in an area irrigated with effluents treated in an intensive sewage treatment facility. In this case a significant accumulation of organic micropollutants was observed in both the soil profiles and groundwater.

14.2.4
Givat Brenner Wastewater Treatment Facility

The wastewater entering the Givat Brenner Reservoir originates from municipal and food industry uses. The treatment includes an anaerobic stage followed by conventional oxidation ponds. After pre-treatment, wastewater flows into an anaerobic pond of about 7000 m² with a residence time of 2–3 days. Then wastewater is directed to an oxidation pond aerated by a turbine, with a residence time of about 10 additional days. From the oxidation pond the effluents are transferred to a polishing pond with about 6 days residence time, and finally to the reservoir, 6 m deep. The removal of micropollutants in this system was assessed by the follow-up of alkyl phenols. At least ten different isomers of alkyl phenols were detected in the Givat Brenner facility. As mentioned above, these substances result from the alkyl phenol ethoxylate detergents, which are of wide household use. The data (Table 14.5) indicate high removal efficiency. The remarkable reduction in the anaerobic pond is interesting, since the most significant degradation occurred during such a short residence time. The combination

Table 14.5. Removal of alkyl phenols[a] in the Givat Brenner treatment facility

Stage of treatment	Concentration (μg l^{-1})	Retention time (days)
Anaerobic pond, inlet	71	2 – 3
Anaerobic pond, outlet	27	
Aerated pond, inlet	27	10 – 14
Polishing pond, inlet	29	6
Reservoir, inlet	16	Variable (months or weeks)
Reservoir, outlet	<0.5	

[a] Octyl phenols, nonyl phenols, dodecyl phenols.

of the benefits of an anaerobic process with those of an aerobic system may offer remarkable potential. The low level of alkyl phenols at the reservoir outlet is also to be noticed.

14.2.5
Distribution of Organosynthetic Chemicals in Agricultural Soils Irrigated with Effluents from the Reservoirs

The distribution of pollutants in soils was examined as a function of the origin of the waters used for irrigation. In this context soil profiles were examined in agricultural areas of similar soil composition, which underwent prolonged irrigation with effluents after storage in wastewater reservoirs. In case A soil sites were examined in Na'an (irrigated by Na'an Reservoir) and in Kfar Menachem (irrigation with effluents from reservoirs in the Judean Hills). The results were compared with control sites, irrigated with groundwater, and with soils where no irrigation was applied. In the examined soils no penetration of pollutants deep into the soil was noticed. Also, no significant differences could be observed in the pollutants distribution in sites irrigated by effluents and by groundwater. These findings were further compared with an additional area irrigated by wastewater, in a location close to Herzlia (case B). The wastewater treatment plant in Herzlia started as an oxidation pond in the early 1960s. Since 1977 an extended aeration treatment plant and infiltration lagoons have replaced the old facility. Hence, two different technologies of effluents treatment could be examined through their effects on soil pollution, when used for irrigation. The results in case B revealed the accumulation of organic micropollutants only in the effluent irrigated site, both deep in soil and also in groundwater (Muszkat et al. 1993a). The list of contaminants included pesticides and industrial organics, which under common conditions should have undergone biodegradation in the soil.

It is possible that the high content of organic matter in the Na'an area provides an adequate basis for a stable microbial population capable of efficient biodegradation of organic xenobiotics. Because of the high practical interest, the above phenomena require additional investigation.

14.2.5.1
Na'an: Organic Micropollutants in Soil at a Wastewater Irrigated Site

At the soil surface relatively high levels of aliphatic hydrocarbons were observed. These do not penetrate deep into the soil. Long chain ($n = 16$–18) carboxylic amides, phenolic antioxidants and polar derivatives capable of penetration into the soil were also observed. These were detected even at the depth of 14 m (Muszkat et al. 1993b). In view of the small extent of pollutants penetration, it must be noted that the high DOC (dissolved organic carbon) levels in this soil (20–50 mg l^{-1}) originate from the organic matter constituents of the soil rather than from organic pollutants. Moreover, the DOC levels in the soil of the Na'an site are much higher than in the site examined in Herzlia which was irrigated by effluents (where DOC levels are in the range of 2–20 mg l^{-1}).

The findings are of special interest for semiarid areas where agriculture is highly dependent on irrigation. Therefore, the subject deserves a more detailed study in order to understand the effect of treatment technologies on the removal of organic pollutants from wastewater, and the efficiency of deep wastewater reservoirs in removing micropollutants in comparison with the degradation achieved in intensive aeration systems.

14.2.6
Degradation Mechanisms

While the observations about the purification capacity of wastewater reservoirs and the improvement in water quality during storage are all well established, the detailed degradation mechanism seems to be obscure. Degradation includes biotic and abiotic transformations. However, due to the complexity of the processes and the scarcity of studies dealing with both biotransformation and heterogeneous phototransformation of organic pollutants in the storage reservoirs, it is difficult to reach definite conclusions about the relative rate of both processes. Besides, the biochemical, photochemical and chemical processes seem to be interdependent. Thus, the photochemical promotion to an excited molecular electronic state may lead to a variety of biochemical reactions.

14.2.6.1
Biotic Transformations

Microorganisms are the most ubiquitous agents of biochemical processes, which determine the fate of organic chemicals in water. In the seasonal wastewater reservoirs intensive microbial activity is maintained, typically featured by high levels of BOD (for example, 40–60 mg l^{-1} in the Na'an Reservoir, Dor et al. 1987; Dor and Raber 1990). The microbial activity results in an increased biodegradation of organosynthetic pollutants. The major reaction mechanisms include oxidation, reduction and hydrolysis.

14.2.6.2
Abiotic Photo-Transformations

UV light has a direct effect on organosynthetics in water media. While algae and bacteria absorb light by photosynthetic mechanisms in the visible region, the energy absorbed by *DOC* (dissolved organic carbon) occurs at the range below 400 nm. Absorption of UV by humic substances results in sensitized formation of reactive oxygen species, free radicals and of other excited fragments. Major reactive species are hydrogen peroxide, H_2O_2, singlet oxygen $^1O_2$ (the excited state of molecular oxygen), and hydroxyl OH· and peroxy ROO· radicals.

Hydrogen peroxide is relatively stable. It may react to form hydroxyl radicals OH· which are capable of initiating a sequence of reactions. The abiotic formation of hydrogen peroxide in lake waters has been studied by Lean et al. (1994). These authors found that the concentration of hydrogen peroxide varies with depth and is affected by the production rate, by the effects of vertical mixing of the water column and by the decay rate. The production was limited to the upper layer till ca. 2 m depth. The H_2O_2 formation was affected by the concentration of dissolved organic carbon. Hydrogen peroxide concentration was highest at the surface and decreased with depth. It varied from 100–400 nM at water surface to ca. 50 nM at the depth of 6 m and more.

Direct Photochemical Reactions of Organic Pollutants in Water

This reaction occurs as a result of direct light absorption by the pollutant, promoting the molecule to the excited electronic state. In such a state the molecule may undergo a variety of chemical reactions. The product of such transformation may further react by either chemical or biochemical processes. Therefore it should be difficult to assess each process, chemical and biochemical, in the presence of the other. The light intensity and the solar spectrum at a given point of earth depend on factors such as geographic location, season, time of the day, weather conditions, air pollution, etc. (Finlayson and Pitts 1986). Sunlight at the surface of earth consists of direct and of scattered light which enters the water body at various angles.

Indirect (Sensitized) Photolysis

Organic chemicals may also undergo indirect photolytic transformation induced by light absorbed by other chromophore molecules present in the system. The most important light absorbers in natural water bodies are chromophores present in humic substances. An excited chromophore would react by energy transfer to either molecular oxygen or to organic chemicals which may serve as acceptors (quenchers). The most important acceptor of an excited chromophore is molecular oxygen. The ground state of molecular oxygen is triplet $^3O_2$. Interaction with an excited chromophore results in promotion of $^3O_2$ to its first excited state which is singlet oxygen $^1O_2$. This excitation requires only low energy (94 kJ mol^{-1}). Therefore, almost all chromophores which absorb in the UV and visible range transfer the light energy to an oxygen molecule. The singlet oxygen produced will then react with organic pollutants. There is much less probability that an excited chromophore will transfer energy directly to an organic pollutant, since the energy required for excitation of an organic pollutant is much higher (Zepp et al. 1985). Organic pollutants may compete with triplet oxygen for light energy, but only in water of low oxygen concentration and with pollutants with low excited state triplet energy (Mill and Mabey 1985). Among such pollutants are delocalized π-electron systems (Zepp et al. 1985).

DOC-UV interactions form reactive species, including hydrogen peroxide, hydroxyl radicals, superoxide, formaldehyde and carbon monoxide (Cooper et al. 1989). These species stimulate additional reactions, and their concentrations may be high enough to be toxic on their own.

In addition to energy transfer, an excited chromophore may undergo chemical reactions, producing reactive species which react with organic pollutants. The major photo-oxidants with respect to transformations of organic pollutants are singlet oxygen $^1O_2$, hydroxyl radicals OH$^\cdot$ and peroxy ROO$^\cdot$ radicals (Zepp and Wolfe 1987; Cooper et al. 1989). In surface waters, hydroxyl radicals may be formed by photolysis of hydrogen peroxide, photolysis of nitrite or nitrate and sensitization of oxygen by excited humic acids.

Zepp and Wolfe (1987) have shown that the sorption of compounds may be important in the sensitized photolysis of organic pollutants in natural waters. The rate grows with an increase in hydrophobicity of the compound, indicating that sorption was a rate determining step. Inaba (1992) assessed the degradation of LAS (linear alkyl sulfonates) in wetlands and has also found that LAS removal in the wetland included sorption on sediment particles and degradation by bacteria. The latter was affected by temperature, and thus the significant process for purification was only adsorption in winter time.

Schnidler and Curtis (1997) studied *DOC*-mediated interactions in aquatic ecosystems. When *DOC* concentration decreases there may be a concern for UV damage. Decrease in *DOC* and increase in transparency caused by climatic warming can increase the exposure of aquatic ecosystems to UV-radiation. In lake waters, colored organic compounds are removed much more rapidly than colorless components probably by a combination of photodegradation and photobleaching.

14.3
Conclusions

Under suitable storage conditions, deep wastewater reservoirs successfully remove organic pollutants and release reclaimed water of much better quality, appropriate for irrigation. Deep seasonal wastewater reservoirs are of special interest due to conspicuous advantages such as the relatively low cost of construction, maintenance, operation and low energy consumption. Besides, they do not depend on activated sludge treatment which by itself requires high investments and operational costs. Coupling the conventional wastewater reservoirs with other treatment systems may further accelerate the removal of organic pollutants and yield high quality effluents. Thus, combination of anaerobic-aerobic treatments would provide the benefits of an anaerobic high removal efficiency at relatively low construction and operational costs. Another promising possibility is the coupling of the existing reservoirs with facilities for solar photocatalytic oxidation. In the latter case treatment may be carried out in the existing lagoons and accelerated by the solar photocatalytic process (Matthews 1993; Muszkat et al. 1995). This may be the method of choice for treatment of polluted water in regions with plenty of sunshine and cheap land. The combination of conventional reservoirs with photocatalysis may greatly enhance the detoxification of effluents.

Acknowledgements

The author acknowledges the assistance of Ms. Lina Bir and Dr. Leonid Feigelson in this study. The work was supported by funding provided by the US-AID (Project N° TA-MOU-

CA13-025), for which the author is grateful. The soil experiments in Na'an and Kfar Menachem were carried out in cooperation with M. Magaritz and D. Ronen, Weizmann Institute of Science, within the frame of GIF Project N° I-103-225.08/89.

References

Abeliovich A (1985) Biological treatment of chemical industry effluents by stabilization ponds. Wat Res 19:1497–1503

Abeliovich A, Vonshak A (1993) Factors inhibiting nitrification of ammonia in deep wastewater reservoirs. Wat Res 27(10):1585–1590

Aharoni A, Kanarek A (1994) The wastewater reclamation system of Natania. Performance of the southern reservoir during 1993 Wat Irrig (In Hebrew) 338:42–45

Cooper W, Zika G, Petasne R, Fischer A (1989) Sunlight induced photochemistry of humic substances in natural waters. Adv Chem Ser 219, Amer Chem Soc, pp 333–362

Cohen Z, Eiden C, Lorber M (1986) In: Garner W Y (ed) Evaluation of pesticides in ground water. ACS Symp Ser 315, Amer Chem Soc, Washington, DC, pp 170–196

Dor I, Raber M (1990) Deep wastewater reservoirs in Israel: empirical data for monitoring and control. Wat Res 24:1077–1084

Dor I, Schechter H, Bromley H (1987) Limnology of a hypertrophic reservoir storing wastewater for agriculture at Kibbutz Na'an. Hydrobiologia 150:225–241

Eren J, Dor I (1985) The limnology of Ma'aleh HaKishon Reservoir. Annual reports submitted to Mekorot, Water Supply Co., Tel Aviv, Israel

Eren J, Dor I (1986) The limnology of Ma'aleh HaKishon Reservoir. Annual reports submitted to Mekorot, Water Supply Co., Tel Aviv, Israel

Fattal B, Puyeski Y, Eitan G, Dor I (1993) Removal of indicator microorganisms in a wastewater reservoir in relation to physico-chemical variables. Wat Sci Tech 27:321–329

Faust BC, Hoigne J (1987) Sensitized photooxidation of alkyl phenol by fulvic acid natural water. Wat Sci Tech 21:957–964

Finlayson BJ, Pitts J (1986) Atmospheric chemistry – Fundamental and exprimental techniques. J Atmosph Chem, Wiley Intersci., New York

Gaffney PE (1976) J Water Poll Fed 48(12):2731–2737

Inaba K (1992) Quantitative assessment of natural purification in wetland for LAS. Wat Res 26:893–898

Juanicó M (1993) Alternative schemes for municipal sewage treatment and disposal in industrialized countries: Israel as a case study. Ecological Engineering 2:101–118

Juanicó M, Shelef G (1994) Design, operation and performance of stabilization reservoirs for wastewater irrigation in Israel. Wat Res 28:175–186

Juanicó M, Ravid R, Azov Y, Teltsch B (1995) Removal of trace metals from wastewater during long-term storage in seasonal reservoirs. Water, Air and Soil Pollution 78:1–17

Lean R, Cooper W, Pick F (1994) Hydrogen peroxide formation and decay in lake waters. In: Aquatic and surface photochemistry, CRC Press, pp 207–221

Lewis M (1991) Chronic and sublethal toxicities of surfactants to aquatic animals: a review and risk assessment. Wat Res 25:101–113

Matthews R (1993) Photocatalysis in water purification. In: Ollis DF, Al-Ekabi H (eds) Photocatalytic purification and treatment of water and air. Elsevier Science Publishers, pp 121–136

Mill T, Mabey B (1985) Photochemical transformations. In: Neely WB, Blau GE (eds) Environmental exposure from chemicals, vol 1, CRC Press, Boca Raton, Fl.

Muszkat L, Bir L, Raucher D, Magaritz M, Ronen D (1993a) Unsaturated zone and groundwater contamination of organic pollutants in a sewage effluents irrigated site. Ground Water 31:556–565

Muszkat L, Magaritz M, Ronen D (1993b) Fate of organic contaminants in water and sediments. Final report GIF Project N° I-103-225.08/88

Muszkat L, Bir L, Feigelson L (1995) Solar photocatalytic mineralization of pesticides in polluted waters. J Photochem Photobiol A: Chem 87:85–88

Pionke H, Glotfelty D (1989) Nature and extent of groundwater contamination by pesticides in an agricultural watershed. Wat Res 23:1031–1038

Schnidler D, Curtis P (1997) The role of DOC in protecting freshwater subjected to climatic warning and acidification from UV exposure. Bio-geochemistry 36, pp 1–8, Kluwer Academic Publishers

Schroder H (1991) Identification of non-biodegradable, hydrophilic organic substances in industrial and municipal wastewater treatment plant. Wat Sci Tech 23:339–347

Schroder H (1992) Polar organic pollutants on their way from wastewater to drinking water. Wat Sci Tech 25:241–248

Watts C, Crathorne B, Crane R, Fiedling M (1981) Development of techniques for isolation and identifi-
    cation of non-volatile organics in drinking water. In: Keith LH (ed) Advances in the identification
    and analysis of organic pollutants in water, vol 1, Ann Arbor Sci Pub, pp 383–398
Zepp R, Wolfe N (1987) Abiotic transformation of organic chemicals at the particle-water interface. In:
    Stumm W (ed) Aquatic surface chemistry. Wiley Interscience, New York
Zepp R, Schlotzhauer P, Sink R (1985) Photosensitized transformations involving electronic energy trans-
    fer in natural waters: Role of humic substances. Env Sci Tech 19:74–81

Chapter 15

Trace Metals

Marcelo Juanicó · Roza Ravid · Yossi Azov · Benjamin Teltsch

15.1
Introduction

A preliminary study on the reduction of the concentration of trace metals in wastewater during its storage in stabilization reservoirs was made by Kaplan et al. (1987) in a reservoir receiving effluents with 180 mg $BOD\,l^{-1}$ from the sewage treatment plant of Beersheva, Israel. These authors reported a reduction of 30–40% in the concentration of Cu, Zn, Pb and Cd. The main limitation of these data is that, being the reservoir a nonsteady-state reactor with seasonal outflow, a reduction in concentration is not a direct measurement of the actual removal of heavy metals.

This chapter presents a quantitative evaluation of the removal of five metals, by two different methods (annual budget and sediment traps), in two stabilization reservoirs operated in series. This information, although preliminary, is the first of its kind available and thus of great importance for the future design and operation of stabilization reservoirs as wastewater treatment reactors.

15.2
Materials and Methods

15.2.1
The Studied Reservoirs and Water Balance

The Northern and Southern Kishon Reservoirs (Fig. 15.1) are located in the centre of the Jeezrael Valley, northern Israel. The reservoirs have a storage capacity of 6 million m³ each and a maximum depth of 10.4 m. The two reservoirs are connected by a pipe located at the bottom level in the eastern side ('passage' in Fig. 15.1), but they can be disconnected by closing the passage. During the covered year the reservoirs were operated in series: inflow was directed to only one of the reservoirs and effluents were released from the other. The direction of the flow was inverted once. The reservoirs receive about 40 000–50 000 m³ d⁻¹ of effluents from the city of Haifa, after treatment in parallel activated sludge and trickling filters. Complete descriptions of the design, operation and performance of the Kishon Complex for wastewater treatment and storage can be found in Rebhun et al. (1987), Weber and Juanicó (1990), and Azov and Shelef (1991). The quality of the effluents entering and leaving the reservoirs during the studied period is characterized in Table 15.1.

Inflows to, and outflows from the reservoirs were measured with continuous recording flowmeters. Water level in both reservoirs was recorded once a day and the volume of effluents within the reservoirs calculated from a volume/depth curve. Rain measurements were taken from a meteorological station located 7 km from the reservoirs. Evapo-

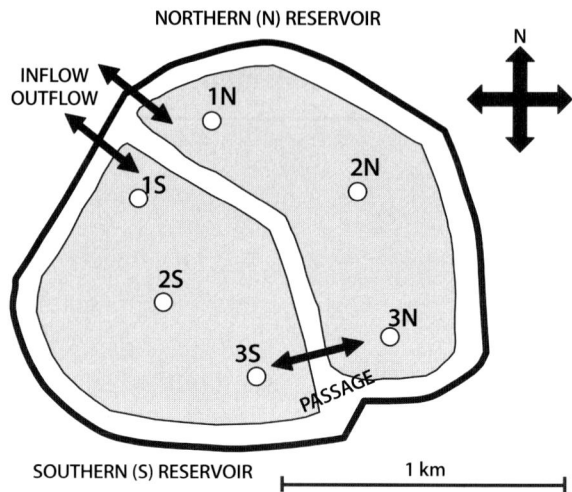

Fig. 15.1. Lay-out of the Kishon Reservoirs and sampling points. The *arrows* indicate that flow can follow one of two possible directions

Table 15.1. Minimum and maximum values of selected parameters in the inlet and outlet of Kishon Reservoirs during the covered period

Parameter	Units	Inlet	Outlet
BOD	mg l^{-1}	20 – 50	3 – 20
COD	mg l^{-1}	80 – 210	45 – 75
MBAS (detergents)	mg l^{-1}	1 – 9	0.2 – 0.8
Faecal coliforms	MPN/100 ml	103 – 106	2 – 104

ration calculations were based upon average multi-annual evaporation rates for the area, increased by 10% based on the calibration of the water balance of the reservoirs and because 1990–1991 was a dry year. Seepage was estimated by means of a simplified function:

Seepage (mm d^{-1}) = K (mm d^{-1}) × (*water level* (m) / 10 m) .

The value of K was estimated as 6 mm d^{-1} for an almost full reservoir 10 m in depth, by calibration of the water balance of the reservoirs. The mean residence time of effluents within the reservoirs, and the percentage of fresh effluents with 5 days or less within the reservoirs (PFE_5), or with 30 days or less within the reservoirs (PFE_{30}), were calculated according to Juanicó and Friedler (1994, and in this volume).

15.2.2
Sampling

Water sampling was performed from 8 May 1990 to 20 May 1991. Inflow was sampled at the inlet of the reservoirs, with a frequency of approximately once every four weeks, totaling 15 samples during the covered period. Outflow was sampled at the outlet with a similar frequency to that of the inflow, but a higher sampling frequency was performed during periods of strong outflow, totaling 18 samples during the covered period. Sampling dates can be seen in Fig. 15.4. Concentrations of metals in the effluents within the two reservoirs at the start and end dates of the sampling period were estimated by taking composite samples from several depths at the six points showed in Fig. 15.1. The amount of metals present in the reservoirs at these two dates was calculated.

The sedimentation rate of metals in the reservoirs was measured with sediment traps located at the six sampling points showed in Fig. 15.1. Each trap has a catching area of 472 cm² and was left in the bottom for a period of 30–40 days, totaling 10 measurements at each point during the covered period. The concentration of metals was determined also in the soil surrounding the reservoirs and used to contruct them.

15.2.3
Analytical Methods

Total concentrations (i.e., inorganically plus organically bound metals) of Cu, Zn, Cr, Pb and Al, in the water and sediment samples, were determined by atomic absorption after wet digestion with nitric and sulfuric acids, following Standard Methods procedures 3030 G and 3500 B (APHA et al. 1989). Dissolved concentrations were determined by filtering through 0.5 micron paper.

15.2.3.1
Reduction in Concentration, Removal and Budget of Metals

Reduction in the concentration (C) of metals was calculated as follows:

Reduction C (%) = [(Outflow C – Inflow C) / Inflow C] × 100 .

The amount of metals entering the reservoirs (*input*) was calculated by multiplying the inflow volume of effluents by the corresponding concentration of metals in the influent at the same time. The amount of metals leaving the reservoirs (*output*) was calculated in the same way. The concentration of metals between two sampling dates was estimated by linear interpolation. It was assumed that metal inputs and outputs due to rain and evaporation were null. The amount of metals lost by seepage was calculated by multiplying the dissolved metal concentrations by the seepage flows. The removal and budget for each metal was calculated as follows:

Removal = Start – End + Input – Output – Seepage ,

where '*start*' is the amount of a given metal in the volume of effluents within the reservoirs at the start of the covered year, and '*end*' the amount which remained within the reservoirs at the end of the year. The removal was also expressed as a percentage of the input.

The amount of a given metal which settled within a trap (*Amount*) was calculated as follows:

$$Amount = \frac{(Total\ concentration - Dissolved\ concentration) \times Sediment\ volume}{2}.$$

The division by 2 is to compensate for the sediments resuspended from the bottom of the reservoir by hydraulic turbulence and settling again within the sediment traps. A resuspension trapping factor of 2 was estimated after Avnimelech and Wodka (1988). The sedimentation rate (*Rate*) was calculated as:

$$Rate = \frac{Amount}{0.0472\ \mathrm{m}^2 \times number\ of\ sampling\ days},$$

and expressed in $\mu g\ m^{-2}\ d^{-1}$. It was assumed that each one of the traps located in sampling points 1N, 1S, 3N, 3S is representative of 1/4 of the area of its respective reservoir, while those located in sampling points 2N and 2S are representative of 1/2 of the area. The quantity of metals deposited in both reservoirs on a given day was obtained by multiplying the sedimentation rate at each point by its respective area, and then summing up the quantities deposited in all six sampling points.

15.2.4
Potential Sources of Errors

The measurements and estimations made in the present work may have several sources of errors:

- Inflow/outflow flowmeters may over- or under-estimate actual flows.
- Multi-annual evaporation rates adjusted to the 1990–1991 conditions were used instead of actual ones.
- Rain measurements were made 7 km from the reservoirs in an area where the rain patterns may change in space.
- Only approximate area/depth and volume/depth curves of the reservoirs were available.
- Seepage was calculated by calibration of the water balance; thus, errors in inflow/outflow, rains, or evaporation may affect the calculation of seepage.
- The concentration of metals in inflow and outflow was analysed in only 15 and 18 samples per year while differences in time were found.
- Sedimentation rates were calculated in only six sampling points while differences in space were found.
- Analytical errors.

15.3
Results and Discussion

15.3.1
Operation of the Reservoirs and Water Balance

The inflow of wastewater effluents to the reservoirs fluctuated between 40 000 and 50 000 $m^3\ d^{-1}$ (Fig. 15.2a). Effluents entered the northern reservoir from May to No-

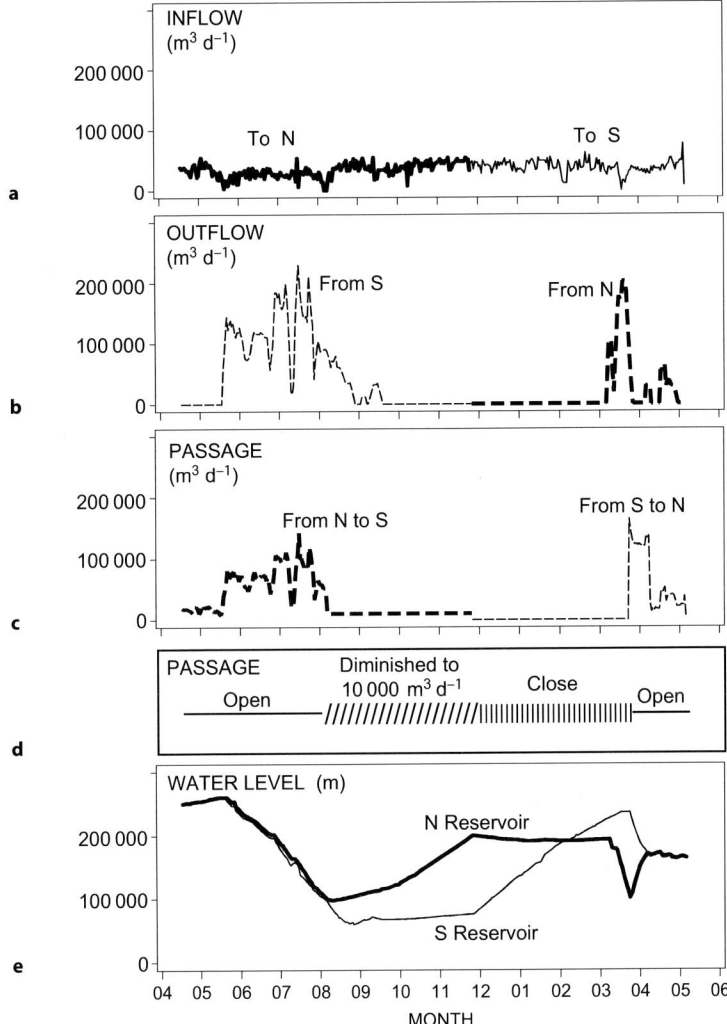

Fig. 15.2. Operational parameters of the Kishon Reservoirs during the year. In December 1990 the direction of the flow was inverted

vember, and the southern reservoir from December to May. The effluents were released during two periods (Fig. 15.2b) always from the second reservoir of the series, depending on the flow direction. The passage between the two reservoirs was fully open from May to August, reduced to 10 000 m³ d⁻¹ from September to December, and then closed till April when it was again fully opened (Figs 15.2c and 15.2d). Figure 15.2e shows that there were conspicuous changes in the water level of both reservoirs as a consequence of the seasonal operational regime. The reservoirs contained 12.1 million m³ of effluents at the 'start' date. Almost 13.6 million m³ of wastewater entered the reservoirs during the covered year, while 14 million m³ of effluents were released for irrigation.

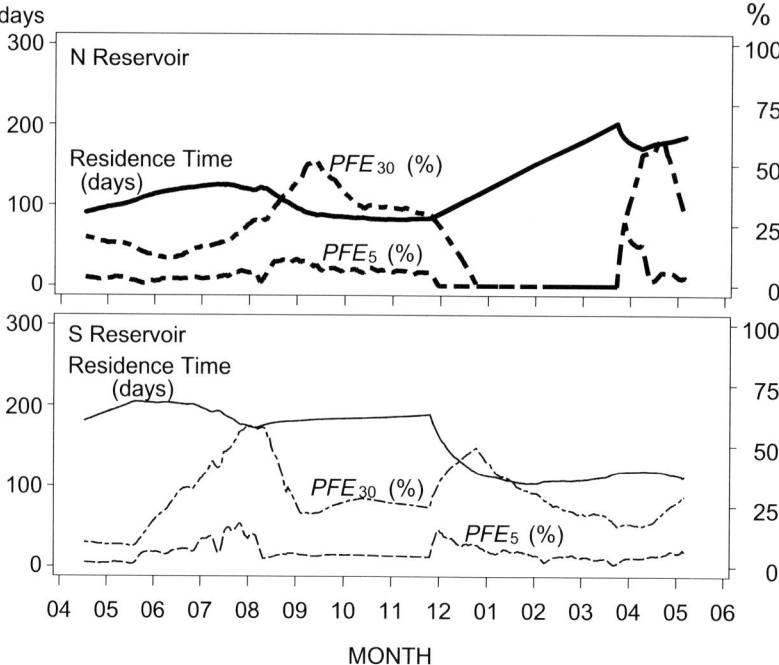

Fig. 15.3. The mean hydraulic residence time of the effluents, and the percentage of fresh effluents with 5–30 days or less within the reservoirs (PFE_5 and PFE_{30}) during the year

Seepage losses summed 2.1 million m³ (about 15% of total inflow), evaporation losses 2.9 million m³ (about 20% of total inflow) and rain gains about 0.5 million m³. The reservoirs contained only 7.2 million m³ of effluents at the 'end' date. The mean residence time of the effluents within the reservoirs was very long and irregular, varying between 80 and 220 days (Fig. 15.3). The other two descriptors of the hydraulic age distribution (PFE_5 and PFE_{30}) also experienced strong changes during the year.

15.3.2
Concentration of Metals in Inflow and Outflow

The concentration of the studied metals in the inflow to the reservoirs was relatively low, already below the maximum values recommended for irrigation waters (Table 15.2). The storage of the effluents in the reservoirs further reduced the mean concentrations, from 20% reduction of Al to 75% reduction of Cr. This reduction brought the concentration of metals in the effluents back to the base level found in the Central and Northern Coastal Plain Aquifers of Israel (Ronen et al. 1975). The Pb concentration in other natural non-contaminated freshwaters is within the range of <1–10 µg l⁻¹ and in rain within the range of 5–30 µg l⁻¹ (Fergusson 1990). The inflow to the reservoirs can thus be characterized as very clean from Pb, even though the storage period further reduced its concentration in the effluents by 35%.

Table 15.2. Concentration of trace metals ($\mu g\,l^{-1}$) in the inflow and outflow of the Kishon Reservoirs, compared with irrigation standards and the base level found in the coastal aquifers of Israel

Metal	Statistic	Reservoirs inlet	Reservoirs outlet	Conc. re-duction (%)	FAO irrigati-on standard[a]	Base level conc. in coastal plain aquifer[b] Central[c]	Northern[d]
Cu	Min	5.5	1.8				
	Mean	16	5.5	65		9	2
	Max	28	20		200		
Zn	Min	45	20				
	Mean	110	40	65		40	50
	Max	230	80		2 000		
Cr	Min	1.6	0.5				
	Mean	12	3	75		4.5	3
	Max	45	10		100		
Pb	Min	1.2	0				
	Mean	6	4	35		3.5[e]	3[e]
	Max	20	20		–		
Al	Min	150	70				
	Mean	500	400	20		–	–
	Max	1 600	1 600		5 000		

[a] *Source:* Leeden et al. (1990) The water encyclopedia;
[b] *Source:* Ronen et al. (1975) The base level is the concentration that is equal or less than the observed in 80% of the wells;
[c] The Central coastal plain aquifer – Israel, 115 wells sampled;
[d] The Northern coastal plain aquifer – Israel, 20 wells sampled;
[e] Pb actual concentrations may be a bit higher than the reported, due to problems in the handling of the samples.

The concentration of metals found in the reservoirs were far below those required to produce acute inhibition of algae or zooplankton. Generally, concentrations above $1\,000\,\mu g\,l^{-1}$ are required to affect the biological community (Moshe et al. 1972). However, the concentrations were not far from the values required to produce chronic effects. Kaiser (1980) quotes that 16% reproductive impairment was found in *Daphnia magna* at concentrations of Cu = 22 $\mu g\,l^{-1}$, Zn = 70 $\mu g\,l^{-1}$, Cr = 330 $\mu g\,l^{-1}$, Pb = 30 $\mu g\,l^{-1}$ and Al = 320 $\mu g\,l^{-1}$.

The concentrations of the five metals varied significantly in time, both at the inlet and at the outlet of the reservoirs (Fig. 15.4). In the case of Cu and Zn, the concentration at the outlet was always lower than that at the inlet, but in the case of Pb and Al the contrary occurred in some occasions.

15.3.3
Input/Output Amounts

The input and output amounts of metals are given in Table 15.3. The minimum input amount corresponded to Pb with only 55 kg yr^{-1} and the maximum to Al with 8 t yr^{-1}.

The daily metal amounts which entered and left the reservoirs during the year are presented in Fig. 15.5. The values are the result of the inflow/outflow concentrations

Fig. 15.4. Concentration of the five metals in the inflow (*solid lines*) and outflow (*dashed lines*) of the reservoirs during the year

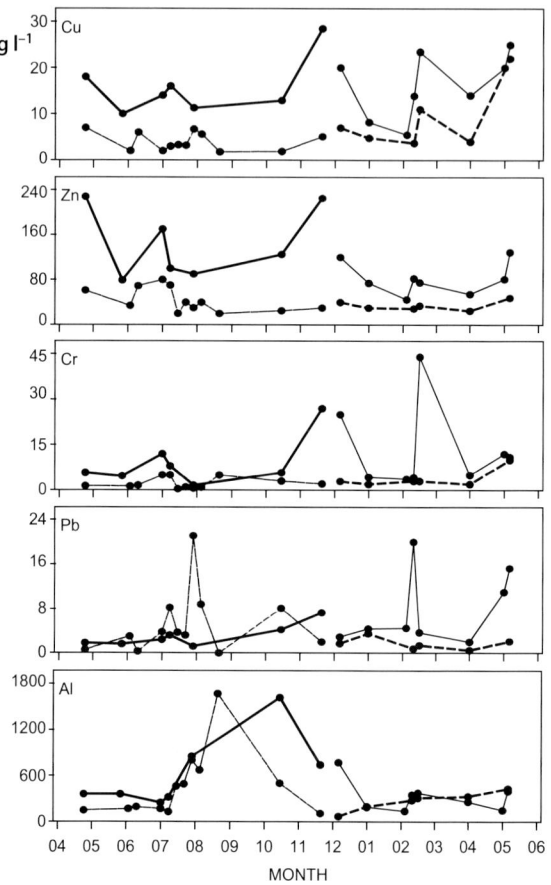

Table 15.3. Annual (378 days) budget of trace metals in the Kishon Reservoirs. The amount within the reservoirs when sampling was *start*ed, the *input* through inflow, *output* through outflow, losses by *seepage*, the amount which remained in the reservoirs at the *end* of sampling, the *removal* (kg) computed from the metal budget, *removal* (%) as a percentage of the input, and the amount *sedimented* as computed from sediment traps

	Start (kg)	Input (kg)	Output (kg)	Seepage (kg)	End (kg)	Removal[a] (kg)	Removal (%) input	Sedimented (kg)
Cu	122	204	62	12	227	24	12	55
Zn	801	1475	624	67	285	1300	88	305
Cr	25	154	35	4	61	79	51	103
Pb	16	55	54	2	13	2.5	5	5
Al	1913	8396	5738	242	1914	2415	29	89000

[a] Removal = start − end + input − output − seepage.

Fig. 15.5. The daily amount of the five metals which entered (*solid lines* and *shaded areas*) and left (*dashed lines*) the reservoirs during the year

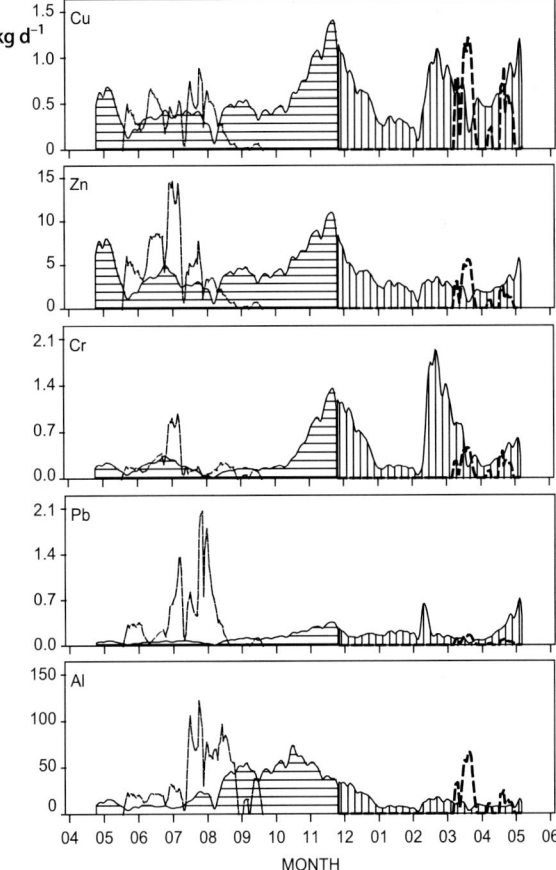

(Fig. 15.4) multiplied by the inflow/outflow volumes (Figs 15.2a and 15.2b). A comparison of Figs 15.2a, 15.4 and 15.5 indicates that as inflow rate is almost constant, metal inputs depend mainly on inflow concentrations. On the contrary, a comparison of Figs 15.2b, 15.4 and 15.5 indicates that as outflow rate is very variable, it strongly affects output amounts. Steady-state flow reactors for sewage treatment such as stabilization ponds and activated sludge have an almost constant outflow and thus the outflow concentration is a good measurement of the performance of the unit. On the contrary, in reactors with controlled discharge such as the seasonal reservoirs, both concentration and outflow rates must be included in the analysis.

15.3.4
Losses by Seepage

Table 15.3 shows that losses of trace metals from the reservoirs by seepage accounted for only 3–6% of the input to the reservoirs, and this is negligible in quantitative terms. However, these figures may gain significance in reservoirs receiving effluents with higher metal concentrations. Two independent works on stabilization ponds (Wolfberg et al. 1980 in Israel; Ghosh et al. 1985 in India) concluded that leaching of metals in dissolved form through the bottom of the pond may be an important vector of metal contamination of groundwater.

15.3.5
Removal and Metal Budget

The storage of effluents in the reservoirs resulted in considerable removal of metals (Table 15.3). The removal percentage was very irregular, from only 5% for Pb to almost 90% for Zn. More than one ton per year of Zn was removed by the reservoirs. Suffern et al. (1981), in a study of metal removal in stabilization ponds, also found that Zn is the metal best removed in these impoundments. The poor Pb removal can be explained by the very low concentration of Pb in the inflow to the reservoirs (Table 15.2) which is within the range of natural unpolluted waters (Fergusson 1990) and the base level found in Israeli aquifers (Ronen et al. 1975).

15.3.6
Sedimentation Rates

The concentrations of the five studied metals in the sediments caught by the sediment traps are far below those found in the soil of the area where the reservoirs are built (Table 15.4). Sigg et al. (1987) in a work on the vertical transport of metals through settling in lakes, found that biological material is the main carrier of settling metals. Thus, the sediments caught by the sediment traps are made mainly of organic matter (bacteria, algae and zooplankton biomass) with a low portion of local soil particles. In the case of Al, the concentration in the soil is so high (70 mg g^{-1}) that even a small fraction of soil particles caught by the traps can affect the final value.

The concentrations of metals in the sediment traps are also lower than those found in the sediments of lakes and estuaries (Table 15.4). The sediment traps were left in the bottom for 30–40 days and thus the sediments caught by the traps were fresh with a considerable amount of organic matter. On the other hand, sediments which are sampled directly from the bottom are older ones, and the concentration of metals

Table 15.4. Mean concentration of metals in the sediment traps, and in the soil of the area surrounding the reservoirs, compared with the concentration in sediments of other water bodies. All values in $\mu\text{g g}^{-1}$ (*Sources:* [a] Fergusson 1990; [b] Lecden et al. 1990; [c] Suffern et al 1981)

	Soil near the reservoirs	Kishon R. sediment traps	Non-polluted arctic lakes[a]	Polluted Michigan lake[a]	40 USA estuaries[b]	Stabilization ponds[c]
Cu	20	1			0.1–22	1 100
Zn	60	5.7			0.7–43	3 100
Cr	70	1.9			2.3–230	350
Pb	0	0.09	10–33	88	0.2–19	230
Al	70 000	1 700			–	–

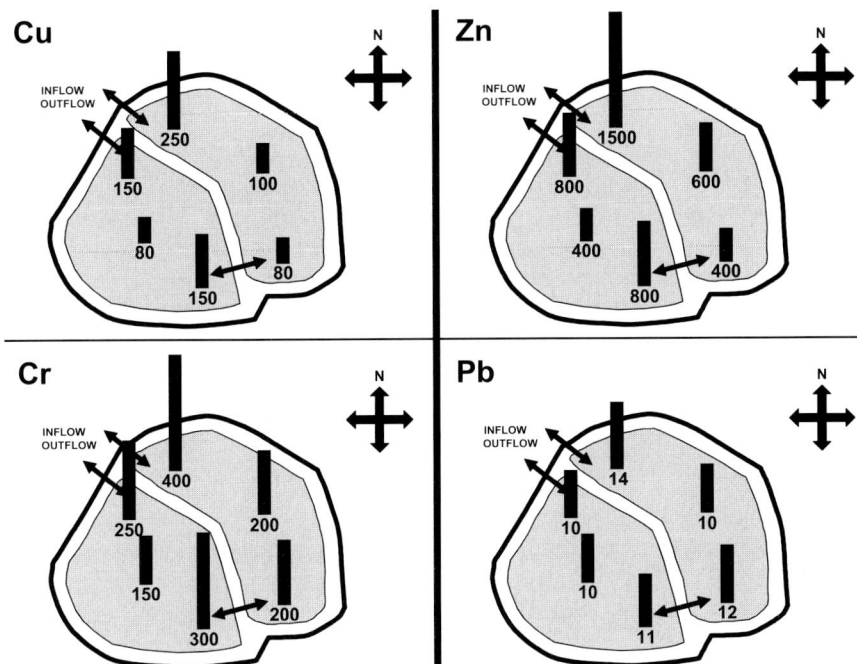

Fig. 15.6. The annual mean sedimentation rates of four of the metals ($\mu g\, m^{-2}\, d^{-1}$) at the six sampling points

in these sediments is increased by the degradation of the organic matter. The stabilization ponds studied by Suffern et al. (1981) (Table 15.4) received metals and *BOD* concentrations similar to those received by the Kishon Reservoirs, but had a residence time of only 6 days. Thus the actual loading of metals in the ponds was more than one order of magnitude higher than the metal loading in the Kishon Reservoirs.

The differences in sedimentation rates between the six sampling points are presented in Fig. 15.6. The sedimentation rates were generally higher at the extreme points of the reservoirs (1N, 1S, 3S) than at the central points (2N, 2S). Points 1N and 1S are located near the inlets to the reservoirs, but an increase of the sedimentation rates during the periods when inflow entered to northern or southern reservoirs was not found. The extreme points are located in areas where there is an accumulation of suspended solids in the water due to wind induced currents (Fig. 15.7). Dominant winds create longshore currents which erode and resuspend sediments from the embankments, transporting the particles along the shores. Longshore currents accumulate the suspended particles at the extreme points of the reservoirs where rip-currents disperse them over relatively large areas. Both the longshore currents and the rip-current dispersion areas indicated in Fig. 15.7 can be easily recognized in air photographs of the reservoirs, with brown patterns contrasting with the green water surface. Proper location of outlet should avoid these areas.

Fig. 15.7. Longshore currents and rip-current dispersion areas of resuspended sediments, generated by dominant winds in the reservoirs

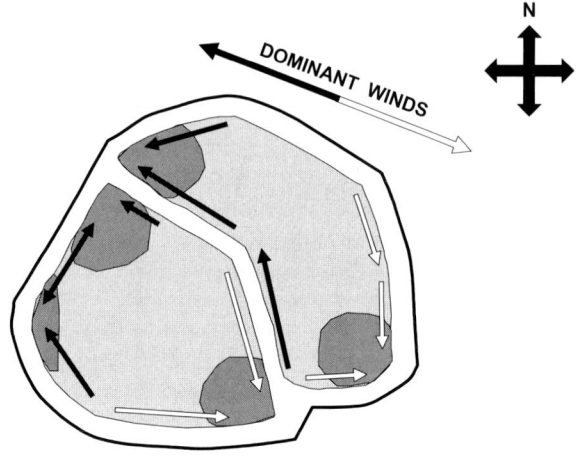

15.3.7
Sedimentation Quantities

Table 15.3 lists the quantity of metals deposited in the bottom of the reservoirs as calculated from the sediment traps data. These numbers do not match the removal calculated from the metal budget. Figure 15.6 shows that there were strong differences in sedimentation rates between different areas within the reservoirs, and thus the selection of only six sampling points was insufficient to obtain a representative sample. This seems to be the main source of error explaining the differences in removal percentages between the two methods. Al is a special case where the sediment traps give an estimate two orders of magnitude higher than those of the metal budget. As mentioned before, this can be explained by the high concentration of Al in the soil of the area where the reservoirs are built (Table 15.4). The settling of even small amounts of soil particles within the sediment traps can result in a very high estimate of the sedimentation rate.

In spite of the differences found between the two methods, the data from the sediment traps indicate that the metals accumulate in the sediment of the reservoirs and that settling is the main or single process involved in the removal of the metals.

15.3.8
Parameters Affecting the Removal of Metals in the Reservoirs

No correlation was found between the operational parameters of the reservoirs (residence time, PFE_5, PFE_{30} and water level) and the concentration of metals in the outflow, the reduction in concentration, the removal, or the sedimentation rates. A comparison of Tables 15.3 and 15.4 indicates that there is no correlation between the reduction in concentration (the difference in metal concentration between inflow and outflow) and the removal (the difference in metal amount between inputs and outputs). Two independent studies in the removal of metals in a series of stabilization ponds (Smillie and Loutit 1982 in New Zealand; Carre et al. 1986 in France) concluded that most metal removal occurred in the first pond of the series with a residence time of few days. These results suggest that metal removal is a quick process and that the residence time obtained in the Kishon stabilization reservoirs (more than 80 days) is much longer than the time required for metal removal. This can explain the lack of correlation between metal removal and the covered range of residence time values.

Sigg et al. (1987), in a work on the vertical transport of metals through settling in lakes, found that biological material is the main carrier of settling metals. The sedimentation rates depend more on the population dynamics of plankton (affected by nutrients-phytoplankton and phyto-zooplankton interactions, temperature, solar radiation, etc.) than on the loading of metals through effluents or the age distribution of the effluents within the lake. In the case of earthen reservoirs with strong slopes like the Kishon ones, shore erosion by wind induced currents also affects the sedimentation pattern.

15.4
Conclusions and Recommendations

In reactors with controlled discharge such as the stabilization reservoirs for the seasonal storage of effluents, both the outflow concentrations and the outflow volumes must be included in the evaluation of the performance of the reactor. A comparison between inflow and outflow concentrations alone does not allow a correct evaluation of the removal of heavy metals or other pollutants.

The concentrations of the five metals were reduced between 20% and 75%, back to the base level found in unpolluted groundwater in Israel. The amounts were reduced by 5% for Pb (2.5 kg), 10% for Cu (24 kg), 30% for Al (2.5 t), 50% for Cr (78 kg), and 90% for Zn (1.3 t).

Settling and accumulation in the bottom was the main or single process involved in the removal of metals from the effluents during storage. No correlation was found between the removal of metals and the operational parameters of the reservoirs.

In reservoirs with a seepage rate of 3–4 mm d^{-1}, losses of metals by leaching can amount to 3–6% of the input. Bottom sealing is recommended in those reservoirs which receive significant amounts of heavy metals and/or other pollutants which may leach through the bottom to groundwater.

The release of bottom sediments in the outflow from the reservoirs means a direct release of trace metals, as well as of other settling pollutants (e.g., faecal coliforms and nematode eggs) and of particles which can clog the irrigation networks. Thus, it is recommended that the effluents be taken from the uppermost water layer, and that strong outflow rates which may drag out part of the sediments by hydraulic turbulence be avoided. It is recommended that the outlet of future reservoirs be located away from the dominant wind axis. In the studied reservoirs the outlets are located in the areas where the easterly wind accumulates resuspended sediments. It is recommended to stop the outflow from these reservoirs when the easterly wind blows.

The main tools available to improve the removal of heavy metals in stabilization reservoirs are not the manipulation of the age distribution of the effluents and/ or the loading of the reservoirs, but the proper location, design and flow rate of the outlet.

Acknowledgements

This work was made within the framework of the Kishon Complex Monitoring Program – Technion, sponsored by the Mekorot Water Company of Israel.

This article is reproduced from Water, Air and Soil Pollution 82:617–633, 1995, with permission of Kluwer Academic Publishers.

References

APHA, AWWA, WPCF (1989) Standard methods for the examination of water and wastewater. 17th edn

Avnimelech Y, Wodka M (1988) Accumulation of nutrients in the sediments of Ma'aleh HaKishon reclaimed effluents reservoir. Wat Res 22(11):1437–1442

Azov Y, Shelef G (1991) Effluents quality along a multiple stage wastewater reclamation system for agricultural reuse. Wat Sci Tech 23(10–12):2119–2126

Carre J, Baron D, Legeas M, Maurin J (1986) Heavy metal, bacterial, and viral contamination of sewage sludges in oxidation ponds. Environ Technol Lett 7(2):119–27

Fergusson J (1990) The heavy metals: Chemistry, environmental impact and health effects. Pergamon Press, 614 pp

Ghosh TK, Kshirsagar SR, Kshirsagar DG, Kumar A (1985) Leachability of heavy metals through waste stabilization ponds. Asian Environ 6(5):29–34

Juanicó M, Friedler E (1994) Hydraulic age distribution in perfectly mixed non steady-state reactors. ASCE J Environ Eng 120(6):1427–1445

Kaiser KLE (1980) Correlation and prediction of metal toxicity to aquatic biota. J Fish Res, Board of Can 37(2):211–218

Kaplan D, Abeliovich A, Ben-Yaakov S (1987) The fate of heavy metals in wastewater stabilization ponds. Wat Res 21(10):1189–1194

Leeden F, Troise F, Todd D (1990) The water encyclopedia. Lewis Publ. Inc., 2nd edn, USA, 808 pp

Moshe M, Betzer N, Kott Y (1972) Effect of industrial wastes on oxidation pond performance. Wat Res 6(10):1165–71

Rebhun M, Ronen D, Eren J (1987) Monitoring and study program of an inter-regional wastewater reclamation system for agriculture. Journal WPCF 59(5):242–248

Ronen D, Kanfi Y, Rebhun M (1975) Trace metals in Israel groundwater: The coastal plain aquifers. Water Commission, Report 75(1)

Sigg L, Sturm M, Kistler D (1987) Vertical transport of heavy metals by settling particles in Lake Zurich. Limnol Oceanogr 32(1):112–130

Smillie RH, Loutit M (1982) Removal of metals from sewage in an oxidation pond system. N Z J Sci, 25(4):371–6

Suffern JS, Fitzgerald CM, Szluha A (1981) Trace metal concentrations in oxidation ponds. J Wat Pollut Control Fed 53(11):1599–608

Weber B, Juanicó M (1990) Variability of effluent quality in a multi-step complex for wastewater treatment and storage. Wat Res 24(6):765–771

Wolfberg A, Kahanovich Y, Avron M, Nissenbaum A (1980) Movement of heavy metals into a shallow aquifer by leakage from sewage oxidation ponds. Wat Res 14(6):675–9

Chapter 16

The Clogging Capacity of Effluents

Benjamin Teltsch

16.1
Introduction

Effluent reservoirs provide a special aquatic environment in which physical and biological processes affect water quality and influence the reservoir effluent characteristics (Dor et al. 1987). The effluent is often applied through drip irrigation systems. Besides the advantages as an efficient irrigation method, drip irrigation with wastewater effluent provides safety in terms of preventing dispersion of bacteria and viruses into the air and on the irrigated crop canopy and produce (Libhaber and Mintzker 1990). Clogging of the drippers is the greatest problem of these systems. The use of filters at the head of the irrigation network sharply improves the network performance, but the filters themselves may clog so frequently that the costs of manual or automatic backwash are too high (Juanicó et al. 1995). Hence, the performance of the drip irrigation systems is highly dependent on the filterability of the irrigation water.

The relationships between the clogging capacity and wastewater reservoir effluents are analysed in this chapter. Different reservoir management strategies including various chemical, operational and biological parameters, and their effect on the clogging capacity of reservoir effluent are herein discussed.

16.2
Drip Irrigation

Drip irrigation systems reduce water use by up to 75%, minimize runoff, and save up to 85% in energy costs, while increasing application efficiency (Verkade and Fitzpatrick 1986). Unfortunately, clogging of drippers can adversely affect both the rate of water application and the uniformity of water distribution (Bucks et al. 1979). Drip irrigation, which uses a series of pipes, tubes, and emitters to apply water more precisely to specific zones around plants, is the method most susceptible to water quality.

The clogging process was found to be related to the dripper design and the degree of filterability of the source water (Gilbert et al. 1980). The low flow rate (1–8 l h^{-1}), the length of the flow path within the dripper (25–960 mm), and the size of the passages in the drippers (0.5–1.5 mm) are among the main reasons for the drippers' clogging (Adin 1986; Adin and Sacks 1991). The clogging process is gradual. Bacterial slime, to which other particles adhere, is the primary cause (Ford 1978; Adin and Sacks 1991). Clogging of the drippers by algae only occurs when they attach themselves to other particles. The resulting sediment contains skeletal remains of zooplankton and whole, live algae.

Adequate water filtration is a primary requirement for reliable dripper operation. Different types of filters are being used. These include strainers, granular deep bed filters, and filters based on a combination of both mechanisms (Light 1993). Since the particle removal mechanisms in the above filter types are different, the causes for filter clogging are also different. Large particles which form a compressed cake are the primary reason for screen filters clogging (Adin and Elimelech 1989; Juanicó et al. 1995), while the causes for clogging in granular, deep bed filters are more complex and relate to grain size, filtration rate, and media depth (Adin and Alon 1993).

16.3
Measurements of the Clogging Capacity of Effluents on Screen Filters

An important step toward efficient irrigation and the saving of water is the analysis of the clogging capacity of reclaimed wastewater, which is defined as the potential of water to clog screen filters. Such analysis requires water quality criteria for determining the clogging capacity during storage in open reservoirs and the identification of components which are responsible for filter clogging. There are no simple or precise quantitative methods to determine the potential clogging capacity of water on filters. The traditional tests of turbidity, total suspended solids, or particle size distribution are indirect. They may only be used after the establishment of correlations between the measured parameters and filter performance for each type of water. Some of the tests can only be performed in a laboratory, and particle qualities may change during the transport of samples.

16.3.1
Filterability Index and Filter Specific Resistance

Adin and Alon (1986) suggested that the filterability index "*I*" based on Boucher's Law (Boucher 1947) can be used as an index for the screen filter resistance. An increasingly frequent filter clogging can be expected with increasing "*I*" values. The resistance of the cake formed on the screen is also an important parameter. McCabe and Smith (1971) define the cake's characteristics in terms of compressibility. Adin and Alon (1986), based on the McCabe and Smith formula, developed the "specific resistance" parameter which is defined as the resistance of a filter cake having a unit weight of dry solids per unit area of filtration surface. The values for these parameters can be derived from pilot experiments.

16.3.2
"ATMMIN" and "ATMLIT"

Ben-Harim and Steinhauer (1986) evaluated the possibility of developing a simple "filterability test" for field measurements and concluded that such a test can be based only on the simulation of screen filters (strainers), rather than on the simulation of granular deep bed filters. The instrument that was developed measures the differential pressure needed to maintain a constant flow of water through a standard screen filter. At the starting point, when the mesh is clean, the differential pressure is set at zero; then, as the screen starts to clog, higher and higher differential pressures are necessary to maintain a constant flow of water through the filter. The instrument was suited for field work on DC 12V current and can be operated from a boat allowing for direct measurements at any water depth in the reservoir.

This test has the following advantages:

- It is a direct test based on the actual clogging of a real screen.
- It is performed in operational conditions similar to those of the full-scale field screen filters (mesh size, filtration velocity, etc.).
- There is no need to transport samples.
- Results are obtained in the field within 10–30 min.
- There exists the possibility of analysing the nature of the clogging matter because it remains in the mesh of the instrument.

The operational advantages of this test have been demonstrated in several works (Teltsch et al. 1989; Teltsch et al. 1991; Teltsch and Leventer 1993; Juanicó et al. 1995). Teltsch et al. (1991) proposed a new quality criterion for reclaimed wastewater used for drip irrigation based on this "filterability test": the clogging capacity of the water. It is expressed as the rate of the increase in differential pressure with time – (atmospheres/minute) × 1 000 – and was called *ATMMIN*. Another parameter is the rate of the increase in differential pressure with filtered volume – (atmospheres/litre) × 1 000 – and was called *ATMLIT*.

These parameters were derived from controlled experiments using synthetic suspensions of clay, algae (*Scenedesmus* sp.) and zooplankton (*Daphnia magna*), and performed with the above filterability test apparatus (Wodowski 1991). The compliance of *ATMMIN* and *ATMLIT* parameters to Boucher's Law was studied by comparing them to the filterability index "*I*" and to the specific resistance. The correlation coefficients of *ATMMIN* and *ATMLIT* with "*I*" were 0.80 and 0.95 respectively (Fig. 16.1), and of *ATMMIN* and *ATMLIT* with the specific resistance were 0.083

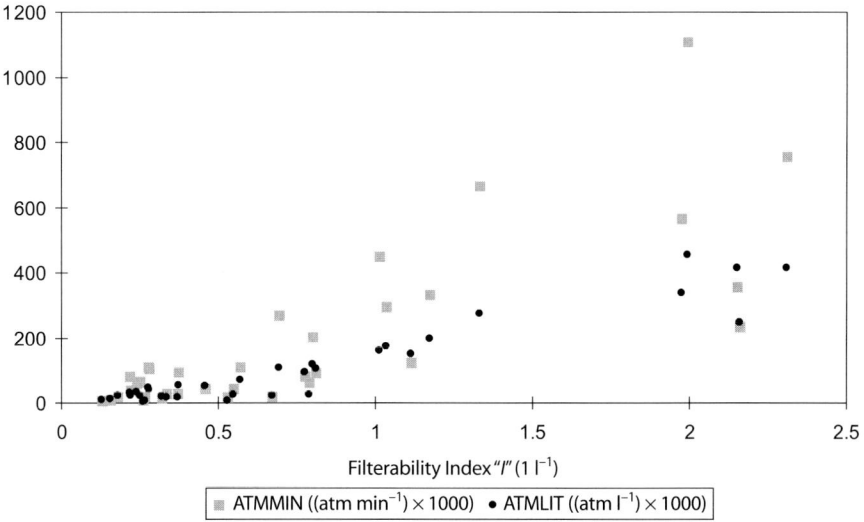

Fig. 16.1. The rate of the increase in differential pressure with time (*ATMMIN*) and the rate of the increase in differential pressure with filtered volume (*ATMLIT*) as a function of filterability index, *I* (derived from Wodovski 1991)

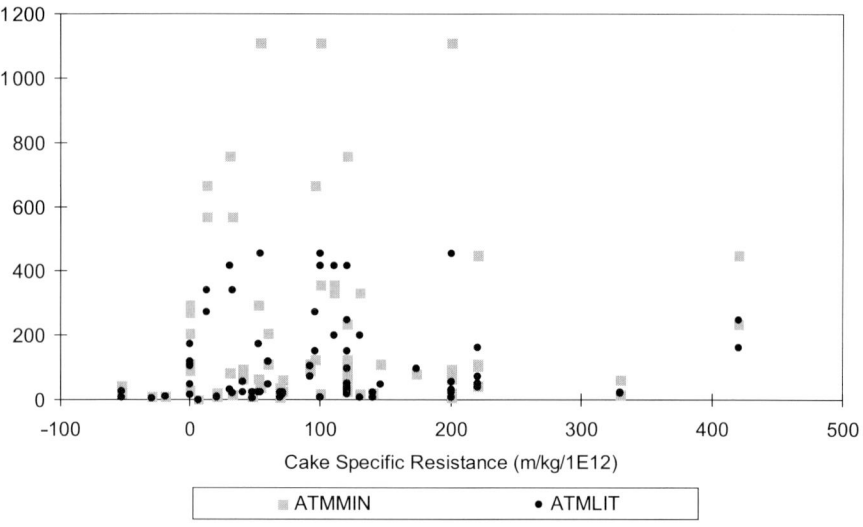

Fig. 16.2. The rate of the increase in differential pressure with time (*ATMMIN*) and the rate of the increase in differential pressure with filtered volume (*ATMLIT*) as a function of cake specific resistance (derived from Wodovski 1991)

and 0.034 respectively (Fig. 16.2). The lack of a linear fit between *ATMMIN* and *ATMLIT* is due to variations in the filter cakes which were produced by the different experimental suspensions. Thus, *ATMMIN* and *ATMLIT* computed from the filterability test can be used for predicting the clogging capacity of the irrigation water in full-scale filter screens. It cannot, however, provide useful information on the filter cake be-haviour.

16.3.3
Clogging Time

In more sophisticated filters, which include automatic backwash mechanisms, the evaluation of the filter performance may be done by determining the time between consecutive backwashes. Terminating the run in the "filterability test" apparatus at a certain differential pressure, which is equal to a clogging event, may create a new measured parameter "clogging time" in minutes. This parameter is very convenient when comparing the "filterability test" results to data on clogging events in full-scale screen filters that has been received from farmers.

16.4
Causes for Screen Filter Clogging

Two simultaneous processes occur during straining (Ives 1960):

- deposition of particles within the pores
- deposition of material on the screen, forming a filter cake. The cake is porous and contains tortuous capillaries, irregular in diameter (Adin and Alon 1986)

Fig. 16.3. The rate of the increase in differential pressure with time (*ATMMIN*) and the rate of the increase in differential pressure with filtered volume (*ATMLIT*) as a function of flow rate (derived from Wodovski 1991)

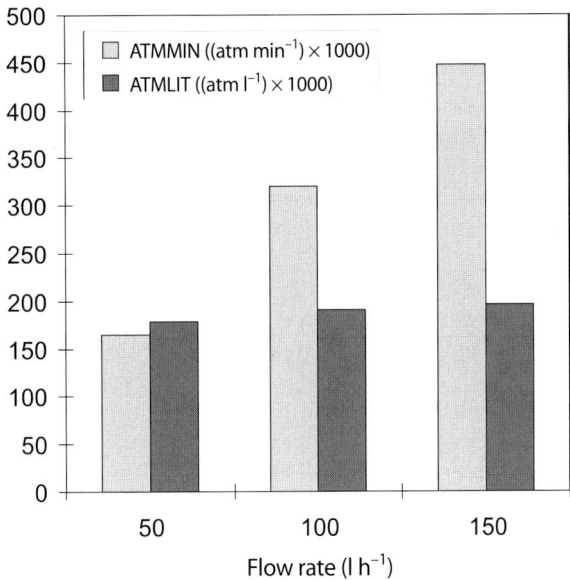

Fig. 16.4. The rate of the increase in differential pressure with time (*ATMMIN*) and the rate of the increase in differential pressure with filtered volume (*ATMLIT*) as a function filter mesh size (derived from Wodovski 1991)

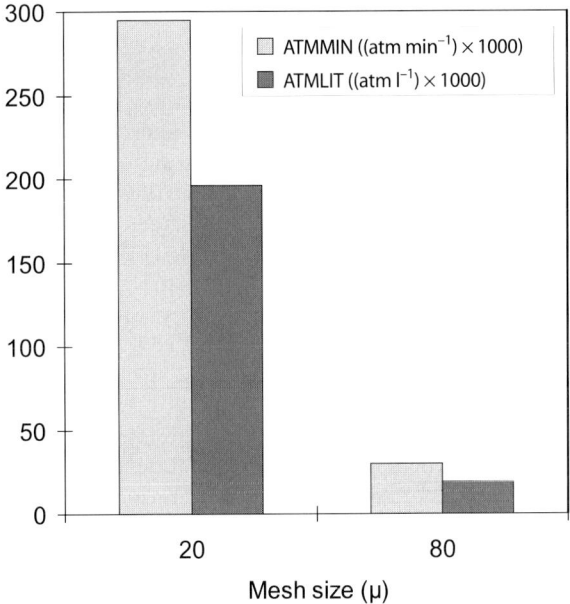

The causes for screen filter clogging can be divide into two related categories:

- causes related to filter operation
- causes related to the effluent

16.4.1
Causes Related to Filter Operation

An increase in the flow rate through the filter (Fig. 16.3; Wodowski 1991) and an increase in filtration velocity (Adin and Alon 1986; Adin and Elimelech 1989) will increase the head loss rate. These observations may be attributed to the fact that, with higher flow rates or filtration velocities, more particles per time unit reach the screen. Therefore the opportunity for bridging and coupling between particles and screen fibers increases. It was also found that the head loss directly correlates to the inlet pressure. This, together with high filtration velocities, increases the intensity of the collisions between the arriving particles and the filtration cake, thus compressing the cake and enhancing the screen clogging. Due to the straining mechanism, it follows that an increase in the mesh size grade is inversely related to the rise in head loss rate, as shown in Fig. 16.4. As the filter porosity increases, the chance for deposition of particles within the pores decreases; the cake forming stage therefore becomes longer, and the rate of head loss decreases.

16.4.2
Causes Related to the Effluent

Since material accumulation on the filter screen tends to reduce the filterability of the screen, higher concentrations of suspended solids in the wastewater effluents will increase the clogging capacity of the water. However, other effluent particle characteristics, such as size, shape, volume, and number distribution, as well as their compressibility and adhesiveness, are important properties when the filterability of the effluents is considered. Typical wastewater reservoir effluent is composed of a majority of very fine particles, less than 10 μ (Adin and Elimelech 1989; Teltsch et al. 1991; Teltsch et al. 1992). These make a notable contribution to the total volume of the particles. Nevertheless, the immediate cause of screen clogging is that of large particles, which form a filter cake on the screen. Juanicó et al. (1995) found that screen filter clogging in reservoir effluents was primarily due to an increase in the number of the large copepod *Mesocyclops oqunnus* (size range 1700 μ) and the cyanobacteria *Spirulina platensis* (size range 700–1000 μ). Zooplankton levels are reported to be highly correlated to filterability test data (Teltsch et al. 1989). In order to confirm the assumption that large particles are the main cause of the poor filterability of the water, filterability tests were performed on samples that had been prefiltered with a 200 m screen filter.

Prefiltration improved filterability fourfold (Teltsch et al. 1989). Adin and Elimelech (1989) also reported that pretreatment by granular filtration before screen filter runs resulted in a radical improvement. Often, the maximum clogging capacity of water on screen filters is found where both the large plankton species and the small particles are present in high concentrations, causing a decrease in filter cake porosity. Adin and Alon (1986) reported that screen clogging occurs even when most of the particles in the suspension are one order of magnitude smaller than the screen pore, due to both a bridging mechanism in the screen and an attachment mechanism in the filter cake, resulting in filter cake thickening. A detachment process can also occur, and, as a result, aggregates become separated from the cake and penetrate the screen to appear in the filtrate, thus forming pores in the filter cake and gradually decreasing the final head loss.

Microbial activity has an important role in the clogging process and organic microbial products act as bioflocculant material. Wastewater reservoirs' effluent containing both sulfides and a low level of dissolved oxygen can be a source for microaerophile bacteria, such as the colorless filamentous sulfur bacteria, *Beggiatoa alba*, which may form slime layers on the filters. Pumping water from an anaerobic layer in the reservoir does not always solve the problem, since air can penetrate in many filter systems especially during backwash.

16.5
Reservoir Management Alternatives

The clogging capacity of the effluents from reclaimed wastewater reservoirs can be effectively reduced by means of biological control (biomanipulation) and/or chemical treatment and/or proper reservoir management practices.

16.5.1
The Biological Approach

Stocking reservoirs (potable and agricultural water) with fish in order to cope with 'nuisance organisms' and to improve the water quality is an attractive approach which is becoming a common practice in Israel (Leventer 1979; Leventer and Teltsch 1990; Teltsch and Leventer 1993).

Reclaimed wastewater reservoirs are ecologically unstable water bodies characterized by outbursts of growth of certain links in the food web, such as phyto- and zooplankton. These cause considerable problems by clogging filters receiving the reservoir effluents. Cyanobacteria, which are liable to clog filters and irrigation systems, are widespread in irrigation reservoirs not populated by fish in the summer season (Dor 1990). Experiments with the filter–feeding fish – silver carp (*Hypophthalmichthys molitrix*), big-head carp (*Aristichthys nobilis*) and their hybrids – conducted in enclosures and large tanks by Sagi (1992; 1993), indicated that although the total concentration of the planktonic chlorophyll a was not always reduced by the filter-feeding fish, the algae community was shifted to that of smaller size species. Blooms of Cyanobacteria occurred only in the enclosures without fish. During periods of bloom, the corresponding concentration of the planktonic chlorophyll a was also significantly higher in the enclosures without fish. Zooplankton concentration in the enclosures with no fish was much higher (up to 10 times or higher) compared with the enclosures with filter-feeding fish. The effect of grazing by fish on attached algae was studied in a secondary effluent reservoir by placing paving-stones, covered or uncovered by plastic nets, on the reservoir bottom at different depths. Bottom-feeding fish, such as the hybrid Tilapia (*Oreochromis niloticus × O. aureus*), could feed on the bentonic biomass grown on the uncovered paving-stones and could not feed on the covered paving-stones. The ratio of chlorophyll a to total organic dry weight (autotrophic index) was calculated from monthly measurements of the total solids accumulated on the paving-stones. The index was lower for the uncovered paving-stones by 50% to 75%, indicating that the bottom-feeding fish consume the attached algae very efficiently. In order to evaluate the feasibility of controlling the clogging capacity of water on screen filter by using fish, other experiments with enclosures were performed in a

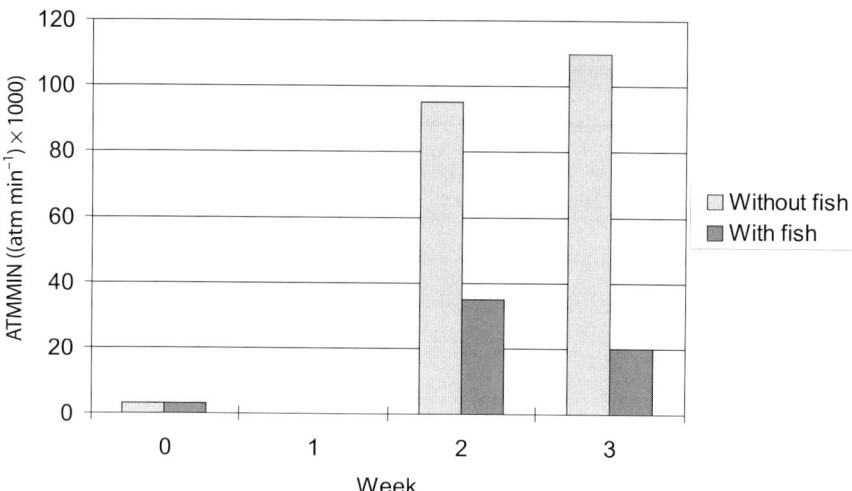

Fig. 16.5. The rate of the increase in differential pressure with time (*ATMMIN*) as a function of fish treatment in 50 m³ enclosures

wastewater stabilization reservoir (Teltsch et al. 1991). Enclosures of 50 m³ size were stocked by eight filter-feeding planktivorous fish silver carp (*Hypophthalmichthys molitrix*). Filterability test measurements, as well as chemical and biological parameters, were carried out over a period of 3 weeks. Figure 16.5 shows the evolution of *ATMMIN* values in the two treatments (with fish and without fish). The clogging capacity at the end of the run was lower in the enclosure with fish than in the enclosure without fish. Chlorophyll a levels showed the same trend – 20 and 60 µg l⁻¹ in the enclosure with fish and without fish respectively. The effect of filter-feeding fish on the algae biomass may be attributed to the antagonistic/synergistic effect of these fish on the algae populations. The fish are able to filter daily a quantity of water which is estimated at five times the order of magnitude of their own volume (Gulati and Van Donk 1989). Large algae and zooplankton, which are larger than the passages of the gill rakers, are swallowed by the fish and pass through their digestive system, thus reducing the algae biomass levels in the reservoirs. However, excessive densities of silver carp may lead to a dramatic increase in the algal standing crop by promoting small algae populations (Opuszynski 1981; Milstein et al. 1988). The development of small algae (ichthyo-eutrophication) is promoted by the silver carp in three ways: directly, by selectively not eating them; indirectly, by eliminating larger cells which compete for nutrients; and by eliminating zooplankton which grazes on small algae (Leventer and Teltsch 1990; Milstein 1992). Therefore, a decrease in the clogging capacity can be achieved in spite of an increase in the algal standing crop.

Sagi (1992) in a survey of 25 reservoirs, varying in size from 0.5 to 60 ha, concluded that in all the reservoirs stocked with filter-feeding fish, the presence of the fish has contributed to a change in the algae population, as well as to a decline in the density of the zooplankton, thus alleviating the problem of water filtration and reducing the clogging in the irrigation systems.

16.5.2
The Chemical Approach

There are relatively few chemicals available for use in reservoirs. Of these, some many have restrictions on their use. Some of the reasons are: 1) there are human and environmental hazards associated with the chemical treatment, and 2) when chemicals are applied directly to water, they seldom remain specifically at the site of application. Residue tolerance in crops and acceptable residue levels in water may be altered over extended periods of time (Hansen et al. 1983). Two groups of chemicals are commonly used to control filter clogging: copper (or other metals) salts and oxidants.

16.5.2.1
Copper Sulfate

This salt of copper highly soluble in water is the most commonly used chemical for the control of undesirable plankton growth in reservoirs. The copper ion affects the algal cell membrane inhibiting growth and interfering with the photosynthesis process resulting in starvation of the algae. Copper also causes deactivation of micro-organisms through interference at the metal-enzyme level, for example, interference in metabolic pathways (Chen 1995). Different species of algae and bacteria have different sensitivities to the cupric ion. The concentration of copper sulfate required to control plankton in water-supply reservoirs varies from as little as 0.05 or 0.10 mg l^{-1} for some organisms to as much as 12 mg l^{-1} for very resistant species (Chancellor 1958). The dissolution rate of copper sulfate is influenced by: application methods (various slug techniques, suspension-dissolution, and continuous methods), crystal size, water temperature, and water flow velocity. Alkalinity, suspended matter, and water temperature are water quality factors affecting the effectiveness of copper sulfate (Hansen et al. 1983). Alkalinity is the principal one. Copper ions react with bicarbonate and carbonate ions in water to form insoluble complexes which precipitate thus reducing the amount of

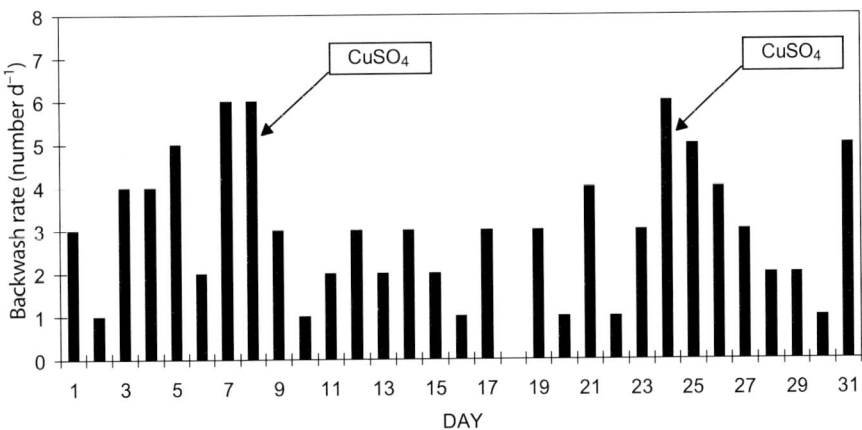

Fig. 16.6. Backwash rate as a function of time and $CuSO_4$ treatment

biologically active copper (Meador 1991). Its effectiveness is significantly reduced when the bicarbonate alkalinity exceeds 150 ppm. The toxicity of copper is also reduced due to complex formation by ligands, e.g., with humic acids (Tubbing et al. 1994).

The application of copper sulfate is used in many wastewater reservoirs as a method for improving water filterability. The effect of copper sulfate treatment in conjunction with an agricultural screen filter with a mesh of 125 μ (M103B, Filtomat Ltd., Israel) was studied (Azov et al. 1991). The backwash rate of a filter receiving water from a secondary effluent reservoir as a function of time, and the effect of copper sulfate treatments of the whole reservoir on the clogging capacity of the water are shown in Fig. 16.6. Treatment of the reservoir with 1 mg l^{-1} of copper sulfate reduced the backwash rate by 66%. This effect is limited in time, since there is regrowth of plankton in the reservoir.

16.5.2.2
Oxidants

Oxidants, especially Chlorine, are used as the cheapest treatment for controlling microbial activity in water used for micro-irrigation (English 1985). However, only a few studies have been undertaken on the use of oxidants in preventing filter clogging. The effect of common oxidants such as Chlorine (Cl_2) and chlorine dioxide (ClO_2), on effluent constituents, which may take part in filter clogging, are extensive.

Algal cell viability and chlorophyll concentration decrease as a result of Cl_2, ClO_2 and ozone (O_3) application to *Scenedesmus* sp. cultures (Sukenik et al. 1987). Similar results were reported by Rav-Acha et al. (1995), who observed that, although the chlorophyll content and replication ability were reduced by more than 90% 20 min after the application of 10 mg l^{-1} Cl_2 or ClO_2, the number of algae cells (*Chlorella vulgaris*) and their structure were not affected by the oxidants. Viability reduction therefore does not necessarily imply elimination of clogging potential. The algicidal efficiency of 1.5 mg l^{-1} Cl_2 was higher for the Cyanobacteria *Phormidium minnesotense* (more than 99% viability inhibition) than for the Chrysophyta *Pleurochloris pyrenoidosa* and the Chlorophyta *Oocystis* sp. (only 35% and 50% inhibition respectively) in cultures isolated from swimming pools (Sommerfeld and Adamson 1982).

Mucilages, wax esters, oils and other food storage products of plankton are released into the water, especially by crushed plankton trapped on the filter (Edyvean and Sneddon 1985). Chlorine has a direct effect on algal surface and cell contents, so that cellular components leak from the cell (Sukenik et al. 1987) and may accumulate on the filter fibers, thus providing an anchorage for blockage. Adding higher doses or stronger oxidants brings further oxidation and cleavage, yielding smaller molecules. This may result in enhancement of microalgal flocculation (Sukenik et al. 1987) and aggregation of suspended particles (Rav-Acha et al. 1995) through a destabilization mechanism.

Experiments which were carried out in order to test the long term performance of commercially available drippers and filters, under controlled operational conditions, were performed (Ravina et al. 1990, 1992). The water source was a storage reservoir with mixed wastewater and fresh water (Teltsch et al. 1989, 1992). The main findings were that filtration at mesh size of 125 μ is not superior to that of 180 μ and that daily chlorination is very effective in reducing clogging problems in the filtering and dripper systems. It was concluded that the aggregates formed in the chlorinated water resulting from the microalgal flocculation and coagulation with organic products were either very compressible, thus passing through the filters, or disintegrated when passing through the filters (Ravina 1992). These results confirm the findings of Hsiung (1980) that the head-loss buildup across the filter was much slower for chlorinated effluent than for unchlorinated effluent, since the flocks in the chlorinated effluent were smaller, lighter, and more fragile.

In some situations, short backwash cycling developed because the backwash failed to wash away the algae material entangled in the filter. Treatment by backwash combined with chlorination is very effective (Ravina et al. 1990), but chlorination with too high doses may cause harm to the drippers, especially to the rubber diaphragm of the self-regulating dripper (Ravina and Sagi 1993).

16.5.3
The Operational Approach

The operational approach includes the timing of the addition of wastewater effluents to reservoirs, changing water level and retention time, periodic drying, etc.

The importance of the timing of effluent input on the clogging capacity of water in a storage reservoir of mixed wastewater and fresh water was discussed by Teltsch et al. (1991) and is presented in Fig. 16.7. The response of algae populations and the clogging capacity of the water in relation to the inflow of effluents to the reservoir become expressed by the increase in chlorophyll a concentrations and *ATMMIN* values. *ATMMIN* values were the worst when plankton growth was maximal, and decreased after inflow of effluents ceased. There is no simple correlation between chlorophyll a and *ATMMIN*, since not only phytoplankton but also zooplankton has an important role in the clogging of the filter. The same data show the importance of the water level in determining the clogging capacity of the water. In shallow reservoirs, a decrease in water level results in a sharp decrease in surface area and volume. Hence, the trophogenic layer and the residence time also decrease, leading to a reduction in the quantities of autotrophic and heterotrophic organisms. This reduction can explain the differences in *ATMMIN* following the input of effluent to full or to an almost empty reservoir. In the studied reservoir, maintaining a low and stable water level in the summers of 1986 and 1987 led to a decrease in residence time and the trophogenic layer and consequently in all biological activity, resulting in lower clogging capacity than in summer 1985 (Teltsch et al. 1989).

The hydraulic parameter which best represents reservoir behaviour and has the highest correlation with effluent quality parameters is the percentage of fresh effluent within the reservoir. This is computed from daily inflow, outflow, and reservoir volume data (Juanicó and Shelef 1994). A correlation analysis was made between *ATMMIN* and the percentage of effluent with 3 days retention in a stabilization reservoir (Teltsch et al. 1991). The clogging capacity of the water increased rapidly by the end of the irrigation season, when the reservoir was almost empty and the percentage of fresh effluent was higher. However, the correlation was not especially high ($r = +0.68$), indicating that the development of clogging particles is not a simple function of nutrients supply but also depends on other factors, such as hydraulic flow patterns, water level, solar radiation, population dynamics, etc.

The timing of the input of reclaimed effluent to the reservoirs can also be used as an operational tool to control the clogging capacity of the water. The separation be-

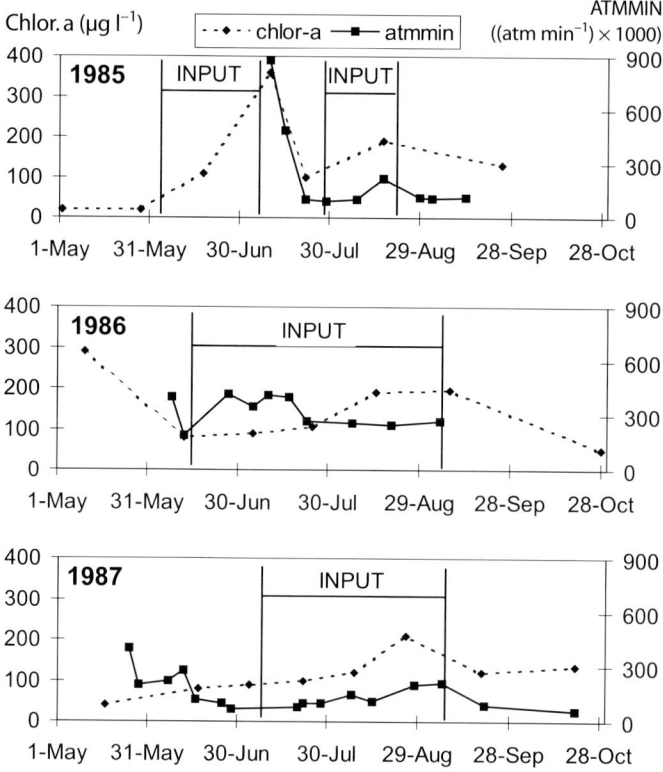

Fig. 16.7. Chlorophyll a to total organic dry weight ratio (Autotrophic Index) of covered and uncovered artificial substrate at 3 m depth on the reservoir bottom on different sampling dates

tween the time of effluent input and the increase in the clogging capacity following the input of the effluent to the irrigation high season will lead to lower clogging capacity in the irrigation water.

16.6
Concluding Comments

The relationship between the drip system and the filtration system is paradoxical: filters are installed in order to protect the drippers from clogging, but this causes the filter itself to become clogged. Therefore, the problem of drip clogging decreases at the expense of a filter clogging increase and vice versa. Various solutions to this situation are being proposed by both the dripper equipment manufacturers and the filtration equipment manufacturers but – undoubtedly – the best solution is to improve source water quality by reducing its clogging capacity. Considering the reservoir as an ecosystem enables the execution of trophic-level manipulations which, in addition to their efficiency in reducing the clogging capacity of the water, do not consume chemicals and do not constitute a risk to the environment.

References

Adin A (1986) Problems associated with particulate matter in water reuse for agricultural irrigation and their prevention. Wat Sci Tech 18 (Split):185–195

Adin A, Alon G (1986) Mechanisms and process parameters of filter screens. J Irrig Drain Engrg, ASCE, 112(4):293–304

Adin A, Elimelech M (1989) Particle filtration for wastewater irrigation. J Irrig Drain Engrg, ASCE, 115(3):474–487

Adin A, Sacks M (1991) Dripper clogging factors in wastewater irrigation. J Irrig Drain Engrg, ASCE, 117(6):813–826

Adin A, Alon G (1993) The role of particle characterization in advanced wastewater treatment. Wat Sci Tech 27(10):131–139

Azov Y, Juanicó M, Shelef G, Kanarek A, Priel M (1991) Monitoring the quality of secondary effluents reused for unrestricted irrigation after underground storage. Wat Sci Tech 24(9):267–275

Ben-Harim I, Steinhauer Z (1986) A new filterability test for water quality control. In: Dubinsky Z, Steinberger Y (eds) Environmental quality and ecosystem stability, vol 3A/B, Bar-Ilan Univ Press, Israel, pp 901–905

Boucher PL (1947) A new measure of the filterability of fluids with applications to water engineering. J Inst Civil Engrg 27:415–421

Bucks DA, Nakayama FS, Gilbert RG (1979) Trickle irrigation water quality and preventive maintenance. Agri Wat Manag 2(2):149–162

Chancellor AP (1958) The control of aquatic weeds and algae. Ministry of Agriculture, Fisheries and Food, Her Majesty's Stationery Office, London, 20, 2s, 6d

Chen J (1995) Copper salts as an algicide/bactericide for reservoirs. New World Water, pp 67–69

Dor I (1990) Deep wastewater reservoirs in Israel: Empirical data for monitoring and control. Wat Res 24(9):1077–1084

Dor I, Schechter H, Bromley HJ (1987) Limnology of hypertrophic reservoir storing wastewater effluent for agriculture at Kibbuz Na'an, Israel. Hydrobiologia 150:225–241

Edyvean RGJ, Sneddon AD (1985) The filtration of plankton from seawater. Filtration and Separation, May/June, pp 184–189

English SD (1985) Filtration and water treatment for micro-irrigation. Proc. 3rd Intern. Drip/Trickle Irrig. Congress. Fresno CA, Amer. Soc. Agric. Eng., St. Joseph, WI, p 50

Ford HW (1978) Bacterial clogging in low pressure irrigation systems. In: Proc. Irrig. Asso., 1978 Technical Conference, Cincinnati, Ohio, Silver Spring, Md., pp 239–244

Gilbert RG, Nakayama FS, Bucks DA, French OF (1980) Emitter clogging and other flow problems. Agri Wat Manag 3(3):159–178

Gulati RD, Van Donk E (1989) Biomanipulation in the Netherlands. In: Fresh water ecosystem and estuarine water. An introduction. Hydrobiol Bull 23:1–4

Hansen GW, Oliver FE, Otto NE (1983) Herbicide manual. U.S. Department of the Interior, Bureau of Reclamation, Denver, Colorado, pp 97–168

Hsiung AK (1980) Chlorine effect on secondary effluent filtration. J Environ Engrg, ASCE, 106(EE3): 649–654

Ives KJ (1960) Filtration through a porous septum: A theoretical consideration of Boucher's Law. J Inst Wat Engineers 17:333–338

Juanicó M, Shelef G (1994) Design, operation and the performance of stabilization reservoirs for wastewater irrigation in Israel. Wat Res 28(1):175–186

Juanicó M, Azov Y, Teltsch B, Shelef G, (1995) Effect of effluent addition to a freshwater reservoir on the filter clogging capacity of irrigation water. Wat Res 29(7):1695–1702

Leventer H (1979) Biological control of reservoirs by fish. Mekorot Water Co., Nazareth-Illit, Israel, pp 16–32

Leventer H, Teltsch B (1990) The contribution of silver carp (*Hypophthalmichthys molitrix*) to the biological control of Netofa Reservoirs. Hydrobiologia 191:47–55

Libhaber M, Mintzker N (1990) Wastewater treatment for agricultural reuse in Israel. Proc. 5th Intern. Conference on Irrig., Tel-Aviv, Israel, pp 203–221

Light C (1993) Taupo water supply treatment pump station microstrainers: The arkal spin klin system. Water and Wastes in N.Z., March, pp 17–19

McCabe WL, Smith JC (1971) Unit Operations of Chemical Engineering, McGraw-Hill Books Co., New York, N.Y., pp 885–904

Meador JP (1991) The interaction of pH, dissolved organic carbon, and total copper in the determination of ionic copper and toxicity. Aquat Toxic 19:13–32

Milstein A (1992) Ecological aspects of fish species interactions in polyculture ponds. Hydrobiologia 231:177–186

Milstein A, Hepher B, Teltsch B (1988) The effect of fish species combination in fish ponds on plankton composition. Aquacult Fish Mgmt 19:127–137

Opuszynski K (1981) Comparison of the usefulness of the silver carp and the bighead carp as additional fish in carp ponds. Aquaculture 25:223–233

Rav-Acha C, Kummel M, Salamon I, Adin A (1995) The effect of chemical oxidants on effluent constituents for drip irrigation. Wat Res 29(1):119–129

Ravina I, Sagi G (1993) Water filtration for controlling cloggings in drip irrigation systems. Palgey Mayim Water Works Association, Yoqne'am, Israel, p 28 (in Hebrew)

Ravina I, Paz E, Sagy G, Yechiely Z, Sofer Z, Lavy Z, Marcu A (1990) Filtration requirements for emitter clogging control. Proc. 5th Intern. Conference on Irrig., Tel-Aviv, Israel, pp 223–233

Ravina I, Paz E, Sofer Z, Marcu A, Shisha A, Sagi G (1992) Control of emitter clogging in drip irrigation with reclaimed wastewater. Irrig Sci 13:129–139

Sagi G (1992) The effect of filter feeding fish on water quality in irrigation reservoirs. Agricult Wat Mgmt 22:369–378

Sagi G (1993) Fish for water quality control in irrigation reservoirs. Pro. 6th Intern. Conference on Irrig., Tel-Aviv, Israel, pp 231–246

Sommerfeld MR, Adamson RP (1982) Influence of stabilizer concentration on effectiveness of chlorine as an algicide. Appl Envir Microbiol 43(2):497–499

Sukenik A, Teltsch B, Wachs AW, Shelef G, Nir I, Levanon D (1987) Effect of oxidants on microalgal flocculation. Wat Res 21(5):533–539

Teltsch B, Leventer H (1993) The use of fish for biological control of supply reservoirs. In: Qian Yi, Hao Jiming, Long Jun (eds) Proc. of Word Congress III on Enging. Environ., Beijing, China, International Academic Publishers, China, pp 397–404

Teltsch B, Ben-Harim I, Eren J, Leventer H (1989) Short term effects of nutrient enrichment on the quality of irrigation water. Wat Res 23(6):719–724

Teltsch B, Juanicó M, Azov Y, Ben-Harim I, Shelef G (1991) The clogging capacity of reclaimed wastewater: A new quality criterion for drip irrigation. Wat Sci Tech 24(9):123–131

Teltsch B, Azov Y, Juanicó M, Shelef G (1992) Plankton community changes due to the addition of treated effluents to a freshwater reservoir used for drip irrigation. Wat Res 26(5):657–668

Tubbing DMJ, Admiraal W, Cleven RFM, Iqbal M (1994) The contribution of complexed copper to the metabolic inhibition of algae and bacteria in synthetic media and river water. Wat Res 28(1):37–44

Verkade SD, Fitzpatrick GE (1986) Trickle irrigation: Is it for you. American Nurseryman, 163(11): 114–116, 118–120

Wodowski I (1991) The filterability test as a means for localization of the causes for screen filter clogging. Thesis presented to The Hebrew University of Jerusalem, in partial fulfilment of the requirements for the degree of Master of Science (in Hebrew)

Chapter 17

Particle Characterization and Filtration

Aver Adin

17.1
Introduction

Open wastewater reservoirs contain a variety of particulates, ranging from ones that have maintained their original nature while passing through the sewerage system and the treatment train to those which emerge in the system due to chemical or biochemical interactions. Some of them are particles which originated in the secondary treatment step, e.g., activated sludge bioflocs that have not settled out at the secondary clarifier. Others are particles that have been developed in the reservoir itself, e.g., algae. Those particulates affect the ecosystem in the reservoir as well as the effluent reuse systems.

Investigations have demonstrated that drip irrigation systems are dependent on virtually particle free water since even colloidal particles can clog dripper pathways (Adin and Sacks 1991). Research work and field observations concerning the performance of drip irrigation systems utilizing either freshwater or wastewater effluents, indicate that the causes of clogging of low rate applicators may be divided into three main categories: 1) suspended matter, 2) chemical precipitation, and 3) bacterial growth (Nakayama et al. 1977). Clogging is actually a combination of some of the aforementioned factors: algae, clay, and corrosion products entrapped within a biological mass cemented with $CaCO_3$ precipitate. It is now clear, however, that the major clogging problems are caused by the presence of suspended particles in the irrigation water (Bucks et al. 1982; Dasberg and Bresler 1985; Adin 1987).

Filtration prevents immediate clogging by large particles. It also plays an important role in protection from smaller particles that cause gradual clogging. Proper design and operation of filters is essential for water reuse in agriculture as well as in industry or municipalities. Particle characteristics are a key factor to successful operations in this area.

Accordingly, this chapter is aimed at providing the reader with an insite on particle characteristics in reservoir influent (secondary) and effluent, which relate to its physical and physicochemical behaviour. The two major filtration processes which are widely applied nowadays in reservoir effluent reuse – mechanical and granular filtration – will be discussed.

17.2
Particle Characterization

Wastewater reservoir effluents contain a variety of particles, often characterized by gelatinous particle shapes (Fig. 17.1) with extremely high filter clogging potential (Adin at al. 1989). This contrasts with activated sludge particles which have a well defined, typically oval shape. It is more difficult to design a filtration process for a mixture of particles due to the unpredictable behaviour within the filter medium.

The differences observed in shape corresponding to different origins of the wastes, are reflected in the particle surface chemical composition as detected by the SEM x-ray analysis (Table 17.1).

Fig. 17.1. Micrograph of parti-cles in wastewater reservoir effluent

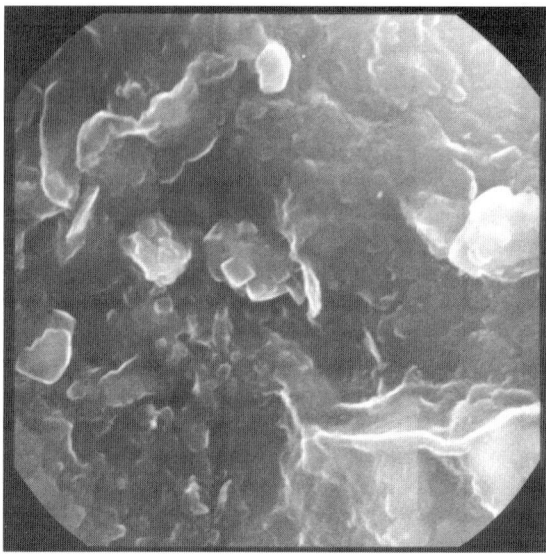

Table 17.1. Particle surface composition in wastewater effluents (after Adin et al. 1989)

Element	Effluent reservoir	Activated sludge	Aerated lagoons	Biological filters
Na	3(4)	3(6)	3(6)	1(3)
Mg	2(4)	4(5)	3(6)	2(3)
Al	4(4)	4(6)	3(6)	2(3)
Si	4(4)	5(6)	6(6)	1(3)
P	2(4)	4(6)	6(6)	3(3)
S	2(4)	5(6)	6(6)	3(3)
Cl	1(4)	5(6)	3(6)	2(3)
K	3(4)	6(6)	5(6)	1(3)
Ca	4(4)	6(6)	6(6)	
Ba	2(4)	4(6)	1(6)	
Fe	3(4)	5(6)	3(6)	
Cu	2(4)	1(6)	1(6)	
Zn	2(4)	3(6)		

Results point out that the particles from the open reservoir are compositionally dominated by Si, whereas the particles from the activated sludge plant are dominated by Cl, Si, and Ca. At the same time, the particles from aerated lagoons and maturation ponds are mostly dominated by P, K, and Ca, which is typical to algal environment. The particle composition of open wastewater reservoirs may change, however, with the season (Dor et al. 1987).

Particles in wastewater effluents are mostly colloidal in nature and negatively charged, thus repelling each other and being stabilised. A common, somewhat qualitative measure of the stabilizing energy is the electrokinetic surface potential, often referred to as ζ (zeta) potential (ZP). ZP measurements of particles originated in wastewater reservoirs and other types of effluents in Israel are presented in Table 17.2. ZP of the particles is negative and varies in the range of –10 mV to –18 mV.

These values indicate that sometimes the repulsive forces among the particles are not as strong as repulsive forces in different types of natural surface water sources and a certain coagulation potential does exist in these systems (Adin et al. 1989).

Biological films and slimes derived from bacteria and algae are among the major biological clogging factors in drip irrigation systems applying wastewater effluents. The individual cells in biofilms are surrounded by extracellular polymers, polysaccharides, and glycoproteins (Jones et al. 1969; Fletcher 1979). Biofilms primarily utilize organic compounds as a source of energy and carbon for cell growth and maintenance.

Such slimy gelatinous deposits of amorphous shape serve as triggers for serious blockage. Particles of a definable shape are found in a matrix of the gelatinous substance and form the primary sediment in the drippers (Fig. 17.2). The primary sediment then acts as a trap for algae and zooplankton limbs that increase the volume of the sediment until they block the water pathway (Adin and Sacks 1991).

Kanaani et al. (1992) and Kanaani (1997) point out that polysaccharides play an important role in biofilm formation and structure. They are formed as a result of bacte-

Table 17.2. Zeta potential (mV) for particles in wastewater effluents

Sampling date	Effluent reservoir	Activated sludge I	Activated sludge II	Biological filters	Aerated lagoons
12/82		−14	−16	−15	−15
1/83			−15	−15	
2/83			−15	−14	
3/83			−10		−12
4/83		14	−13		−10
5/83		−16	−13		−11
6/83	−14				
9/83	−16	−14		−15	−14
10/83	−16	−16		−16	−18
11/83	−17	−14		−15	−13

Fig. 17.2. Biofilm entrapping particles in a dripper (from Adin and Sacks 1991)

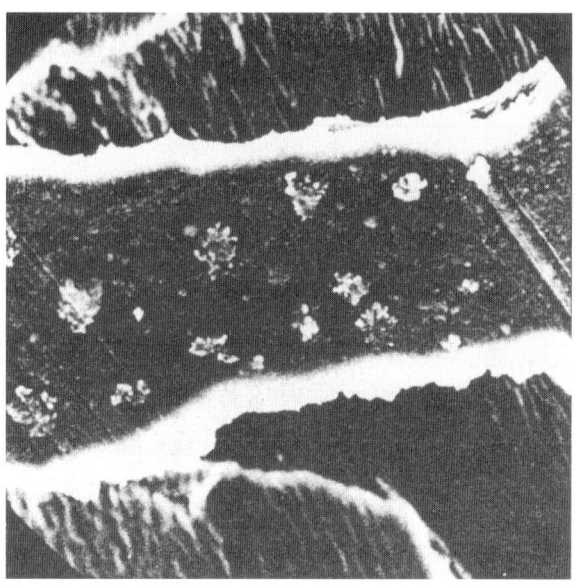

rial activity when microorganisms grow, multiply, and produce extracellular components while being attached to a wall. Suspended clays and other components in wastewater can interact with polysaccharides and cause clogging. These investigators found out that adsorption was increased with the increase in ionic strength of the solution. Adsorption was significant at low pH values (5.7). A significant effect on particle size distribution was marked in the presence of polysaccharides. After equilibrium, time-volume distribution of clay particles was shifted to higher values. In view of these results, polysaccharides have the ability to make good flocculation for clay minerals at considerable ionic strength.

The binary interaction of clay-polysaccharide was postulated in the above mentioned research from the clay-polysaccharide rheology system data, zeta potential, volume size distribution and IR spectrum for the complex clay-polysaccharide. Polysaccharides were shown to develop aggregate flocs which participate, to a greater or lesser extent, in the overall rheological properties of clay-polysaccharide complexes. The ability of polysaccharides to develop floc is seen as an important factor to be considered in studies on clogging phenomena in irrigation with effluents. In the IR studies, it was observed that an interaction between polysaccharides and clay occurs, carried out through a flat configuration of polysaccharide in which the intermolecular hydrogen bonds are broken. This takes place between free montmorillonite "clay" carboxyls and free oxygen present on both the clay surface and the polymer surface, by dipole-dipole bonds on hydrogen bonds.

The presence of hydrophobic organics affect significantly the binary interaction. Humic acids could accelerate binding with polysaccharide to form block polymer, which causes earlier clogging, as implicated by filter run time. When polysaccharide was induced to the media of soil coated with humic acid, clogging occurred and strongly depended on polysaccharide concentration.

The ability of polysaccharide to develop flocs is seen as a factor to be considered in studies of clogging phenomena in wastewater reuse systems. As a result of IR spectroscopy analysis, together with rheological tests of polysaccharide-clay interaction, a mathematical model is proposed to predict biofilm formation which is proportional to the extent of clogging. The advantage of the model is its ability to predict potential clogging caused by growth of biofilm in a designed effluent irrigation network, according to data calculated from total polysaccharide in a given system (Kanaani 1997).

Algae can be considered a special group of particles of a high indirect clogging potential within the size range of a few to 60 μm. One of the prominent findings with respect to clogging mechanism by algae is that this type of clogging develops only when initial deposits of minerals or a gelatinous material occurred previously. Algae particles are trapped and compressed into the primary sediments in the dripper. Adin and Sacks (1991) also conclude that:

- Clogging of emitters using reservoir effluents is caused primarily by suspended solids in the water but they do not necessarily initiate the clogging process. The clogging rate is more affected by particle size than by particle-number density.
- The sediment buildup begins with the deposition of amorphous slimes, to which other particles adhere. The algae caused dripper clogging only when they attached themselves to other particles. That happened when their concentrations were above 10^5 cells ml^{-1} for continuous period of a few weeks.
- Filtration without chemical pretreatment prevents immediate clogging by large particles. Rapid granular filtration has an important role in protection from clogging due to effective removal of particles with irregular shapes.
- Clogging potential may be decreased by modifying the emitter internal design and by chemical pretreatment with oxidants and flocculants.

Rav-Acha et al. (1995) demonstrated that the aggregation of suspended particles caused by the introduction of oxidants such as chlorine or chlorine dioxide, as expressed by the shifts in both particle size and particle volume distributions toward larger particles, was in accordance with the decrease in the absolute value of the suspended particles zeta potential, which indicated a destabilization effect due to a decrease in the electrostatic repulsion between the particles. Both oxidants affected very markedly the algal viability as expressed by its chlorophyll content and replication ability, the effect of ClO_2 was the strongest.

Tchobanoglous and Eliassen (1970) suggested that PSD (particle size distribution) in an activated sludge effluent was bimodal. On the other hand, PSD data for aqueous particulates larger than 1 μm in many freshwater and wastewater systems are known to be modelled with a two-parameter power law distribution function (Kavanaugh et al. 1980) where the exponent provides an estimate of particulates contribution by size to the total particulate number, the surface area, the volume and the light scattering coefficient. When the power values exceeds 3 it indicates that the smaller size fraction dominates and vice versa. Adin et al. (1989) did not observe any bimodal distribution in effluents, neither from reservoirs nor from trickling filters, aerated lagoons or activated sludge. The distribution in those effluents was either of the power-law or the exponential type.

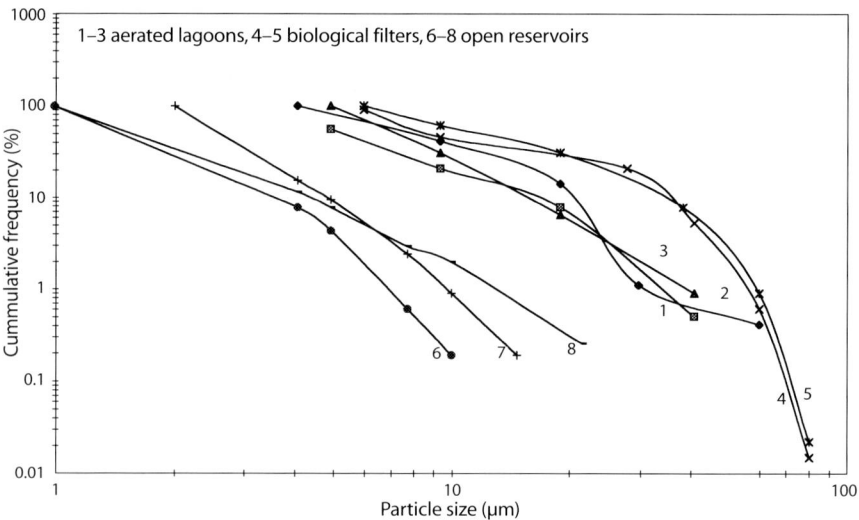

Fig. 17.3. PSD for different effluents (from Adin et al. 1989)

PSD for different types of effluents at different dates are plotted in Fig. 17.3 in terms of cumulative particle number frequency vs. particle size (Adin et al. 1989). For an open reservoir it can be observed that about 90% of all particles are below 5 μm whereas about 99% of all particles are below 10 μm.

The curves presented in Fig. 17.3 were fitted by exponential and power law functions via least square correlation analysis, resulting in correlation factors (r^2) greater than 0.88. In most cases the exponential function described the distributions slightly better than power law function. The difference was not so radical however, as to rule out the use of the latter.

It has been demonstrated that PSD in wastewater effluents relates to a power law function (Adin and Elimelech 1989; Alon and Adin 1994) given by the following equation, for particles larger than 1 μm:

$$\frac{dN}{d(d_p)} = A(d_p)^{-\beta} \quad , \tag{17.1}$$

where N is the number of particles in the size interval of the particle, d_p is the average particle size of the interval and A and β are empirical constants. Each group size is ruled by different mechanisms. Particles smaller than one micrometer are transported by diffusion. Larger particles are transported by gravity. Transport of relatively large particles may also be dominated by interception or they may be retained by interstitial straining. Yao et al. (1971) showed that particles of 1–2 μm in size have minimal opportunity for removal, since transport mechanisms of these particles within the filter bed are less efficient.

Particle volume distribution (PVD) can be calculated from PSD data assuming a spherical shape with a diameter equal to the logarithmic average of the particle diameter measured in each measured size range. While particle number increases inversely to particle size, PVD resembles a bell shaped curve. The obvious explanation to it is that since particle volume is a third power of particle size, the total volume of the small particles is small in spite of their larger number, while the total volume of the larger particles is small due to their smaller number. Such a calculation may help determine solids volume entrapped within a filter bed or a dripper between predetermined particle sizes.

Process theories generally assume uniform suspended particles. That assumption is incorrect for wastewater effluent suspensions, where particle size may range from several nanometers up to more than 100 μm. Transport equations express strong dependency of transport mechanism efficiency on particle size. In order to know on which mechanism to focus, it is necessary to determine the relevant size group of particles in the wastewater. PSD measurement can be applied in mathematical formulation (replacing the parameter representing the average particle size).

17.3
Mechanical Filtration

Different types of filters are currently being used to cope with the clogging problem. Many of them are based on straining mechanisms whose filter (strainer) pores are smaller than most of the particles to be strained. This differentiates them from granular filters which operate on the principle of in-depth filtration. The more sophisticated filters, by incorporating automatic cleaning mechanisms, relieve the farmer of time-consuming cleaning work. Most problems associated with such filters derive from incomplete back-flushing which causes clogging phenomena that can only be corrected by individual filter servicing. Irrigation water often contains considerable amounts of suspended matter (silt, algae, etc.), which affects the proper functioning of the flush mechanisms. This may result in too frequent back-flushing and also may lead to complete pore clogging (Adin and Alon 1986).

Two main categories of filters may be defined on the basis of filtration mechanisms: 1) strainers, in which straining is the main filtration mechanism (also referred to as mechanical or surface filtration), e.g., screen filters; and 2) granular filters, also deep bed, rapid, or high rate filters, e.g., sand filters. Here the suspended particle size is smaller than the filter grain size and the removal is dominated by physical-chemical mechanisms.

The straining process is based on the principle that the pores of the medium are smaller than the particle diameters. Figure 17.4 presents an SEM photograph of a filter screen partly clogged by activated sludge particles. It is observed that an enhanced screen filter clogging may be caused by particles adhering to the corners of the pores and peripheral wires, or to the fewer formerly settled larger particles. It provides evidence to the claim that screen nominal pore size has nothing to do with the actual size of particles that are removed by it (Adin et al. 1989).

A mathematical expression of exponential growth in pressure losses as a function of volume filtered through a steel screen is known as Boucher's Law (Boucher 1947); it assumes the form

$$H = H_0 e^{IV} , \tag{17.2}$$

where H_0 = the pressure loss across a clean screen, V = the volume filtered, and I = a filterability index, related to the suspended material content of the water. I increases

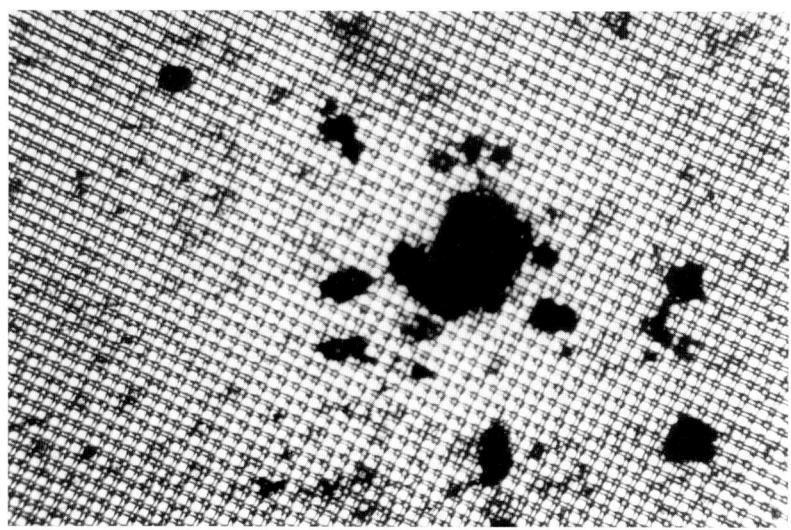

Fig. 17.4. Particles adhering to an 80 μm filter screen effluents (from Adin et al. 1989)

in value as water becomes more difficult to filter. Increase in head loss rate is caused by: 1) increase in suspended material, 2) decrease in pore diameter in various screens, and 3) increase in filtration velocity.

This formula is the result of two simultaneous processes (Ives 1960): 1) the deposition of material within the pores of the screen and the filter cake which obstructs the passage of the suspension, and 2) the deposition of material at a constant rate on the surface of the screen. An extension of the pores is made by the formation of capillaries, and a porous filter cake is formed, the pores of which correspond to the pores upon the filter screen. The fouling rate of a filter depends to a considerable extent upon the distribution of the capillary diameters and entrapped particle dimensions.

It should be noted that Boucher's formula applies solely to the connection between the pressure drop through the screen itself and the flow rate (or volume filtered) without considering a specific filter configuration. In addition, it does not provide a definite idea about the behaviour of the cake formed on the screens of the strainers. The filterability index, I, can be used as an index for the filter and cake resistance. With increasing I values, more frequent filter clogging can be expected. Thus the derivation of the filterability index from pilot experiments can be possibly used for predicting clogging rates in full-scale filter screens. It cannot, however, provide useful information as to the expected effluent quality and the filter cake behaviour (Adin and Alon 1986). An important cake parameter is its compressibility, which is particularly important when the rinsing of the filter is to be considered. Cake filtration theory may serve for that purpose.

A filter can usually serve a wide range of flow rates (Dickey 1961). As the flow rates passing through a given filter increase, the work cycle (i.e., length of time until backflushing) is shortened. Experimental and practical evidence indicates that increasing the flow rate in a granular medium may bring about a detachment process (Adin

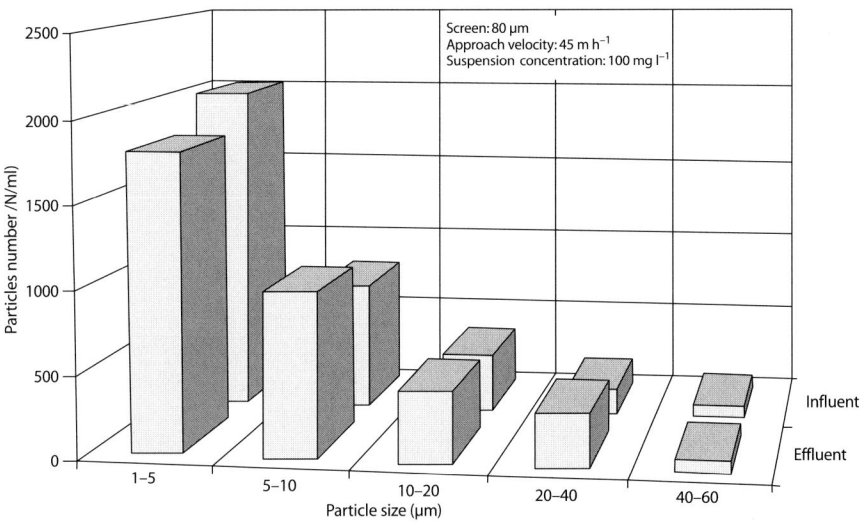

Fig. 17.5. Particle size range in influent and effluent of a strainer (from Adin and Alon 1986)

et al. 1979; Adin and Rebhun 1977). This is caused by extended hydraulic shear forces on the particles in a granular medium. Similar forces act on the filter cake in straining as described by Hosseini and Rushton (1979). Evidence of particle release from strainers in effluent irrigation lines was demonstrated by Oron et al. (1982). It has been postulated by Tien and Payatakes (1979) that the detachment phenomenon in strainers is frequent and contributes to the relative inefficiency of this process, particularly in its initial stage. Colloidal particles which do not appear in effluents from granular filters may be present in effluents from straining due to this phenomenon. Thus it can be concluded that it is impossible to establish a clear and universal optimum for the operation of screen filters.

Figure 17.5 shows that although most particles in the influent to a 80 μm filter do not exceed 20 μm (smaller than the pore diameter), the dominant removal mechanism was straining. This does not correspond to the proper definition of straining, where the particle diameter ought to exceed the pore diameter. Although many researchers stress the advantage of granular filtration over straining in filtering colloidal and submicronic particles (Adin 1979; Rothwell 1978), the possibility of retaining such particles cannot be excluded in the cake formed by straining.

A profile of the particles shows a numerical decrease in the 5 μm range (Fig. 17.3), where particles smaller than 1 μm are contained. From here it can be concluded that the bridging phenomenon, along with the formation of 20 μm aggregates and their subsequent detachment from the screen into the filtrate, increases the number in the effluent suspension. The existence of bridging in strainers is mentioned by Ives (1975) and Dickey (1961), although no mention has been made concerning the formation of aggregates in the filtrate.

The detachment phenomenon prevails toward the end of the run (Adin and Alon 1986). In this case, the build-up of particles on the cake competes with detachment of particles from the cake (sometimes labeled "channeling phenomenon"), thus causing the decrease in head loss rate. The detachment phenomenon may cause clogging in drip irrigation system components that are placed after the filters, and in the drippers themselves. Evidence for this was received from farmers and others who attempted to evaluate the reasons for clogging by field observations.

This type of behaviour could be observed by microscopic examination of the aggregates in the effluent. They are compacted amorphous particles typically formed on a filter cake, as compared to the original particles in the suspension.

17.4
Granular Media Filtration

Granular deep-bed filtration plays a major role in the protection of system components vulnerable to clogging in agricultural and industrial water reuse projects (Adin 1987). It is commonly agreed that suspended solids provide the major cause of serious clogging problems in filters and low-rate applicators that serve in effluent transport schemes (Adin 1979). Suspended solids cause rapid pressure drops and flow disturbances in screen filters commonly used for the protection of sensitive appurtenances. However, filter design in this field still lacks a sound scientific basis.

Filtration through porous media is a process which is based in principle on the capture of particles, rather than on removal of masses of solids. The main tool for evaluation of filter performance in wastewater filtration practice has been, so far, the removal ratio of *TSS* and sometimes of turbidity.

In conventional water treatment by granular filters, removal of suspended solids is usually improved by the use of a finer medium, deeper filter bed, or a lower filtration rate, while headloss buildup is increased by a finer medium, deeper filter bed, or a higher filtration rate (Ives and Sholji 1965; Adin and Rebhun 1974; Ives 1980). In filtration of wastewater effluents, however, these relationships may not hold. Departures from these generalizations have been reported during the last two decades in works in the literature dealing with wastewater effluent filtration.

Fitzpatrick and Swanson (1980) reported that *TSS* removal efficiency and filtration rate were inversely related. Tchobanoglous and Eliassen (1970) have shown that the filtration rate had little effect on *TSS* removal. According to Tebbutt (1971) and Bench et al. (1981), increasing the filtration rate does not appear to reduce the removal of *TSS*. In secondary effluent filtration, filtrate quality is less dependent on filtration rate and influent suspended solids concentration compared to water treatment (Baumann and Cleasby 1974).

The size of filter media has little effect on *TSS* removal, but does significantly affect head-loss build-up (Baumann and Huang 1974). Tebbutt (1971) found that suspended solids could be removed independently of media grain size in the range of 1.0–2.5 mm, but there was some evidence of suspended solids breakthrough for larger media. Tchobanoglous and Eliassen (1970) concluded that removal of suspended solids in secondary activated sludge effluent is primarily a function of the bed grain size, showing a significant improvement in *TSS* removal with finer media.

Adin and Elimelech (1989) noted, however, that particle removal was improved with media grain size getting larger in the range of 0.7–1.2 mm. Small size media filters (0.45–0.55 mm) are actually surface straining devices, resulting in an exponential head-loss build-up and uneconomical filter runs (Bench et al. 1981). In filtration of second-

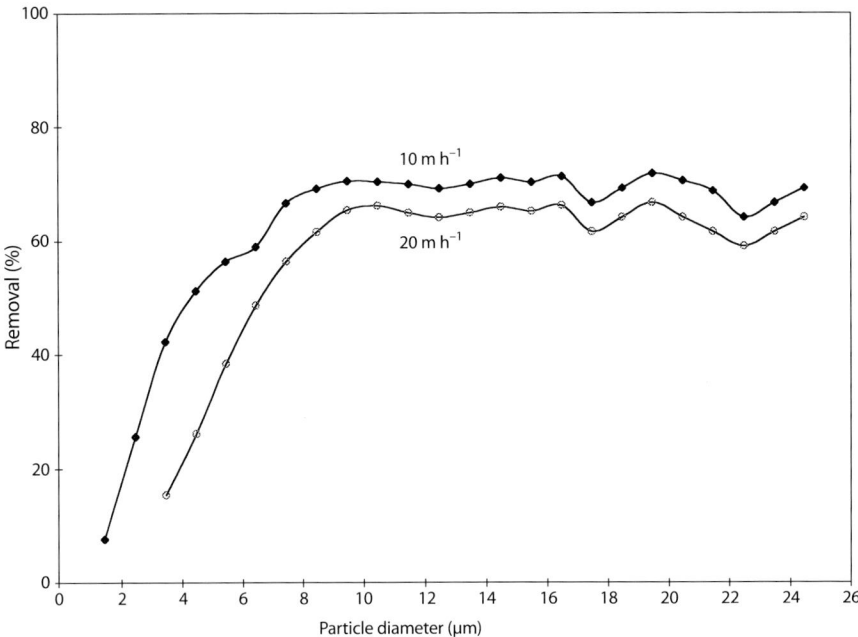

Fig. 17.6. Removal ratio vs. particle size for different filtration rates. Wastewater reservoir effluents, effective grain size = 0.7 mm, bed depth = 150 mm (from Adin and Elimelech 1989)

ary effluents media of at least 1.2 mm effective grain size is required and coarser media is preferred if appropriate backwash is to be provided. Increased media depth may not compensate for coarser media in achieving filtrate quality (Baumann and Cleasby 1974).

Comparison of filter removal efficiency as a function of particle size for different filtration rates shows a lower efficiency for the smaller particles, with a tendency of the removal curve toward a minimal efficiency in the 1–2 μm range (Adin and Elimelech 1989). Figure 17.6 presents particle removal efficiency of reservoir effluents as a function of particle size. Better removal efficiency of particles at the lower filtration rate for all particle sizes was observed. The difference in percentage became smaller as particle size increased. Removal efficiency increased significantly with increase in particle size up to 9–10 μm. At higher filtration rates, an instantaneous phenomenon of more particles in the filtrate than in the effluent was observed in the size range of 1–3 μm. Particles larger than 10 μm were removed with a constant removal efficiency of about 80%, most probably by the layer formed on the medium surface.

The effect of media grain size on particulate removal for the reservoir effluent was also studied with interesting results: larger grain sizes clearly showed better removal efficiency of very fine particles (1–10 μm). For particles larger than 10 μm (up to 60 μm), there was no difference in the rate of particle removal. This finding was confirmed by PSD measurements in influent and filtrate in the filtration of activated sludge. The rate of head loss buildup was much higher for the finer media although there was some evidence of surface removal in the coarser medium.

Experimental results corresponded well to the power-law distribution function for both filters influent and filtrates. Considering the most basic form of the function (appears in more detail later in the text), the power-law exponent absolute value was larger for the coarse media filtrate than for the finer media (e.g., 2.01 vs. 1.80) pointing at larger number of particles smaller than 10 μm in the latter. In general, the results indicate an advantage of the coarse bed to the fine bed by better removing particles larger than 10 μm and by retaining larger volume of suspended material. This phenomenon can be attributed to several factors: pore geometry, grain surface roughness, diffusion and gravity deposition mechanisms, and interstitial hydraulic gradients.

A conventional tertiary treatment includes chemical coagulation followed by flocculation, sedimentation, filtration, and disinfection. In case of in-line filtration, equipment and chemical costs are minimized. In cases where the PSD is dominated by particles greater than 1 μm, the use of a suitable cationic polyelectrolyte at a minimum dosage can help optimize the process. At low suspended solids loadings or in cases where the PSD is dominated by submicron particulates, use of alum at optimum doses together with a polyelectrolyte as a flocculant aid can result in an effective and low cost filtration.

The filtration process can be described macroscopically by three sets of mathematical expressions – continuity (mass balance) equation, kinetic equation, and head loss development expression (Adin and Rebhun 1977). The continuity equation is given by:

$$v\left(\frac{\partial C}{\partial x}\right)_t + \left(\frac{\partial \sigma}{\partial t}\right)_x = 0 \quad , \tag{17.3}$$

where x and t are bed depth and time, C is the concentration of particulates in the water, v is the approach velocity, and σ is the specific deposit (mass of particulate per volume of media). A kinetic equation to be solved with the above expression is given by:

$$\frac{\partial \sigma}{\partial t} = K_a vC(F - \sigma) - K_d \sigma J \quad . \tag{17.4}$$

Here K_a is the attachment coefficient and K_d the detachment coefficient. F is the theoretical filter capacity (mass/volume) and J is the hydraulic gradient. Head loss through the granular media as a function of the specific deposit based on Shektman's expression is:

$$\frac{J_0}{J} = \left(1 - \left(\frac{\sigma}{\varepsilon_0 \gamma}\right)^{0.5}\right)^3 \quad , \tag{17.5}$$

where J_0 is the initial hydraulic gradient, ε_0 is the clean bed porosity and γ is the solid concentration in the deposit. The change of suspension concentration with depth can be expressed for the single collector model as (Amirtharajah 1988):

$$-\left(\frac{\partial C}{\partial x}\right)_t = \frac{\eta(\sigma)(1 - \varepsilon_0)C}{d_C} \quad , \tag{17.6}$$

where $\eta(\sigma)$ is the single collector efficiency and is a function of the specific deposit and d_C is the grain diameter. η can be written as:

$$\eta = \eta_0 \alpha \quad , \qquad\qquad\qquad (17.7)$$

where η_0 being the clean bed single collector efficiency factor and accounts for all transport properties, and α is the collision efficiency factor which is a function of physicochemical parameters including particle attachment strength.

Another calculated factor is the Ives Filterability Number (Ives 1980):

$$F.N. = \frac{H\dfrac{C}{Co}}{Vt} \quad , \qquad\qquad (17.8)$$

where C is average filtrate quality (NTU), Co filter influent quality (NTU), H head-loss (m), V filtration rate (m h⁻¹), and t filtration time (h). The lower the filterability number, the better is the filtration performance.

Size density and attachment strength of filter deposits produced by the interaction of effluents with alum, alum-polymer aids and cationic polymers as primary flocculants and the filter media during contact filtration, were calculated and compared using the above mentioned filtration models (Cikurel et al. 1996). Conclusions have been drawn as follows:

- Secondary effluent filtration without flocculant addition was relatively ineffective.
- Effective alum dosage domains for contact filtration are of up to 10 mg l⁻¹. For the same filtration conditions (1.1 mm grain size and 10–15 m h⁻¹ approach velocity) better removal efficiencies were obtained when the alum dosage was 10 mg l⁻¹ as compared to a dosage of 20 mg l⁻¹. This data is in agreement with the findings that in the 5.5–8.5-pH range, at an alum dosage greater than 15 mg l⁻¹, the prevailing flocculation mechanism is sweep coagulation giving way to a bulky floc.
- While the alum seems to work more through adsorption and charge neutralization, the anionic polymer operates via bridging. The high molecular weight polymer affected filter performance better than the low molecular weight polymer, giving way to stronger alum-polymer-particle deposit. Low detachment values characterize this mechanism. The detachment mechanism seems to have more control over the process in effluent filtration than in surface water filtration, where attachment mechanisms have more effect, probably due to the lack of bio-destabilizers.
- The effect of grain size was more significant at lower approach velocities (5–10 m h⁻¹) and less significant at higher velocities (15–25 m h⁻¹). In some cases the improvement observed in filtration with greater media size could be rationalized by the higher interstitial approach velocity at the lower grain size, which caused a higher detachment of the already attached floc.
- High molecular weight (2×10^6) branched chain polyacrylamides, at any charge density, and at as low as 0.5 mg l⁻¹ dosages can be more effective than 10–20 mg l⁻¹ alum and can be comparable in efficiency to (alum + polymer) causing only mild head-loss increase for low influent particulates loading.

- Very low molecular weight (20 000) polyamines are not effective even at dosages up to 7 mg l⁻¹. Medium molecular weight high cationic polyamines are effective at dosages >5 mg l⁻¹. The filtration efficiency increased proportionally to the applied dosage as did the head loss.
- The high molecular weight polyacrylamides' performance was not significantly affected by the change in approach velocities. Changing media size from 1.1 mm to 1.5 mm improved the filtration efficiency for v = 10 m h⁻¹.
- The low molecular weight, linear polyamines destabilize particulates by patch neutralization and in a minor proportion by bridging. The high molecular weight branched chain polyacrylamides work by bridging and by interception of particulates by the previously deposited, ramified, polymer-particulate deposit.

17.5
General Conclusions

1. Effluents from wastewater reservoirs contain a variety of particles often characterized by gelatinous particle shape. The particles are compositionally dominated by Si, their content may change, however, with the season. They are mostly colloidal in nature and negatively charged, with somewhat repulsive zeta potential ranging –14 mV to –17 mV.

2. Biological films and slimes, derived from bacteria and algae, are among the major biological clogging factors in drip irrigation systems applying wastewater reservoir effluents. Such deposits serve as triggers for serious blockage. Particles of a definable shape form the primary sediment in the drippers. The primary sediment then acts as a trap for algae and zooplankton limbs that increase the volume of the sediment until they block the water pathway.

3. The dominant mechanism in mechanical filters is straining out particles of a diameter that exceed the screen pore diameter, the possibility of retaining colloidal particles cannot be excluded in the cake formed by straining. A detachment phenomenon prevails toward the end of the run. The detachment of relatively large aggregates may cause clogging in drip irrigation system components that are placed after the filters, and in the drippers themselves.

4. Filtration without chemical pretreatment prevents immediate clogging by large particles. Rapid granular filtration has an important role in protection from clogging due to effective removal of particles with irregular shapes. Secondary effluent filtration without flocculant addition is relatively ineffective, however, in removing colloidal particles. Effective alum dosage for contact filtration ranges 10–20 mg l⁻¹. An addition of 0.05–0.1 mg l⁻¹ of low anionic high molecular weight polymer strengthens the alum-particle bond. High cationic, medium to high molecular weight polymers may perform equally or better than alum as primary flocculants, at doses as low as 0.5 mg l⁻¹, at different charge densities. In high particle loadings, the differences among the various treatments are less pronounced

5. Particle size distribution (PSD) can serve as a powerful tool for understanding filter behaviour and discrepancies in case of filtration under various conditions. It may be applicable for advanced wastewater filter design. PSD should be taken into consideration in filtration models, rather than relying on a single representative particle diameter of questionable value.

References

Adin A (1979) Problems and advanced methods in filtration for low rate applicators. Proceedings of the International Conference on Water Systems and Applications, Israel Center of Waterworks Appliances, Israel V–3, pp 1–9

Adin A (1987) Problems associated with particulate matter in water reuse for agricultural irrigation and their prevention. Wat Sci Tech 18:185–195

Adin A, Rebhun M (1974) High rate contact–flocculation filtration with polyelectrolytes. J American Water Works Association, 66(2):109–117

Adin A, Rebhun M (1977) A model to predict concentration and head–loss profiles in filtration, J American Water Works Association, 8:444–451

Adin A, Alon G (1986) Mechanisms and process parameters of filter screens. J Irrig Drainage Eng, ASCE, 112(4):293

Adin A, Elimelech M (1989) Particle filtration in wastewater irrigation. J Irrig Drainage Eng, ASCE, 115(3):474–487

Adin A, Sacks M (1991) Dripper clogging factors in wastewater irrigation. J Irrigation and Drainage Eng, ASCE 117:6:813–826

Adin A, Baumann ER, Cleasby JL (1979) The application of filtration theory to pilot plant design. J AWWA 71:1:17–27

Adin A, Rubinstein L, Zilberman A (1989) Particle characterization in wastewater effluents in relation to filtration and irrigation, filtration and separation 26(4):284–287

Alon G, Adin A (1994) Mathematical modelling of particle size distribution in secondary effluent filtration. Wat Envir Res 66(6):836–841

Amirtharajah A (1988) Some theoretical and conceptual views of filtration. J American Water Works Association, 80:36–46

Baumann ER, Huang JYC (1974) Granular filters for tertiary wastewater treatment. J Water Pollution Control Federation 46(8):1958–1973

Baumann ER, Cleasby JL (1974) Wastewater filtration: Design considerations, technology transfer, Series N° EPA–625/4–74–007a: USEPA

Bench B, Middlebrooks E, George D, Reynolds J (1981) Evaluation of wastewater filtration. Water Quality Series, N° UWRL/Q–81/01, Utah Water Research Laboratory

Boucher PL (1947) A new measure of the filterability of fluids with applications to water engineering. J Inst Civ Engrs 27:415–423

Bucks DA, Nakayama FS, Warrick AW (1982) Principles, practices and potentialities of trickle irrigation. In: Hillel D (ed) Advances in Irrigation, vol 1, San Diego, California, Academic Press, p 219

Cikurel H, Rebhun M, Amirtharajah A, Adin A (1996) Wastewater effluent reuse by in–line flocculation filtration process. Wat Sci Tech 33:10–11:203–211

Dasberg S, Bresler E (1985) Drip irrigation manual. International Irrigation Information Center (IIIC), Volcani Center for Agricultural Research, Beit–Dagan, Israel

Dickey GD (1961) Filtration. Reinhold Publishing Co., New York, N.Y.

Dor I, Schechter H, Bromley HJ (1987) Limnology of hypertrophic reservoir storing wastewater effluent for agriculture at Kibbutz Na'an, Israel, Hydrdrobiologia 150:225–241

Fitzpatrick JA, Swanson R (1980) Evaluation of full–scale tertiary wastewater filters. EPA–600/2–80–005: USEPA

Fletcher M (1979) The attachment of bacteria to surfaces in aquatic environments. Adhesion of micro-organisms to surfaces. London, U.K., Academic Press

Hosseini M, Rushton A (1979) Shear effects in cake formation mechanisms. Filtration and Separation 16 (Sept–Oct):456–460

Ives KJ (1960) Filtration through a porous septum: A theoretical consideration of Boucher's Law. J Institute of Water Engineers 17:333–338

Ives KJ (1975) Capture mechanisms in filtration, The Scientific Basis of Filtration. In: Ives KJ (ed), Netherlands, Noordhoff–Leyden

Ives KJ (1980) Deep bed filtration: Theory and practice, filtration and separation 17:2:157–166

Ives KJ, Sholji I (1965) Research variables affecting filtration. J Environmental Engineering Division, ASCE, vol 91, N° SA4, Proc. Paper 4436, pp 1–18

Jones HC, Roth IL, Sanders WM (1969) Electron microscope study of slime layer, J Bacteriol 99(1):316–325

Kanaani YM (1997) Interaction of particulate and dissolved matter with inorganic surfaces in conjunction with water reuse. PhD thesis, The Hebrew University of Jerusalem (in Hebrew)

Kanaani YM, Adin A, Rav–Acha C (1992) Biofilm interactions in water reuse system: Adsorption of polysaccharides to kaolin. Wat Sci Tech 26:3/4:673–682

Kavanaugh MC, Tate CH, Trussell AR, Treweek G (1980) Use of particle size distribution measurement for selection and control of solid/liquid separation processes. In: Kavanaugh MC, Luckie JO (ed) Particles in Water, Advances in Chemistry Series. Washington, DC, American Chem Soc p 305

Nakayama FS, Bucks DA, French OF (1977) Reclaiming partially clogged trickle emitters. Transactions of the ASAE, pp 278–280

Oron G, Ben-Asher J, Demalach Y (1982) Effluent in trickle irrigation of cotton in arid zones. J Irrig Drainage Division, ASCE, 108(N° IR2, June):115–126

Rav–Acha C, Kummel M, Salomon I, Adin A (1995) The effect of chemical oxidants on effluent constituents for drip irrigation. Wat Res 29:1:119–129

Rothwell E (1978) Fabric filter failures. Filtration and Separation 15(Nov–Dec):586–593

Tien C, Payatakes AC (1979) Advances in deep bed filtration. J American Institute of Chemical Engineering 4(Sept):737–748

Tchobanoglous G, Eliassen R (1970) Filtration of treated sewage effluent. J Sanitary Engineering Division, ASCE, 96, SA2, Proc Paper 7210, pp 243

Tebbutt HY (1971) An investigation into tertiary treatment by rapid sand filtration. Wat Res 5:3:81–92

Yao KM, Habibian MT, O'Melia CR (1971) Water and wastewater filtration: Concepts and applications. Environ Sci Tech 5:11:1105

Chapter 18

Satellite Remote Sensing of Water Quality

Chaim Braude · Nissim Ben Yosef · Inka Dor

18.1
Introduction

Satellite remote sensing is a well established tool for the classification and monitoring of water quality in big water bodies such as oceans and lakes. Satellite sensed data has been successfully used to monitor concentrations of chlorophyll a, dissolved organic matter and suspended solids (Alfoldi 1982).

Inland wastewater reservoirs differ from these big water bodies in both size and water composition. Wastewater reservoirs are much smaller in size, and normally contain much higher concentrations of organic matter and pigments in the water. Such small hypertrophic reservoirs with heavy organic loads have not been studied in detail from satellite images.

The concentrations of many water constituents in wastewater reservoirs are up to four orders of magnitude higher than in typical ocean or lake waters. Water quality in these reservoirs is characterized by high concentrations of suspended solids, organic matter (yellow matter), green algae, and purple bacteria.

The feasibility of satellite monitoring of water quality under these conditions was studied by the analysis of the French satellite, SPOT, multispectral images. The radiance values of the SPOT channels were taken from a sample of about 100 water reservoirs in images from three different dates.

A detailed statistical analysis of the reservoir images was carried out in order to classify the reservoirs according to their water quality. Two wastewater reservoirs and one clean drinking water reservoir served as ground truth. Water samples were taken from these three reservoirs for laboratory analysis.

A principal components analysis (PCA) was performed on the reservoirs' images. The principal components vectors were studied in terms of water quality and composition.

A physical model based on the radiative transfer equation and the optical properties of the water constituents and atmosphere was developed to support the interpretation of the results of the image analysis. The model simulates the radiance measured by the satellite as a function of water quality, atmospheric conditions, sun and satellite location and imaging system. Variations in water quality are modelled by the concentrations of three characteristic water components: chlorophyll a found in algae, bacteriochlorophyll a found in purple bacteria, and dissolved organic matter generally called yellow matter.

18.2
SPOT Images of Wastewater Reservoirs

18.2.1
Deep Wastewater Reservoirs

Wastewater reservoirs are complex chemical and biological systems. The reservoirs are typically 6–10 m deep and may be mixed or stratified. They are hypertrophic water bodies which contain very high organic loads. These reservoirs have been classified into three major categories (Dor and Raber 1990):

i.   *Chlorella reservoirs.* The water is dominated by the small green *Chlorella* algae. Most of the water is urban sewage. They have a moderate organic load and compensation depth (1% of surface light) is around 1 m.
ii. *Euglena reservoirs.* The water is dominated by the bigger flagellate *Euglena* algae. Most of the wastewater contains manure from dairy farms. The organic load is high and compensation depth is around 50 cm.
iii. *Photosynthetic bacteria reservoirs.* The water is dominated by the red *Thiocapsa* bacteria which cause very high turbidity. The organic load is very high and there is a lack of dissolved oxygen. Compensation depth is only 5–10 cm.

18.2.2
Ground Truth

Three reservoirs served as ground truth for this study. The Eshkol Reservoir contains clean drinking water (coded NN1). The Na'an Reservoir (coded SR1) and Tel-Adashim Reservoir (coded NA2) contain wastewater and carry heavy organic loads. Reservoir SR1 represents a *Chlorella* type reservoir, and NA2 represents a *Thiocapsa* bacteria type reservoir. Samples from these reservoirs were taken for laboratory analysis on dates close to the satellite overpass. A summary of the ground truth measurements is given in Table 18.1.

Table 18.1. Water quality parameters of the ground truth reservoirs

Reservoir	SPOT overpass	Sampling date	Turbidity (NTU[a])	TSS ($mg\,l^{-1}$)	Chloroph. a ($\mu g\,l^{-1}$)	Bchl. a ($\mu g\,l^{-1}$)
NN1	03-09-92	03-09-92	1.8	3.1	2.9	~0
	04-10-92	04-10-92	1.4	3.2	1.5	~0
	07-12-92	07-12-92	2.3	3.2	2.6	~0
SR1	03-09-92	nd	–	–	–	–
	04-10-92	08-10-92	106	169	805	–
	07-12-92	07-12-92	97	111	514	3 146
NA2	03-09-92	17-09-92	276	300	1 599	1 960
	04-10-92	04-10-92	228	227	2 314	5 953
	07-12-92	09-11-92	195	262	1 165	7 494

[a] *NTU* = Normalized Turbidity Units.

18.2.3
Reservoir Classification

The database for this study was created from SPOT images of Israel from September, October and December 1992. A sample of about 100 water reservoirs of different types taken from each of the images served as the database for this study. These include drinking water, irrigation, and wastewater reservoirs as well as fish ponds and a sewage treatment plant. The spectral satellite measurements from the reservoirs were statistically analysed in several ways to see whether a classification of clean and polluted reservoirs can be achieved. For all analysis the satellite gray level images were converted to true radiometric radiance units using the SPOT Absolute Calibration Coefficients.

Three methods of classification were examined as tools for the evaluation of the quality of the water in the reservoirs. First the radiance values of the three SPOT bands were studied. These values were then transformed to chromatic coordinates to enhance the spectral difference between the reservoirs. Finally a principal components analysis was used to find the base of spectral vectors which best describe the spectral variation of the reservoir sample.

18.2.4
Radiance Values

The three SPOT spectral bands are XS1 (0.5–0.59 µm), XS2 (0.61–0.68 µm), and XS3 (0.79–0.89 µm). The values of measured radiance for all reservoirs are highest in band XS1 and lowest in band XS3. This is mainly due to the intensity of the atmospheric path radiance in these bands. The calculated path radiance for the NN1 area for the October image are 39, 29 and 12 watt m^{-2} str^{-1} $µm^{-1}$ for the XS1, XS2, XS3 bands respectively. All atmospheric optical calculations were done using the US Airforce LOWTRAN7 code for atmospheric optical modelling.

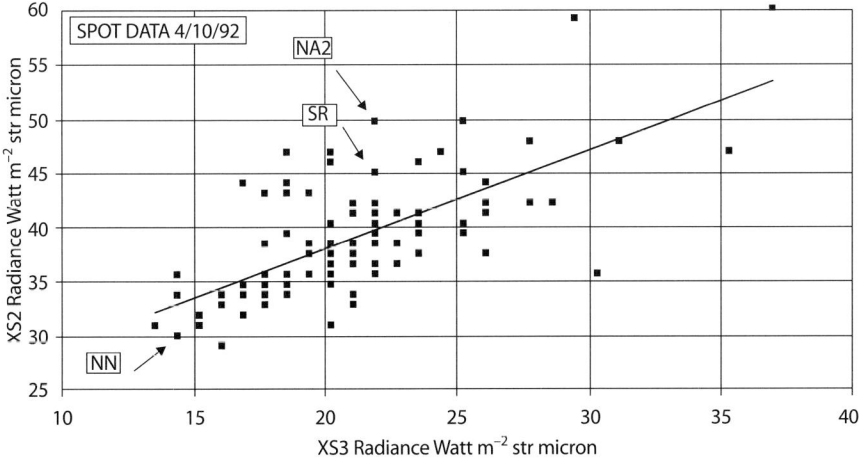

Fig. 18.1. Band *XS2* vs. band *XS3* of 99 reservoirs and the linear regression

A strong correlation is found between the three spectral bands. This is due to mutual variations in path radiance in all three bands and to the fact that dirtier water generally reflects more light all across the spectrum than does clean water. This correlation is described by a linear regression line plotted through the sample. The variations in the spectral distribution of the radiation from the different reservoirs are expressed by the amount by which the spectral values deviate from the regression line.

The location of the reservoirs relative to the regression lines is found to represent their water quality. The clearest indication for water quality is the XS2 vs. XS3 regression, as seen in Fig. 18.1. The cleaner reservoirs are found below the regression line, and the wastewater reservoirs above it. This result is consistent on all three dates studied. This classification is found to be valid for reservoirs located in areas with different atmospheric conditions, such as SR1 and NA2. The regression line describes most of the atmospheric variation, and the deviation from the regression line is mainly related to the water reflected radiation.

18.2.5
Chromatic Coordinates

In order to enhance the differences in the spectral distribution of the radiation over the absolute radiance measured, the satellite bands were transformed into chromatic coordinates (Pratt 1978; Bukata et al. 1983), using the following transformation:

$$(L_3, L_2, L_1) \rightarrow (X, Y, Z) \qquad L_k = \text{radiance in band } k \quad ,$$

$$X = \frac{L_3}{L_1 + L_2 + L_3}; \quad Y = \frac{L_2}{L_1 + L_2 + L_3}; \quad Z = \frac{L_1}{L_1 + L_2 + L_3};$$
$$\Rightarrow X + Y + Z = 1 \tag{18.1}$$

These are normalized coordinates which remove the factor of the total radiation and show the partial contribution of each band to the total radiance.

The transformed coordinates remain highly correlative, in spite of the fact that the transformation is not linear. The ground truth reservoirs NN1, NA2 and SR1 fall at the opposite extremes of the range of the values of the Y and Z coordinates calculated for the entire sample. Therefore it appears that most of the variation in the Y and Z values is associated with variations in the water reflectance. The cleaner reservoirs have a higher Z value than those containing sewage water – i.e., they look greener in true colours. The wastewater reservoirs have a higher Y value than the clean reservoirs – i.e., they look more red in true colours. This is consistent with ground observations of water colour. The variations in the X coordinate are not fully explained by the variations in water quality. The X coordinate is derived from the XS3 band in the near-IR which is more sensitive to variations in atmospheric conditions and vegetation cover; furthermore, this spectral band penetrates less into the water body.

Since the transformed X, Y, Z coordinates retain a high degree of correlation it was found useful to examine the amount of deviation of the reservoirs from the regression line. This results in an improved classification of water quality. Reservoir NA2 had an extremely high purple bacteria population in December which caused a high volume reflectance in the XS3 band on this date. This change is very well seen in the deviation of NA2 from the Y-X regression for this month, as seen in the regression in Fig. 18.2. This representation shows the difference between the two wastewater reservoirs more clearly, where NA2 is more polluted than SR1.

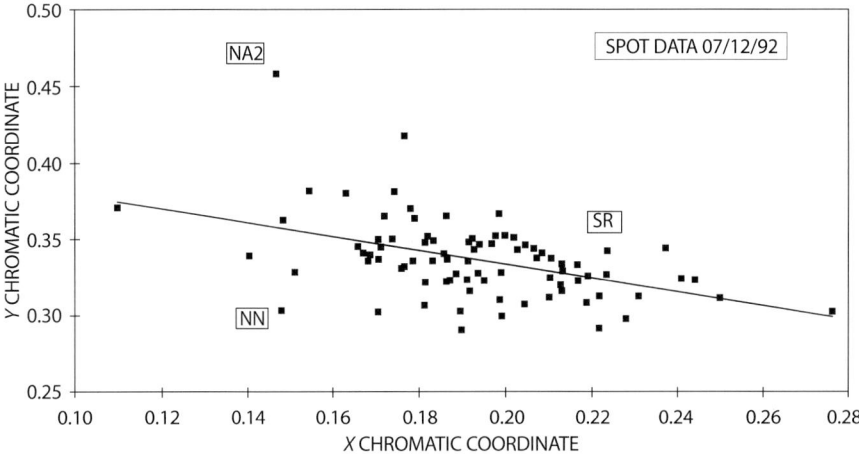

Fig. 18.2. Y chromatic coordinate vs. the X chromatic coordinate and the linear regression

Fig. 18.3. Histograms of Y coordinate deviation (for each reservoir) from the Y vs. X linear regression

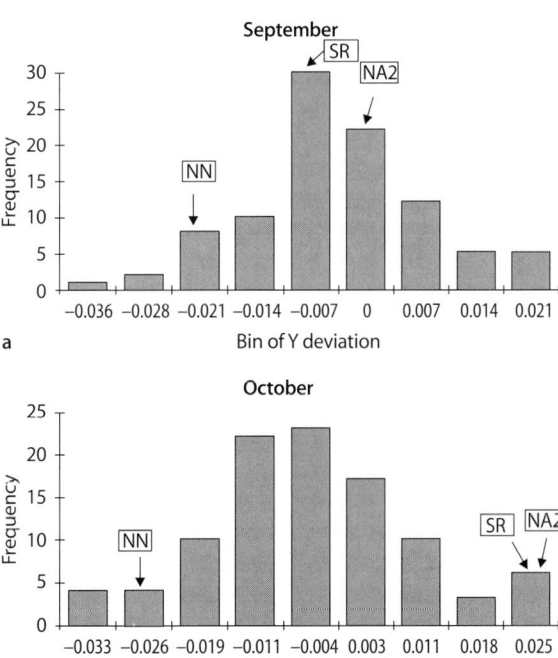

These results are consistent for all three dates studied. The location of the ground truth reservoirs in the sample is given in the set of histograms in Fig. 18.3 which show the deviations of the *Y* coordinate from the *Y-X* regression line. In September the clean and polluted reservoirs are separated, but not to the extremes of the distribution, in part due to poor atmospheric conditions. In October there is excellent separation between clean and polluted reservoirs. There is no clear separation between NA2 and SR1. In December NA2 is very polluted and is clearly separated from SR1 and the rest of the sample.

This type of classification fails in several marginal cases such as the oxidation ponds of a sewage treatment plant. Typical wastewater reservoirs such as NA2 and SR1 are 6 to 10 m deep and very turbid, and thus no sunlight is reflected from the bottom. The oxidation ponds are much shallower, 1–2.5 m deep and some have a reflecting sandy bottom. Light reflected from the bottom modifies the typical spectral distribution of the volume reflected radiation.

18.2.6
Atmospheric Effects

The greatest difficulty in quantitatively evaluating satellite radiometric measurements of ground targets is the correction of atmospheric radiation. Atmospheric conditions determine target illumination, optical transmittance of the path and path radiance.

The NN1 and NA2 reservoirs are located close together in a rural environment. SR1 reservoir is located closer to big urban centres. The reservoirs are therefore imaged under different atmospheric conditions. The calculated pertinent atmospheric parameters are given in Table 18.2, together with the actual satellite measurements. All atmospheric data are for the October image and radiance values are in watt $m^{-2} str^{-1} \mu m^{-1}$.

The effect of the spectral ground albedo of the reservoir's environment on the path radiance is significant. This effect was included in the calculation of the path radiance using a special algorithm to calculate the average surface albedo around each reservoir for each SPOT band.

Table 18.2. Measured radiance and calculated atmospheric factors (LOWTRAN 7 code)

Atmospheric parameter	XS1	XS2	XS3
NN1 radiance	43.5	30.0	14.3
NA2 radiance	54.6	49.8	21.9
Path radiance	39.0	29.0	12.0
Atm. transmittance	0.53	0.60	0.68
SR1 radiance	50.0	45.1	21.9
Path radiance	39.0	28.0	15.0
Atm. transmittance	0.57	0.64	0.75

18.3
The Spectral Radiance Model

18.3.1
The Optical Properties of Hypertrophic Wastewater Reservoirs

Deep wastewater treatment reservoirs are similar to ordinary shallow wastewater oxidation ponds but are much deeper, typically 6–10 m deep. These reservoirs carry very high organic loads and their optical properties are totally different from those of bigger and cleaner water bodies. The optical density of the water in these reservoirs is very high as the concentrations of most organic water constituents and pigments is orders of magnitude higher than in oceans or lakes. A comparison of typical values of several standard water quality parameters is given in Table 18.3.

These differences prevent the direct application of remote sensing models developed for oceans and lakes to these reservoirs. The high concentrations of pigments give a stronger spectral signature, yet they are strongly masked by the high concentrations of suspended solids and dissolved organic matter. The water includes large populations of photosynthetic bacteria, such as the purple bacteria *Thiocapsa*, which contain the reddish bacterial pigment not normally found in clean water bodies.

The high optical density and the depth of the water prevents any interaction of the reservoir bottom with the incoming solar radiation.

The individual pixel area within a big water body is located in a uniform environment of neighboring pixels of similar waters. In the case of a small inland reservoir the immediate environment is very different; it may for example be forest or sand dunes. This different spectral albedo of the target area affects the path radiance. Therefore special attention must be given to atmospheric correction of radiance measured from reservoirs located in different environments.

18.3.2
Model Structure and Inputs

The model calculates the optical interaction of the solar radiation all the way from the sun to the satellite detector, including the paths through the atmosphere, and the interaction with the water volume and interface. The atmospheric effects of transmit-

Table 18.3. Optical properties of different water bodies (modified from Dor et al. 1991)

Parameter	Typical value in oceans/lakes	Typical value in wastewater reservoirs	Units
Turbidity	<30	50 – 300	NTU
TSS (Total Suspended Solids)	<30	50 – 300	$g\,m^{-3}$
Chlorophyll a	0 – 100	200 – 500	$mg\,m^{-3}$
Bacteriochlorophyll a	~0	500 – 7 500	$mg\,m^{-3}$

tance and path radiance for each image were calculated for the specific sun and satellite geometry in which the image is taken.

Full optical modelling of a water body requires the knowledge of the spectral absorption and scattering cross sections of all the water constituents, and their particle size distributions. Based on a realistic database available for this study the optical behaviour of the water body was modelled using the optical properties of several dominant water constituents. Water quality in the model is characterized by chlorophyll a contained in the green algae *Chlorella*, bacteriochlorophyll a contained in the purple bacteria *Thiocapsa*, and spectroscopic measurements of yellow matter from wastewater samples. The following notation for the concentrations of the three modelled water components for the rest of this chapter:

- Chlorophyll a: µg l^{-1}
- Bacteriochlorophyll a: µg l^{-1}
- Yellow matter: Na'an units

The density of the yellow matter in the Na'an Reservoir was taken as a unit for comparison between different water bodies and denominated "Na'an unit".

The interaction of radiation with the air-water and water-air interface is treated by the standard Fresnel equations. The volume reflectance of the water body was calculated as a function of water component concentrations and their spectral cross sections. The radiative transfer equation is solved using the quasi-single scattering approximation (Gordon 1973). This approximation assumes that all the forward scattered radiation continues to propagate in the original direction, and all the back scattered is extinct. This approximation is most accurate when the backscattering coefficient is small compared to the forward scattering coefficient. It is assumed in the model that the water depth is infinite, which is equivalent to assuming no radiation reaches the bottom. This is normally true in deep turbid reservoirs. Radiation reflected from the water body is divided into surface reflected radiation which is reflected at the air-water interface, and volume reflected radiation which is spectrally modulated by the water components.

The total upwelling radiation leaving the water body in the direction of the satellite is given by:

$$L_{up}(\phi, \varphi) = [r_{air \rightarrow water}(\phi) \cos^2(\phi)]L_{sky}(\phi, \varphi + \pi) + R_{vol}(\phi, \varphi)I_0(\theta, \varphi + \pi) \quad . \qquad (18.2)$$

The angles ϕ and φ define the satellite to target zenith and azimuth angles respectively. The angles ϕ_w and φ_w define the respective propagation angles inside the water. r and t are the reflectivity and transmittance at the air-water interface. I_0 is the direct solar irradiance incident on the water surface and the volume reflectance is given by:

$$R_{vol}(\phi, \varphi) = \frac{L_{vol}(\phi, \varphi)}{I_0} = t_{air \rightarrow water}(\theta) \, t_{water \rightarrow air}(\phi_w) \frac{n_{air}^2}{n_{water}^2} \frac{\sigma(\pi - \theta_w - \phi_w)}{c\left(1 + \dfrac{\cos(\phi_w)}{\cos(\theta_w)}\right)} \quad , \qquad (18.3)$$

where c (in cm^{-1}) is the spectral extinction coefficient and σ (in cm^{-1} str^{-1}) is the spectral differential scattering coefficient, and n is the index of refraction.

Adding the atmospheric effects of path radiance and atmospheric transmittance the signal reaching the satellite is:

$$L_{sat}(\lambda) = T_{sat}(\lambda)L_{up}(\lambda) + L_{path}(\lambda) \quad , \tag{18.4}$$

where
- L_{sat} = radiance at the satellite detector
- L_{path} = path radiance
- T_{sat} = transmittance of the target to satellite path

Finally, adding the spectral response of the satellite imaging system, the radiance measured by the satellite in band k is (CNES and SPOT IMAGE 1991):

$$L_k = \frac{\int_{\lambda_1}^{\lambda_2} L_{sat}(\lambda)S_k(\lambda)d\lambda}{\int_{\lambda_1}^{\lambda_2} S_k(\lambda)d\lambda} \quad , \tag{18.5}$$

where
S_k = detector spectral response
L_k = spot equivalence radiance (in watt $m^{-2}\,str^{-1}\,\mu m^{-1}$)
k = 1, 2, 3

18.3.3
The Optical Properties of the Water Constituents

The absorption and scattering coefficients of natural water were taken from published literature (Smith and Baker 1981). They are well known and will not be given here.

Dissolved organic matter appears to be the only water component (except pigments) that has a significant absorption coefficient in the SPOT spectral range (Davis-Colley and Vant 1987). These are substances similar to humic acid normally found in the water either as a suspension or in a dissolved state. The absorption of the yellow matter was modelled as an exponential decay as a function of wavelength (Davis-Colley and Vant 1987), where the absorption coefficient given by:

$$a(\lambda) = a_0 e^{A(\lambda_0 - \lambda)} \quad . \tag{18.6}$$

The scattering is inversely proportional to the wavelength (Davis-Colley and Vant 1987), and it is given by:

$$b(\lambda) = b_0 \lambda^{-1} \quad . \tag{18.7}$$

The coefficients a_0, b_0 and A were numerically fitted to empirical spectroscopic measurements of samples taken from the SR1 reservoir. A comparison of the measured and computed optical densities of a sample of wastewater is given in Fig. 18.4. The absorption peaks deviating from the calculated graph are due to the chlorophyll and bacteriochlorophyll pigments in the water.

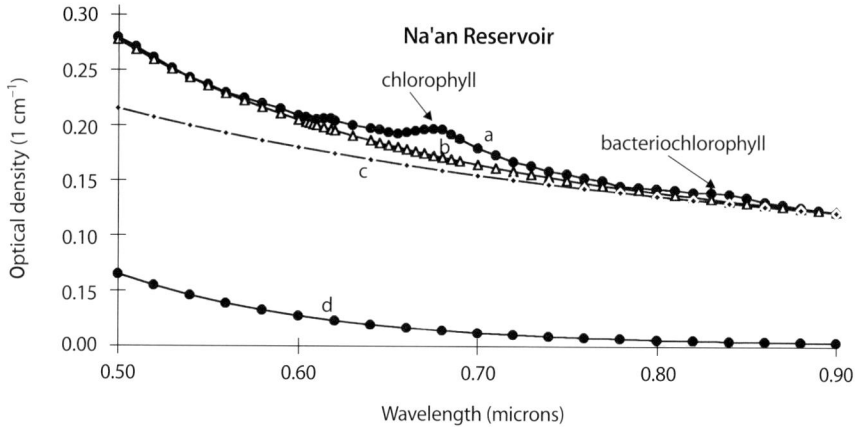

Fig. 18.4. Optical properties the yellow matter in wastewater for the Na'an (SR1) Reservoir. Curve *a* measured optical density, *b* calculated optical density, *c* calculated scattering extinction coefficient, *d* calculated absorption coefficient

The optical properties of the green algae *Chlorella* were taken from published spectroscopic measurements (Privoznik et al. 1978). The major absorption lines are around 0.67 µm and in the 0.4–0.5 µm range, thus giving the algae their green colour. Scattering efficiency generally drops at higher wavelengths, except where affected by absorption lines.

The optical properties of the *Thiocapsa* bacteria were not found in the remote sensing literature. The dominant pigments in these bacteria are green bacteriochlorophyll a and red carotenoid pigment spirilloxanthin. The bacteria is round in shape and optically similar to the algae, but is smaller in size and contains a different pigment. The extinction efficiency of the bacteria was calculated from spectroscopic measurements of their absorption lines *in vivo*. The calculation is based on scattering theory for small spheres near narrow absorption lines (van de Hulst 1957). Bacteriochlorophyll a has strong absorption lines around 0.805 µm and 0.829 µm. Other pigments in the *Thiocapsa* absorb strongly in the 0.5–0.6 µm range.

Water quality was defined by standard parameters used as a measure of water quality for water resource management. The algae population is defined by the chlorophyll a concentration. The photosynthetic bacteria population is defined by the bacteriochlorophyll a concentration. The organic load is defined by the amount of yellow matter, which is normalized to a given sample taken from reservoir SR1. This sample had an extinction coefficient of 0.28 cm^{-1} at $\lambda = 0.5$ µm.

18.3.4
Model Results

The calculated spectral volume reflectance for four characteristic types of reservoirs is given in Fig. 18.5. Clean and very clean represent drinking quality water. *Chlorella* and *Thiocapsa* represent respective types of wastewater reservoirs. The absorption lines of chlorophyll a around 0.67 µm and of bacteriochlorophyll a around 0.805 µm

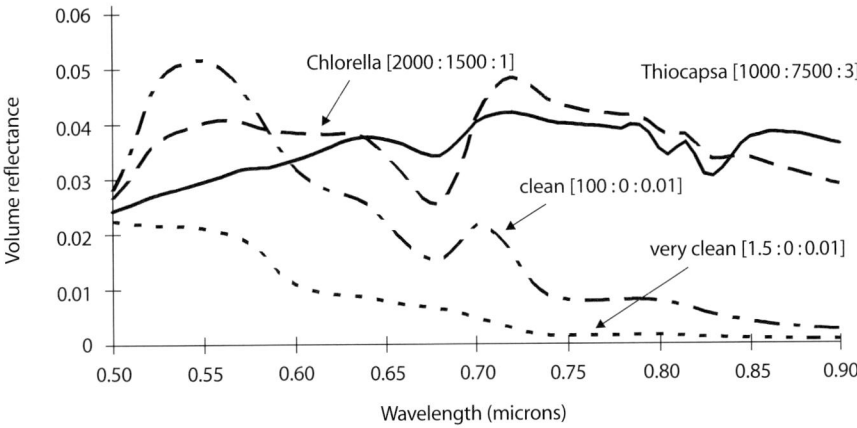

Fig. 18.5. Simulated volume reflectance for different reservoir types

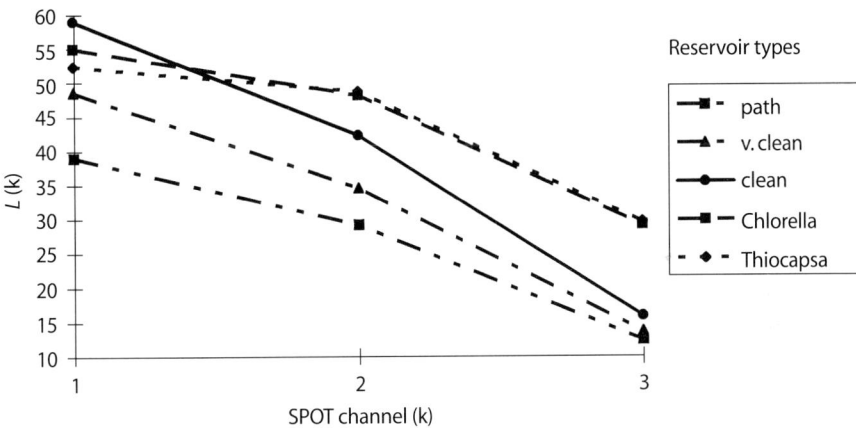

Fig. 18.6. Simulated SPOT radiance for the different reservoir types

and 0.829 μm can be identified on the reflectance graphs. The volume reflectance was calculated for an identical sun and satellite configuration for all four examples.

The model simulation of the radiance measured by SPOT from the different types of reservoirs is given in Fig. 18.6. The calculated path radiance is added to illustrate its dominant part in the total radiance measured by the satellite. The graph shows that the polluted reservoirs reflect considerably more light in the XS2 and XS3 bands than clean reservoirs. The polluted reservoirs also reflect more light in the XS1 band than a perfectly clean reservoir, but slightly less than a clean reservoir with a small algae population. The results of the simulation are consistent with the actual measurements from the satellite images: bands XS2 and XS3 show a clearer classification of water quality than band XS1.

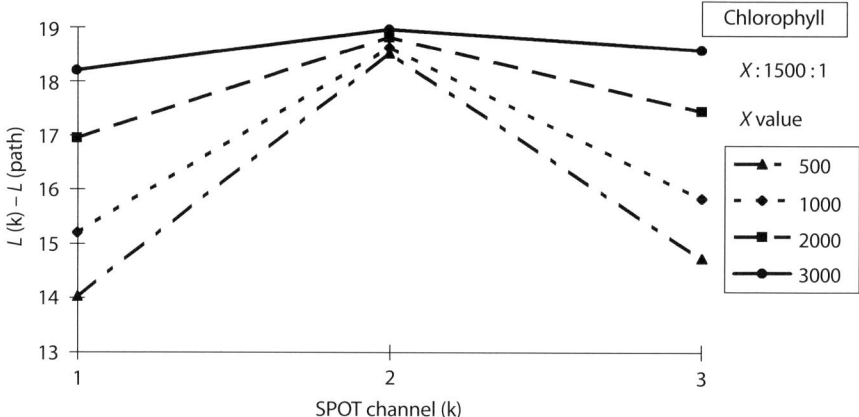

Fig. 18.7. SPOT simulated radiance at different chlorophyll a concentrations (x), for water compositions $x : 1500 : 1$. Where x = chlorophyll a concentration in mg l^{-1}, the second number represents the bacteriochlorophyll a concentration in mg l^{-1}, and the third number represents the yellow matter concentration in "Na'an units" (extinction coefficient = 0.28 cm^{-1})

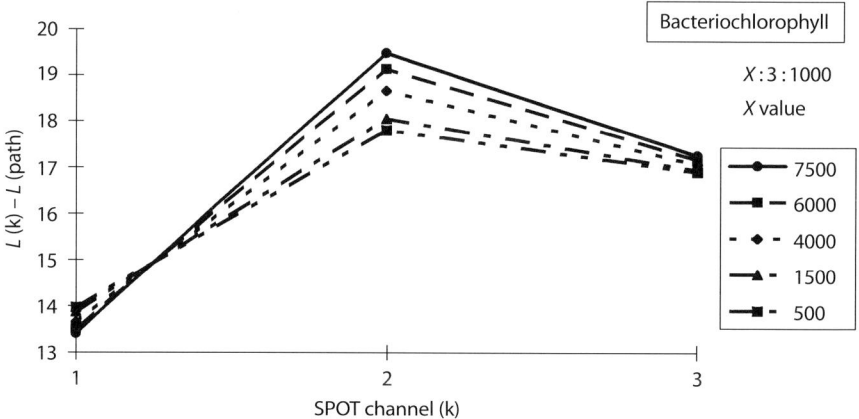

Fig. 18.8. SPOT simulated radiance at different bacteriochlorophyll concentrations (x), for water compositions $1000 : x : 3$

Figures 18.7–18.9 show the calculated effect of changes in the concentrations of the chosen water constituents on the SPOT bands. For clarity the results are shown omitting the path radiance, i.e., L[satellite] – L[path]. An increase in the concentration of chlorophyll a causes an increase in the reflected radiation in bands XS1 and XS3 due to high scattering in the 0.5–0.6 μm range, and very little absorption beyond 0.7 μm. Chlorophyll a has little effect on band XS2, where the absorption peak at 0.67 μm and the scattering peak around 0.7 μm cancel each other in the spectral range of this band.

An increase in the concentration of bacteriochlorophyll a causes a significant increase in the reflected radiation in band XS2 due to low absorption and high scatter-

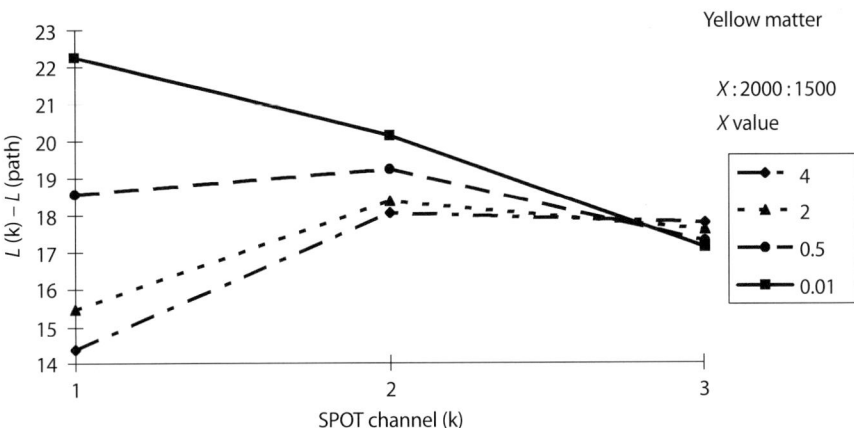

Fig. 18.9. SPOT simulated radiance at different yellow matter concentrations (*x*), for water compositions 2 000 : 1 500 : *x*

ing in this region. The smaller increase in the reflected radiation in band XS3 is due to scattering being masked by the bacteriochlorophyll a absorption lines around 0.8–0.85 μm. The strong absorption of other pigments in the *Thiocapsa* bacteria in the 0.5–0.6 μm region cause increased bacteria concentrations to reduce the amount of radiation reflected in band XS1.

Higher concentrations of yellow matter cause a considerable decrease in reflected radiation in band XS1 due to the strong absorption of the dissolved organic matter in this band. A lesser decrease in band XS2 is due to the exponential decay of absorption with the wavelength. In band XS3 higher concentrations increase the reflection slightly because the scattering of the yellow matter rather than absorption becomes the dominant factor in this region.

18.4
Principal Components Analysis

18.4.1
Principal Components

PC transformation is a statistical tool for transforming data measured in one spectral base, the satellite bands in this case, to a new spectral base which best describes the information content of the image (Green 1978). This technique has been successfully applied to a wide range of multispectrally sensed data. A *PC* analysis using the covariance matrix was made on the three SPOT bands for each of the three dates studied. The covariance matrix calculated for the reservoir sample from the October image is:

$$M_{covariance} = \begin{bmatrix} 27.9 & 25.6 & 14.3 \\ 25.6 & 31.6 & 16.9 \\ 14.3 & 16.9 & 18.8 \end{bmatrix} .$$

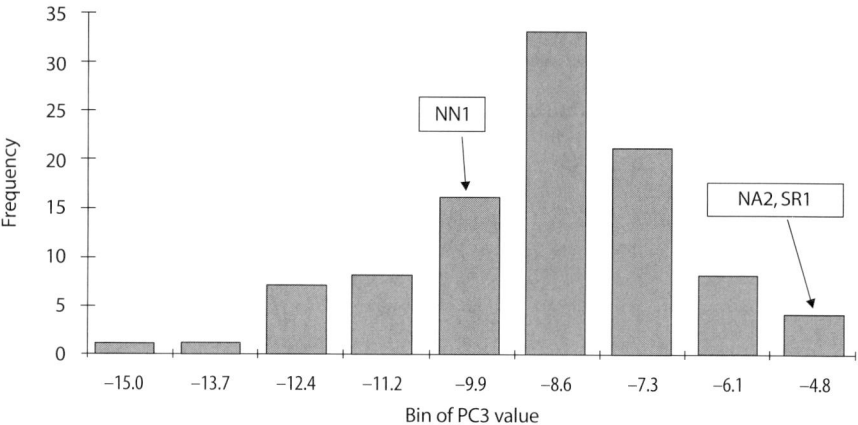

Fig. 18.10. Histogram of $PC3$ vector values. After the principal components of the reservoirs multispectral images are calculated, the spectral composition of each is represented as a linear superposition of the three principal components. The relative weight of the third principal component is called the $PC3$ vector value

The eigenvalues and the eigenvectors calculated from the matrix are:

$$\text{eigen values}: \; \lambda_1 = 65.8 \; ; \; \lambda_2 = 8.6 \; ; \; \lambda_3 = 3.8$$

$$\text{eigen vectors}: \; PC1 = \begin{bmatrix} 0.61 \\ 0.67 \\ 0.43 \end{bmatrix} ; \; PC2 = \begin{bmatrix} -0.44 \\ -0.16 \\ +0.88 \end{bmatrix} ; \; PC3 = \begin{bmatrix} -0.66 \\ +0.73 \\ -0.19 \end{bmatrix} .$$

The eigenvectors are the new PC vector base, and the eigenvalues represent the information content of each PC vector.

A classification of the reservoirs by the new base gives good separation between the different reservoirs. The best separation is achieved in the classification by the least significant vector $PC3$. A histogram of the $PC3$ contribution for October is given in Fig. 18.10. The clean and polluted reservoirs fall at the extreme opposite tails of the distribution. The values of $PC1$ contribution also give a reasonable classification of the reservoirs. In December there was a sharp increase in the *Thiocapsa* population in reservoir NA2. This gave a very strong signal on the $PC3$ histogram for that month, which is not given here.

Plate 18.1a shows the images of all the studied water bodies using the original multispectral data (XS1, XS2, XS3) in an RGB display. Plate 18.1b shows the images of the reservoirs after PC transformation, displayed in RGB according to the key: $PC1$ = Blue, $PC2$ = Green, $PC3$ = Red. One notices that the PC display resolves "better" the various reservoirs. Notice, in particular, the separation of the four reservoirs bellow the NN1 reservoir.

18.4.2
Spectral Interpretation of the PCA

In order to understand the meaning of each PC vector, its content in terms of the original SPOT bands has to be examined. The SPOT band composition of the PC vectors is given in Fig. 18.11.

Plate 18.1 a. SPOT images of all the water reservoirs in the sample using the original data in RGB display

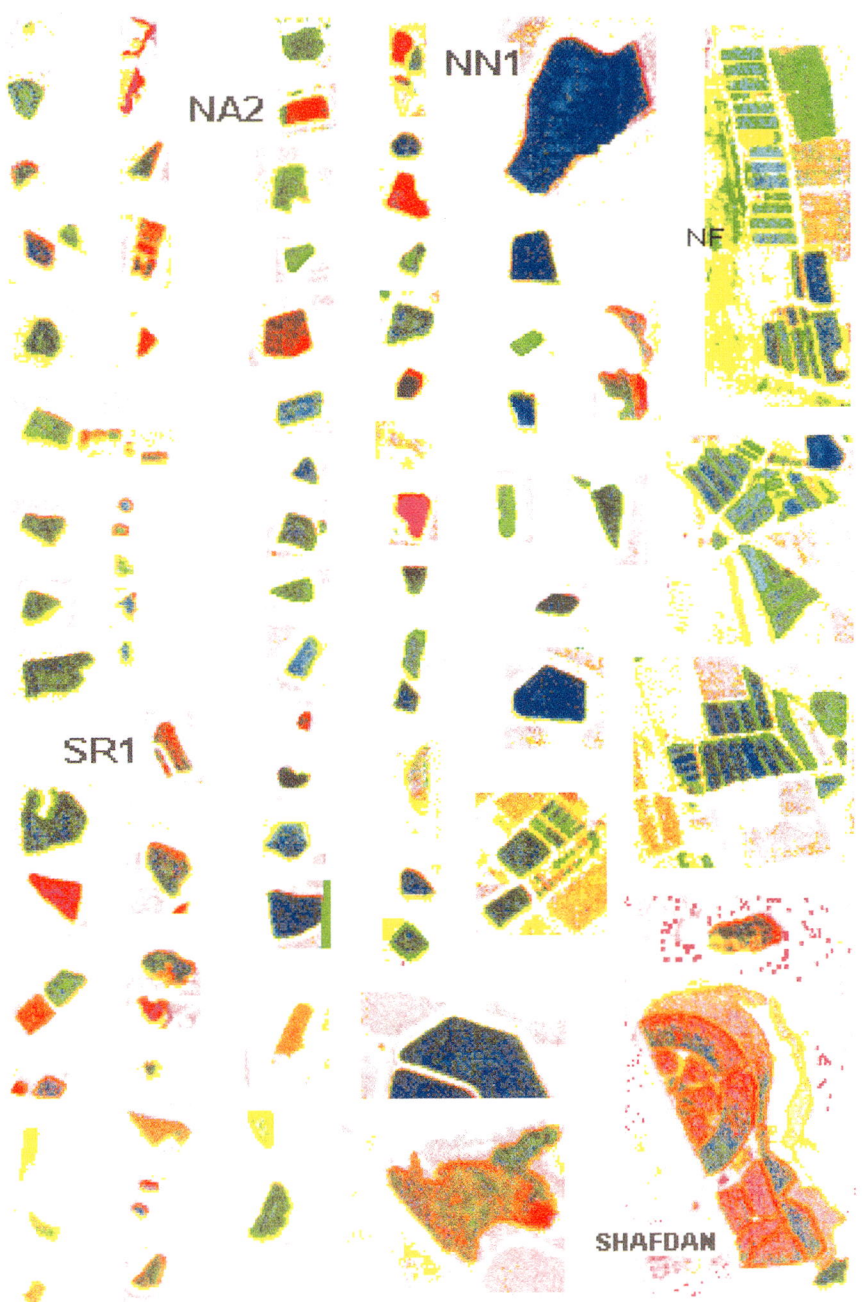

Plate 18.1 b. SPOT images of all the reservoirs after PC transformation, in RGB display

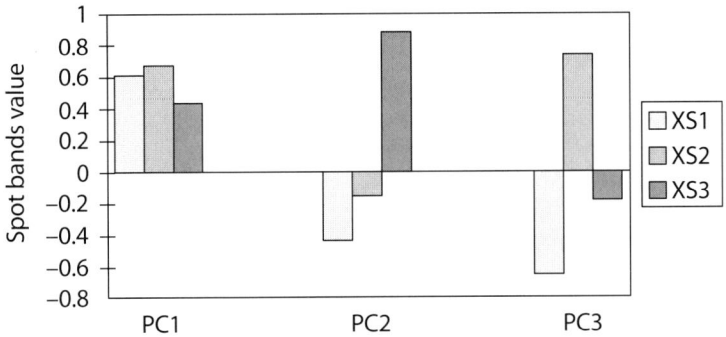

Fig. 18.11. Principal component vectors calculated from the October reservoir sample

The first vector, PC_1, contains most of the information but is spectrally flat. It appears to represent the total radiation intensity, mainly affected by atmospheric conditions and the total amount of light reflected. The low value of XS3 in PC_1 is consistent with the decrease in the path radiance and in the volume reflectance of pure water at the longer wavelengths.

The second vector, PC_2, mainly represents the green to near-IR contrast. This contrast appears to be mainly related to chlorophyll content as can be seen from variation in volume reflectance in Fig. 18.5, especially in the difference between the very clean and clean reservoirs which differ only in chlorophyll content. Figure 18.7 also demonstrates that band XS2 is not significantly affected by variations in chlorophyll content. The correlation between the second PC and green chlorophyll was also observed in multispectral images of natural ground terrain (Agassi et al. 1992). Higher concentrations of yellow matter tend to reduce overall spectral contrast due to increased scattering. This masking by the yellow matter makes the classification of reservoirs by PC_2 more ambiguous.

The third vector, PC_3, mainly represents the green to red contrast. This contrast appears to be mainly related to yellow matter and bacteriochlorophyll concentrations. Higher concentrations of yellow matter cause an increase in scattering, which increases overall volume reflectance but reduces spectral contrast. Therefore the absolute value of PC_3 decreases with higher concentrations of yellow matter as is seen in the histogram in Fig. 18.10. As seen in Fig. 18.9, higher concentrations of bacteriochlorophyll considerably increase the green to red contrast.

18.4.3
Interpretation of Model Results by Principal Components

The calculated PC vectors for the ground truth reservoirs and the model simulations appear in Fig. 18.12. Model results are generally consistent with the satellite measurements, except for PC_2 in the heavily loaded reservoirs. The effect of small changes in the concentrations of each water component depends on the concentrations of all three components. To illustrate this the effects of small changes in each of the three water components and the atmosphere were simulated for each reservoir

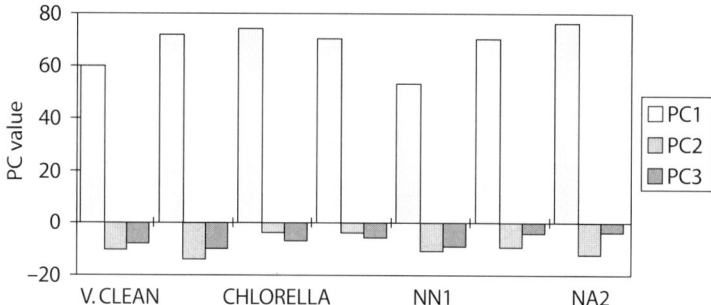

Fig. 18.12. Principal component values for the different measured and calculated reservoir types

type given in Fig. 18.6. The changes were treated as partial derivatives and are given as:

$$\text{delta ratio} = \frac{\dfrac{\Delta PC_i}{PC_i}}{\dfrac{\Delta X_i}{X}} \qquad \text{where} \quad \begin{aligned} PC_i &= PC \text{ vector value}\\ X_i &= \text{water component concentration} \end{aligned} \qquad (18.8)$$

The delta ratio defined above was calculated for a + 10% change in the concentrations of each water component. The calculated delta ratio which is used to explain the effect of each water component in terms of the PC vectors is presented in Figs 18.13a–e.

Chlorophyll a. Additional chlorophyll a in the water is found to always cause an increase in $PC1$ as the scattering coefficient of chlorophyll is higher than its absorption coefficient, thus causing an increase in the overall water reflectance. Chlorophyll a gives a strong increase in scattering in the green region. Therefore it increases the green to red and green to near-IR contrast in cleaner waters, as seen in the increase of $PC2$ and $PC3$ in the clean and very clean reservoir types. In more turbid waters higher concentrations of chlorophyll tend to reduce the green to near-IR contrast, as seen in Fig. 18.7 which describes a *Chlorella* type reservoir. This can be explained by the already high scattering efficiency of yellow matter present in the water. The scattering from the algae has a negligible effect, and thus their absorption becomes the dominant effect. The increase in $PC3$ caused by higher chlorophyll concentrations is explained by the model results in Fig. 18.7, which shows that XS1 significantly increases while XS2 remains nearly constant.

Yellow matter. In very clean waters the increase in scattering from yellow matter causes an increase in the overall reflected radiation. This causes $PC1$ to increase at higher concentrations of yellow matter. In more turbid waters, as in the clean type reservoir, the absorption of the yellow matter becomes significant and reduces the total radiation reflected.

Pure water reflects more light in the green region and very little light in the red and near-IR regions. They thus have a high spectral contrast as seen in $PC2$ and $PC3$. Yellow matter behaves in the opposite way: it strongly absorbs in the shorter wavelength and much less toward the longer wavelengths where scattering becomes more

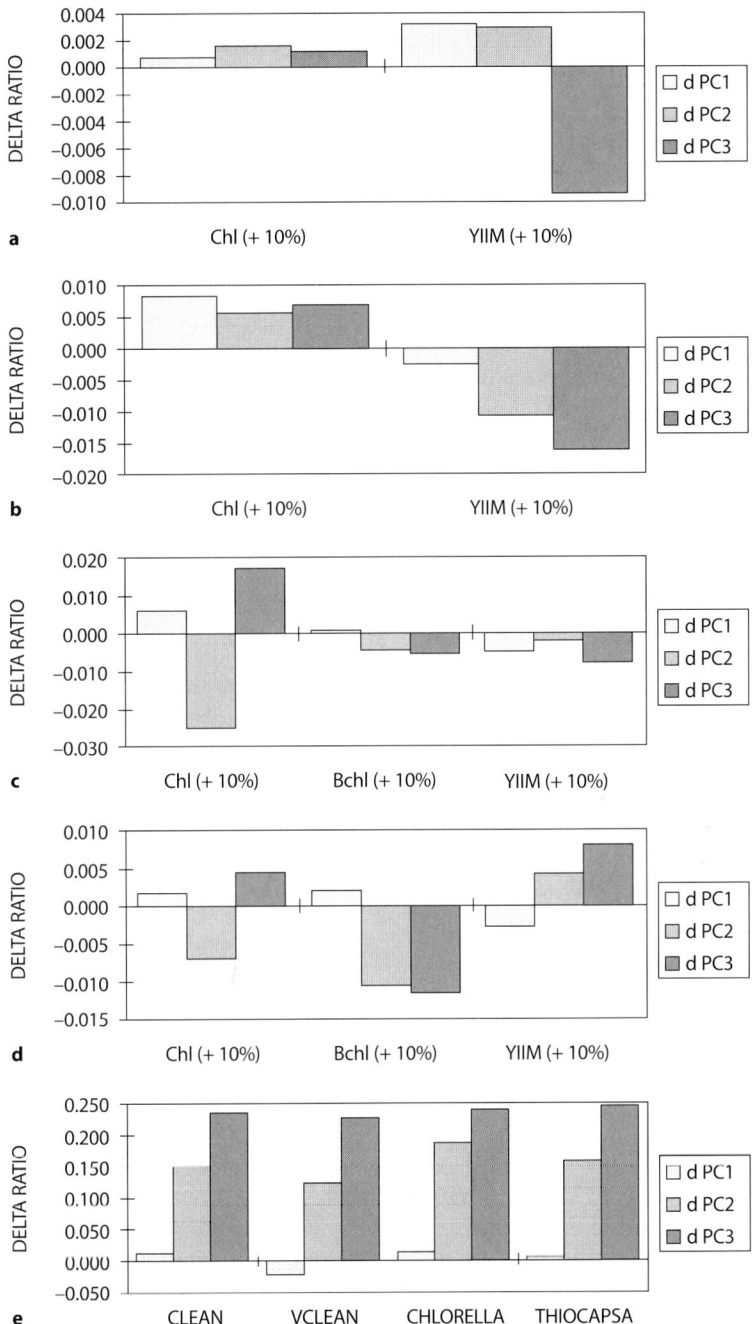

Fig. 18.13. Calculated delta ratios for the *PC* vectors in different types of reservoirs. **a** very clean type reservoir; **b** clean type reservoir; **c** *Chlorella* type reservoir; **d** *Thiocapsa* type reservoir; **e** variation under different atmospheric conditions

dominant (see Fig. 18.4). Thus yellow matter tends to decrease spectral contrast as seen in the negative delta ratios of $PC2$ and $PC3$ of the clean and *Chlorella* type reservoirs. The increase in spectral contrast in the *Thiocapsa* type reservoir may be due to the saturation of the scattering efficiency at very high concentrations of yellow matter. Thus absorption becomes the significant factor which increases spectral contrast, as seen in the increase of bands XS1 and XS2 in Fig. 18.9.

Bacteriochlorophyll a and spirilloxanthine. The above pigments are not found in clean unpolluted waters where the *Thiocapsa* bacteria do not develop. Therefore the effect of variations in these pigments is tested only for the wastewater reservoirs.

The effect of these pigments is equivalent in both types of wastewater reservoirs, but is much stronger in the more heavily loaded *Thiocapsa* type reservoir. $PC1$ increases at higher bacteria concentrations due to an increase in scattering. The strong absorption in the green region causes a decrease in the radiation reflected in band XS1. As band XS1 has the highest radiation values, this reduction causes a decrease in green to red and green to near-IR spectral contrast. Therefore $PC2$ and $PC3$ decrease at higher concentrations of the *Thiocapsa* pigments.

Atmospheric conditions. The four reservoir types were modelled under the atmospheric conditions of the NN1 and SR1 reservoirs. The differences are given in Fig. 18.13e as the ratio:

$$\frac{\Delta PC_i}{PC_i} = \frac{PC_i[\text{SR1}] - PC_i[\text{NN1}]}{PC_i[\text{NN1}]} \ . \tag{18.9}$$

It is clearly seen that $PC2$ and $PC3$ are far more sensitive to atmospheric conditions than XS1.

18.5
Summary and Conclusions

Statistical analysis of SPOT images of small inland water bodies shows a clear classification between clean water reservoirs and wastewater reservoirs. It is more difficult to differentiate between different levels of pollution in wastewater reservoirs, but major changes in the biochemical condition of the water can be detected. Reservoirs containing effluents reflect overall more radiation than clean reservoirs and have a flatter spectral distribution. The amount of reflected radiation in the near-IR band (XS3) is proportionally greater for wastewater reservoirs than for clean water reservoirs.

Radiance values for all three SPOT bands are found to be highly correlative in the sample of about 100 reservoirs studied on all three dates. The deviation of an individual reservoir from the linear regression line of the entire sample is found to be a good indicator of water quality, particularly the deviation of XS2 from the XS3 vs. XS3 regression. This effect is enhanced when the sample is studied after it is transformed to normalized spectral coordinates.

A principal components analysis of the reservoir sample yields three new principal component vectors (PC), each of which can be given a meaning in terms of water composition. $PC1$ contains most of the spectral information. It represents the total radiation intensity and is spectrally flat. *PC2* represents the green to near-IR contrast. This contrast appears to be mainly related to chlorophyll a content. *PC3* represents green to red contrast, which is mainly related to yellow matter and *Thiocapsa* pigments concentrations. The differential variations of the *PC* vectors with respect to differential variations in water component concentrations vary greatly for different types of reservoirs. This is due to the overall differences in the optical properties among these environments and tradeoffs between component absorption and scattering. Atmospheric conditions are found to have a strong effect on the *PC* vectors, especially *PC2* and *PC3*.

Simulations of the effects of parametric variations of water composition were run using the optical model for the reservoir's volume reflectance. Increased chlorophyll concentrations increase reflection in bands XS1 and XS3, particularly in clean waters. Increased *Thiocapsa* pigments concentrations increase reflection mainly in band XS2 and a decrease in band XS1. High concentrations of yellow matter strongly reduce reflection in band XS1. Due to increased absorption in band XS2 and the high scattering efficiency of yellow matter, it causes an overall flattening of the radiance spectral distribution in waters with heavy organic loads. It is therefore more difficult to detect small variations in water quality in wastewater reservoirs.

SPOT images were found to contain significant information related to water quality in a wide range of small inland reservoirs. These images can serve as a useful tool for monitoring water quality, as a good classification of water quality can be achieved. The images can also assist water management of hypertrophic wastewater reservoirs, as the spectral distribution is related to the ecological state of the reservoir. The analysis of minor differences between wastewater reservoirs is more difficult and would require a wider database for the optical modelling of the water. Higher spectral resolution, available with other satellites with more spectral bands such as LANDSAT, is likely to improve the ability to differentiate between different qualities of water in hypertrophic reservoirs.

References

Agassi E, Ben Yosef N, Wietz A (1992) Spectral correlation of natural ground terrain images at the spectral range of 0.4–1.05 μm. Proceeding of the 8th Meeting on Optical Engineering in Israel, held in Tel-Aviv, Israel, on 14–16 December 1992, SPIE Volume 1971, pp 160–169

Alfoldi T (1982) Remote sensing for water quality monitoring. In: Johannsen CJ, Sanders JL (eds) Remote sensing for resource management. Soil Conservation Society of America, Ankey, Iowa, pp 317–328

Bukata RP, Bruton JE, Jerome JH (1983) Use of chromaticity in remote measurements of water quality. Remote Sensing of the Environment 13:161–177

CNES, SPOT IMAGE (1991) SPOT users handbook, vol 2, ed 01 rev 02, Toulouse, France, SPOT IMAGE

Davis-Colley RJ, Vant WN (1987) Absorption of light by yellow substance in freshwater lakes. Limnology and Oceanography 32:416–425

Dor I, Raber M (1990) Deep wastewater reservoirs in Israel: Empirical data for monitoring and control. Wat Res 24(9):1077–1084

Dor I, Ben Yosef N, Adin A (1991) Monitoring water quality in deep wastewater reservoirs using remote sensing: Calibration of the method. Third Research Report to NCRD and BMFT, School of Applied Science and Technology, The Hebrew University of Jerusalem

Gordon HR (1973) Simple calculation of diffuse reflectance of the ocean. Applied Optics 12:2803–2806

Green PE (1978) Analyzing multivariate data. The Dryden Press, Hinsdale, Illinois

Pratt WK (1978) Digital image processing. John Wiley & Sons, New York

Privoznik KG, Daniel KJ, Incropera FP (1978) Absorption, extinction and phase function measurements for algal suspensions of *Chlorella pyrenoidosa*. Quantitative Spectroscopic and Radiative Transfer 20:345–352

Smith RC, Baker KS (1981) Optical properties of the clearest natural waters (200–800 nm). Applied Optics 20:177–184

Van de Hulst HC (1957) Light scattering by small particles. John Wiley & Sons, New York, pp 172–193

Chapter 19

Experiences Outside Israel

Marcelo Juanicó · Eran Friedler

Israel is today the leading country in the research and use of wastewater reservoirs, but wastewater reservoirs were neither invented in Israel nor is Israel the only country to use them. Other countries already used wastewater reservoirs much before their massive introduction in Israel in the seventies, and many countries continue or are starting to use them nowadays.

This chapter intends to make a short review of published literature on wastewater reservoirs outside Israel. Unfortunately, it seems that much information is published in journals and/or languages which are out of our reach. For example, we received just one reference in English on wastewater reservoirs in China, but the reference suggests that there may be more Chinese experiences on the issue. The same situation may happen with Japan (Sano 1983a, 1983b) and other countries. We have received isolated and informal information from Mexico on the 'Endhó' wastewater reservoir (200 million m³!) in the State of Hidalgo, and on the existence of other wastewater reservoirs, but we failed to find reliable information on them.

There is a spectrum of different names to call wastewater reservoirs both outside and inside Israel. Some of the differences are just semantic, others refer to impoundments which are similar but not equal to those used in Israel. Some of the terms that can be found in literature are:

- stabilization reservoirs
- deep storage reservoirs
- detention reservoirs
- controlled discharge ponds
- storage ponds
- holding ponds
- deep stabilization ponds
- storage lagoons

A review of the experience with wastewater reservoirs outside Israel must differentiate between:

1. Wastewater reservoirs developed by independent research/professional groups in other countries;
2. Wastewater reservoirs designed abroad by Israeli consultants.

19.1
Independent Experience

19.1.1
Canada and the United States

19.1.1.1
Controlled Discharge Ponds for River Discharge

Controlled discharge ponds have been used in Canada and the Unites States for many years to treat sewage and allow the discharge of effluents into rivers in proper time, i.e., to avoid the ice cover season (Canada, Northern USA), or the low stream flow season (Southern USA). In this second type of holding pond (also called 'hydrograph controlled release' ponds, Middlebrooks 1995) discharge is performed when the receiving stream flow rate is adequate to receive the effluent. This practice produces a significant increase in the instream dissolved oxygen level (Tapp 1976).

These 'controlled discharge ponds' are made of 2–3 cells (a few have 4–5 cells), working in series or parallel, 2–2.5 m deep, which receive raw sewage. The basic criteria for the design of these ponds are described in an old EPA report (EPA 1975) quoted by Middlebrooks (1983, 1990):

> "*The unique features of controlled discharge ponds are long-term retention and periodic controlled discharge usually once or twice a year. Ponds of this type have operated satisfactorily in the north-central US using the following design criteria:*

- *Overall organic loading: 22–28 kg $BOD_5 ha^{-1} d^{-1}$;*
- *Depth of water column: not more than 2 m for the 1st cell, not more than 2.5 m for subsequent cells;*
- *Hydraulic retention time: At least 6 months of storage above the 0.6 m liquid level (including precipitation), but not less than the period of ice cover;*
- *Number of cells: at least 3 for reliability, with piping flexibility for parallel or series operation.*

> *The operation and maintenance manual must include instructions on how to correlate pond discharge with effluent and stream quality.*
> *Isolate the cell to be discharged … by valving off the inlet line.*" (this is a batch operation).

The prediction of *BOD* in the effluents is calculated by a simple formula (Middlebrooks 1983):

$$\frac{C_e}{C_0} = e^{-k_p t} \quad ,$$

where:
- C_e = effluent BOD_5 in mg l^{-1}
- C_0 = influent BOD_5 in mg l^{-1}
- e = base of natural logarithms
- k_p = plug flow first order reaction rate, d^{-1}
- t = hydraulic residence time in days

A survey of almost 100 controlled discharge systems in Michigan, Minnesota and North Dakota, USA (Pierce 1974) indicates that the most probable effluent quality for 3-cell systems is $BOD = 17$ mg l^{-1}, $TSS = 30$ mg l^{-1}, and faecal coliforms < 400/100 ml.

Half of the 1750 sewage treatment plants operating in Canada in 1985 were based on waste stabilization ponds, many of them with controlled discharge to avoid discharge to rivers during the ice cover season. In the colder regions sludge accumulated in the lagoons during the winter, and removal of sludge was necessary each 5–10 years (Environment Canada 1985).

Hatcher (1980) developed a model to optimize the design of sewage treatment plants which includes a 'wastewater storage pond.' The study is oriented to communities which have to discharge treated wastewater in streams that have low stream flow seasons (Georgia, USA) and face the problem of the strict wastewater quality requirements for discharge during these dry periods. The reservoir holds lower quality wastewater during the dry season for subsequent release when natural stream flows are higher, or for release to a wastewater irrigation system. The model allows the optimization of the reservoir sizing, and includes a simple function to estimate *BOD* degradation during storage.

19.1.1.2
Storage Lagoons for Wastewater Irrigation

The Muskegon County Wastewater Management System receives both domestic and industrial sewage, from sources including 14 municipalities, a paper mill, 3 chemical companies, an engine manufacturing plant, a metal casting and plating firm, and over 150 smaller industries, totaling 40 million $m^3 yr^{-1}$ of sewage. Sewage is treated in aerated lagoons followed by 'storage lagoons'. These storage lagoons are only 3.5 m deep. A study made by Frykberg et al.(1977, 1978) shows that (like in wastewater reservoirs in Israel) the behaviour of the impoundments and water quality were affected by the wastewater input regime more than by seasonal factors.

Funderburg et al. (1978) performed an interesting study in Texas, USA, on the removal of polivirus (Polivirus I – Chat) in wastewater holding ponds. The experiments were performed in outdoor tanks 150 cm in diameter and with depths from 45 to 225 cm, with both primary treated and secondary plus chlorination treated effluents. Results indicate that five days of batch operation are enough to remove 99% of viruses in summer (water temperatures 17–25° C), but 15 days are needed to remove the same percentage in spring (water temperatures 17–25° C, the same as in summer), and 25 days in winter (water temperatures 4–12° C). There was no virus detectable in the water column after 7 days in summer, after 35 days in spring and after more than 42 days in winter. The inactivation of viruses was slower in primary treated effluents than in secondary treated ones. The inactivation of viruses in the deeper ponds was slower than in the shallow ones. The rate of inactivation was positively correlated with high *p*H values and chlorophyll concentration. All environmental data indicate photosynthetic related process as one of the main causes of virus removal. The concentration of viruses in the sediment increased sharply during the first 6 days and then continued without changes during five weeks, indicating that settling (possible associated with bioflocculation) was also one of the processes related to virus removal. Although the pilot size of the tanks imposes some limitations to the results, this work was one of the pioneering ones in revealing the potential capacity of batch operated ponds for the removal of pathogens. Kott (1975) and Kott et al. (1978) obtained similar results in full scale shallow holding ponds in Israel, and Liran (1990), Liran et al. (1994) and

Table 19.1. Removal of *BOD* in the ponds and reservoirs of Rancho Murieta	*BOD* (mg l^{-1})
Raw sewage	175
Effluent from 5th aerated lagoon	12 – 16
Effluent from wastewater reservoirs	4 – 6

Juanicó (see Chapter 5) extended these preliminary conclusions to full-scale deep storage reservoirs.

The Rancho Murieta sewage treatment and reuse system is one of the latest published systems for sewage treatment and reuse in golf course irrigation (Fuog et al. 1995). The system is made of five aerated lagoons in series followed by two deep wastewater reservoirs in parallel operated in continuous flow. The reservoirs have a maximum depth of about 16 m and a combined storage capacity of 930 000 m³. Strict Californian standards for golf course irrigation require further dissolved air flotation, rapid sand filtration, and disinfection after the reservoirs. The removal of *BOD* in the ponds/reservoirs is shown in Table 19.1.

19.1.1.3
Recreational and Landscape Wastewater Reservoirs and Similar Water Bodies

There are several impoundments of this kind in the USA. The first to be commissioned seems to be the Indian Creek Reservoir (Porcella et al. 1972), an artificial reservoir created to receive tertiary treated effluents after phosphorus removal with a *BOD* < 5 mg l^{-1}. Storage in the reservoir sharply improved effluent quality and the reservoir developed a diversified biological community. Although the original objective of the reservoir was to discharge effluents downstream for irrigation, it became a prominent recreational facility in Alpine County (Adams et al. 1983).

Lake Balboa in Los Angeles and the wetlands at the Sepulveda Basin Wildlife Refuge are other well known projects where treated wastewater is stored for recreational purposes (WR-News 1992, 1994).

19.1.1.4
Phosphorus Removal in Batch Operated Lagoons by Chemical/Floculant Treatment

Surampalli et al.(1995) present a review of experiences with 32 storage-lagoon systems in Minnesota and Michigan, USA, in precipitating P with aluminium sulfate, lime, and ferric chloride. These authors based their work on two previous similar studies in Canada which reached the following conclusions:

- Batch chemical treatment of seasonal lagoons achieved total P effluent of less than 1 mgl^{-1}.
- Effluent quality from batch treated lagoons was comparable to or better than that achieved by conventional secondary treatment.
- Alum and ferric chloride applications produced consistently high quality effluents while lime applications were not as effective in removing P.

- Batch chemical treatment is feasible for existing lagoon treatment systems which have adequate retention time for winter storage and is also effective in removing algae from lagoon wastewater if the chemical dosage is sufficient.

Based on these studies, the Province of Ontario designed and operated 20 full-scale municipal lagoon treatment systems using alum to precipitate P in seasonal discharge ponds.

The survey of Surampalli et al. arrived to conclusions similar to those of the Canadian surveys:

- The technology of adding requisite amounts of alum, ferric chloride and polymer in the secondary cells of facultative lagoons (the storage pond) ... removed P to less than 1 mg l^{-1} in most of the systems evaluated.
- In these lagoon systems (some of them operated for 20 years) no sludge build-up problems were found which affected the effluent quality.
- Secondary benefits noted from such chemical additions were lower *BOD* and *TSS* levels in the effluents.

19.1.2
Spain

The development of wastewater reservoirs in Spain is led by two independent groups:

- Rafael Mujeriego from the Universidad Politécnica de Catalunya y Lluis Sala from The Consorcio de la Costa Brava, on two wastewater reservoirs at the Mas Nou Golf Course, located on the Mediterranean coast of Catalunya.
- A research team from the Universidad de Murcia, on two small wastewater reservoirs, one located in Cartagena and one in Murcia, on the Mediterranean coast of Murcia.

19.1.2.1
The Reservoirs of the Mas Nou Golf Course

The sewage treatment and reuse system is described in Mujeriego and Sala (1991), Sala et al. (1994) and Sala and Millet (1997), plus several local Catalunyan publications.

The Mas Nou golf course relies on reclaimed wastewater for irrigation of its 34 ha of playing grounds. Pumping for irrigation varies from a maximum of 1 800–1 900 m^3 d^{-1} in July to a minimum of 0–160 m^3 d^{-1} in December.

The secondary treated effluents from the municipal treatment plant (conventional activated sludge) are chlorinated and elevated 300 m to the golf course where they are stored in two wastewater reservoirs in series located within the golf course. The reservoirs were designed to fulfil three main purposes: 1) aesthetic enhancement, 2) storage of irrigation water, and 3) additional treatment of the irrigation water. The reservoirs are operated as continuous-flow reactors. Each reservoir is equipped with a water jet to promote water circulation and aeration, while contributing to visual enjoyment of the impoundments. The main characteristics of the system and effluent quality are presented in Table 19.2.

The quality of effluents in the reservoirs follows an annual cycle, with good quality during the winter when hydraulic retention time within the reservoirs is long, and

Table 19.2. Main design, operation and effluent quality characteristics of the two reservoirs at Mas Nou golf course (modified from Mujeriego and Sala 1991, and Sala et al. 1994)

Parameter	Units	Inflow from treatment plant	Reservoir 1	Reservoir 2
Max volume	m^3		13 300	21 000
Max depth	m		4.5	5.5
Retention time	days		<10 – 500	40 – 600
pH	pH units	7.3 – 8	7.8 – 9.2	7.9 – 9.7
Conductivity	µS cm^{-1}	700 – 6 500	1 190 – 2 440	1 470 – 1 930
Alkalinity	mg l^{-1} CaCO$_3$	140 – 390	130 – 300	150 – 210
SAR	SAR units	3.6 – 19	6 – 16	8 – 15
NO$_3$-N	mg l^{-1}		0.15 – 19	0.03 – 4
NO$_2$-N	mg l^{-1}		0.04 – 11	0.01 – 0.7
NH$_4$-N	mg l^{-1}	6 – 70	0.05 – 40	0.02 – 12
N-organic	mg l^{-1}		1.5 – 4	1.5 – 4
PO$_4$-P	mg l^{-1}	2 – 14	0.3 – 11	0.2 – 8
Chlorophyll a	µg l^{-1}	0	10 – 730	3 – 1 450
Faecal coliforms	cfu/100 ml	$6 \times 10^3 - 5 \times 10^6$	0 – 40 000	0 – 2 500

degradation of water quality during the irrigation season as a consequence of reduction in the hydraulic retention time. This cycle is typical of reservoirs operated as continuous-flow reactors (see Chapter 5).

The main problems faced by from this system have been:

- Periodic failures of the chlorination system have produced peaks of pathogen concentration in the effluents (the problem is overcome by improving the performance of the chlorinator).
- Periodic break-down of algae results in low oxygen contents in the reservoirs (the problem is overcome by increasing water circulation and seeding of reservoirs with algae rich water from other impoundments).
- The SAR and salinity of the effluents from the sewage treatment plant is sometimes too high.

A more sophisticated operation of the reservoirs in order to optimize water quality, nutrient recycling, and energy consumption has been lately proposed by Sala and Millet (1997).

19.1.2.2
The Cartagena Reservoir

This 'deep stabilization pond' has a volume of about 45 000 m³ and a depth of 5 m. Moreno et al. (1988) took field samples during one year and developed a simulation model, comparing the actual measurements with model predictions. A tracer study indicated that the pond is perfectly mixed. The pond removed COD by 60%, BOD by 80%, and faecal coliforms by four orders of magnitude. Unfortunately, the authors give no data on the operation of the pond/reservoir (flow rates, hydraulic retention time, whether the input of effluents to the pond was stopped or not, etc.) except the single statement that the volume of the pond did not change during the study period. Thus, the interpretation of field data and model predictions is not possible.

19.1.2.3
The Murcia Reservoir

This system is described by Berná et al. (1986), Soler et al. (1991), Llorens et al. (1992, 1993) and Torres et al. (1997).

The reservoir is located on the campus of the University of Murcia and receives sewage from a residential complex plus several university centres. This reservoir has a volume of 15 000 m³ and a depth of 8 m. Both inlet and outlet are located in the surface layer.

The studies by Berná et al. (1986) and Soler et al. (1991) cover two periods:

1. Filling, November–December 1984 followed by batch operation till July 1985
2. Filling, September–December 1985 followed by batch operation till July 1986

The work of Berná et al. (1986) addresses mainly the changes in physico-chemical parameters. The effluents entering the reservoir had a $BOD = 180$ mg l⁻¹ and faecal coliforms 1 000 000/100 ml during the covered period. The reservoir presented thermal stratification during the spring-summer. The water column was anaerobic during the filling period and for one month after inflow valving-off. The concentration of oxygen increased during the batch period reaching supersaturation. BOD concentration was reduced from 180 mg l⁻¹ to <50 mg l⁻¹ (period 1) and to less than 25 mg l⁻¹ (Period 2) after 40 days of batch operation. The concentration of pathogens (total and faecal coliforms, Streptococcus and Clostridium) decreased much faster in the surface layers than at bottom. Most pathogens were not detectable at the surface after 30–50 days of batch operation.

The work of Soler et al. (1991) stresses the changes in the planktonic community. *Chlamydomonas* algae dominated during the early batch operation under heavy loading and poor oxygen conditions. Small chlorophytes (*Golenkinia, Chlorella* and *Scenedesmus*) dominated the second algae community to develop in the reservoir. Later on, grazing by rotifers and ciliates favoured the development of a third bloom dominated by the small Cyanophyta *Synechococcus*. Two blooms of photosynthetic bacteria were identified: one dominated by the sulfur red *Thiocapsa* and *Chromatium*, which coincided with the bloom of *Chlamydomonas*, and one dominated by the green photosynthetic bacteria *Chlorobium* at and below the thermocline when the concentration of H_2S reached 95 mg l⁻¹.

Llorens et al. (1992, 1993) studied the system from October 1986 to December 1987. Inflow BOD was 350 mg l⁻¹ in average. The impoundment was operated as a continuous-flow reactor with continuous inflow and fixed volume, with a mean hydraulic retention time of almost 90 days; thus, it must be considered a deep stabilization pond and not a 'reservoir.' The system was strongly stratified during the summer. Strong primary productivity exhausted the nutrients in the epilimnion in June, resulting in a decrease of algae concentration and clear waters. Stratification did not allow nutrient from hypolimnion to reach the upper layers. The pond provided 70–90% removal of COD, 85–95% removal of BOD, and 1–3 orders of magnitude faecal coliform removal.

Torres et al. (1997) analysed the hydraulic performance of the reservoir by means of tracer experiments with sulfo-rhodamine B in periods of continuous flow at constant volume (the impoundment behaves as a deep pond, not a reservoir). The experiments indicated an active volume of 70% of the theoretical volume in winter (no stratification) and of only 22% (!) of theoretical volume in summer (stratification). In summer, the tracer did not reach the water column below 2 m depth.

19.1.3
China

The single article on wastewater reservoirs in China we succeeded in finding an international journal is that of Baozhen (1987). A summary of the article is herein transcribed.

"Qiqihar City is situated at 123°30'E and 47°52'N. It is characterized by a continental monsoon climate with a mean annual temperature of 3.2° C, which ranges from an extreme high of 40° C to an extreme low of –39.5° C. The mean annual rainfall is 419 mm, most of it from June to August. Mean annual evaporation is 1 430 mm. The period of sub-zero temperatures with water freezing lasts for about five months from December to April.

The sewage treatment and storage system – put into operation in 1970 – is made of a single lagoon surrounded by earth dams lined with concrete. The total area is about 800 ha (9 km long by 0.5 to 2 km wide) but present use covers only 200 ha in summer and 450 ha in winter.

The system receives 150 000 $m^3 d^{-1}$ of raw sewage. There is zero discharge during the five freezing months in order to protect the river, which has no fish deaths now. During the summer the lagoon receives more solar energy and establishes a suitable ecosystem for effective treatment and multi-purpose utilization of the wastewater.

Specific type of anaerobic and facultative bacteria and fungi are capable of effectively decomposing refractory organic compounds such as lignin and synthetic detergents with typical removal efficiencies of 95% and 80% at 20° C respectively.

Algae were represented by 97 species and zooplankton by about 45 species. *Cladocera* appeared in large numbers in the shallow water along the shore with a population density of 100–1000 individuals/l. The grazing pressure by *Cladocera* sharply decreased algae and bacteria populations.

The lower reach of the lagoon (near the outlet) had clear water and submerged aquatic plants (*Chara*, *Cerataphyllum* and *Hydrilla*). Several species of fish were found in the middle and lower reaches, dominated by *Carassius auratus* (crucian carp) which grows naturally in the lagoon. Benthic fauna was dominated by Chironomid larvae. The middle and lower reaches of the lagoon attracted thousands of gulls and terns, and hundreds of wild ducks and various other birds.

Hundreds of ducks and geese had been raised in trials on the lower reach of the lagoons in the past years where they grew much faster than those raised in clear water ponds.

Table 19.3. Quality of the influent and effluent in the Qiqihar Wastewater Reservoir (modified from Baozhen 1987)

Parameter	Units	Influent	Effluent
pH	pH units	6.2 – 7.3	7.5 – 8.1
COD	mg l^{-1}	300 – 350	30 – 60
BOD$_5$	mg l^{-1}	125 – 160	2.5 – 10
TSS	mg l^{-1}	240 – 310	14 – 30
Lignin	mg l^{-1}	16 – 20	0.2 – 0.6
DBS (synthetic detergents)	mg l^{-1}	0.7 – 1.0	0.05 – 0.12
Phenol	mg l^{-1}	0.01 – 0.16	0.001 – 0.002
Cyanide	mg l^{-1}	0.01 – 0.06	0.001 – 0.002
Mercury (Hg)	μg l^{-1}	0.25 – 0.64	0.03 – 0.08
Zinc (Zn)	μg l^{-1}	3 300	40
Chromium (Cr)	μg l^{-1}	400 – 1 000	70 – 90
Copper (Cu)	μg l^{-1}	150 – 500	13 – 80

It is evident from the observed results (Table 19.3) that the lagoon removed more species of pollutants at higher efficiency than conventional secondary treatment facilities, particularly in terms of COD and refractory compounds. Moreover, its capital and operational costs are much lower than that of conventional secondary treatment.

The purified wastewater is pumped to paddy fields for irrigation, thus producing bumper harvests while saving greatly on chemical fertilizers.

The lagoon is very simple. It lacks pretreatment facilities, particularly sedimentation basins, thus resulting in the deposition of a large quantity of sludge, silting up the upper reach of the lagoon, leading to the occurrence of a large area of septic and odorous sediment and a progressive decrease in available volume of the lagoon. The lagoon is irregular in shape and not divided into several cells, with several stagnant and/or dead areas.

The lagoon has been lately reconstructed according to the authors suggestion, including the construction of primary conventional treatment facilities, and dividing the lagoon into several cells acting in series: anaerobic pond, facultative pond, aerobic fish pond, duck and goose pond."

19.1.4
Germany

Felgner and Sandring (1983) describe a series of experiments on wastewater storage in tanks of 300 m³ each and 2.2 m depth in Dresden.

The tanks were filled from December to March and then inflow was stopped. Wastewater was released for irrigation during June-July-August. Quality of wastewater is presented in Table 19.4.

The authors stressed the importance of taking water for irrigation from the surface layer where water quality is better, and proposed the construction of two reser-

Table 19.4. Quality of wastewater in the outflow from the tanks – surface layer water (modified from Felgner and Sandring 1983)

Parameter	Units	Inflow to tanks	Batch operation		Continuous-flow operation	
			End of filling (in-flow is stopped)	Start of irrigation	End of filling (in-flow continues)	Start of irrigation
BOD$_5$	mgl^{-1}	210	113	18	114	24
Diss. oxygen	mgl^{-1}	0	1	>18	1	>18
Faecal coliforms	ml^{-1}	1×10^9	1×10^8	0	10 000	100

Table 19.5. Quality of wastewater in the surface layer of the 1 000-m³ reservoir at the beginning and end of the winter storage (modified from Indelicato et al. 1995)

Parameter	Units	Start of storage (February)	End of storage (May)
Dissolved oxygen	mg l^{-1}	0	19
COD	mg l^{-1}	325	150
TSS (180)	mg l^{-1}	70	42
Total P	mg l^{-1}	2.2	0.8
N-Kjeldahl	mg l^{-1}	50	15
NH$_4$-N	mg l^{-1}	48	6
NO$_2$-N	mg l^{-1}	0	0.04
NO$_3$-N	mg l^{-1}	1	1
Total coliforms	MPN/100 ml	4.6×10^6	1.6×10^3
Faecal coliforms	MPN/100 ml	4.6×10^6	2.4×10^1
Faecal streptococci	MPN/100 ml	8.6×10^5	2.0×10^1

voirs to be operated alternatively in order to obtain better effluent quality (a sequential batch operation).

19.1.5
Italy

Indelicato et al. (1995, 1996) describe the experience gained in Sicilia, Italy with a 1 000 m³ capacity wastewater storage reservoir for agricultural irrigation. Other reservoirs up to 50 000 m³ capacity are being operated in the region.

The small earthen reservoir is 3.2 m deep. It was filled with untreated sewage on February 1994 and then operated as a batch reservoir (no inputs nor outputs) for 90 days. The reservoir was not stratified and totally anaerobic till April, when a thermocline started to form, separating the reservoir into an anaerobic hypolimnion and an aerobic epilimnion.

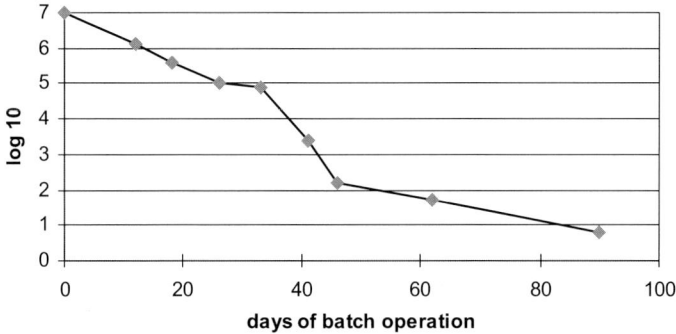

Fig. 19.1. Decay of faecal coliforms in the 20-cm surface layer of a 1 000-m³ reservoir under winter conditions (February) in Sicilia, Italy

Table 19.5 describes effluent quality at the beginning and end of the batch period. Figure 19.1 shows the evolution of faecal coliforms decay during the storage period. Preliminary data from a 30 000 m³ reservoir indicate similar results.

19.2
Wastewater Reservoirs Designed Abroad by Israeli Consultants

Israeli firms have designed many wastewater reservoirs abroad. The three examples herein presented intend only to illustrate potential approaches to different problems.

19.2.1
Chile – The Santiago Poniente Experimental Treatment Plant

This sewage treatment and reuse plant was designed as an experimental plant to obtain optimal design parameters in a region with less solar radiation and lower temperatures than Israel. The objective was to produce treated wastewater for agricultural irrigation. A summary of the system design is herein presented, based on a more detailed description by Libhaber (1995).

19.2.1.1
Inflow

The design sewage inflow was for 17 000 m³ d⁻¹. Inflow *BOD* is about 340 mg l⁻¹.

19.2.1.2
Outflow

Water demand for agricultural irrigation in the area of Santiago is divided into an irrigation season from August to April with maximum water demand in January, and a nonirrigation season from May to July. Thus winter storage of effluent is required to optimize reclaimed water supply in summer.

19.2.1.3
Required Effluent Quality

Quality requirements for agricultural irrigation were defined as:

- faecal coliforms < 1 000/100 ml
- *BOD* < 20 mg l^{-1}

19.2.1.4
Conceptual Solution and Process Design

The treatment plant is divided into two modules:

- Module 1 is made of four anaerobic ponds in parallel, four low energy aerated lagoons in parallel, one facultative pond, and one wastewater reservoir with a storage capacity of 1 million m³ operated in continuous flow (see Chapter 5 for description of operating regimes).
- Module 2 was similar, but the aerated lagoons were located before the anaerobic ponds for experimental purposes.

The plant was provided with many options which are not found in 'normal' treatment plants but that were necessary for experimental purposes: eight small anaerobic ponds and aerated lagoons (instead of 4–5), anaerobic ponds of changing volume, recirculation facilities within the anaerobic ponds and between the anaerobic ponds and the aerated lagoons, *p*H control units, flowmeters in each of the units, flow bypass for each unit, the possibility to close batch any unit or increase/decrease its hydraulic loading, etc.

The plant was commissioned in 1993.

19.2.2
Morocco – The Ben Slimane System

This sewage treatment and reuse plant was designed to provide reclaimed water for the irrigation of a golf course.

19.2.2.1
Inflow

The design sewage inflow was from 3 000 m³ d^{-1} at the commission of the project up to 5 500 m³ d^{-1} in 10 years. Most sewage is domestic, with an average *BOD* of about 325 mg l^{-1}, and faecal coliforms of 10^7–10^8/100 ml.

19.2.2.2
Outflow

Water demand for the irrigation of the golf course follows an annual curve reaching zero during the rainy winter and a maximum in July (driest summer month). Thus, winter storage of effluent is necessary to supply the water demand during the summer intensive irrigation.

The accumulated flow of effluents during the whole year exceeds the annual water demand of the golf course. Thus, some of the effluent must be released to a nearby stream during the winter.

19.2.2.3
Required Effluent Quality

Quality requirements for golf course irrigation were defined as:

- faecal coliforms < 100/100 ml
- $BOD < 20$ mg l^{-1}

Quality requirements for discharge in the stream (which today receives the full raw sewage of the town) were defined as:

- faecal coliforms < 10^6/100 ml
- $BOD < 40$ mg l^{-1}

19.2.2.4
Conceptual Solution and Process Design

The conceptual solution (Fig. 19.2) separates the treatment plant in two different parts:

1. The first part is designed as a function of sewage inflow and the quality requirements for discharge into the stream. It includes the following process units:
   - Anaerobic ponds. Main task: removal of suspended organic matter.
   - Aerated lagoons (low energy). Main task: removal of dissolved organic matter.
   - Facultative ponds. Main task: further removal of organic matter and reduction of faecal coliforms by 1–2 orders of magnitude in order to reach the quality requirements for discharge of the effluent to the stream during the winter.
2. The second part is designed as a function of the quantity and quality requirements for the irrigation of the golf course. It is made of four stabilization reservoirs (the number may be increased in the future) operated as sequential batch reactors in parallel. Main task: further removal of organic matter, reduction of faecal coliforms by 5 orders of magnitude (before chlorination), removal of toxic/refractory compounds, and storage.

The system also includes pre-treatment of raw sewage by screen bars and grit chambers, and post-treatment disinfection by chlorination (optional).

The system was commissioned in 1997.

19.2.3
India – Wastewater Reservoirs in Gujarat State

The following text is a summary of a commercial proposal prepared by an Israeli consulting firm for the design and construction of wastewater reservoirs in the Gujarat

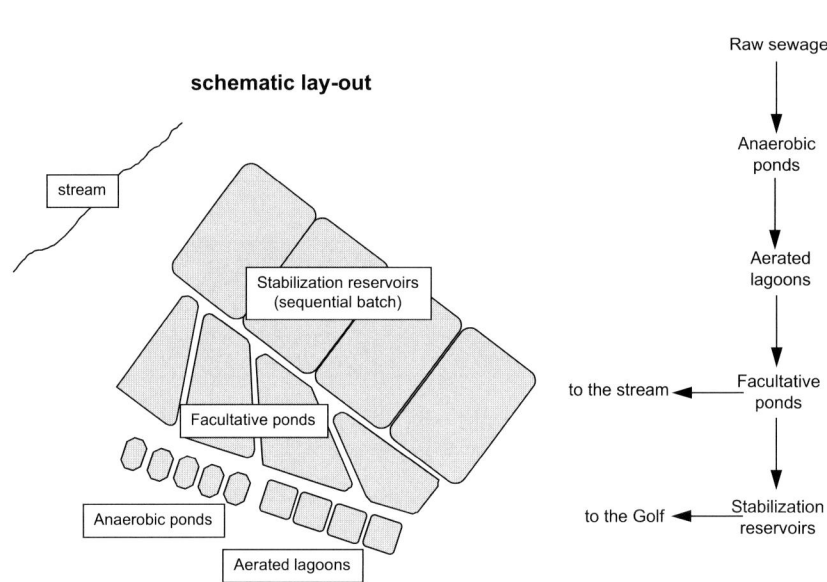

Fig. 19.2. Flow chart and schematic lay-out of the treatment plant of Ben Slimane, Morocco

State, India. It is herein presented to illustrate the rational behind this kind of sewage treatment and reuse project which includes – as many wastewater reuse projects – the use of wastewater reservoirs.

19.2.3.1
Wastewater Reservoirs for Combined Treatment and Reuse in Gujarat State, India (by Y. Shevah)

Introduction
The options for reuse of wastewater in Gujarat State were studied by TAHAL Consulting Engineers Ltd. in 1996, as part of the Water Resources Planning for this State. Among the various ways of increasing water potential, reuse of municipal wastewater was considered to be the most convenient water source. The estimated potential of 1.9 million $m^3 d^{-1}$ of sewage would allow irrigation of almost 1 million ha, releasing scarce water resources for other potable uses and bringing in valuable returns and higher crop production. Domestic wastewater collection and treatment is already practiced in 20 towns of Gujarat State (40% of the town and 85% of the urban population), and further expansion of wastewater treatment and disposal systems is planned as water supply is extended to other communities in the state. Against this background, reuse of effluents after adequate treatment is recommended in a regular and controlled manner, and within the general framework of public health protection. The conceptual design of a staged development plan for treatment and reuse of domestic effluents, as an ultimate disposal system and a viable water resource are herein described.

Methodology

The conceptualized plan was evolved through a systematic methodology and planning procedure, including collection and analysis of existing data, and field trips to water works, wastewater treatment plants and irrigation schemes. Supplementary surveys related to physical and socio-economic issues were also conducted.

Project Background Data

Location and Population. Gujarat State is situated on the west coast of India. The boundaries of the state extend to the Arabian Sea in the west, Pakistan and Rajasthan in the north, Madhya Pradesh and Maharashtra in the south. The geographical area of the state is about 196 000 km² (6.1% of the total area of India) and the total population is 42 million (1991 census).

Domestic Water Supply and Wastewater Generation. Gujarat State experiences water shortages in most of its districts, with the exceptions of the southeastern part of the state. The level of water supply in the major cities is about 120–150 l per capita per day, provided mainly from rivers and supplemented with groundwater sources. Wastewater is discharged through stormwater outlets into rivers and the sea. The estimated potential of sewage generation by the year 2025 is given in Table 19.6.

The domestic and industrial water consumption in Gujarat State, up to the year 2025, will amount to about 6.3 MCM d^{-1} or about 2 300 MCM yr^{-1}. Based on the above figures, the potential generation of sewage effluents to the year 2025 will amount to about 3.3 MCM d^{-1} ≈ 1 200 MCM yr^{-1}. Sea discharge of raw sewage and effluents is becoming less and less acceptable and such practice will be allowed only after tertiary treatment in the future, making sea disposal less favourable and more expensive than biological treatment followed by land disposal. In semi-arid areas, such as Gujarat, land disposal carries additional benefits by the utilization of the effluents as a source of water for irrigation.

Conceptual Design

Design Quantities. The immediate reuse potential was evaluated based on the current situation in Ahmedabad and in Mehsana District, representing large, medium, and small towns. The reuse potential in the two districts was estimated at about 600 MCM yr^{-1} by the year 2025.

Sector	Domestic water supply		Sewage potential 2025	
	MCM d^{-1}	MCM yr^{-1}	MCM d^{-1}	MCM yr^{-1}
Urban	3.0	1 080	2.4	860
Rural	2.5	900	0.5	180
Industrial	0.8	300	0.4	150
Grand total	6.3	2 300	3.3	1 200

Table 19.6. Domestic water supply and sewage generation potential

System Design. Currently, most of the irrigation schemes are based on surface water and storm flow water ponds, filled during the monsoon period for use over the dry periods. These storage systems can be incorporated into a sewage treatment and reuse system by diverting treated effluents from a treatment plant or a water course, into an existing system or to a newly developed irrigation scheme. For small and medium towns (less than 100 000 people) a scheme comprising sedimentation ponds, followed by a series of oxidation and storage ponds discharging into existing or newly built storage ponds is envisaged.

The Design of the Treatment and Reuse Module. The proposed scheme would be based on a typical module of 1 MCM capacity, to serve an equivalent population of 25 000. The module can be duplicated according to the size of the population of specific towns. The module consists of the following components:

- Primary sedimentation ponds
- Large detention reservoir
- Conveyance/pumping system of effluents to irrigation plots
- On-farm irrigation system

Based on the irrigation and cropping practices of Gujarat, a net area of about 110 ha can be irrigated, assuming a gross water demand of 10 000 $m^3 ha^{-1} yr^{-1}$, for the irrigation of suitable cash and industrially processed cash crops. Discharge of excess effluents is avoided by providing adequate storage volume for the detention of effluents not used during the wet season. Under the cropping pattern of Gujarat State, a storage volume equal to 30% of the total yearly supply is required. The added seasonal storage provides for added purification and quality improvement of the effluent.

The proposed system is applicable to areas where the effluents can be diverted to specific locations, having suitable cropping system and the areas to be irrigated are well-defined, away from shallow aquifers used for drinking water supply.

Cost Estimates and Cost Allocation
The costs of construction were estimated for a module of 1 MCM capacity (Table 19.7).

Cost of implementation will amount to about 24 million Rs for a module of 1 MCM. The cost will be shared between the three beneficiaries namely:

- The local authority
- The state
- The farmer

The first beneficiary (local authority) is provided with an acceptable form of treatment and disposal, avoiding the uncontrolled discharge of effluents into estuaries and water bodies. The second beneficiary (the Gujarat State) is generating a water source for irrigation thus alleviating pressure on existing water resources. The third beneficiary (the farmer) benefits from a regular supply of water for irrigation. For a town with an equivalent population of 100 000 people, with no treatment facilities, the capital requirements will amount to about 100 million Rs (US$3 million) shared between the local authority (40%) the state (40%), and the farmers (20%), as shown in Table 19.8.

Table 19.7. Construction costs of a combined treatment and reuse system; capacity 1 MCM

System component	Cost (million Rs)
Treatment/detention – storage lagoons	13.8
Conveyance and irrigation system (90 000 Rs ha^{-1} × 110 ha)	10.0
Total	23.8

Table 19.8. Capital and operation and maintenance (O&M) cost allocation for a 1-MCM yr^{-1} plant capacity

System component	Cost allocation (× 1 000 Rs)		Farmer	Total
	Local authority	State		
Capital Cost				
Treatment/storage system	8 300	5 500	–	13 800
On-farm irrigation	–	5 000	5 000	10 000
Total	8 300	10 500	5 000	23 800
Annual O&M				
Capital recovery (5%, 40 years)	800	–	580	1 380
O&M (1% of capital)	–	–	238	238
Total	800	–	818	1 618

According to the above formula, the investment cost of a treatment and reuse system will be shared by the local authority and the state, while the capital cost for the on-farm system will be borne by the farmer who will also share with the local authority the annual operation and maintenance costs of the combined treatment and reuse system (Table 19.8). The state will assist in financing the construction of the treatment and reuse facilities, while the annual operation including capital costs will be negotiated between the local authorities and the farmers.

Discussion

The proposed concept provides a solution for treatment and reuse of domestic effluents as a safe land disposal and additional potential water source, demonstrating a system where reuse of effluents can be a safe disposal system for the municipal authorities, while providing a significant increase in crop production and income to the farmers. The system is suitable for countries of the semi-arid regions, where land disposal of wastewater via irrigation should be preferred to sea or river discharge, using a low cost technology that can be easily implemented under well defined conditions. The proposed system demonstrates an effective way of managing wastewater treatment and reuse through participation of the various sectors including the farming community, which is the ultimate user of the treated effluents. A major consideration in adopting the system is the large tracts of land needed for the construction of large long detention reservoirs. Consequently, the application of this treatment scheme is limited to small and medium-sized cities (less than 500 000 inhabitants) and where land is not a limiting factor.

A responsible authority is required to coordinate and supervise the implementation plans and to monitor the various conditions set forth by the environmental protection regulations, including the preparation of relevant studies and overseeing the implementation in the field and the resulting impacts on health, ground-water, drainage, water supply and agricultural management.

References

Adams V, Cleave M, Cofer J, Middlebrooks J (1983) Indian Creek Reservoir sediments. Wat Res 17(11): 1709–1712

Baozhen W (1987) Development of ecological wastewater treatment and utilization systems (EWTUS) in China. Wat SciTech 19(1–2):51–63

Berna L, Torrella F, Soler A, Saez J, Llorens M, Martinez I (1986) Estudio de la autodepuracion microbiologica y fisico-quimica de las aguas residuales por lagunaje profundo. Anales de Biologia 10 (Biologia General 2):49–59

Environment Canada (1985) Sewage lagoons in cold climates. Technical Services Branch, Environmental Protection Service, Environment Canada, Report EPS 4/NR/1, 89 pp

EPA (1975) Wastewater treatment ponds. EPA 430/9–74/001

Felgner G, Sandring G (1983) Wastewater storage – a way to ensure wastewater treatment and utilization over the whole year. Wasserwirtsch Wassertech, WWT, 33(9):321-323 (in German)

Fuog R, Giberson K, Lawrence R (1995) Wastewater reclamation at Rancho Murieta, California; Golf course irrigation with upgraded pond effluent meeting California's strictest requirements for wastewater reuse. Wat Sci Tech 31(12):399–408

Funderburg S, Moore B, Sorber C, Sagic B (1978) Survival of poliovirus in model wastewater holding ponds. Progr Wat Tech 10(5–6):619–629, Discussion pp 1071–1072

Frykberg W, Goodnight C, Meier P (1977) Muskegon, Michigan industrial-municipal wastewater storage lagoons, biota and environment. EPA-600/3-77-039, 78 pp

Frykberg W, Meier P, Goodnight C (1978) Storage lagoons come to life. Water Wastes Eng 15(9):138–140

Hatcher K (1980) Finding design criteria and operating schedules for a wastewater storage pond. Procc Symp Surface Water Impound, Univ Minnesota, vol II, pp 1209–1226

Indelicato S, Barbagallo S, Cirelli G (1995) Change in wastewater quality during seasonal storage. Procc Conf Natural and Constr Wetlands WW Treatment and Reuse, Perugia, pp 195–204

Indelicato S, Barbagallo S, Cirelli G, Zimbone S (1996) Reuse of municipal wastewater for irrigation in Italy. Procc 7th Internat Conf Water and Irrigat, Tel Aviv, pp 210–221

Kott Y (1975) Wastewater conditioning for irrestricted irrigation. Technion, Lab of Environ Eng, Report 013-630, 50 pp (in Hebrew)

Kott Y, Ben-Ari H, Betzer N (1978) Lagooned, secondary effluents as water source for extended agricultural purposes. Wat Res 12:1101–1106

Liran A (1990) The removal of indicator microorganisms in stabilization reservoirs as a function of the operational regime. MSc Thesis, Technion – Israel Institute of Technology, 174 pp (in Hebrew)

Libhaber M (1995) The treatment design of the Santiago Poniente treatment plant as a pilot plant. Proceedings of the Seminar on Sewage Treatment and Reuse, World Bank, Santiago de Chile, 120 pp (in Spanish)

Liran A, Juanicó M, Shelef G (1994) Bacteria removal in a stabilization reservoir for wastewater irrigation in Israel. Wat Res 28(6):1305–1314

Llorens M, Saez J, Soler A (1992) Influence of thermal stratification on the behaviour of a deep wastewater stabilization pond. Wat Res 26(5):569–577

Llorens M, Saez J, Soler A (1993) Primary productivity in a deep sewage stabilization pond. Wat Res 27(12):1779–1785

Middlebrooks E (1983) Municipal wastewater stabilization ponds – Design manual. EPA report EPA/625/1-83/015, 326 pp

Middlebrooks E (1990) Waste stabilization ponds. In: Reed S (chairman) Natural systems for wastewater treatment. WPCF Manual of Practice FD-16, pp 159–185

Middlebrooks E (1995) Upgrading pond effluents: An overview. Wat Sci Tech 31(12):353–368

Moreno D, Medina M, Moreno J, Soler A, Saez J (1988) Modelling the performance of deep waste stabilization ponds. Wat Resour Bull 24(2):377–87

Mujeriego R, Sala L (1991) Golf course irrigation with reclaimed wastewater. Wat Sci Tech 24(9):161–172

Pierce DM (1974) Performance of raw waste stabilization lagoons in Michigan with long period storage before discharge. In: Middlebrooks E, Falkenborg D, Lewis R, Ehreth D (eds) Upgrading wastewater stabilization ponds to meet new discharge standards (Symposium Proceedings, Aug. 21–28), National Environmental Research Center, EPA, Cincinnati, Ohio, pp 89–136

Porcella DB, McGauhey PH, Dugan, Gordon L (1972) Response to tertiary effluent in Indian Creek Reservoir. J Wat Pollut Contr Fed 44(11):2148–61

Sala L, Millet X (1997) Aspectos basicos de la reutilizacion de las aguas residuales regeneradas para el riego de campos de golf. Consorci Costa Brava Publishers, 126 pp

Sala L, Garcia J, Mujeriego R, Hernandez M (1994) Phytoplankton studies in hypertrophic lakes used for irrigation. Verh Internat Verein Limnol 25:1983–1988

Sano I (1983a) Evaporation of organic compounds from wastewater holding ponds. Akushu no Kenkyu 12(56):6–8 (in Japanese)

Sano I (1983b) Offensive odor control of an industrial wastewater storing lake. Akushu no Kenkyu, 12(57):1–8 (in Japanese)

Soler A, Saez J, Llorens M, Martinez I, Torrella F, Berna L (1991) Changes in physico-chemical parameters and photosynthetic microorganisms in a deep wastewater self-depuration lagoon. Wat Res 25(6):689–695

Surampalli R, Banerji S, Pycha C, Lopez E (1995) Phosphorus removal in ponds. Wat Sci Tech 31(12):331–339

Tapp J (1976) Wastewater storage – simulation of instream effects. ASCE Environ Eng Journal 102(EE6):1151–1159

Torres J, Soler A, Saez J, Ortuno J (1997) Hydraulic performance of a deep wastewater stabilization pond. Wat Res 31(4):679–688

WR-NEWs (1992) Bulletin of the Office of Water Reclamation, City of Los Angeles, vol. 3(3)

WR-NEWs (1994) Bulletin of the Office of Water Reclamation, City of Los Angeles, vol. 5(2)

Part II

Case Studies

Chapter 20

The Na'an Reservoir

Inka Dor · Hanna Schechter · Hether Bromley

Reproduced with permission from Hydrobiologia 150:225–241 (1987)

20.1
Introduction

Israel is a semi-arid country with very limited water resources. However, the needs of the growing population, increasing urbanization, industry, and agriculture, have required the development of new, unconventional water resources, which include the construction of large seasonal reservoirs collecting effluents and runoff water.

Among the various kinds of reservoirs, most problematic are those collecting undiluted wastewater. In these reservoirs the wastewater undergoes primary and secondary treatment before being stored, and after a prolonged retention period of about 9 months, is used for irrigation of industrial crops such as cotton or corn for animal feed. Reservoirs of the above type are multipurpose systems serving as tertiary treatment plants for improvement of water quality, and as a source of water rich in nitrate and phosphate fertilizers for agriculture. Such systems require very little operational control and maintenance, and are very popular in Israel.

A peculiarity of these reservoirs is their considerable depth (about 10 m) combined with high levels of organic and mineral nutrients. This richness places them in a category of highly eutrophic or hypertrophic ecosystems. Updated knowledge about extremely enriched lakes and reservoirs is collected by Barica and Mur (1980). Hypertrophy is characterized by a high concentration of biomass, wide fluctuations in physicochemical and biological parameters, and anaerobic periods, causing environmental nuisances. The obvious paradox in the operation of wastewater reservoirs is that they must supply water of defined standards (State of Israel 1976), and thus remain constantly oxygenated and free of nuisances, while being inherently unstable and supposedly balancing closely at the limit of deterioration.

The present interdisciplinary investigation provides detailed data on the seasonal changes in physico-chemical, biological and sanitary variables in Na'an Reservoir, one of the most carefully designed and operated reservoirs in Israel. We hope that our study will contribute to the further optimization and development of this important technology in Israel and elsewhere while throwing more light on the peculiarities of the hypertrophic water bodies in our region.

20.2
Materials and Methods

20.2.1
Structure and Operation of the Reservoir

The reservoir was constructed in 1977 in central Israel (Fig. 20.1). The reservoir collects wastewater effluent from the neighboring city of Ramla, which constitutes the major source of reservoir water. Minor amounts (up to 8% monthly) come from Kibbutz Na'an and sporadically also from excess clean water supplied by Mekorot Water

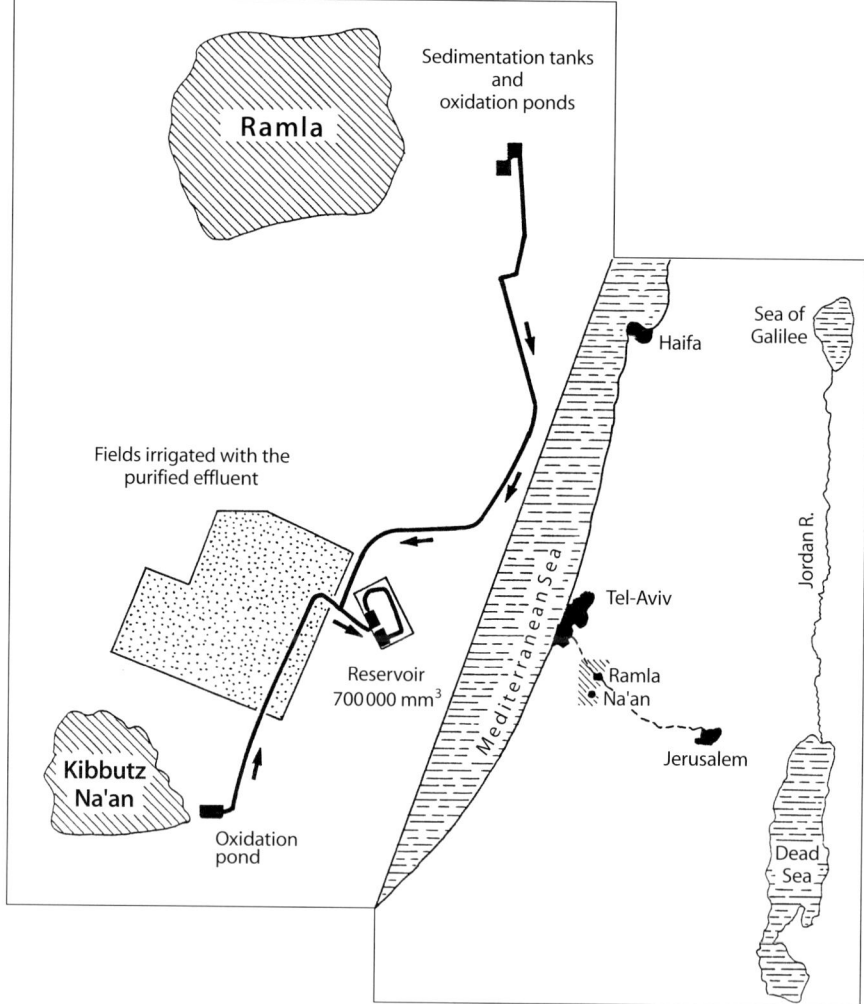

Fig. 20.1. Location of Na'an Reservoir

Fig. 20.2. Structure of Na'an Reservoir with the oxidation ponds and sampling point

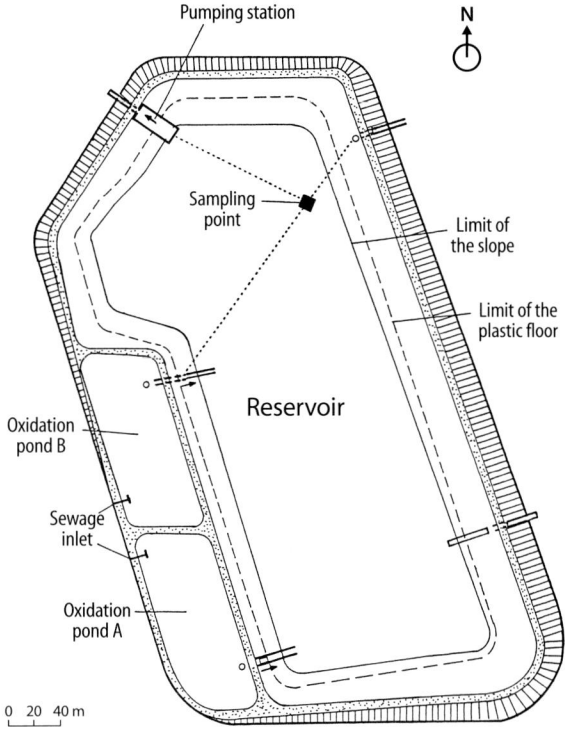

Company. After the primary and secondary treatment, effluents enter two shallow, 5 000 m³ oxidation ponds adjacent to the reservoir (Fig. 20.2). Both oxidation ponds collect wastewater effluent from Ramla, but while pond A is fed exclusively from this source, pond B also receives the effluent from the kibbutz and the clean water, thus undergoing wide fluctuations in water quality. The filling cycle begins in September and lasts until June. During this period the wastewater enters the reservoir continuously via the oxidation ponds. In June the filling is discontinued and the water is pumped for irrigation until the reservoir is empty at the end of August.

20.2.2
Field Sampling and Measurements

Field trips were usually performed once a month during 1979/1980. Sampling and field work were carried out from a wooden raft anchored in the middle of the reservoir (Fig. 20.2), far from the inlet and from the pumping station.

Temperature and dissolved oxygen were measured in the water column using a YSI Oxygen Meter 58 with a self-stirring electrode; pH was determined with the Radiometer 29 pH Meter; light penetration was recorded using either Bruno Lange GmbH Luxmeter or Lambda Li-Cor 185 Quantum Meter.

Readings of the water column were taken every 10 cm in the surface meter layer and every 50 cm in the deeper water. Profiles were usually recorded 2–4 times in the daylight hours. Changes in the night content of dissolved oxygen in the upper water layer were recorded four times: in April, June, July, and November. Night profiles of temperature, dissolved oxygen, and pH were taken in July during hot and calm weather.

Primary productivity was measured using the oxygen method (Vollenweider 1969). Duplicate pairs of 260-ml bottles were filled with the water sampled from 0, 0.1, 0.2, 0.5 and 1.0 m and suspended from a floating bar at the corresponding depths. In a few cases the measurements were also done in deeper water. The initial and final oxygen content in the bottles was measured using YSI Model 54 Oxygen Meter with an electrode fitting the bottle opening. During the measurements the water was stirred with a magnetic bar on a battery-operated stirrer. Exposure duration was usually about 2–3 h in winter and only 1 h in spring and summer. In summer 3–4 runs of experiments were set at different times while in winter a single experiment was performed around midday.

Water for laboratory analyses was sampled either with a transparent Nansen bottle or by pumping. Samples of about 10 l were collected every 50 cm in the upper 1.5 m and every 100 cm in the deeper water. The samples were transported as soon as possible to the laboratory, kept overnight at 4° C, and processed the next day.

20.2.3
Laboratory Analyses

Phytoplankton was concentrated by centrifugation from the sub-samples of 10–50 ml, identified and counted in a haemocytometer. Chlorophyll a was measured after centrifugation of 100–400 ml of water, followed by extraction with boiling 90% methanol according to Talling and Driver (1963). Zooplankton, concentrated from 2 l of the water, was identified and counted under a stereoscopic microscope. Heterotrophic bacteria were counted on nutrient agar (Standard Plate Count); faecal coliforms and Salmonella were determined according to APHA et al. (1975). The following analyses were performed according to APHA et al. (1975): *COD, BOD, TSS*, ammonia, total nitrogen, organic nitrogen, total phosphate, and ortho-phosphate.

20.3
Results

Day profiles of temperature throughout the year are given in Fig. 20.3, showing the summer stratification and winter homothermy. Seasonal changes of midday temperatures at the surface and bottom are summarized in Fig. 20.4. Night profiles of the temperatures recorded in July showed that during very calm weather thermal stratification remained stable despite some cooling of the surface water (Fig. 20.10). Day profiles of dissolved oxygen are illustrated in Fig. 20.5 and diurnal curves of dissolved oxygen at various depths of the reservoir in July 1980 are in Fig. 20.6. Seasonal fluctuations of dissolved oxygen at a depth of 0.2 m are given in Fig. 20.7. Night levels of dissolved oxygen in the upper water in different seasons are presented in Fig. 20.8 and night profiles of dissolved oxygen during the summer stratification in Fig. 20.10. The pH in the reservoir ranged between 6.6 and 7.5 in the deep water and increased up to 9.0 in the euphotic zone during the summer peak of photosynthesis (Fig. 20.9). Figure 20.11 summarizes the seasonal fluctuation of the compensation level. The gross primary productivity in the surface water measured between 09.00 and 14.00 varied within

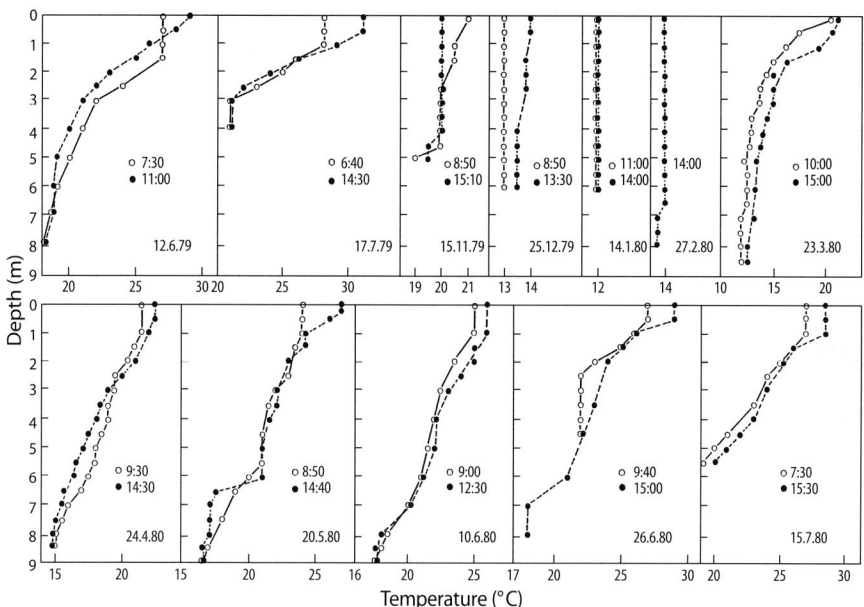

Fig. 20.3. Day profiles of temperature

Fig. 20.4. Summary of the midday temperature at the reservoir surface and bottom

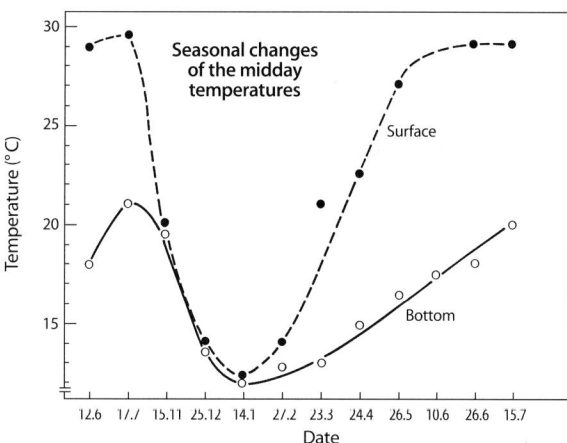

the range of 1.3–7.7 mg O_2 $l^{-1}h^{-1}$ (Fig. 20.12). Maximal values exceeding 7 mg l $^{-1}$ were
recorded in March and again in June 1980, while the minimal value of 1.3 mg O_2 $l^{-1}h^{-1}$ was
recorded in July 1980. Table 20.1 provides a summary of the mean rates of primary
productivity and respiration for the upper 0.5 m water layer.

Total cell counts of algae in the reservoir were in the range of 10^{-5}–10^6 cells ml^{-1} and
only once in July 1980 dropped to 10^4 cells ml^{-1} (Table 20.2). The algal community was
composed of several species, but *Chlorella vulgaris* predominated in all seasons and

Table 20.1. Mean rates of primary productivity and respiration for the upper 0.5 m, measured between 09:00–13:00 h (mg $O_2 l^{-1} h^{-1}$)

Date	GPP	NPP	R
12.06	1.7	1.2	0.5
17.07	1.2	0.5	0.7
15.11	1.5	−0.3	1.8
25.12	1.4	0.5	0.9
14.01	1.1	0.6	0.4
27.02	1.8	−0.2	2.0
23.03	3.6	2.6	0.9
24.04	1.0	0.7	0.3
20.05	2.2	1.4	0.8
10.06	3.2	2.7	0.4
26.06	4.9	3.4	1.4
17.07	0.9	0.3	0.6

GPP: Gross Primary Productivity;
NPP: Net Primary Productivity;
R: Respiration.

Table 20.2. Concentration of phytoplankton, percentage of *Chlorella vulgaris* and chlorophyll a in the upper 0.5 m of the reservoir

Date	Phytoplankton (cells ml^{-1})	*Chlorella* (%)	Chlorophyll a (µg l^{-1})
12.06	2.2×10^5	92	181
17.07	4×10^5	89	295
15.11	1.1×10^5	91	246
25.12	1.5×10^5	88	347
14.01	3×10^5	95	180
27.02	1×10^6	97	403
23.03	3.1×10^6	98	843
24.04	4×10^4	82	34
20.05	1.1×10^5	91	243
10.06	1×10^6	99	347
26.06	7.2×10^5	97	225
15.07	2.6×10^4	76	58

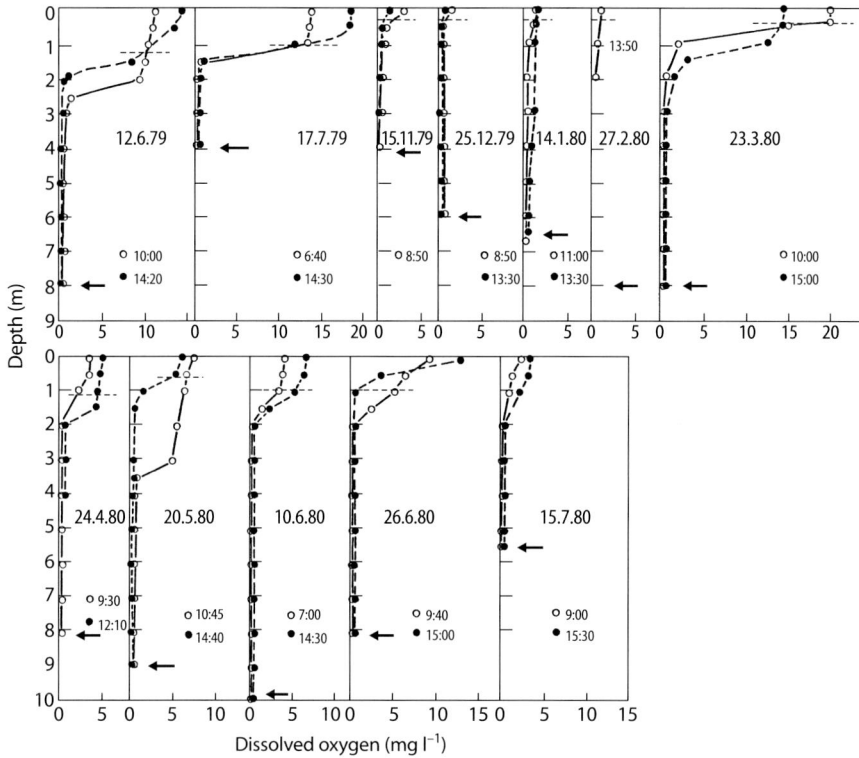

Fig. 20.5. Day profiles of dissolved oxygen. *Pointers* mark actual depth of the reservoir and *horizontal broken lines* indicate limits of the photic zone

constituted 76–99% of the total cell count. Next in importance were the following algae which sporadically increased in numbers: *Scenedesmus dimorphus* (8% of total cell count in June), *Actinastrum hantzschii* (4% in July), *Tetraedron muticum* (6% in July) and *Selenastrum minutum* (2% in March). Other species appeared in negligible numbers. Blue-green algae appearing at the end of the summer did not exceed 10^3 colonies ml^{-1} in the case of *Microcystis*, or 10^3 filaments in the case of *Oscillatoria* or *Spirulina*.

The chlorophyll a content fluctuated most of the time around a few hundred μg l^{-1} (Table 20.2). Vertical distribution of the algae and chlorophyll a are given in Fig. 20.13. The prominent drop in numbers of algal cells in April in the upper 2–3 m evidently resulted from the excessive grazing by Rotifers, and further fluctuations in the concentrations expressed the dynamic balance between the phyto- and zooplankters. In the oxidation pond the photosynthetic community was predominately composed of the flagellate algae *Euglena* sp. and *Chlamydomonas* sp.

Zooplankton was first evident in March (1980), represented by less than 11 specimens per l of the rotifer *Epiphanes senta*. The rotifer fauna increased rapidly in surface waters with *Brachionus angularis* dominant in April (ca. 9 000 spec. l^{-1}) and *B. calyciflorus* dominant in May (ca. 10 000 spec. l^{-1}). In June, numbers decreased with

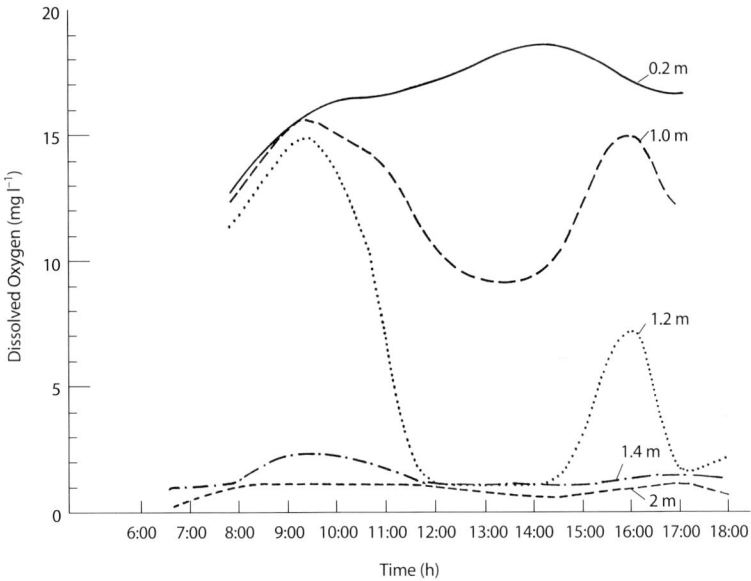

Fig. 20.6. Diurnal curves of dissolved oxygen, July 1979, Na'an

B. angularis (2 600 spec. l⁻¹) and *B. budapestinensis* (1 700 spec. l⁻¹) dominant, while in July, *Hexarthra* sp. was dominant (55 spec. l⁻¹) (Fig. 20.13). The crustacean fauna was first represented in April by a few copepodites of *Microcyclops minutus*, which increased in numbers, peaking in June (342 spec. l⁻¹). Cladocera appeared and peaked in May, represented by *Moina micrura* (488 spec. l⁻¹), which was replaced in July by the large cladoceran *Daphnia magna* (25 spec. l⁻¹) (Fig. 20.14).

Vertical distribution of Rotifera is summarized in Fig. 20.13 and that of the dominant crustacea in Fig. 20.14. Rotifers and *Moina* peaked at the 0.5–1.5 m level, where dissolved oxygen did not drop below 1.5–2 mg l⁻¹; *Microcyclops* was limited to the more oxygenated water, having at least 3–4 mg l⁻¹. The limit of survival of the zooplankton was at a depth corresponding approximately to 0.2–0.3 mg O_2 l⁻¹.

In addition to the animals mentioned above, Protozoa, Hydracarina, Chironomidae, Dityscidae and Corixidae were sporadically observed in the reservoir. However these taxa were not investigated.

Composition of the raw sewage feeding the system and the progressive changes in the effluent quality indicated in the oxidation ponds and in the reservoir are summarized in Table 20.3. For most of the parameters investigated, the mean water quality characterizing raw sewage, oxidation pond, and reservoir remained quite stable throughout the year. The extent of changes in *BOD*, ammonia, organic nitrogen and total phosphate occurring in the reservoir (as compared to the oxidation pond) depended on season (Table 20.3, last column). The relative and total removal of *BOD* within the system is illustrated in Fig. 20.15. In the cold season 74–82% of the total *BOD* removed by the whole system was already decomposed during the stages of pre-treatment (sedimentation and oxidation ponds), while 18–26% of the removal occurred during the

Table 20.3. Changes of chemical composition (mg l^{-1}) during the warm and cold season in the Na'an system (mean values)

Source	Raw sewage		Oxidation pond		Change in O.P. (%)		Reservoir		Change in R. (%)	
Season	Warm	Cold	Warm	Cold	Warm	Cold	Warm	Cold	Warm	Cold
COD	935	762	354	296	−62	−61	193	207	−46	−31
SD	182	194	108	33			115	87		
BOD f	326	513	59	151	−82	−71	56	60	−5	−61
SD	202	138	17	14			24	20		
NH$_3$-N	24	20	14	26	−42	+28	15	19	+7	−28
SD	14	2.7	12	8			7.1	1.9		
Tot. N	51	52	32	47	−37	−9	29	35	−10	−26
SD	16	21	14	8			6.5	3.8		
Org. N	26	36	19	21	−28	−43	13	16	−28	−23
SD	5	17	3.2	12			5.8	4		
Tot. P	8.0	12.0	5.6	8.2	−30	−32	5.6	6.7	0	−19
SD	3.3	2.6	5.8	1.8			2.9	2.0		
Ortho. P	5.0	3.5	2.5	3.0	−50	−15	4.2	2.8	+68	−7
SD	1.2	0.6	1.0	0.4			1.2	0.9		
TSS	329	253	163	123	−51	−52	98	78	−51	−32
SD	108	98	156	29			66	48		

SD: standard deviation; *O.P.:* oxidaton pond; *R.:* reservoir; *Warm season:* March–July;
Cold season: November–February.

retention of the water in the reservoir. In the warm season the percentage of removal in the reservoir was either in the same range (April–May) or much lower (March, July).

Standard plate counts (SPC) of bacteria are summarized in Table 20.4, and concentration of coliform bacteria and *Salmonella* are given in Tables 20.5 and 20.6 respectively.

20.4
Discussion

In every aquatic ecosystem the physico-chemical factors and the biomass reciprocally affect each other. However in deep, oligotrophic water bodies, these interrelations are masked by scarcity of the biomass and complexity of the biotic system. By comparison, in wastewater reservoirs, where biomass is highly concentrated at each trophic level and the community of the organisms is rather simple, the interactions between biomass and environment are powerful and clear. In the following discussion of the limnology and sanitary performance of the wastewater reservoir, an attempt is undertaken to analyse interactions at various levels and determine their effect on water quality.

The reservoir studied is monomictic, behaving like other deep water bodies in this region. The upper and lower water mass separated by the summer stratification differed considerably: the epilimnion was warm and rich in oxygen (up to 250% saturation at midday), while the hypolimnion was cool and poorly oxygenated (1–10% saturation); the completely mixed water column in winter was cold and poor in oxygen.

The above extremities of the oxygen regime are derived from intense biological activities and are characteristic for water bodies rich in organic and inorganic nutrients. Under such conditions the concentrated biomass of algae, bacteria, and zooplankton affects oxygen content directly trough photosynthesis and respiration, and

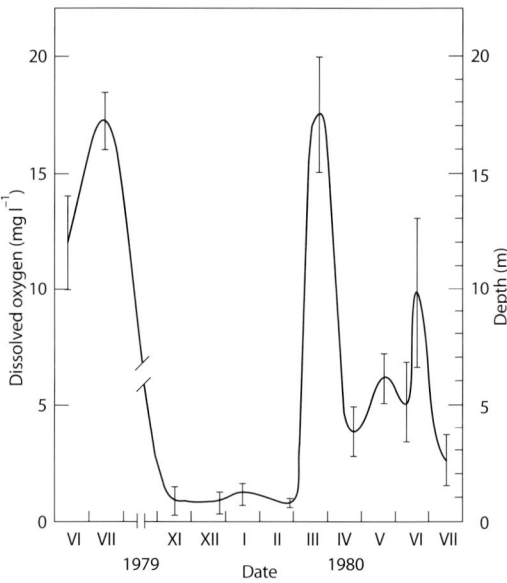

Fig. 20.7. Seasonal fluctuations of daytime dissolved oxygen in the surface water

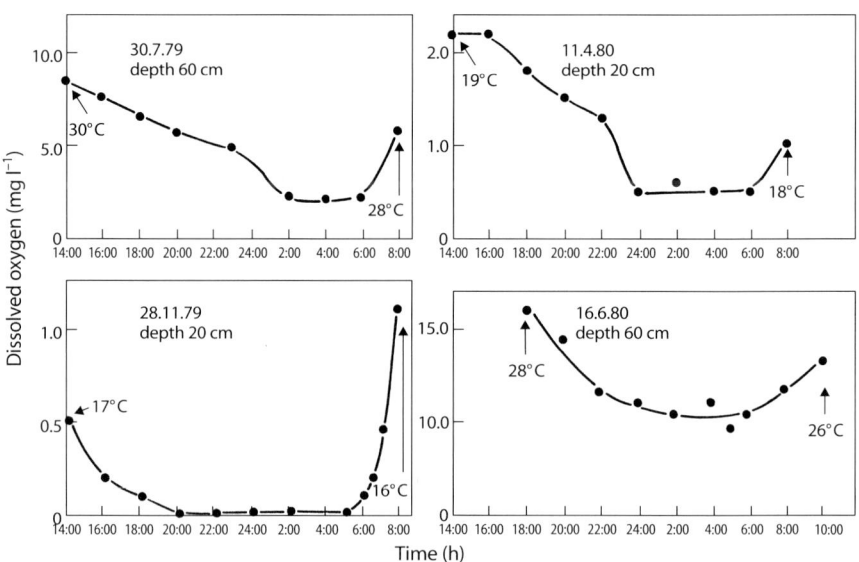

Fig. 20.8. Night record of dissolved oxygen

indirectly by controlling water turbidity and thus affecting the depth of photic zone where oxygen is produced by algae.

Diurnal oxygen levels in the upper water mass fluctuated seasonally, but after a temporary decline a rapid recovery always followed (Fig. 20.7). Anaerobic conditions at the reservoir surface were rare and short-lived, only occurring during the nights, oxygenation being rapidly restored in the morning (Fig. 20.8). The deep water mass was anaerobic for varying time intervals during November and February; however, the anaerobic gases were evidently oxidized in the upper aerobic layer, because nui-

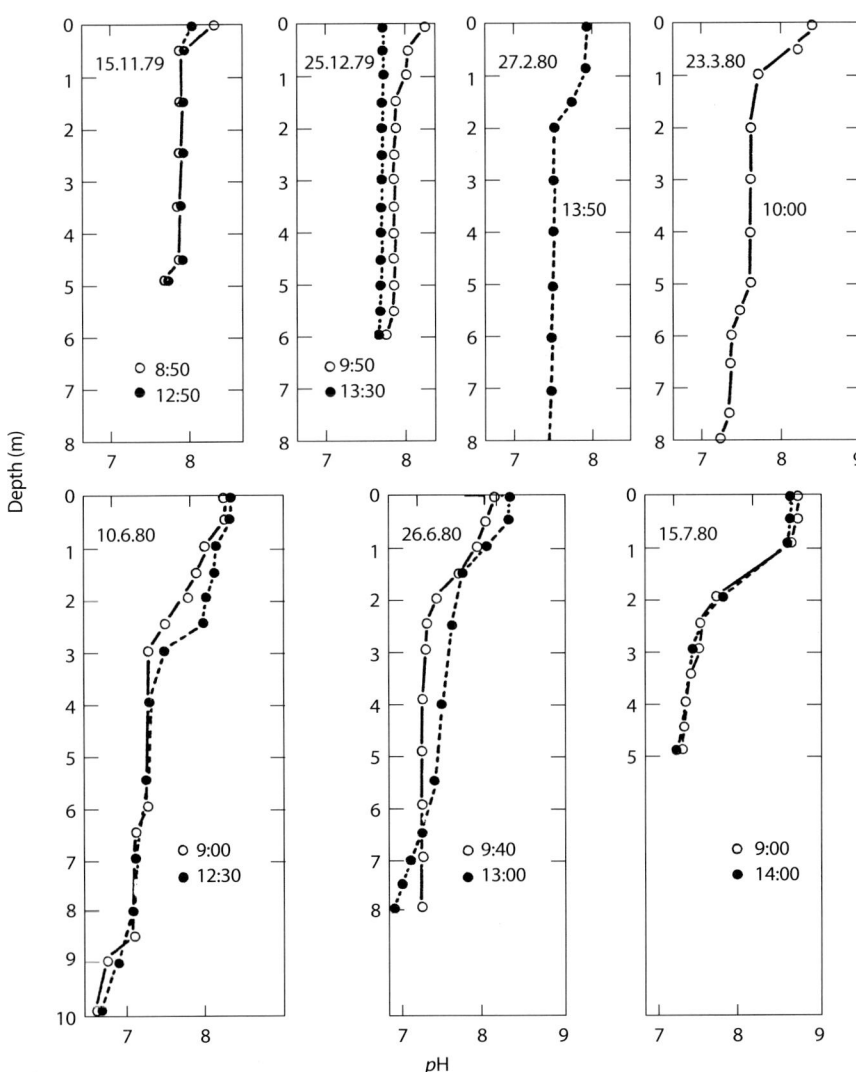

Fig. 20.9. Day profiles of pH

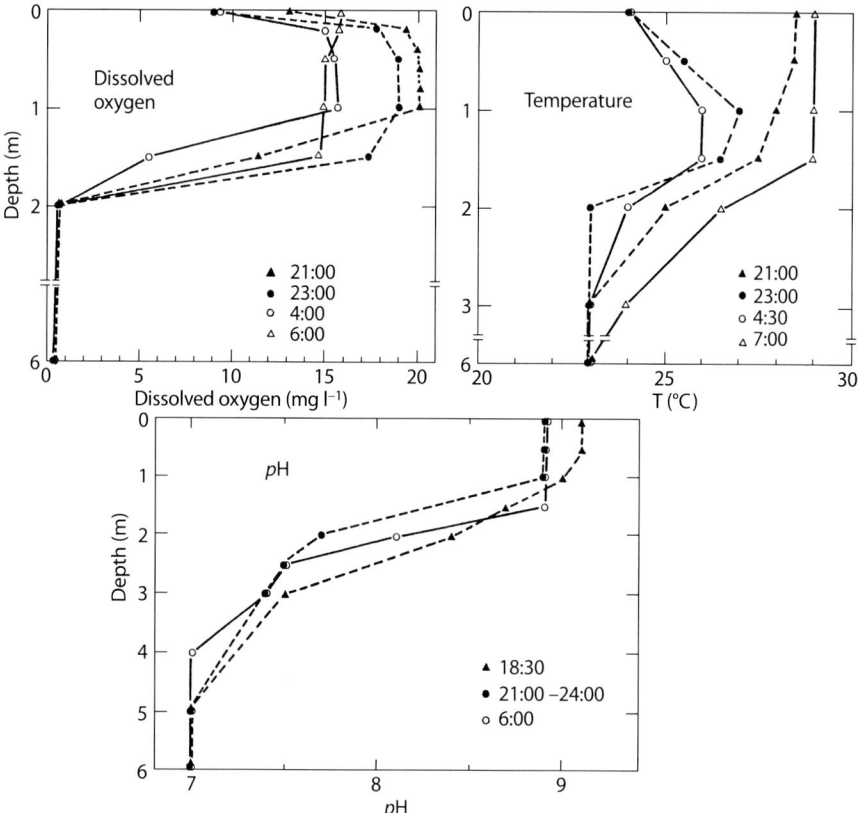

Fig. 20.10. Night profiles of temperature, dissolved oxygen and pH, 10–11 July '80

sance odours did not occur. The 0.2–0.5 mg l⁻¹ of dissolved oxygen still present in the hypolimnetic water during the summer stagnation most probably originated from the incidental overturns occurring during some windy nights, when normal cooling of the surface water augmented by the waves induced a vertical mixing.

The thermocline depth during the summer stratification correlates well with the level of compensation and the depth of oxycline (Fig. 20.16). When both the thermocline and the boundary of the photic zone are located at similar depths, the photosynthetic oxygen naturally accumulates within the same layer and the oxycline overlaps the level of compensation. In this case from each one of the following measurements: profiles of temperature, dissolved oxygen or light, and the depth of the thermocline, oxycline and thickness of the euphotic zone can be extrapolated.

Progressive changes in the level of compensation exhibited a fairly regular seasonal pattern (Fig. 20.11). During the winter mixing the water was gray, visibly turbid, and suspensions contained silt and organic debris in addition to the algae and bacteria. At this stage the compensation level fluctuated narrowly between 0.20–0.3 m. In March at the onset of stagnation, suspended solids consisted mainly of unicellular green al-

Fig. 20.11. Seasonal changes in compensation level (1% of surface light)

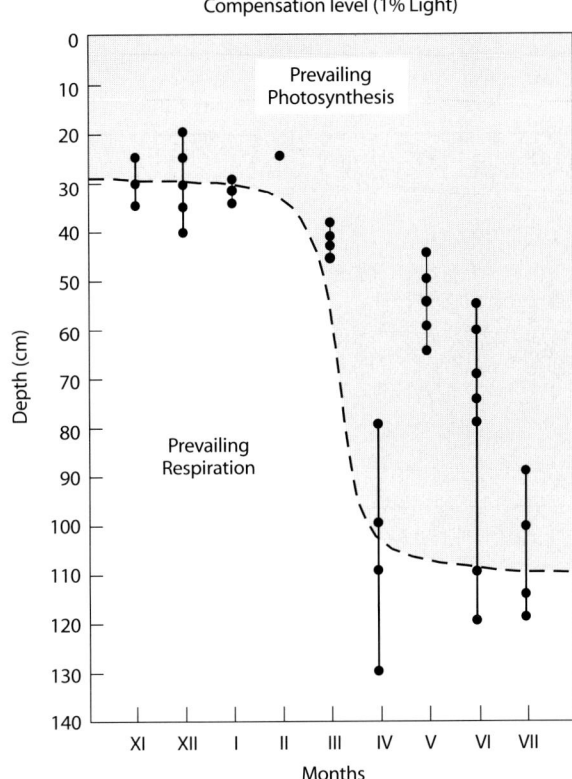

Compensation level (1% Light)

Table 20.4. Standard plate count of bacteria (cells ml⁻¹) and % of removal in Na'an system

Date	Raw sewage	Oxidation pond	Removal in O.P. (%)	Reservoir (surface)	Removal in R. (%)	Total removal (%)
17.07[a]	1.8×10^6	5.3×10^5	–	2.4×10^5	–	–
15.11	5.1×10^7	1.2×10^6	–97.6	3.9×10^6	+225	–92.3
25.12	2×10^7	5×10^6	–75.0	3×10^6	–40.4	–85.0
14.01	5.6×10^7	1.6×10^7	–71.4	3.3×10^6	–79.4	–94.1
27.02	4.5×10^8	2.7×10^7	–94.0	3.5×10^7	+29.6	–92.2
23.03	2×10^8	8.7×10^7	–56.5	2×10^6	–97.7	–90.0
24.04	5.5×10^7	2.9×10^6	–94.7	4.2×10^5	–85.5	–99.2
20.05	5×10^8	1×10^8	–80.0	8.8×10^7	–12.0	–82.4
26.06	2.3×10^8	3.4×10^6	–98.5	5×10^5	–85.3	–99.8
15.07[a]	4.9×10^7	6×10^4	–	4.4×10^5	–	–

[a] In July raw sewage supply to the system was disconnected;
O.P.: oxidation pond; R.: reservoir.

Table 20.5. Counts of coliforms (cells 100 ml^{-1}) and % of removal in Na'an system

Date	Raw sewage	Oxidation pond	Removal in O.P. (%)	Reservoir (surface)	Removal in R. (%)	Total removal (%)
17.07[a]	2.5×10^4	20	–	44	–	–
15.11	2.4×10^6	4.6×10^4	−98.1	4.3×10^3	−90.7	−99.8
25.12	4.5×10^8	4.7×10^6	−99.0	4.6×10^6	0	−99.0
14.01	2.4×10^8	2.4×10^7	−90.0	2.3×10^4	−99.9	−99.9
27.02	1.5×10^9	2.4×10^6	−98.8	4.0×10^4	−98.3	−99.9
23.03	–	9×10^3	–	5.1×10^2	–	–
24.04	2.3×10^8	1.5×10^3	−99.9	2.4×10^2	−84	−99.9
20.05	2.4×10^8	4.5×10^6	−98.1	2.4×10^4	−99.5	−99.9
26.06	9×10^6	2.3×10^3	−99.9	4.3×10^2	−81.3	−99.9
15.07[a]	1.1×10^8	1.1×10^4	–	3.6×10^4	–	–

[a] In July supply of the raw sewage to the system was disconnected;
O.P.: oxidation pond; R.: reservoir.

gae (10^6 cells ml^{-1}), which produced quite high turbidity, limiting the photic zone to some 0.40 m depth. A deepening of the euphotic zone to some 1.0–1.4 m in April resulted from the explosive development of the rotiferan filter feeders which decimated algae (from 10^6 ml^{-1} in March to 10^4 ml^{-1} in April, Fig. 20.13) and bacteria (from 10^6 ml^{-1} in March to 10^5 ml^{-1} in April, Table 20.4). In addition, the calm weather promoted sedimentation, further improving transparency. However the summer fluctuations of the compensation level were rather large (Fig. 20.11) expressing the constantly changing balance between the concentration of suspended unicells and grazers.

Primary productivity exhibited comparatively little seasonal variation: the maximal rate of oxygen production in summer (7 mg l^{-1}h^{-1}) was higher only by a factor of three compared to the rate in winter (2 mg l^{-1}h^{-1}). Evidently, the environmental variables like nutrients, light and temperature were satisfactory throughout the year for *Chlorella vulgaris*, the dominant alga. Such suppression of periodicity and permanent algal blooming are typical for hypertrophic conditions (Leentvaar 1980). Similar biological stability with respect to both the algal species composition and cell concentration was recently reported by Abeliovich (1982) in the deep Ram Reservoir, also in Israel.

However, despite quite similar rates of primary productivity in all seasons, the total supply of photosynthetic oxygen was low owing to the limitation of the photic zone. In addition, the excessive vertical mixing in this season distributed the oxygen produced near the surface throughout the water column, and its mean concentrations remained low, dropping below saturation level. Consequently, during the winter overturn the supply of atmospheric oxygen became more important than the photosynthetic source. In contrast, during the summer stagnation and particularly during very calm weather, the role of diffusion must be negligible (Golterman 1975) and algal photosynthesis supplied most of the oxygen available in the system.

Table 20.6. Counts of *Salmonella* (cells l^{-1}) and % of removal in Na'an system

Date	Raw sewage	Reservoir	Total removal (%)
15.11	1 000 S. anatum S. hadar S. tennenssee	10 S. anatum	99.9
25.12	10 200 S. irumu S. montevideo S. blockley	1 000 S. irumu S. hadar	90
24.04	5 050 S. newport S. carmel	30 S. newport	99.4
26.06	24 020 S. typhi-murium S. negev S. sentenberg S. infantis	0	100
15.07[a]	6 500 S. newport S. heidelberg S. infantis	20 S. sp.	

[a] In July supply of the raw sewage to the system was disconnected.

The limited diversity of algal community indicates that the conditions in the reservoir were highly selective and the question is, which factor or factors control the penetration of additional species into this water? *Chlorella vulgaris*, the alga dominant in the reservoir, is also widely distributed in wastewater all over the world and it must have some selective advantages over other algae, like resistance to wide fluctuations in *DO* and *p*H and a high rate of reproduction to compensate losses during the excessive grazing by zooplankton. In addition, strains of *Chlorella* are known as heterotrophs, able to utilize a variety of organic molecules (Pipes and Gotaas 1960). This obviously helps in surviving through periods of low light in turbid water, particularly during the winter mixing. This characteristic, however, places *Chlorella* in the position of a competitor of heterotrophic bacteria living in the same medium. Besides the direct competition for organic molecules, antagonistic effects between the sewage bacteria and *Chlorella* also exist. According to our previous experimental studies (Dor and Benzion 1980; Benzion and Dor 1981), numerous strains of heterotrophic bacteria occurring naturally in the wastewater inhibit growth of *Chlorella* by excreting toxic by-products into the medium. Despite this suppressing effect, *Chlorella* is able to grow at a slower rate even in the presence of a highly concentrated population (10^8 cells ml^{-1}) of the inhibitory bacteria. These still unidentified toxic products may be responsible for the elimination of the other, less resistant algal species. Such a selective, inhibitory effect of sewage bacteria on the algae was also assumed by Fitzgerald (1969), who demonstrated that adding wastewater to algal cultures inhibited *Microcystis* while leaving *Chlorella* unaffected. On the other hand, *Chlorella* is also known to excrete fatty acids (Pratt 1942; Spoehr et al. 1949), which repress respiration and growth of numerous

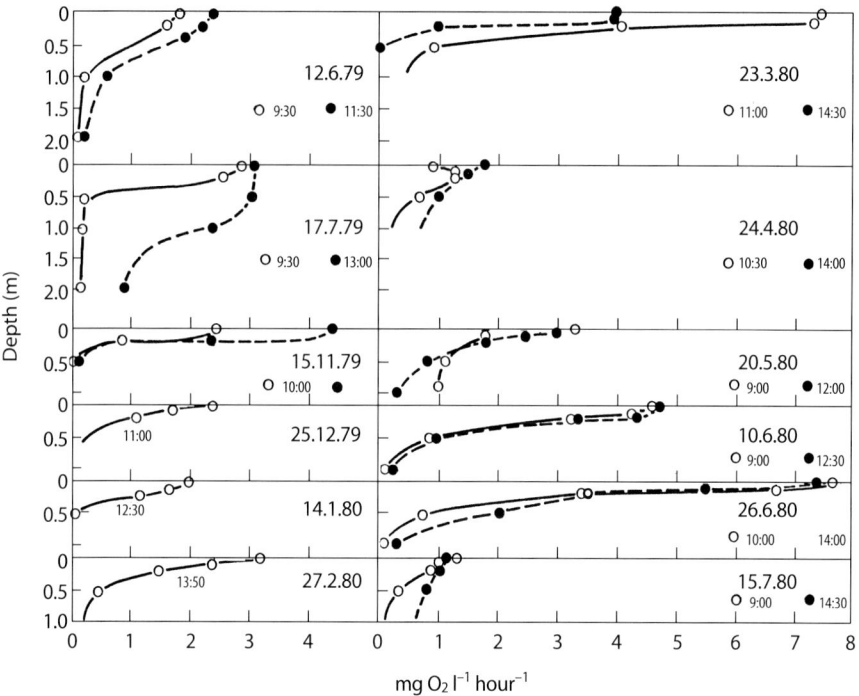

Fig. 20.12. Gross primary productivity

bacterial strains isolated from sewage (Davis and Gloyna 1970; Dor 1978). The above results lead us to the conclusion that only algal species resistant to bacterial by-products and able to repress bacterial competitors for space and nutrients, can grow in environments rich in organic matter. It is most probable that the bacterial community undergoes a similar competitive selection, but this phenomenon has not yet been investigated.

The competitive and antagonistic relationships between heterotrophic algae and bacteria do not exclude their mutual dependence on metabolic by-products, i.e., bacterial dependence on photosynthetic oxygen and algal dependence on respiratory CO_2 and other nutrients derived from bacterial decomposition. The complementary relationships between algae and bacteria in wastewater were recognized a few decades ago and have been incorrectly called "symbiosis" (Oswald 1973). Whatever the interpretation, the intricate relationships between both groups of microorganisms obviously play a central role in the supporting process of self purification in the open wastewater systems.

Chlorella vulgaris has an additional advantage, being a small unicellular organism freely suspended in the water column, thereby creating a moderate self-shading effect. By contrast, flagellate algae like *Euglena* or *Chlamydomonas* and blue-greens like *Microcystis* or *Spirulina*, form dense scum on the water surface, preventing light penetration and photosynthesis in the deeper water mass.

Progressive improvement of water quality resulting from microbial activity and sedimentation is usually expressed in terms of removal of dissolved and suspended organic matter and enteric bacteria as a function of time (Dor et al. 1976). The changes recorded seasonally in the oxidation pond and in the reservoir reflect the self purificatory performance of the system and are of particular interest. (Table 20.3).

The mean BOD in the oxidation pond was much higher during the cold season (151 mg l^{-1}) than in the warm (59 mg l^{-1}). This difference may be explained by variations in seasonal temperatures, which affect the rate of bacterial decomposition (12–17° C and 25–32° C during the cold and warm season respectively). Nevertheless, the BOD of the reservoir remained almost uniform in all the seasons. This unexpected lack of seasonal variation of BOD within the reservoir cannot be easily explained. As seen from Table 20.3, the reservoir is quite efficient in removing BOD during the winter circulation, when the whole body of water is well mixed. Evidently winter temperature ranging between 12–17° C and oxygen content of only 0.8–1.5 mg l^{-1} were sufficient to promote bacterial decomposition, which removed some 61% of the BOD entering the reservoir.

By contrast, at temperatures ranging 25–32° C and with an ample supply of O_2 from algal photosynthesis, the relative removal of BOD in the summer was poor and did not exceed 5%. The question is, why was the efficiency so much lower during the warm season and why did a further decrease of the BOD (below 60 mg l^{-1}) not occur? One assumption might be that most of the biodegradation during the warm season had already been accomplished in the oxidation pond and the residual organic matter entering the reservoir was not easily decomposable. This explanation, however, cannot be accepted, because the results of the BOD test itself, carried out in the laboratory, indicated that the organic matter present in the reservoir was decomposed during 5 days at 20° C. Despite this, it remained almost unchanged in the reservoir during the prolonged retention time. More quantitative data on the subject is needed before any conclusive explanation can be provided.

The removal of bacteria in the oxidation pond was consistent in all seasons. By contrast, changes in SPC in the reservoir were extremely irregular, showing decreases of varying extent and sometimes a marked increase in bacteria numbers. The bacterial increase recorded in the reservoir during November and February possibly resulted from amounts of organic particles resuspended by the vertical mixing, which provided an additional substrate for bacterial growth. The maximal decline in bacteria numbers (97.7% in March) coincided with the onset of stratification and development of a heavy bloom of *Chlorella*. As filter feeders were still scarce at that time, the disappearance of heterotrophic bacteria must result from adverse environmental conditions as discussed above.

Removal of coliforms in the oxidation ponds showed no seasonal change (98.1–99.9%), indicating that the rate of die-off of these bacteria was quite stable and time dependent. The increase in coliform numbers in the reservoir indicated in July 1980, when the sewage flow to the system was disconnected, must have resulted from faecal droppings of migrating aquatic birds. A somewhat lower rate of removal of *Salmonella* in winter (96% in December as compared to 99–100% in warmer months), probably indicates the preservatory effect of lower temperatures for these bacteria. The fact that the performance of the oxidation pond in removal of various bacteria was more consistent than the performance of the reservoir may be explained in several ways: a) an oxidation pond is a simple system which displays uniform activity, and bacterial num-

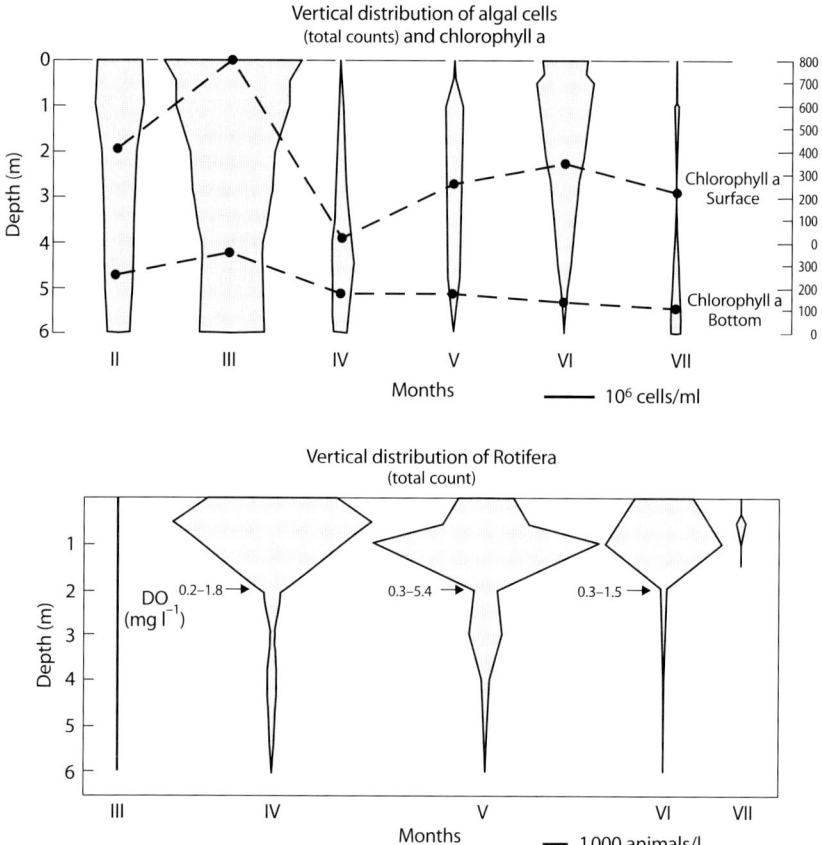

Fig. 20.13. Vertical distribution of algae, chlorophyll a and Rotifera

bers in it depend directly on the concentration of available organic matter; by comparison the reservoir is much more complex and bacterial numbers are additionally affected by filter feeders and relationships with the dominant *Chlorella*; b) in an oxidation pond a constant die-off of the sewage bacteria occurs, while in the reservoir limnetic species develop, which do not decline in direct relation to the decrease of sewage pollutants.

Only a limited number of zooplankton species appeared in the reservoir, all of them ubiquitous inhabitants of waters rich in organic matter. The explosive appearance of the rotifers in April was evidently related to the algal bloom, which developed earlier and created favourable conditions by providing abundant food and oxygen. Grazing by rotifers considerably improved transparency in the water column (Figs 20.11 and 13), so increasing the depth of the photosynthetic zone producing oxygen and stabilizing the environmental performance of the reservoir. In addition, the oxygenated upper layer provided water for irrigation without causing nuisance odours. By contrast, in the absence of grazers, the self-shading effect of the algae reduced the photic zone to

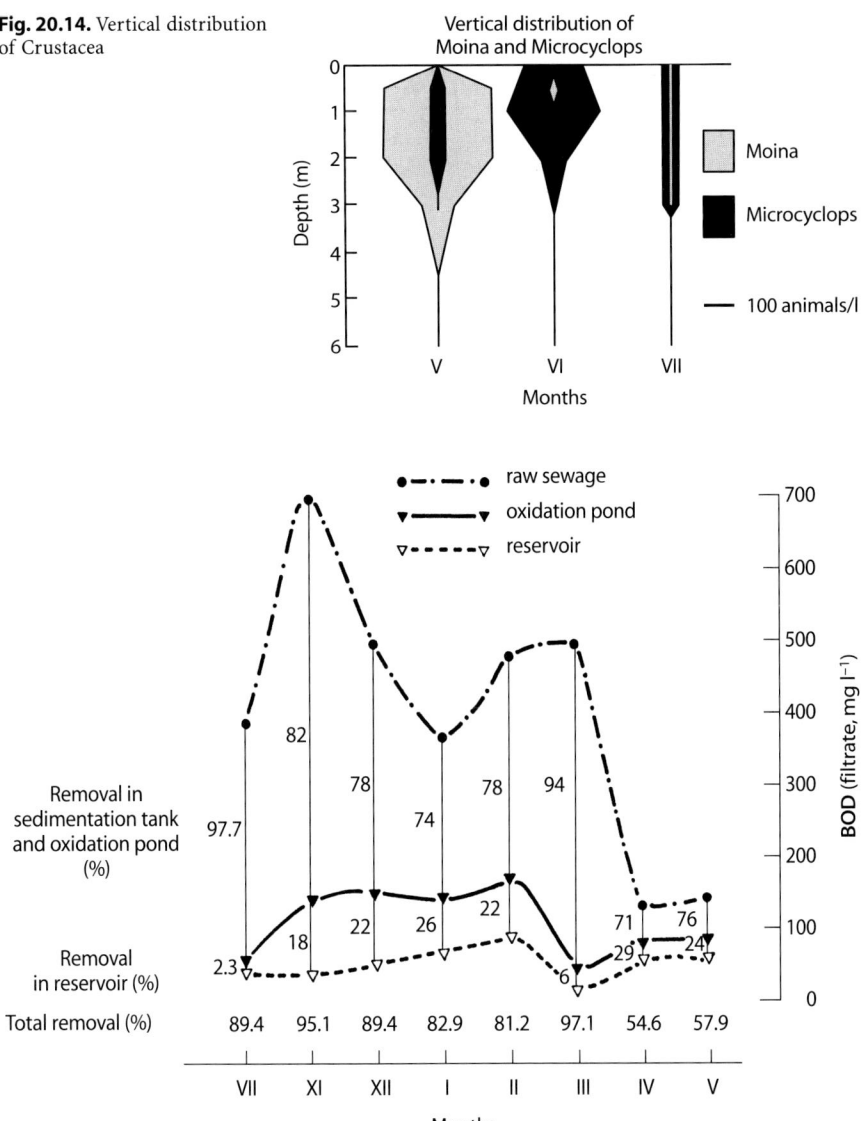

Fig. 20.14. Vertical distribution of Crustacea

Fig. 20.15. Relative and total removal of *BOD* in Na'an system

30–40 cm only (Fig. 20.11), and during stratification, the oxygenated layer became limited to the same depth (Fig. 20.16). Under such conditions anaerobic incidents are more likely.

Vertical distribution of the zooplankton was evidently related to the availability of dissolved oxygen in the water column and the limit for most species was at a concentration of some 0.2–0.3 mg l^{-1}, representing a depth of 0.5–1.5 m (Fig. 20.13).

Fig. 20.16. Relationship between the depth of thermocline[a], oxycline[a] and compensation level[b]; [a] the upper boundary; [b] 1% of surface light

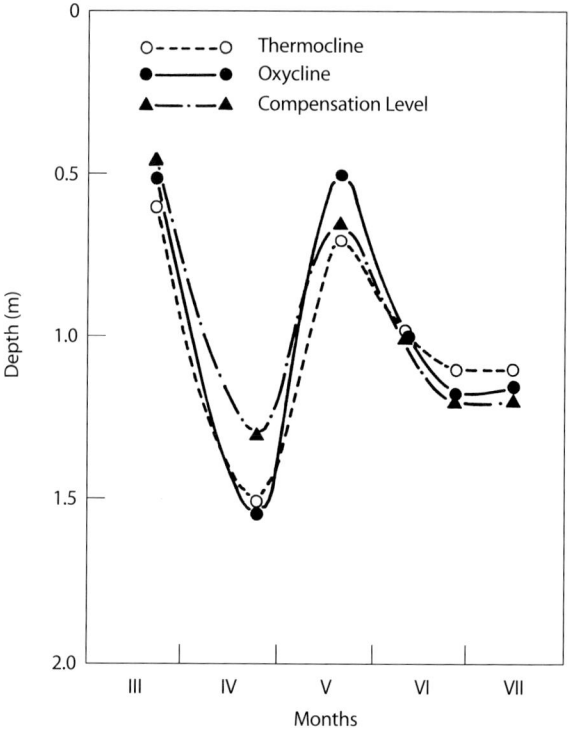

The upper limits for various parameters expressing the trophic level of Na'an Reservoir were as follows: *BOD* – 80 mg l⁻¹; chlorophyll a – 800 mg m⁻³; gross primary productivity – 14 g O_2 m⁻² d⁻¹ equivalent to 5.4 g C m⁻² d⁻¹. According to these characteristics the reservoir should be classified as polysaprobic (according to its organic matter content) and polytrophic (according to its mineral nutrients content) (Sladecek 1979), or more generally as hypertrophic. The last term is usually applied to the mineral enrichment, but frequently mixotrophic conditions prevail, when some organic matter is introduced together with the nitrogen and phosphate (Leentvaar 1980).

According to Barica and Mur (1980), such highly enriched bodies of water easily desequilibrate and undergo periodic crashes of zooplankton and algal blooms, resulting in anaerobic catastrophes. During a one-year study the Na'an Reservoir never deteriorated to produce obnoxious anaerobic conditions. With the transitory states between oxygen super and undersaturation, it in fact displayed a well balanced dynamic stability. It seems that the high enrichment with organic matter provided a measure of stabilization to this ecosystem.

The greater ability of sewage ponds to adjust to environmental stress as compared with oligotrophic waters was documented by May (1974), Petersen (1975), and further by Uhlmann (1980). The above regulatory potential of hypertrophic waters rich in organic matter, although recognized, was not however adequately explained. The following hypothesis is here proposed: a hypertrophic ecosystem (rich in mineral nutrients), which also receives an input of organic matter, increases bacterial activity and becomes selective toward algae (see former discussion of algal-bacterial relations). Above a certain level of organic enrichment, only a few, specialized algal species remain, which are facultative heterotrophs and resistant to bacterial toxins. These algae continue production of photosynthetic oxygen even under multiple environmental stresses. The respective bacterial community probably undergoes a similar selection toward forms more resistant to high DO, pH and algal excretions. Both groups, algae and bacteria, are resistant to environmental fluctuations, closely coupled by competition (for organic molecules), mutual suppression (excretory products of both groups) and support (exchange of metabolites). At this stage, sensitivity of the ecosystem to the environmental factors (light, temperature, pH, DO) greatly declines and dynamic equilibrium is supported chiefly by the biotic interactions between the dominant organisms. The reduced, specialized community seems to be more resistant than the more diversified communities of clean water bodies.

If this hypothesis is proved to be right, the unavoidable conclusion is that "mixotrophic" conditions may be more favourable than purely hypertrophic ones. The practical recommendation for the operation of wastewater reservoirs would be that a certain optimal level of organic matter is required in order to promote the specialized community of bacteria, algae, and zooplankton, which, owing to high resistance, supports a long term dynamic stability of such a system, while avoiding deterioration into obnoxious conditions.

Acknowledgements

M. Dinour of Kibbutz Na'an is sincerely thanked for the cooperation and help provided during the field trips. Participation of L. Metzger in the early phase of the study is much appreciated. J. Schechter is acknowledged for the identification and counts of *Salmonella*. Our special thanks are extended to A. Verechson, who organized the field work and provided skillful technical assistance. Finally we wish to thank F.D. Por for reading the manuscript and for valuable criticism. The study was supported by a grant from the Water Commissioner, The Ministry of Agriculture, Israel.

References

Abeliovich A (1982) Biological equilibrium in a wastewater reservoir. Wat Res 16:1135–1138
APHA, AWWA, WPCF (1975) Standard methods for the examination of water and wastewater. Washington, DC, 1193 pp
Barica J, Mur LR (1980) Hypertrophic ecosystems. Dev Hydrobiol 2:348 pp
Benzion S, Dor I (1981) Bacterial inhibition of algal growth in photosynthetic sewage treatment systems. Proc 12th Sc Conf Israel Ecol Soc, pp 249–258
Davis EM, Gloyna EF (1970) Bactericidal effects of algae on enteric organisms. US Dep. Interior, Federal Water Quality Administr, Water Poll Contr Res 18050 DOL 03/70:132
Dor I (1978) The effect of *Scenedesmus* and *Chlorella* on heterotrophic bacteria in polluted waters. Verh Int Ver Limnol 20:1930–1933
Dor I, Benzion S (1980) Effect of heterotrophic bacteria on the green algae growing in wastewater. In: Shelef G, Soeder CJ (eds) Algae biomass production and use. Elsvier, pp 421–429
Dor I, Schechter H, Shuval HI (1976) Biological and chemical succession in Nahal Soreq: A free flowing wastewater stream. J Appl Ecol 13:475–489
Fitzgerald GP (1969) Some factors in the competition or antagonism among bacteria, algae and aquatic weeds. J Phycol 5:351–359
Golterman HL (1975) Physiological limnology. Dev Water Science 2:489

Leentvaar P (1980) Comparison of hypertrophy on a seasonal scale in Dutch inland waters. In: Barica J, Mur LR (eds) Hypertrophic ecosystems. Dev Hydrobiol 2:45–55

May RM (1974) Stability in ecosystems: Some comments. In: Proc Intern Congr Ecol The Hague Dr Junk W, The Hague, Netherlands, p 67

Oswald WJ (1973) Productivity of algae in sewage disposal. Solar Energy 15:107–117

Peterson CH (1975) Stability of species and of community for the benthos of two lagoons. Ecology 56:958–965

Pipes WO, Gotaas HB (1960) Utilization of organic matter by *Chlorella* grown in sewage. Appl Microbiol 8:163–169

Pratt R (1942) Studies on *Chlorella vulgaris*, some properties of the growth inhibitor formed by *Chlorella* cells. Amer J Bot 29:142–148

Sladecek V (1979) Algal tests and the ratio of saprobic vs. trophic levels. In: Marvan P, Pribil P, Lohotsky S (eds) Algal Assays and Monitoring Eutrophication. E. Schweizerbart'sche Verlagsbuchhandlung, Stuttgart, pp 235–237

Spoehr HA, Smith JHC, Strain HH, Miller HW, Hardin GJ (1949) Fatty acids antibacterials from plants. Carnegie Ins Publ 586, Washington, DC, 67 pp

State of Israel, Ministry of Health (1976) Proposed regulation for irrigation with treated sewage effluent. Jerusalem

Talling JF, Driver D (1963) Some problems in the estimation of chlorophyll a in phytoplankton. Proc Conf Primary Prod Measurements, Marine and freshwater. Hawaii 1961, US Atomic Energy Comm TID, pp 142–146

Uhlmann D (1980) Stability and multiple steady states of hypereutrophic ecosystems. In: Barica J, Mur LR (eds) Hypertrophic ecosystems. Dev Hydrobiol 2:235–247

Vollenweider RAA (1969) Manual on methods for measuring primary production in aquatic environments. IBP Handbook N° 12, Blackwell 213 pp

Chapter 21

The Geta'ot Reservoir

Marcelo Juanicó

21.1
The Physical Environment

21.1.1
Introduction

The main objective of this section is to describe ephemeral and permanent stratification in a small reservoir. It also describes other important parameters such as the operation of the reservoir, age structure of the effluents, and the reservoir's physical environment as a whole. The reservoir has an inflow with a mean BOD of 80 mg l^{-1}, and a mean surface organic loading of about 30 kg BOD ha^{-1} d^{-1}, which determines the development of facultative conditions: i.e., an aerobic epilimnion and an anaerobic hypolimnion.

An early paper by Eren (1978) on the Kfar Hasidim Reservoir described a reservoir without stratification. Other ulterior works presented in local workshops and similar informal fora (most of them unpublished) quoted the existence of stratification in some of the reservoirs, generally based on single temperature profiles made at noon or during the morning. The lack of temperature profiles taken around the clock made it difficult to discriminate between ephemeral temperature gradients formed during the day and actual permanent stratification. This created a controversy on the actual existence of permanent stratification in this kind of reservoir in Israel. The controversy must be solved, because stratification is a key factor in hypertrophic warm impoundments (Fry 1987), deeply affecting several limnological processes from oxygen regime to water quality and the plankton-bacteria community.

21.1.2
Methods

21.1.2.1
The Geta'ot Reservoir

The sewage treatment and storage unit of the Kibbutz Lohamei HaGeta'ot is located on the coastal area of northern Israel. It is made up of two anaerobic ponds in parallel, followed by one facultative pond and an earthen reservoir in series (Fig. 21.1). The unit receives the domestic sewage of the rural community (600 inhabitants) plus the wastewater from a vegetarian food factory and a cow stable, totaling 12 000 m^3 of effluent per month. There are no fish in the reservoir. The unit was designed several years ago, and the growth of both settlement and factory has exceeded the original

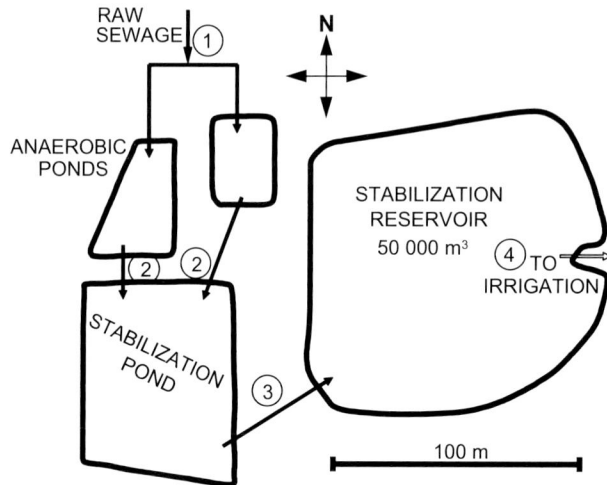

Fig. 21.1. Layout of the Geta'ot sewage treatment and storage unit, and sampling points (encircled numbers *1* to *4*)

Table 21.1. Design and operation of the Geta'ot treatment and storage unit. Only one of the anaerobic ponds may be in operation while sludge from the second pond is dried and removed

Parameter	Units	Anaerobic ponds	Facultative pond	Stabilization reservoir
Volume	m^3	230 each	4400	50000 max
Area	m^2	175 each	3450	11500 max
Depth	m	2	1.3	5.5 max
Inflow	$m^3 d^{-1}$	400	400	355[a]
Volumetric hydraulic loading	$m^3 m^{-3} d^{-1}$	0.85–1.7	0.09	0.02 $(0–0.07)$[c]
$PFE_{0.5}$[b]	%	34–55	4.5	<1
PFE_5	%	>100	35	7 $(0–30)$[c]
PFE_{30}	%	>100	85	35 $(2–100)$[c]
MRT	days	0.6–1.2	11	95 $(25–160)$[c]
Inflow's BOD	$mg\,l^{-1}$	580	370	80
Outflow's BOD	$mg\,l^{-1}$	370	80	25 $(5–100)$[c]
Volumetric organic loading	$g\,BOD\,m^{-3}\,d^{-1}$	500–1000	35	1 $(0–6)$[c]
Surface organic loading	$kg\,BOD\,ha^{-1}\,d^{-1}$	7000–14000	440	30 $(0–60)$[c]

[a] Some losses occur due to evaporation and seepage;
[b] *PFE*: the percentage of fresh effluents with 0.5, 5, and 30 days within the impoundment;
[c] These values strongly fluctuate during the year. The first value represents the annual mean. Values within brackets are minimum and maximum.

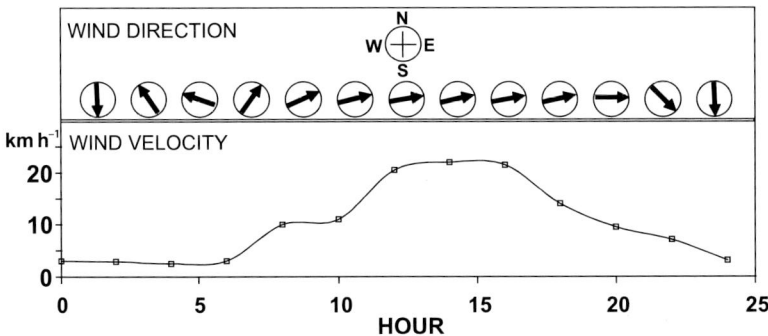

Fig. 21.2. Wind direction and velocity on a typical summer day, in the coastal area of Northern Israel

design. Table 21.1 presents main design and operational parameters of the ponds and the reservoir. The outlet of the reservoir is made of a flexible 10"-pipe hanging from a raft, which maintains the opening 0.7 m below surface, accompanying the changes in water level.

21.1.2.2
Climatic Characteristics of the Area

The coastal area of Israel has two conspicuous seasons. The winter is rainy, with mild temperatures (6° C minimum average, 18° C maximum average), a mean solar radiation of about 1250 Ws cm^{-2} d^{-1}, and an irregular wind regime. The summer is dry (no rain at all), hot (20° C minimum average, 33° C maximum average), with a mean solar radiation of about 2300 Ws cm^{-2} d^{-1}, and a regular wind regime with a strong breeze blowing from 10:00 to 18:00 (Fig. 21.2). A detailed description of the wind regime in the area can be found in Milstein et al. (1992).

21.1.2.3
Sampling and Analyses

The operational parameters of the reservoir (input/output flow rates, water level, volume, water area) were taken from input/output flowmeters, water level scale, and volume-area/depth curves. Meteorological data (radiation, clouds, rain, wind speed and direction) were recorded at the reservoir and complemented with data from the Israel Meteorological Service.

Field measurements and water sampling for analysis were performed in the pipes which connect the different impoundments of the unit (Fig. 21.1) during one hydrological year from September 1986 to September 1987. In the reservoir, sampling was performed at its deepest point, near the outlet, totaling 18 sampling days and 88 surveys (on some sampling days several surveys were performed at different hours). The sampling strategy was designed following Ibanez (1976) and Milstein (1984). On each sampling day profiles of temperature, pH, dissolved oxygen, REDOX potential, electrical conductivity (EC), and light penetration (photosynthetic active radiation – PAR)

were taken in the reservoir with Yellow Spring, El-Hama and LICOR field instruments. Data were collected at eight different times: noon, two hours before sunset, sunset, two hours after sunset, midnight, two hours before sunrise, sunrise, and two hours after sunrise. EC was standardized to 25° C with a correction of 0.8% per degree (APHA 1989). Direct light (surface oriented sensor) and reflected light (bottom oriented sensor) were measured along the water column and expressed as a percentage of incident light above surface. Secchi disk visibility was also measured. Water samples were taken at four depths in the water column. Alkalinity and ammonia were analysed following APHA (1989). Ultraviolet absorption (UV) of filtered samples was measured at 254 and 545 nm and its difference (254–545 absorption) utilized as an estimate of the concentration of dissolved organic matter in the water (Michail and Idelovitch 1980).

21.1.2.4
Data Handling and Analysis

Data were introduced in a general database on stabilization reservoirs in Israel (Shelef and Juanicó 1988), from which they were handled and analysed. The calculation of the percentage of fresh effluents with 30 or less days within the reservoir (PFE_{30}), PFE of other number of days (0.5, 5, etc.) and the mean residence time of effluents (MRT) was made with the algorithms described in Juanicó and Friedler (1994). Statistical Analysis System software, documented in SAS (1985) was utilized.

21.1.3
Results and Discussion

21.1.3.1
Operation of the Reservoir during the Studied Period

Wastewater is available all the year round while irrigation is performed only during the dry summer. The reservoirs have an empty-full-empty cycle and thus their operation is not steady-state flow. Typical reservoirs are almost empty by the end of the irrigation season (September) when most of the effluents within them are relatively fresh (or new) and the percentage of fresh effluents (PFE) is then high. Water level starts to rise when the irrigation season ends and water outflow is cut off, and continues to increase during the winter until spring when the reservoir is full and PFE is very small. When the new irrigation season starts, the strong outflow drops water level while new effluents continue to enter, raising PFE again.

Figure 21.3 indicates that the Geta'ot Reservoir followed this general trend, but it was full of effluents by early February, and then inflow was discontinued (except for a small input in late February to compensate for evaporation and seepage losses). Inflow to the reservoir was irregular during the year, partially because the reservoir was full from February to June and only sporadic inputs were possible, and partially because the pumping system from the stabilization pond to the reservoir works by pulses. The percentage of effluents with 30 or less days within the reservoir (PFE_{30}) reached almost 100% by early October, then started to decrease dropping to about 5% in March–April, and increased again during the summer. PFE_5 followed a similar trend but changed quicker with the changes in inflow. Surface organic loading was

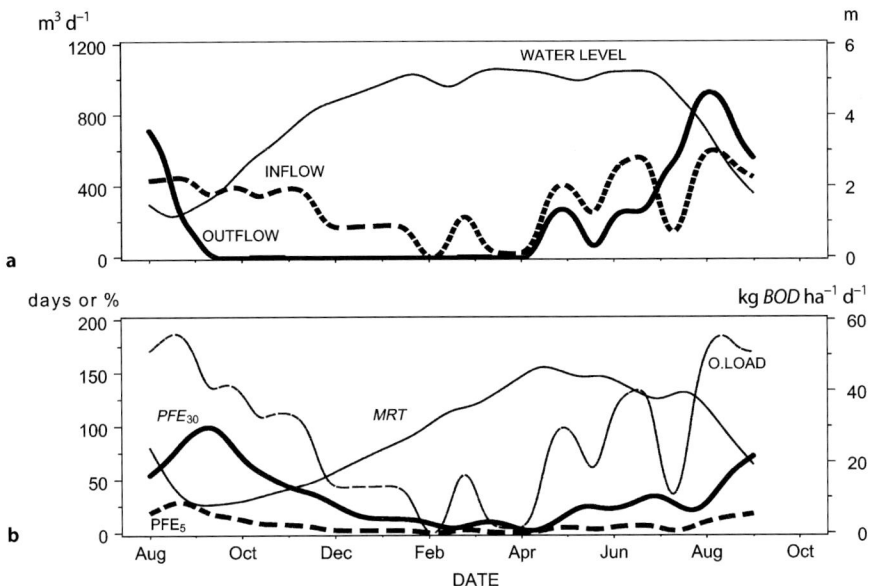

Fig. 21.3. Operation of the Geta'ot Reservoir during the studied period. **a** Changes in water level as a function of inflow and outflow. **b** The percentage of fresh effluents with 5 and 30 days within the reservoir (PFE_5 and PFE_{30}, %), the mean residence time of effluents within the reservoir (MRT, days), and the surface organic loading (kg $BOD\,ha^{-1}\,d^{-1}$)

30 kg $BOD\,ha^{-1}\,d^{-1}$ in average (Table 21.1) but fluctuated even quicker than PFE_5 with the changes in inflow and the area of the reservoir (the latter a function of water level). The mean residence time followed an almost inverse but smoother trend than PFE_{30}, varying between 30 and 150 days with a mean of 95 days.

Evaporation losses amounted to 17 000 m³ during the studied period (about 17% of total inflow to the reservoir in the same period), with a mean evaporation rate of 3 mm d^{-1} in winter and 6 mm d^{-1} in summer.

21.1.3.2
Stratification

Two kinds of gradients were recognized in the reservoir (Fig. 21.4): ephemeral thermoclines formed during the day only, and permanent thermo- and chemoclines were detected around the clock. Ephemeral thermoclines were formed at about 0.5 m below surface from August to October when water transparency was minimal with an average Secchi disk visibility of 20–25 cm, and at 1 or 1.5 m below surface in December and from April to July when Secchi disk visibility was deeper. Ephemeral thermoclines were not detected during winter months. At the beginning of the sampling period (August–September 1986), water level was very low and no permanent stratification was found. A permanent thermo-chemocline was formed at 2.5 m above bottom (2.7 m below surface) in the following spring (mid-May 1987). It maintained its elevation above bottom unchanged during most of the summer in spite of the decreas-

Fig. 21.4. Depth of the thermoclines during the studied period. *Dashed line* indicates water level. '=' refers to ephemeral shallow thermoclines formed during day hours only. Measurements were not continuous, thus lacking '=' does not necessary mean no ephemeral thermocline and *continuous lines* '—' refer to permanent thermoclines detected around the clock

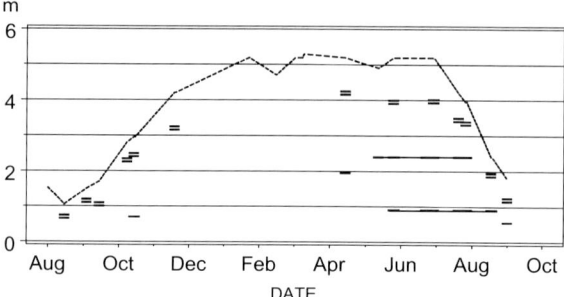

ing water level, until the cline was only 1.5 m below surface by mid-August, when it broke down. A second, deeper thermo-chemocline was formed at 0.9 m above bottom by early June, and also maintained its elevation unchanged in spite of the decreasing water level, until the cline was only 1.5 m below surface by early September. Thus, the minimal epilimnion height before stratification breaks down in this reservoir seems to be 1.5 m.

Figure 21.4 indicates that the outlet, hanging from a raft 0.7 m below surface, took effluents mainly from above the permanent thermo-chemoclines washing out the epilimnion. The decrease in water level was due only to a decrease in epilimnion height, except by the end of the irrigation period (mid-September), when water level was very low and further decrease was due to a drop in hypolimnion height. Felgner and Sandring (1983) performed experimental storage of wastewater in earthen tanks 2 m deep and 300 m³ volume in Germany, recommending taking effluents out only from the upper meter of the tank, in order to avoid the release of sulfide rich waters from the bottom. In the Geta'ot Reservoir water from the deep layer also contained sulfides and the same recommendation is valid.

Figures 21.5 and 21.6 show the three clines (one ephemeral and two permanent) found in June, July and August. The ephemeral thermoclines presented maximum temperatures at surface between 12 and 17 h. During the night heat losses to atmosphere inverted the temperature profile forcing the convective mixing of the whole epilimnion which was again homogeneous at sunrise. On the contrary, the permanent thermo-chemoclines were not affected by the daily temperature changes; the differences in temperature and electrical conductivity between hours in deep layers noticeable in Figs 21.5 and 6 are sampling artefacts due to seiches, errors in the determination of the sampling depth from a rolling boat, and errors in the standardization of EC to 25° C. Surface water temperature increased by 3–5° C during the day. This ephemeral temperature gradient added to the deeper permanent ones, totaling a difference of 10 degrees between surface and bottom waters, which produced a density gradient of 2.4 kg m⁻³ over a depth of 4–5 m. By comparing Figs 21.2 and 5, it can be seen that the period of strongest breeze (between hours 11 and 18) matches that of maximal temperature gradient in the ephemeral epilimnion thermocline.

Figure 21.6 shows that, besides the thermoclines, there were strong chemoclines with a pronounced electrical conductivity gradient totaling a difference of about 500 μmhos cm⁻¹ between surface and bottom. Table 21.2 shows that the EC gradients were due, at least partially, to the increase with depth of alkalinity, ammonia and dis-

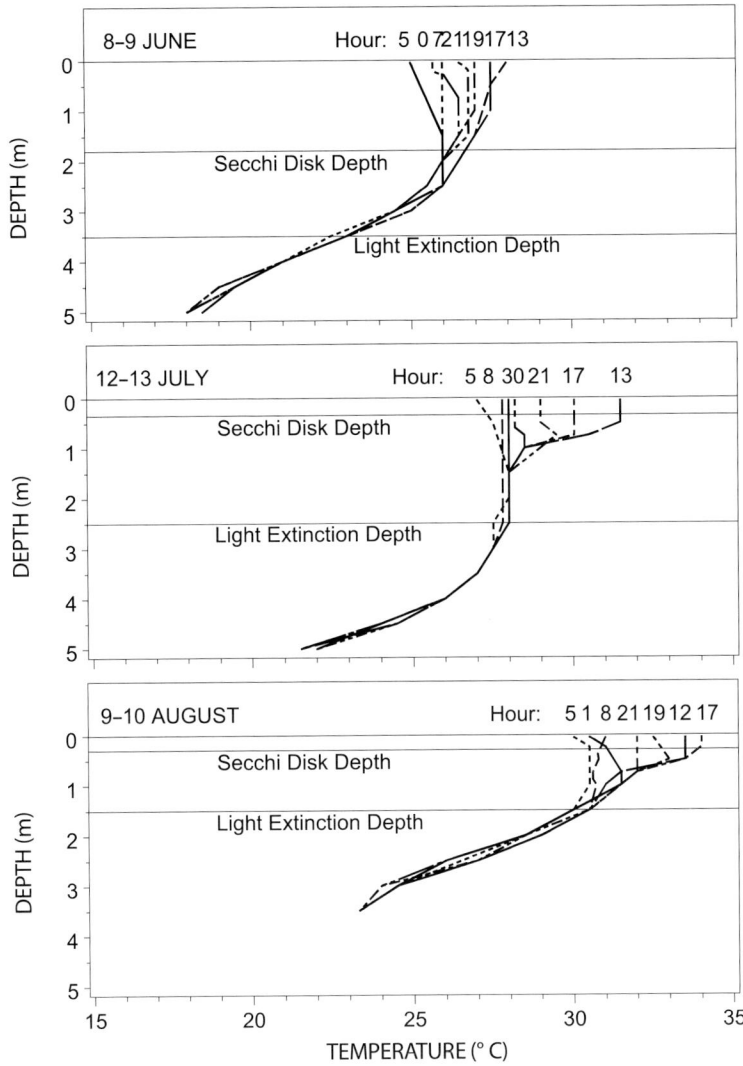

Fig. 21.5. Round-the-clock profiles of temperature in June, July and August/87. The horizontal lines indicate the depth of Secchi disk visibility and of light (*PAR*) total extinction (0% of surface light)

solved organic matter (UV absorption). Patil et al. (1973), in a study of a warm area stabilization pond, also found an increase with depth in the concentration of bicarbonates and ammonia. Bokil and Agrawal (1977) established that even in very shallow stabilization ponds (40–50 cm deep) there may be a build-up of alkalinity and organic matter gradients. Ulhman (1980a) also quotes the hypolimnetic accumulation of dissolved solids as increasing the density gradients in stratified stabilization ponds. This accumulation of dissolved solids would be mainly due to the $Ca(HCO_3)_2$ resultant from

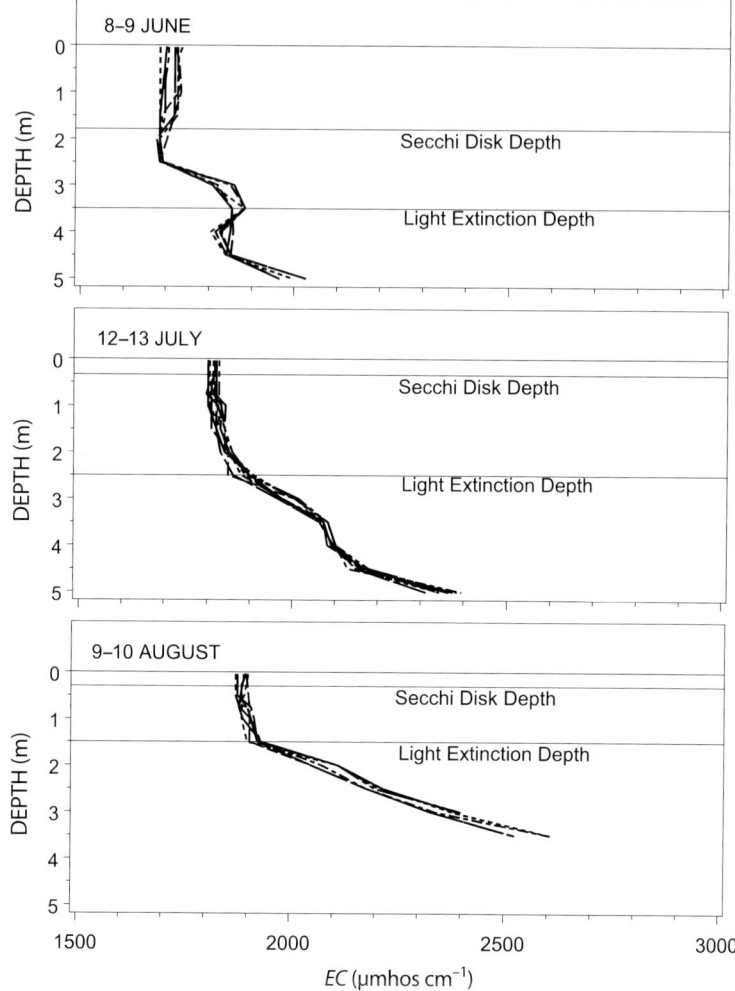

Fig. 21.6. Round-the-clock profiles of electrical conductivity (standardized to 25° C) in June, July and August/87. The *horizontal lines* indicate the depth of Secchi disk visibility and of light (*PAR*) total extinction (0% of surface light)

the anaerobic degradation process in the water column of the hypolimnion. Abeliovich (1982) quotes a conspicuous increase of ammonia with depth in a wastewater reservoir in Southern Israel. In the photic epilimnion of the Geta'ot Reservoir, algae activity reduced ammonia, alkalinity, and dissolved organic matter. In the aphotic hypolimnion, the release of anaerobic degradation products from the organic sediments and in the water column itself had the opposite consequence. Uptake of NH_4^+ by algal photosynthesis (with concomitant reduction in alkalinity) in the epilimnion, and the release of NH_4^+ by the degradation of organic matter in the sediment (increasing alkalinity) are known processes quoted in the classic literature (Stumm and Morgan 1981).

Table 21.2. Vertical chemical gradients in the Geta'ot Reservoir during the summer months

Sampling date	Depth (m)	EC ($\mu mhos\,cm^{-1}$)	Alkalinity ($mg\,l^{-1}$) as $CaCO_3$	N-NH$_3$ ($mg\,l^{-1}$)	UV absorb. 254–545
8-9 June	0	1 700	530	17.3	0.26
	2.5	1 700	540	17.7	0.27
	3.5	1 900	670	33.0	0.44
	4.5	1 850	720	36.6	0.46
	5	2 000			
12-13 July	0	1 800	560	15.7	0.30
	1.7	1 850	575	17.2	0.28
	3.5	2 100	700	33.8	0.60
	4.5	2 250	750	48.4	0.87
	5	2 350			
9-10 August	0	1 900	570	16.2	0.43
	1	1 900	575	16.3	0.41
	2	2 050	640	24.0	0.44
	3	2 400	880	55.6	0.92
	3.5	2 550			
Average of June–August	Inflow to the reservoir	1 900	720	45	0.97

In hypertrophic impoundments such as stabilization ponds, intensive fish ponds, wetlands and wastewater reservoirs, these processes are so intensive that the resulting density gradients may determine the persistence of stratification. In the Geta'ot Reservoir a comparison of alkalinity, ammonia, and UV in the inflow with the vertical gradients within the reservoir (Table 21.2) suggests that the biogenic gradient was further amplified by the entrance of new influent directly to the hypolimnion.

Two previously published works quote no permanent stratification in stabilization reservoirs in Israel. Eren (1978) found no stratification in a 6.5 m deep reservoir receiving effluents with a mean *BOD* of 60 mg l^{-1} in Northern Israel. A temperature gradient of 1 or 2° C was formed between surface and bottom during windless periods, but it was quickly broken when wind started to blow. Argaman et al. (1988) performed a tracer study on the hydraulic flow pattern of two 10 m deep reservoirs in series receiving effluents with a mean *BOD* of 60 mg l^{-1}. They found that the changes in tracer concentration in the outflow (C tracer curve and calculation of active volume, Levenspiel 1972) indicated almost complete mixed conditions within the reservoir (no stratification). Meanwhile, other authors quote the existence of permanent gradients. Abeliovich (1982) found permanent stratification in June–July in a 6 m deep reservoir receiving effluents with a mean *BOD* of 180 mg l^{-1} in arid Southern Israel, with a temperature gradient of 5° C between surface and bottom. The epilimnion was totally washed out by the strong outflow during the irrigation season, exposing the hypolimnion to air with the consequent release of sulfide and ammonia. Dor et al. (1987) also found permanent stratification during summer months in a 9 m deep reservoir receiving effluents with a mean *BOD* of 100 mg l^{-1} in Central Israel. Soler et al. (1991) and Torres et al. (1997) in a study of an 8 m deep reservoir receiving effluents with a mean *BOD* of 100 mg l^{-1} in Spain under Mediterranean climatic conditions, also found permanent marked stratification during the summer months. Dor and Raber (1990), in a study of twelve reservoirs receiving effluents of a wide range of qualities in Central Northern Israel, found both stratified and nonstratified reservoirs. As the measurements were sporadic (1–3 samples per year) and made only during daylight hours, it was not possible to determine whether stratification was permanent or ephemeral. Those reservoirs receiving effluents with higher BOD tended to be stratified, thus these authors postulated that higher BOD results in higher turbidity and consequent trapping of solar energy within the upper water layer accompanied by build-up of strong temperature gradients over small depths. From the five studied reservoirs previously quoted in this paragraph, the two reservoirs receiving effluents with a BOD of 60 mg l^{-1} (Eren 1978; Argaman et al. 1988) presented no permanent stratification, while the three receiving effluents with a BOD between 100 and 180 mg l^{-1} (Abeliovitch 1982; Dor et al. 1987; Soler et al. 1991; Torres et al. 1997) were stratified. The findings of the present work fully confirm the presence of permanent stratification in the small Geta'ot Reservoir receiving effluents with a BOD of 80 mg l^{-1}. In this reservoir, the depth of the ephemeral thermoclines was minimal when Secchi disk visibility was also minimal and PFE was maximal (August to October, Figs 21.3 and 21.4).

The reservoirs for fish culture and irrigation in Israel receive relatively clean freshwater but are fertilized with chicken manure and heavily loaded with fish and feed pellets (although organic loading is less than in wastewater reservoirs). Reservoirs with and without stratification have been found. Water for irrigation is taken from the hypolimnion in these reservoirs, and then the decrease in water level during the irrigation season is due to a decrease in hypolimnion height. In this case, stratification breaks down when the hypolimnion height reaches 1.5 m (Milstein et al. 1992, 1993a, 1993b).

The existence of similar hypertrophic reservoirs with and without stratification in Central-Northern Israel suggests that the build-up of the density gradients in these impoundments does not depend only on physico-mechanical parameters; i.e., temperature, solar radiation, wind, and reservoir size and depth. Other parameters related to the degree of organic loading of the reservoir and the resulting biological activity also seem to play an important role, e.g., the trapping of solar radiation in the upper layers due to the high concentrations of phytoplankton and organic matter, and the biogenic chemical gradients. During field work performed by the author in several wastewater reservoirs in Israel, it was possible to observe that reservoirs receiving heavy BOD loading have much smaller waves than the lightly loaded ones. Two parameters which may explain this phenomenon are: 1) a reduction in the wind/effluent friction coefficient derived from the formation of surface films of detergents and oils (these films are easily observable in heavily loaded reservoirs), and 2) the viscosity of the effluents derived from the high concentration of dissolved organic matter. These two parameters would also play an important role in the establishment of stratification.

The artificial break-down of stratification is known to have several benefits in polluted lakes and heavily loaded ponds (Brown 1983; Krambeck 1986; Stauffer 1987; Chang and Ouyang 1988; Szyper and Lin 1990; Hofmann and Krambeck 1991; Llorens et al. 1992). This technique, applied to wastewater reservoirs with permanent and/or ephemeral stratification, will surely improve the oxygen regime and water quality of the reservoirs. Liran et al. (1994) pointed out that stratification control will also improve the removal of pathogens (faecal coliforms) by exposing the whole water column to photosynthetic activity, high pH, and solar radiation.

21.1.3.3
Temperature and Salinity Changes During the Year

In winter the temperature of both bottom and surface waters was about 12–13° C (Fig. 21.7). In summer bottom waters were 25° C, while surface waters reached 30° C at sunrise and 34° C at noon. Water temperature reached its maximum peak in August, two months after the peak of solar radiation. Water temperature started to rise in spring, but this increase was slower (and started later) at the bottom than at the surface, due to stratification. This surely delayed the beginning of the anaerobic digestion of organic sediments, because the minimum temperature required for an effective anaerobic process is 15° C, and even above this minimum the anaerobic process presents a high dependence on temperature. Thus, the bulk of sediment digestion occurred during a short period of 3–4 months in late summer and fall, with negative effects on water quality (release of H₂S and dissolved organic matter from the sediments to the water column). Artificial break-down of stratification would lead to the increase of water temperature at bottom earlier in spring, distributing the digestion of sediments over a longer period.

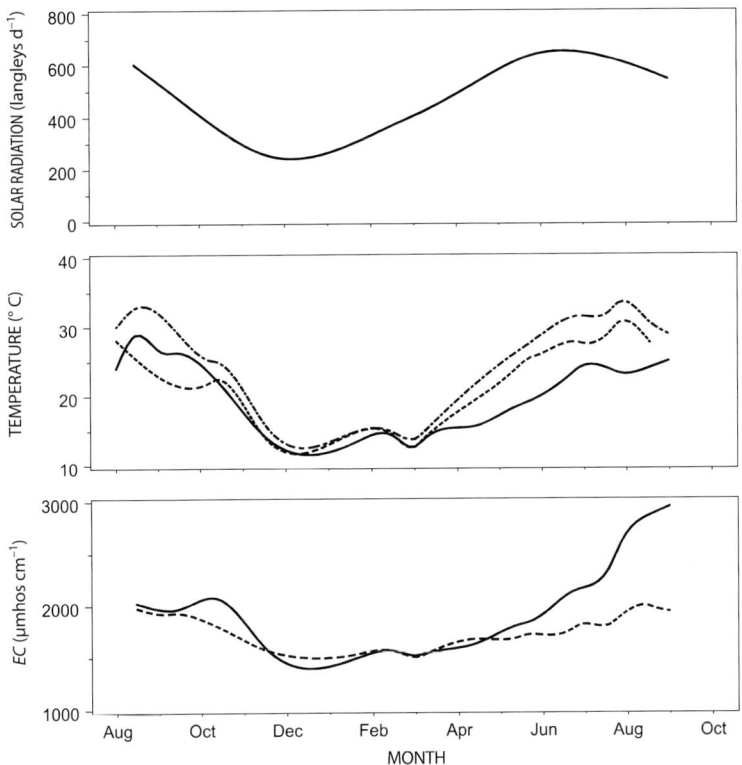

Fig. 21.7. Solar radiation above surface, water temperature and electrical conductivity changes during the year. Temperature: *Point-dashed* and *dashed lines* represent surface layer at noon and sunrise respectively. *Continuous lines* represent bottom layer (all hours). EC: *Dashed line* represents surface layer (all hours) and *continuous line* represents bottom layer (all hours)

Electrical conductivity ranged from 1 500 μmhos cm⁻¹ in the whole water column in
winter, to 2 000 μmhos cm⁻¹ at surface and almost 3 000 μmhos cm⁻¹ at the bottom in late
summer. The degradation of organic sediments during the summer was already quoted
as a cause of the increase in electrical conductivity (through an increase in Alkalinity).
The increase of organic loading to the reservoir in summer (Fig. 21.3) also results in more
biogenic electrical conductivity. Besides, winter effluents contain some proportion of rain
waters which reduce their salinity by dilution, while there is no rain at all during the summer months. The mean evaporation rate is only about 3 mm d⁻¹ in winter, while it reaches
6 mm d⁻¹ during the summer. The combination of these several parameters (more
biogenic EC, higher organic loading, higher salinity and evaporation) determines that
the electrical conductivity of the effluents within the reservoir increases during the irrigation season (summer). Figure 21.7 shows a reduction of the difference in temperature
and electrical conductivity between surface and bottom in mid-July. This indicates a small
breakdown of stratification (which surely was only partial and short-lived because the
gradients were reduced only a bit without vanishing, and increased again later on).

A comparison of Figs 21.3 and 21.7 shows that the annual cycles of temperature and
electrical conductivity apparently coincide in time with the annual operational cycle
of the reservoir (PFE, organic loading, mean residence time, etc.). However, this anal-
ogy is only superficial. For example, the decrease in water temperature starts in August, while the decrease in PFE_{30} starts in late September. Temperature reaches its
minimum in December, while PFE_{30} reaches it two months later.

21.1.4
Conclusions

Wastewater reservoirs in Israel go through three different (only partially overlapping)
annual cycles: 1) a seasonal cycle of solar radiation with maximum values in June, 2) a
seasonal cycle of water temperature with maximum values in August, and 3) an operational cycle determined by the changes in water demand for irrigation (changes
in water level, organic loading, residence time, the percentage of fresh effluents within
the reservoir) with minimum water level in late September.

The studied reservoir presents both permanent (deep) and ephemeral (shallow)
stratification during the summer months in spite of the strong regular breeze. Permanent density gradients are due to both thermal and biogenic chemical gradients.

Outflow from the upper layer washes out the epilimnion of the reservoir. Stratification
breaks down when the decrease in water level reduces the height of the epilimnion to 1.5 m.

There are both permanently stratified and nonstratified wastewater reservoirs in Israel.

The findings of this work suggest that stratification in hypertrophic impoundments
depends not only on the factors generally considered in stratification studies (e.g.,
density gradients, wind, and impoundment depth and size), but also on parameters
related to the chemical composition of hypertrophic waters (e.g., high turbidy which
traps solar radiation in the upper layers, biogenic chemical gradients, the viscosity of
hypertrophic waters, and reduced wind/water friction coefficient).

Stratification retards the warming-up of bottom waters in spring, concentrating
the anaerobic digestion of organic sediments into a short period of 3–4 months, with
negative effects on water quality. The study of artificial destratification is recommended
as a tool to improve water quality and the performance of this type of reservoir.

21.2
Changes in Water Quality

21.2.1
Introduction

The main objective of this section is to identify the principal processes which determine the changes in water quality within the Geta'ot wastewater reservoir. Most hypertrophic impoundments are man-made, purposely constructed to fulfil technological objectives: waste stabilization ponds, intensive fish ponds, artificial wetlands, etc. The limnology of hypertrophic impoundments is still poorly known, while the identification of their principal limnological processes is required as a basis for a better design and operation (Uhlmann 1990).

21.2.2
Methods

Only 'epilimnion' and 'hypolimnion' samples are herein analysed and compared; during the periods without stratification, samples taken from the upper 1/3 of the water column were considered 'epilimnion' and those from the lowest 1/3 were considered 'hypolimnion'. Samples for chemical analyses were taken at sunrise and sunset. It was assumed that sunrise values are the result of night processes (dominated by respiration), while sunset values are the result of the day processes (dominated by photosynthesis).

Data were analysed by means of factor analysis solved through principal components. Details on the use of this method to analyse environmental data can be found in Seal (1964) and Milstein (1993).

21.2.3
Results and Discussion

21.2.3.1
Inflow to the Reservoir

Table 21.3 describes the quality of the inflow to the reservoir. The values listed in the table are characteristic of effluents from stabilization ponds, with high values for *TSS* and chlorophyll a derived from the high concentration of phytoplankton, and *TSS/BOD* and *COD/BOD* ratios of about 3 : 1. The average *BOD* entering the reservoir was 80 mg l^{-1}, determining a mean surface organic loading of 30 kg *BOD* $ha^{-1} d^{-1}$.

21.2.3.2
Epilimnion Versus Hypolimnion

The whole water column was aerobic and almost homogeneous during the overturn periods, but conspicuous differences in water quality between epilimnion and hypolimnion were noticeable once permanent stratification was established. Table 21.4 shows that the hypolimnion had, on average, waters with higher alkalinity, electrical

Table 21.3. Quality of the effluents entering the reservoir from the stabilization pond

Parameter	Units	Mean	Min	Max	N° of samples
COD	mg l^{-1}	260	185	320	11
BOD	mg l^{-1}	80	40	110	11
TSS	mg l^{-1}	270	60	600	15
MBAS	µg l^{-1}	1350	300	2500	15
Alkalinity	mg l^{-1}	700	550	850	16
Turbidity	J.U.	60	30	250	16
UV absorption		0.97	0.55	1.39	16
Chlorophyll a	µg l^{-1}	959	0	2500	14
N-NH$_4^+$	mg l^{-1}	45	20	60	16
N-NO$_2$	mg l^{-1}	0	0	0	12

COD: Chemical Oxygen Demand; BOD: Biological Oxygen Demand, TSS: Total Suspended Solids; MBAS: Methyl Blue Active Substances (anionic surfactants); UV: Ultraviolet absorption at 254 nm minus absorption at 545 nm, utilized as an indicator of the concentration of dissolved organic matter in water (Michail and Idelovich 1980).

conductivity, detergents, sulfides, and ammonium. The epilimnion was richer in dissolved oxygen and chlorophyll a, and had a higher REDOX potential.

The changes in water quality during the year were conspicuously different in the epilimnion and the hypolimnion. The two layers behaved like entirely different water bodies. Thus, a separate analysis of the changes in water quality was made for each layer.

21.2.3.3
Factor Analysis of Epilimnion Data

The analysis identified three factors which together account for about 70% of the overall variability of the epilimnion water quality data (Table 21.5).

FACTOR 1 represents external loading to the epilimnion. It explains 33% of the overall variability of epilimnion water quality data. It is bipolar, contrasting two groups of variables: one pole includes turbidity, alkalinity, UV absorption (dissolved organic matter), TSS (total suspended solids, most of them organic matter in this reservoir), MBAS (detergents) and chlorophyll a, and the other pole includes pH and nitrite. When one of the poles (groups of variables) is high the other is low and vice versa. The FACTOR follows an annual cycle with peaks in fall (Fig. 21.8). There are no conspicuous differences between sunrise and sunset. A correlation analysis between FACTOR 1 and several operation and climatic variables indicated that FACTOR 1 is mainly correlated with the operation of the reservoir, represented by the percentage of fresh effluents (PFE), water level and the area/volume ratio (Fig. 21.8). The FACTOR was not correlated with climatic parameters (water temperature, day length and solar radiation). Figure 21.9 presents the annual cycle of four of the variables included in FACTOR 1. It can be seen that the operation of the reservoir determined a deterioration of water

Table 21.4. Differences in water quality between epilimnion and hypolimnion

Parameter	Layer	Mean	Min	Max	N° of samples
EC (μmhos cm^{-1})	EPI	1746	1487	1992	25
	HYPO	1892	1490	2411	26
Secchi Disk (cm)	EPI	23	10	195	25
	HYPO				
S$^-$ (mg l^{-1})	EPI	0.00	0.00	0.00	22
	HYPO	0.83	0.00	5.28	23
MBAS (μg l^{-1})	EPI	474	213	744	23
	HYPO	613	299	1232	24
N-NH$_4^+$ (mg l^{-1})	EPI	20.9	8.3	37.0	25
	HYPO	33.6	9.1	58.4	26
Alkalinity (mg l^{-1})	EPI	578	500	687	25
	HYPO	687	504	900	26
UV absorption	EPI	0.42	0.26	0.79	25
	HYPO	0.62	0.34	1.03	26
Turbidity (J.U.)	EPI	24.1	3.0	82.0	25
	HYPO	24.5	7.3	54.0	26
TSS (mg l^{-1})	EPI	50	2	330	24
	HYPO	51	10	345	25
Chlorophyll a (μmg l^{-1})	EPI	164	0	1307	24
	HYPO	140	0	1195	25
DO (mg l^{-1})	EPI	9.6	0	25	24
	HYPO	0.8	0	6	25
REDOX (E mV)	EPI	−65	−270	71	25
	HYPO	−171	−350	34	26
N-NO$_2^-$ (mg l^{-1})	EPI	1.1	0	4.3	22
	HYPO	0.1	0	0.7	23
pH	EPI	8.0	7.0	8.7	24
	HYPO	7.3	6.5	8.2	25

quality by the end of the irrigation season: UV absorption and turbidity were high in fall when the reservoir was almost empty and the inflow was poorly diluted in the small water volume of the reservoir (high *PFE* in Fig. 21.8). In contrast, these two parameters were low in spring when the reservoir was full and the inflow was diluted in a large volume of good quality effluents with a long retention time (low *PFE*). The other parameters of the positive pole (*MBAS*, alkalinity, *TSS* and chlorophyll) followed a similar trend. pH followed a mirror cycle (it belongs to the negative pole of the FACTOR), with low values when loading to the reservoir was high and the dominant process was degradation of organic matter, and higher values when loading was low and the concentration of organic matter within the reservoir decreased. Transformation of NH_4^+-N to NO_2-N was possible only in spring-summer with relatively clean waters and high oxygen concentration, and when epilimnion water temperature started rising (Fig. 21.10). In July the concentration of dissolved oxygen became subject to strong fluctuations due to increased loading and photosynthetic activity, with almost anoxic

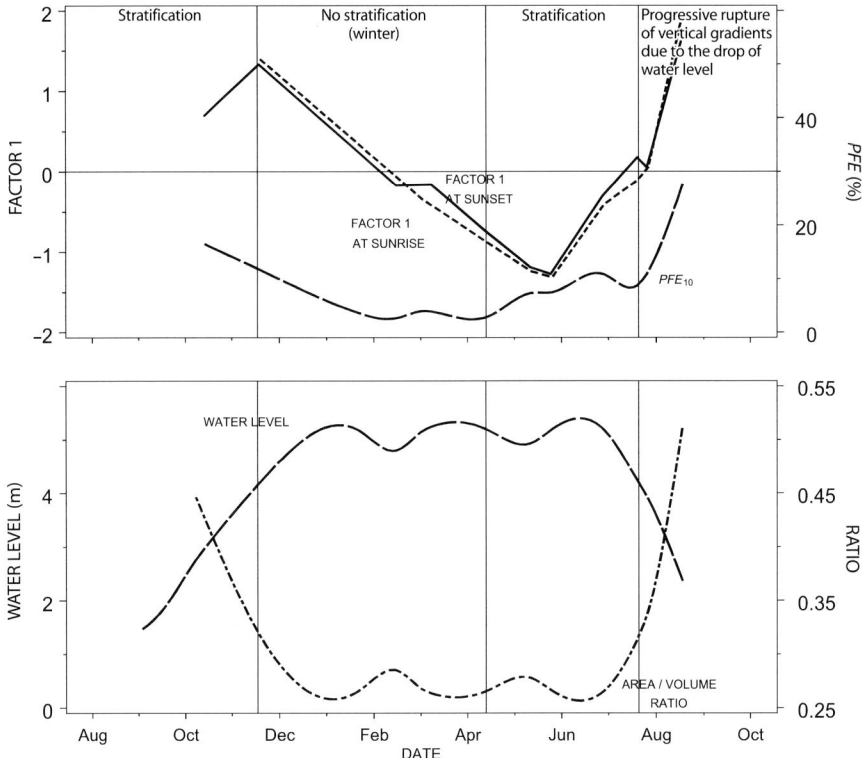

Fig. 21.8. Epilimnion. FACTOR 1, PFE_{10} (percentage of fresh effluents with 10 or less days within the whole reservoir: epi- + hypolimnion), water level of the reservoir, and ratio area/volume of the reservoir, vs. date

conditions at sunrise and oversaturation at sunset (Fig. 21.11), and nitrification was stopped. FACTOR 1 indicates that the main source of variability of water quality in the epilimnion was external loading due to the operation of the reservoir. The relationship between inflow amounts and the volume of effluents within the reservoir resulted in high loadings by the end of the irrigation season in fall, decreasing during the winter filling period and reaching a minimum by spring-summer when the reservoir was full.

FACTOR 2 represents the seasonal climatic changes affecting the photosynthesis/respiration relationship. It explains a further 20% of the overall variability of epilimnion water quality data, and it is also bipolar (Table 21.5). FACTOR 2 contrasts electrical conductivity (*EC*), *TSS*, chlorophyll a, *p*H and dissolved oxygen, with UV, alkalinity and ammonium. The FACTOR follows an annual cycle with peaks in July–August (Fig. 21.10). In the first section of this chapter, it was concluded that wastewater reservoirs in Israel undergo three partially overlapped cycles: a solar radiation one with peak in June, a water temperature one with peak in August, and an operational one with peak in late September. A correlation analysis between FACTOR 2 and several operational and climatic variables indicated that FACTOR 2 is mainly correlated to the climatic variables such as water temperature, day length (number of light hours),

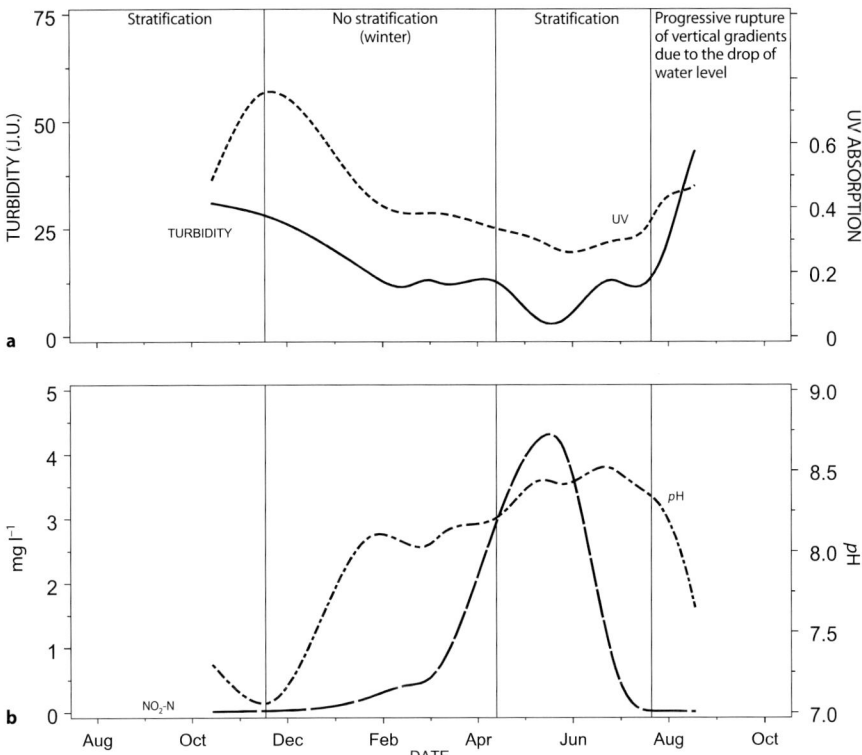

Fig. 21.9. Epilimnion. Four of the variables included in FACTOR 1 vs. date. **a** Positive pole variables. **b** Negative pole variables. The *lines* represent the mean of sunrise and sunset values

and solar radiation. The FACTOR was not correlated with variables describing the operation of the reservoir. The FACTOR has lower values at sunrise than at sunset. Figure 21.9 shows that FACTOR 2 at sunrise (when its values are the result of nocturnal processes, mainly respiration) is more related to water temperature, while at sunset (when its values are the result of day processes, mainly photosynthesis) it is more related to day length. The concentration of *TSS* in the epilimnion of hypertrophic environments is closely related to algae biomass (herein measured as chlorophyll a) and the two parameters are closely related in FACTOR 2.

High values of *p*H and dissolved oxygen are the result of the photosynthetic activity of this algae biomass. On the contrary, electrical conductivity seems to have no relationship with the photosynthetic activity; *EC* appears in the FACTOR because the same parameters which affect photosynthesis and respiration (water temperature, day length and solar radiation) also affect evaporation and the consequent increase in the concentration of salts in the epilimnion. UV, alkalinity and ammonium decrease with photosynthetic activity due to algae uptake. Figure 21.11 shows the annual cycle of two of the variables included in FACTOR 2 (and not included in FACTOR 1). NH_4^+-N concentration increased during the winter with cold waters due to the low uptake by al-

Table 21.5. Epilimnion. Factor analysis of physico-chemical data. Only large coefficients included in the table

Factor	Factor 1 External loading (operational changes)	Factor 2 Photosynthesis/respiration (seasonal changes)	Factor 3 Oxido/reduction (day/night changes)
Explained Variance (%)	33	20	15
Accumulative (%)	33	53	68
Parameters[a]			
EC	–	0.70	–
Secchi disk	–	–	0.42
MBAS	0.58	–	0.47
$N-NH_4^+$	–	–0.37	0.54
Alkalinity	0.78	–0.42	–
UV absorption	0.70	–0.46	–
Turbidity	0.92	–	–
TSS	0.68	0.64	–
Chlorophyll	0.54	0.54	–
DO	–	0.55	–0.51
REDOX	–	–	–0.76
$N-NO_2^-$	–0.77	–	–
pH	–0.69	0.61	–

[a] S^- not included in the analysis because it was always zero in the epilimnion.

gae and lack of nitrification. In spring, the rise in water temperature and solar radiation resulted in higher uptake by algae and the transformation to NO_2 (Fig. 21.9). The concentration of dissolved oxygen increased in spring (at both sunrise and sunset) due to a combination of clear waters with almost no oxygen demand and low concentration of algae due to overgrazing by Cladocera (*Daphnia longispina*), which determined almost no oxygen production during the day and almost no respiration demand during the night. Secchi disk, which had an annual mean value of 0.5 m, reached almost 2 m in spring. Clean waters due to *Daphnia* grazing in spring have been described in other wastewater reservoirs in Israel (Pano 1975; Eren 1978) and in other hypertrophic impoundments in many places (Uhlmann 1980a, 1980b). There was a conspicuous increase in algae biomass from July on. At sunrise, dissolved oxygen dropped to almost zero, due to a combination of algae respiration during the night and increased *BOD* resulting from high water temperatures. But at sunset high concentrations of dissolved oxygen were measured due to intensive photosynthetic activity during the day hours. FACTOR 2 indicates that the second main source of variability of water quality in the epilimnion was the seasonal cycle of climatic parameters (changes in water temperature and hours of light during the year), affecting the relationship between photosynthesis and respiration processes.

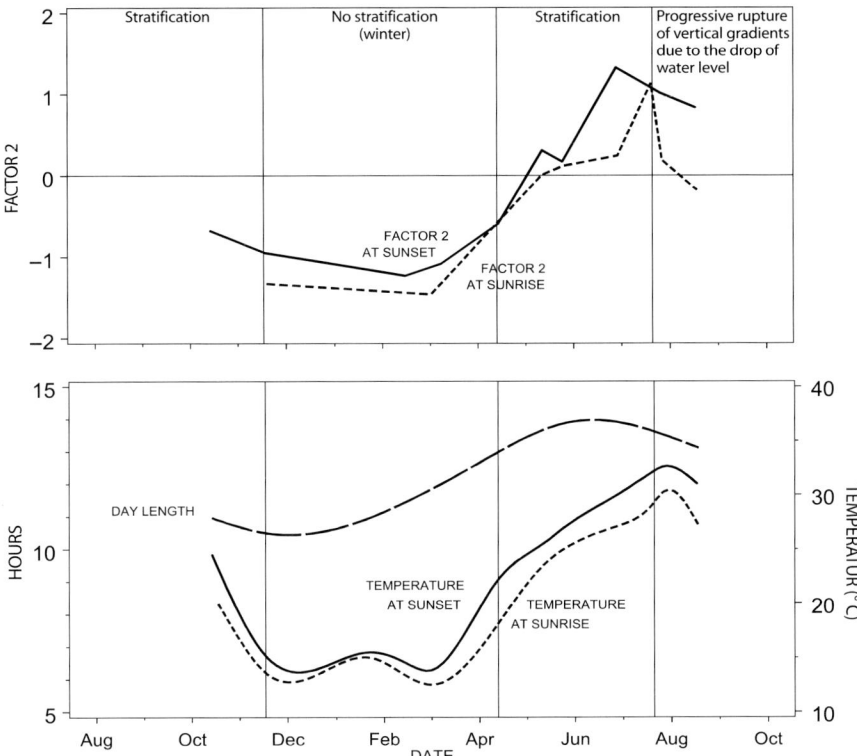

Fig. 21.10. Epilimnion. FACTOR 2, day length and water temperature vs. date

FACTOR 3 represents the differences day/night affecting the oxido-reduction processes. It explains a further 15% of the overall variability of epilimnion water quality data, and it is also bipolar (Table 21.5). It contrasts REDOX and dissolved oxygen, with NH_4^+-N, *MBAS* and Secchi disk. A correlation analysis between FACTOR 3 and several operational and climatic variables indicated that it is correlated only with sampling hour (sunrise or sunset). The FACTOR has different values at sunrise and sunset (Fig. 21.12), which cross at the equinoxes (late December and late June). Sunrise values are conspicuously lower than sunset ones in spring. Spring epilimnion was characterized by clean waters with very low concentration of algae due to both low organic loading and *Daphnia* grazing. Oxygen demand by respiration and degradation of organic matter was low during the night, while low water temperatures determined a good solubility of atmospheric oxygen in water. Thus, REDOX values were relatively high at sunrise. During the day production of dissolved oxygen by photosynthesis was negligible, while the increase of surface water temperature during the day hours in spring (see Section 21.1) determined a poorer solubility of atmospheric oxygen. The conditions were inverted during summer and fall. Higher organic loadings, together with higher solar radiation and water temperatures, determined conspicuous photosynthesis and respiration rates. Intensive night respiration resulted in low REDOX

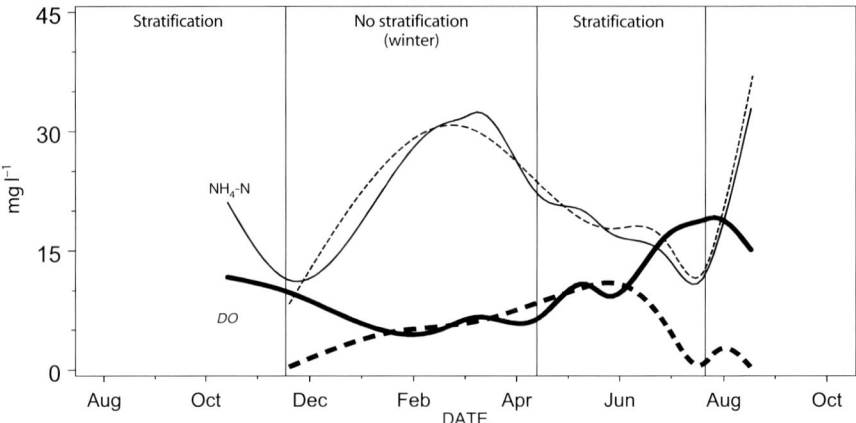

Fig. 21.11. Epilimnion. Two of the variables included in FACTOR 2 vs. date. *Wide lines:* dissolved oxygen (positive pole variable); *Narrow lines:* NH$_4^+$ -N negative pole variable); *Dashed lines:* sunrise; *Continuous lines:* sunset

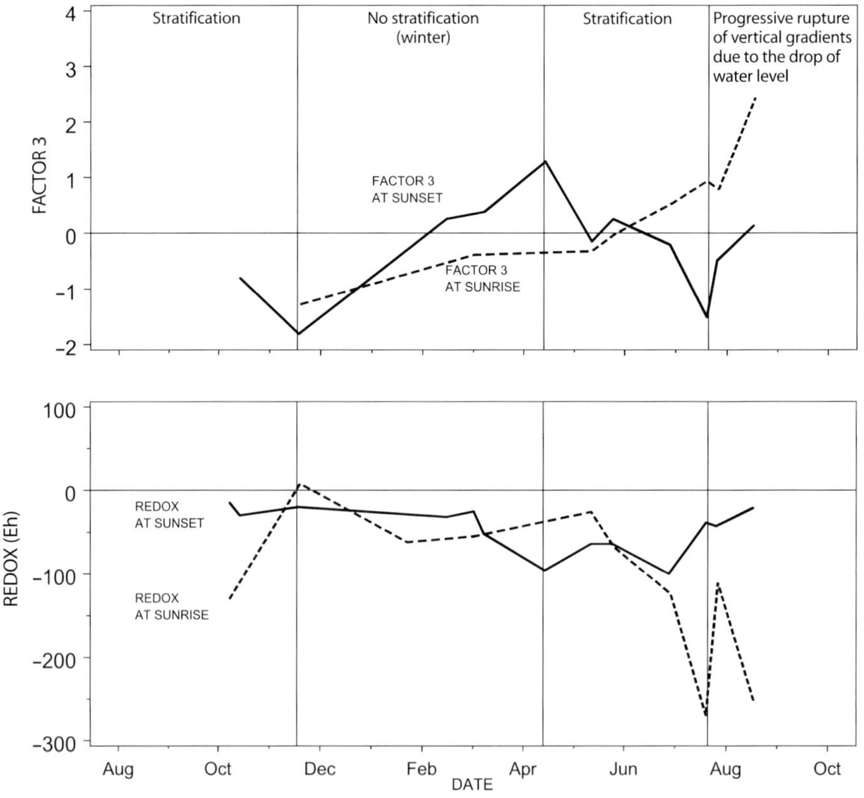

Fig. 21.12. Epilimnion. FACTOR 3 and REDOX vs. date

values at sunrise. Intensive photosynthetic production of dissolved oxygen during the day resulted in high REDOX values (Fig. 21.12) and oversaturation of dissolved oxygen (Fig. 21.11) at sunset. FACTOR 3 indicates that the third main source of variability of water quality in the epilimnion is the difference night/day affecting the REDOX processes.

21.2.3.4
Factor Analysis of Hypolimnion Data

The analysis identified two factors which together account for about 70% of the overall variability of the hypolimnion water quality data (Table 21.6).

FACTOR 1 represents organic loading like in the epilimnion, but due to both external and internal sources. It explains 54% of the overall variability of hypolimnion water quality data, and it is bipolar. It is very similar to FACTOR 1 of the epilimnion, grouping almost the same variables within its two poles. It also follows an annual cycle (Fig. 21.13), but the correlation analysis indicated that it is correlated to both external loading to the reservoir (*PFE* in Fig. 21.8), and hypolimnion water temperature. The hypolimnion receives, besides the external loading related to the input of effluents from the stabilization pond, an internal loading due to the release of several kinds of compounds from bottom sediments (soluble organic carbon, soluble organic nitrogen,

Table 21.6. Hypolimnion. Factor analysis of physico-chemical data. Only large coefficients included in the table

Factor	Factor 1 External and internal loading (operational and seasonal changes)	Factor 2 Inputs of chlorophyll into the hypolimnion
Explained Variance (%)	54	20
Accumulative (%)	54	74
Parameters[a]		
pH	−0.95	−
REDOX	−0.75	−
DO	−0.72	0.50
$N-NO_2^-$	−0.82	0.50
EC	0.93	−
MBAS	0.79	−
Alkalinity	0.92	−
$N-NH_4^+$	0.70	0.67
S^-	−	−
UV absorption	0.78	−
Turbidity	0.55	−
TSS	0.74	−0.41
Chlorophyll	−	−0.95

[a] Secchi disk not included in the analysis of hypolimnion data.

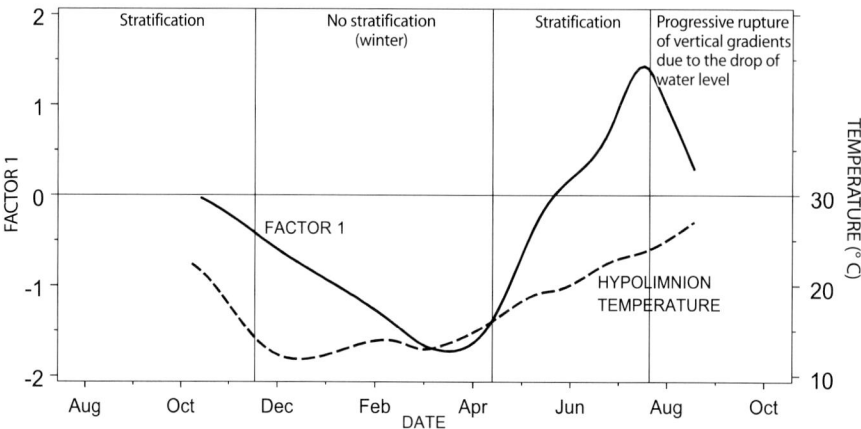

Fig. 21.13. Hypolimnion. FACTOR 1 and water temperature vs. date

ammonium, ortho-P, etc.). The release of these compounds from the sediments depends on the amount of sediments and the intensity of the anaerobic digestion (Avnimelech et al. 1983), the latter being a function of the temperature of the hypolimnion water (Fig. 21.13). During the winter months, external loading was low due to the small inputs of wastewater, and the dilution of these small inputs into a large volume of old effluents within the almost full reservoir. Internal loading was also low in winter because water temperatures at bottom were below 15° C and anaerobic digestion of sediments was minimal. The lack of stratification in winter also reduced the internal loading of the hypolimnion because good vertical mixing diluted the materials released by the sediments in the whole water column. During the summer months, external loading increased in the whole reservoir (increase in *PFE*, Fig. 21.8). The release of compounds from the sediments also increased due to the rise of water temperature at the bottom and consequent higher anaerobic digestion of sediments. Besides, the materials released by the sediments were diluted only within the hypolimnion layer during the summer, due to the establishment of permanent stratification. The importance of stratification in raising the internal loading can be seen in Fig. 21.13: the breakdown of stratification in fall resulted in lower values of FACTOR 1 in spite of the rising values of *PFE* (Fig. 21.8). Figure 21.14 presents the annual cycle of four of the variables related to FACTOR 1 in the hypolimnion. It can be seen that relatively high pH and NO_2 values were typical of the winter months when the whole water column was mixed, but these values dropped very quickly with the establishment of stratification. The concentration of metabolites released by the anaerobic processes in the hypolimnion also increased the alkalinity (and related *EC*) (Fig. 21.14). FACTOR 1 indicates that the main source of variability of water quality parameters in the hypolimnion was loading, due to both external and internal sources.

FACTOR 2 represents the inputs of chlorophyll into the hypolimnion. It explains a further 20% of the overall variability of hypolimnion water quality data. It is bipolar, contrasting chlorophyll and *TSS*, with NH_4^+-N, NO_2 and dissolved oxygen (Table 21.6). FACTOR 2 has peaks in April and August, with minimum values in June and in the fall

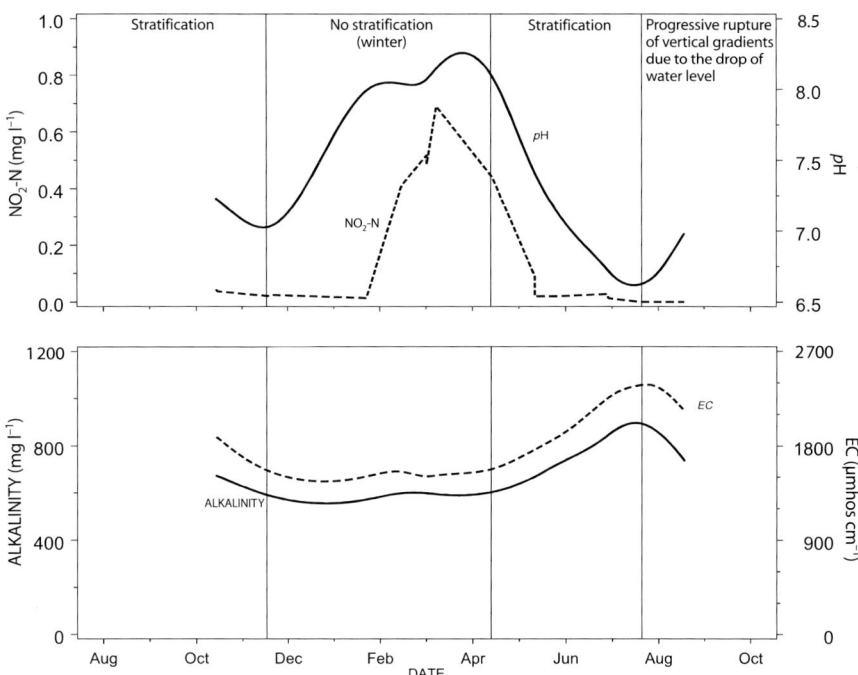

Fig. 21.14. Hypolimnion. Four variables related to FACTOR 1 vs. date

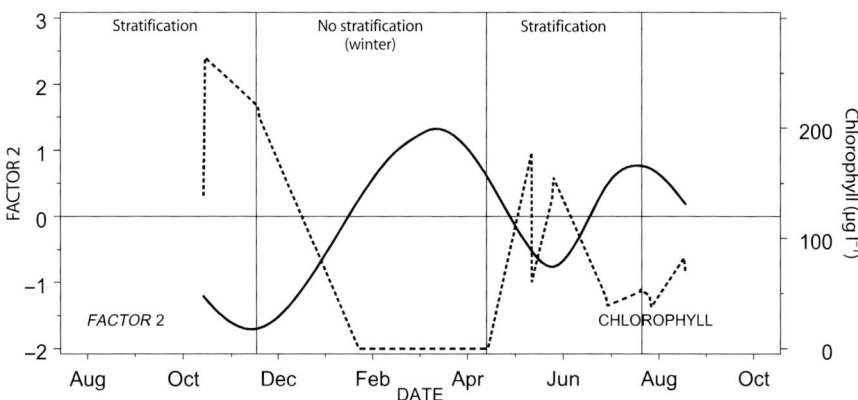

Fig. 21.15. Hypolimnion. FACTOR 2 and chlorophyll a vs. date

(Fig. 21.15). The peaks are determined mainly by the changes in the concentration of chlorophyll a (a coefficient of 0.95, Table 21.6) as can be seen in Fig. 21.15. The high concentration of chlorophyll in fall can be explained by the lack of stratification and low water level in this time of the year, when the mixing of the water column brought

algae-rich epilimnion waters close to the bottom. The peaks of hypolimnion chlorophyll during the summer, in a stratified and almost full reservoir, can be explained by
the entry of chlorophyll rich effluents from the stabilization pond directly into the
hypolimnion. New inflow was pumped into the reservoir when the irrigation season
(and the withdrawal of effluents from the reservoir) started. The reservoir was almost
full at this time of the year, hypolimnion was high (2 m above bottom), and thus the
inlet discharged the chlorophyll rich inflow from the stabilization pond into the hypolimnion. FACTOR 2 indicates that the second main source of variability of quality
data in hypolimnion is the entry of algae in this layer. This would be due to changes
in the relative depth of inlet discharge, and also to the rupture of stratification and
consequent mixing of the water column.

21.2.3.5
Control of Water Quality in Hypertrophic Impoundments

The analysis herein performed points out that the two main phenomena determining
water quality in this kind of reservoirs are: 1) stratification, and 2) the operation of
the reservoir. These two phenomena can be controlled in order to improve water quality. The effects of stratification can be overcome partially by withdrawing the effluents only from the epilimnion of the reservoir (a common practice in Israel), but it is
possible to go beyond this simple approach: stratification can be broken down by several methods which are known to have beneficial effects on water quality in lakes and
reservoirs (Brown 1983; Krambeck 1986; Stauffer 1987; Chang and Ouyang 1988; Szyper
and Lin 1990; Hofmann and Krambeck 1991; Llorens et al. 1992; Liran et al. 1994). The
depth of the inlet is another controllable parameter. The operation of the reservoir
can be changed by cutting off the entry of new effluents to the reservoir during the
irrigation season, deriving the effluents to another reservoir for summer storage.

The third phenomenon in importance is the climatic seasonal changes affecting the
photosynthesis/respiration relationship in the epilimnion and the internal loading in
the hypolimnion. Most of this source of variability is out of control. However, artificial
control of stratification would allow an earlier increase of temperature in bottom waters, thus spreading the anaerobic digestion of the sediments from spring to fall, in contrast with the present situation which delays the beginning of digestion to late summer.

The fourth important phenomenon affecting water quality within the reservoir (the
differences day/night affecting REDOX processes) can also be partially controlled in some
cases, by withdrawing effluents from the reservoir at the hours when water quality is better.

The effect of seasonal climatic changes on the water quality in the studied reservoir was small compared with the effects of controllable phenomena (wastewater inputs, stratification, etc.). The reservoir is located at a relatively low latitude ($33°N$), and
it may be assumed that in higher latitudes the effect of seasonal climatic changes will
be stronger. This may be not so. Dor et al. (1987), in a study of the limnology of another wastewater reservoir in Israel, concluded that hypertrophic conditions lead to
the development of a close relationship between algae and bacteria combining competition, mutual suppression, and mutual support. These authors think that under these
conditions, the sensitivity of the community to environmental changes declines.
Uhlmann (1980b) states that the ability to adjust to strong perturbations (such as
changes in operational parameters) is much higher (faster adjustment) in hypertrophic

systems than in highly diversified oligotrophic ones. He maintains that hypertrophic systems drift relatively slowly with the seasonal variations. This author proposes that the high biomass and high metabolic rates characteristics of hypertrophic systems are responsible for this phenomenon. The findings of the present work on the Geta'ot wastewater reservoir support these earlier hypotheses, by confirming that the parameters which affect the trophic characteristics of the system (external and internal loading, stratification) are more important in determining the behaviour of the system than the seasonal climatic changes. In a recent work on two hypertrophic reservoirs for fish culture in Israel, Milstein et al. (1994a,b) also found that variability of water quality is more affected by operational parameters (fish stocking and water level) than by the seasonal climatic changes. All these considerations suggest that the control of water quality in hypertrophic impoundments may be possible even at higher latitudes.

21.3
Conclusions

The variability of water quality in reservoirs is the result of several different and interrelated processes which occur simultaneously in time and space. Factor analysis of a complex data matrix obtained by sampling at different hours and depths during the year allowed the identification of the main processes responsible for water quality variability in the studied warm hypertrophic wastewater reservoir.

Stratification led to the formation of an aerobic epilimnion and an anaerobic hypolimnion. The two layers behaved as different water bodies.

The main source of water quality variability in the epilimnion was external loading due to changes in the operation of the reservoir (wastewater inputs related to reservoir volume). The second main source was the seasonal climatic changes affecting the photosynthesis/respiration relationship. The third source in importance was the differences between night and day processes affecting the REDOX potential.

The main source of water quality variability in the hypolimnion was also loading, but in this case it was due to both external loading (wastewater inputs) and internal loading (release of compounds from the sediments). The second main source of variability was related to inputs of chlorophyll into the hypolimnion, which may be external (inlet discharges of chlorophyll rich inflow into the hypolimnion) or internal (stratification break down which brings chlorophyll from the upper layers to the hypolimnion).

The effect on water quality, of phenomena which are out of control (seasonal climatic changes) was small compared with the effect of controllable phenomena (wastewater inputs, stratification, etc.). This indicates that from the limnological point of view, the control of water quality in hypertrophic impoundments is a feasible technology.

Acknowledgements

Mr. Zvi Sarig from Kibbutz L. HaGeta'ot provided essential operational information and logistic support. This research was supported by a grant from the National Council for Research and Development, Israel, and the Ministry for Science and Technology, Germany. This article was previously published by Juanicó (1994, 1995) in the journal Int. Revue ges. Hydrobiol. 79(3):423-436 and 80(3):415-428 respectively, and it is herein reproduced with permission from Wiley-VCH Verlag, Berlin.

References

Abeliovich A (1982) Biological equilibrium in a wastewater reservoir. Wat Res 16(7):1135–1138

APHA (1989) Standard methods for the examination of water and wastewater. American Publ Health Ass, 17th edn

Argaman Y, Redlich E, Juanicó M, Rom D (1988) Distribution of residence time in the reservoirs of the Kishon Complex. In: The Kishon Complex monitoring program, Fifth Annual Rep, pp 111–141, Technion (in Hebrew)

Avnimelech Y, Yamamoto M, Menzel R (1983) Evaluating the release of soluble components from sediment. J Environ Qual 12(1):86–91

Bokil S, Agrawal G (1977) Stratification in laboratory simulations of shallow stabilization ponds. Wat Res 11(12):1025–1030

Brown I (1983) Destratification of water storage reservoirs to assist water treatment. Water Supply, 1 (2–3, World Water Supply) 11:8–15

Chang W, Ouyang H (1988) Dynamics of dissolved oxygen and vertical circulation in fish ponds. Aquaculture 74:263–276

Dor I, Raber M (1990) Deep wastewater reservoirs in Israel: Empirical data for monitoring and control. Wat Res 24(9):1077–1084

Dor I, Schechter H, Bromley H (1987) Limnology of a hypertrophic reservoir storing wastewater effluent for agriculture at Kibbutz Na'an, Israel. Hydrobiologia 150:225–241

Eren J (1978) Succession of phyto- and zooplankton in a wastewater storage reservoir. Verh Internat Verein Limnol 20:1926–1929

Felgner G, Sandring G (1983) Abwasserspeicherung – ein Weg zur Sicherung der ganzjährigen Abwasserverwertung und -reinigung. Wasserwirt Wassertech 9:321–323

Fry J (1987) Functional roles of the major groups of bacteria associated with detritus. In: Moriarty D, Pullin R (eds) Detritus and microbial ecology in aquaculture. ICLARM Conference Proc 14:420 pp, Manila

Hofmann W, Krambeck H (1991) Manipulation of the thermocline in a eutrophic lake: The response of the zooplankton. Verh Internat Verein Limnol 24:786–790

Ibanez F (1973) Choix d'une strategie d'echantillonnage par la theorie des jeux pour l'etude in situ d'organismes planctoniques. Union des Oceanographes de France 5(3):18–25

Juanicó M (1994) Limnology of a warm hypertrophic wastewater reservoir in Israel. I. The physical environment. Int Reveue ges Hydrobiol 79(3):423–436

Juanicó M (1995) Limnology of a warm hypertrophic wastewater reservoir in Israel. II. Changes in water quality. Int Reveue ges Hydrobiol 80(3):415–428

Juanicó M, Friedler E (1994) Hydraulic age distribution in perfectly mixed non steady-state reactors. ASCE J Environ Eng 120(6):1427–1445

Krambeck H (1986) Zur hypolimnischen Belüftung von Gewässern mit der "Ökopumpe Ploen". GWF-Wasser/Abwasser 127(3):137–140

Levenspiel O (1972) Chemical reaction engineering. 2nd edn, John Wiley & Sons, 578 pp

Liran A, Juanicó M, Shelef G (1994) Bacteria removal in a stabilization reservoir for wastewater irrigation in Israel. Wat Res 28(6):1305–1314

Llorens M, Saez J, Soler A (1992) Influence of thermal stratification on the behaviour of a deep wastewater stabilization pond. Wat Res 26(5):569–577

Michail M, Idelovitch E (1980) Monitoring of gross organics in water and wastewater. Report Tahal, Water planning for Israel, Tel Aviv, 45 pp

Milstein A (1984) Sampling strategy in a very variable environment. Crustaceana, Suppl 7:336–343

Milstein A (1993) Factor and canonical correlation analyses: Basic concepts, data requirements and recommended procedures. In: Prein M, Hulata G, Pauli D (eds) Multivariate methods in aquaculture research: Case studies of tilapias in experimental and commercial systems. ICLARM Studies and Reviews 20:24–31

Milstein A, Krambeck H, Zoran M (1992) Effects of wind and depth on stratification in reservoirs for fish culture and field irrigation. Limnologica 22(4):375–384

Milstein A, Zoran M, Krambeck H (1993a) Water quality variability in a shallow (4 m) reservoir for simultaneous fish farming and field irrigation. Limnologica 23(4):392–402

Milstein A, Zoran M, Barsadschi D, Krambeck H (1993b) Water quality variability in a deep (8 m) reservoir for simultaneous fish farming and field irrigation. Limnologica 23(4):403–413

Milstein A, Zoran M, Krambeck H (1994a) Water quality variability in a shallow (4 m) reservoir for simultaneous fish farming and field irrigation. Limnologica 24(1):71–81

Milstein A, Zoran M, Barsadschi D, Krambeck H (1994b) Water quality variability in a deep (8 m) reservoir for simultaneous fish farming and field irrigation. Limnologica 24(1):82–92

Pano A (1975) Storage of wastewater and floodwater in Sarid and Mizra Reservoirs. Report by Tahal Co., Tel Aviv, (in Hebrew), 26 pp

Patil H, Dodakundi G, Rodgi S (1973) Succession and stratification of organisms in sewage in an oxidation pond. Indian J Exp Biol 11(4):318–320

SAS (1985) Statistical analysis system, version 5, reference manuals, SAS Inst., USA

Seal H (1964) Multivariate statistical analysis for biologists. Methuen, London, 160 pp

Shelef G, Juanicó M (1988) Establishing design and operation criteria of wastewater effluent seasonal reservoirs. Technion, Final Report, vol II – Database, 172 pp (in Hebrew)

Soler A, Saez J, Llorens M, Martinez I, Torrella F, Berna L (1991) Changes in physico-chemical parameters and photosynthetic microorganisms in a deep wastewater self-depuration lagoon. Wat Res 25(6): 689–695

Stauffer G (1987) Effects of oxygen transport on the areal hypolimnetic oxygen deficit. Wat Resour Res 23(10):1887–1892

Stumm W, Morgan J (1981) Aquatic chemistry, John Wiley & Sons Publ, 2nd edn, 780 pp

Szyper J, Lin C (1990) Techniques for the assessment of stratification and effects of mechanical mixing in tropical fish ponds. Aquaculture Eng 9:151–165

Torres J, Soler A, Saez J, Ortuño J (1997) Hydraulic performance of a deep wastewater stabilization pond. Wat Res 31(4):679–688

Uhlmann D (1980a) Limnology and performance of waste treatment lagoons. Hydrobiol 72:21–30

Uhlmann D (1980b) Stability and multiple steady states of hypereutrophic ecosystems. In: Barica J Mur L (eds) Develop Hydrobiol 2:235–247

Uhlmann D (1990) Anthropogenic perturbation of ecological systems: a need for transfer from principles to applications. In: Ravera O (ed) Terrestrial and aquatic ecosystems; perturbation and recovery. E. Horwood Publ, pp 47–61

Chapter 22

The Ma'aleh HaKishon Reservoir

Jacob Eren

22.1
Introduction

The Ma'aleh HaKishon Reservoir (MKR), with a storage capacity of 12.5 million m³, is the largest wastewater reservoir in Israel. Its construction was completed in 1984 and it started to receive wastewater in the spring of that year. The MKR is part of the Kishon wastewater reclamation project which conveys secondary treated sewage effluents from the sewage treatment plant of Metropolitan Haifa to the Jeezrael Valley. The Kishon wastewater reclamation project adds to the valley 15–20 million m³ water per year, which enables the irrigation of 3 000–4 000 ha. Cotton is the most common crop in this area, but when cotton prices are low some other crops are grown. The scheme of the Kishon wastewater reclamation project is presented in Fig. 22.1.

The Kishon wastewater reclamation project includes an extensive monitoring programme, consisting of chemical, biological, and microbial examinations along the system, from the Haifa sewage treatment plant up to the farmers' irrigation systems. The limnology of the MKR was investigated as part of the monitoring programme. The reservoir was also studied in a framework of several research projects addressing different aspects of the behaviour of this type of hypertrophic water body; part of this work is presented in the following pages.

Fig. 22.1. The Kishon irrigation project

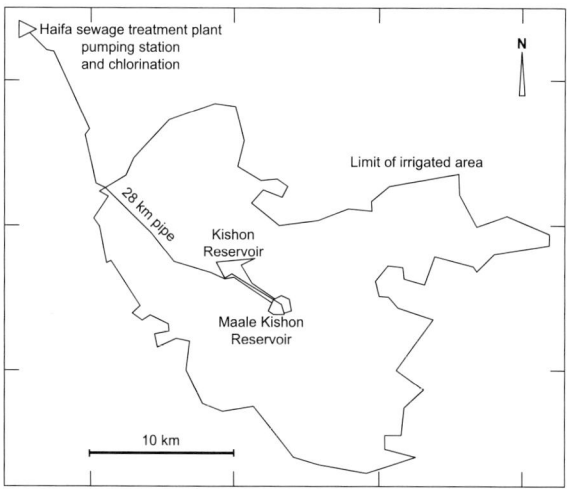

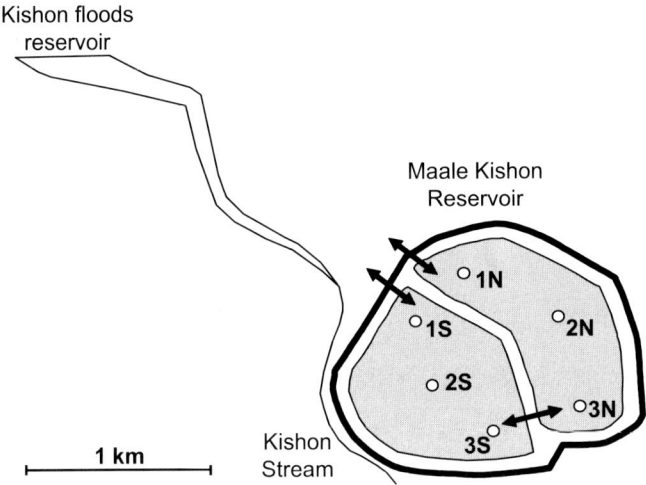

Kishon floods
reservoir

Maale Kishon
Reservoir

1N

1S

2N

2S

3N

1 km

Kishon
Stream

3S

Fig. 22.2. The Ma'aleh HaKishon Reservoir

22.2
Description of the Reservoir

The reservoir is located in the central part of the Jeezrael Valley in northern Israel next to the Kishon stream, and upstream from the Kishon floodwater reservoir which was constructed in 1953. The area of MKR is 130 ha and its volume is 12.5 millions m³. The maximal depth is 11 m and the average depth is 9.5 m.

The reservoir is divided into two approximately equal parts (see Fig. 22.2) to provide a longer storage period and to enable cleaning the reservoir bottom whenever it is required. Each part has its own inlet and outlet and can be operated separately. The normal operation regime is for the water to enter one basin, then flow to the second basin, and from there to the irrigation system. In the past, the water from MKR was supplied for irrigation through the Kishon floodwater reservoir, but since 1994 it can be pumped directly. The walls of the reservoir are constructed from local clay. The bottom is made of compacted clay and it is almost impermeable.

22.3
Operation of the Reservoir

During the first 11 years of operation (1984–1995), the reservoir supplied 114 million m³ of reclaimed water for irrigation. On the average, about 12 million m³ were supplied annually, except for the years 1992 and 1993, which were unusually rainy and there was no demand for irrigation water.

The irrigation season in the Jeezrael Valley runs from April till October. Cotton, the most common crop cultivated there, is irrigated during only three months from June to August. Therefore, the operational regime of the reservoir is divided in two periods: a filling period (only input of secondary treated wastewater into the reser-

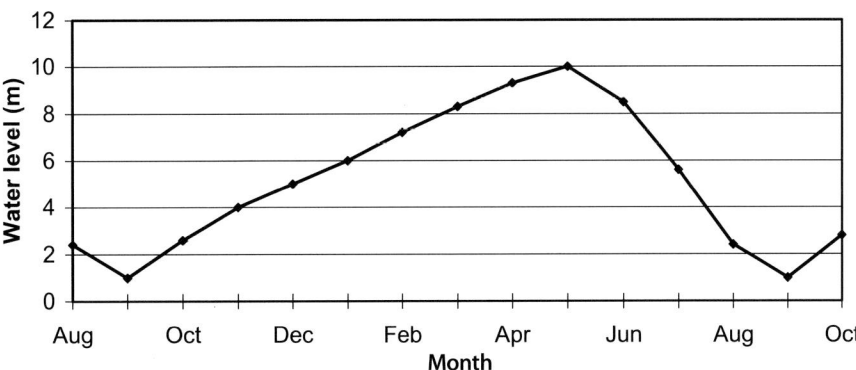

Fig. 22.3. The annual fluctuation of water level in the Ma'aleh HaKishon Reservoir

voir) during fall, winter and early spring, and a withdrawal period (both input and strong output for irrigation) during the summer. The reservoir is full in May and almost empty by the end of the irrigation season. The annual cycle of the water level is presented in Fig. 22.3.

22.4
The Biological Cycle in the Reservoir

The Ma'aleh HaKishon Reservoir receives secondary treated wastewater which contains high concentrations of nutrients such as nitrogen and phosphorus. Under such conditions and with plenty of light, there is massive development of algae during the fall, when the reservoir is at low water level. At this stage the reservoir functions as a stabilization pond. The most common phytoplankton species present at this stage are those frequently found in highly eutrophied water bodies: *Euglena*, *Phacus* and *Chlamydomonas*. As the water level rises in early winter, there is a gradual change in the algal population, and *Chlorella* and other green algae become the dominant phytoplankters. Simultaneously, but mainly in early spring, the development of algivorous zooplankton, mostly *Rotifera* and *Cladocera*, starts. The grazing activity of the zooplankton causes a decline in the density of the algae and so, in the early summer, the reservoir is at its lowest turbidity and concentration of suspended solids. As the water is withdrawn for irrigation during the summer, additional changes occur in the composition of the biomass. After depletion of the algae, there is a decrease in the zooplankton population, followed by new development of algae, this time mostly Cyanobacteria.

22.5
Removal of Organic Material

Stabilization reservoirs serve as an additional secondary treatment system and partially as a tertiary treatment system. When properly operated, the degradation of organic material continues in the reservoir mostly by heterotrophic bacteria. As in stabilization ponds, the oxygen required for this process is supplied by the photosynthetic activity of algae. In

Table 22.1. Removal of *BOD* in
the Ma'aleh HaKishon Reservoir
(average values from 1985–1995)

	BOD
Inlet concentration (mg l^{-1})	33.1
Outlet concentration (mg l^{-1})	7.5
Removal (%)	77

Table 22.2. Removal of heavy metals in the Ma'aleh HaKishon Reservoir

	Cd	Cr	Cu	Zn
Inlet concentration (µg l^{-1})	2.4	4.7	20.2	130
Outlet concentration (µg l^{-1})	1.0	1.7	6.7	46
Removal (%)	58	64	67	65

the MKR, the average removal of organic material expressed as *BOD* is about 65%. Table 22.1 presents the concentrations of *BOD* at the inlet and the outlet of the MKR. It should be mentioned that the composition of the organic matter changes during storage in the reservoir. The influent to the reservoir contains mostly organic components of activated sludge and trickling filter effluents. The organic matter that leaves the reservoir is mostly algal cells which developed during the storage period in the reservoir.

22.6
Removal of Heavy Metals and Detergents

Due to intense algal photosynthetic activity there is an increase of *p*H values up to 9–9.5 in the reservoir. Under such conditions, part of the dissolved metal ions form less soluble compounds and settle to the bottom of the reservoir. During the season 1985 – 1986, a survey of heavy metals was performed and the results are presented in Table 22.2. An additional study on removal of heavy metals was conducted during 1990–1991 (Juanicó et al. 1995) and the removal was similar to the earlier determinations.

Detergents are mostly removed by bacterial degradation. In Israel, like in many other countries, there is a gradual increase in use of biodegradable detergents and as a consequence, a higher percentage of inflow detergents are removed in the reservoir. Comparison between years 1985–1986 and 1995–1996 is presented in Table 22.3.

22.7
Nitrogen, Phosphorus and Potassium

There were no significant changes in the concentration of potassium during the storage period in MKR. On the other hand there was a considerable decrease in the concentration of other fertilizers, e.g., nitrogen and phosphorus. Table 22.4 presents data on the removal of nutrients during 1995–1996. The phosphorus depletion is due to precipitation of insoluble phosphates which are formed when the *p*H rises as result of intense photo-

Table 22.3. Removal of detergents in the Ma'aleh HaKishon Reservoir

	1985–1986	1995
Inlet concentration ($\mu g\,l^{-1}$)	958	1490
Outlet concentration ($\mu g\,l^{-1}$)	556	220
Removal (%)	42	85

Table 22.4. Removal of nutrients (fertilizers) in the Ma'aleh HaKishon Reservoir

	N	P	K
Inlet concentration ($mg\,l^{-1}$)	55	6.1	24.7
Outlet concentration ($mg\,l^{-1}$)	18	4.5	24.4
Removal (%)	67	27	1

Table 22.5. Removal of bacterial indicators in the Ma'aleh HaKishon Reservoir (median values of 62 series of samples during 2 years)

	Total coliforms	Faecal coliforms
Inlet concentration/100 ml	24000	2200
Outlet concentration/100 ml	130	16
Removal (%)	99.5	99.3

synthetic activity (Avnimelech and Wodka 1988). The mechanism of nitrogen decrease in the reservoir is more complicated. Most of the nitrogen reaches the reservoir as ammonia. When the pH values increase above 9, part of the ammonia is lost to the atmosphere by stripping. In the reservoir most of the ammonia is oxidized to nitrate (Azov and Tregubova 1995) and then part of it is denitrified with loss of available nitrogen.

22.8
Suspended Solids

The removal of suspended solids in MKR is quite similar to the removal of *BOD*. There is about 65% decrease in total suspended solids, while the removal of mineral suspended solids is lower. The main factors that affect the rate of the removal are the residence time in the reservoir, temperature and wind regime.

22.9
Enteric Microorganisms

The long residence time of wastewater in the reservoir affects the survival of enteric microorganisms. On the average, a reduction of 99.9% in the concentration of coliform bacteria and faecal coliform bacteria has been measured. Table 22.5 presents data on the removal of bacterial indicators during 1984–1986, and Fig. 22.4 shows the number of the bacteria at the outlet of the reservoir before disinfection till 1995. The disinfection usually removed about 90% of the remaining bacteria.

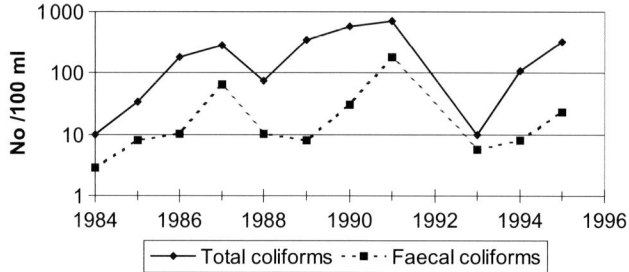

Fig. 22.4. Indicator bacteria in the outlet of Ma'aleh HaKishon Reservoir, before chlorination

The die-off of the enteric indicator bacteria can be explained by several mechanisms:

1. Exposure to irradiation
2. Grazing by filtering zooplankton
3. Bacteriophages
4. Predation by protozoa
5. Bacteriocidic excretions by algae
6. Rapid changes in concentration of dissolved oxygen

22.10
Discussion

Stabilization reservoirs are a relatively new development in wastewater treatment, and they are starting to be considered an integral part of wastewater reclamation projects in arid and semiarid regions. The original purpose was to store the excess wastewater and floodwater from the rainy winter to the dry summer. The first surveys of the reservoirs indicated improvement of water quality, in some cases comparable to partial tertiary treatment. Many of these studies were performed in the MKR. The author (Eren 1987) emphasized the importance of the storage period on the performance of the reservoir. Later on, Juanicó and Shelef (1994) suggested using the parameter of percentage of fresh effluents in the reservoir (*PFE*) in addition to the mean residence time as a factor in the design of stabilization reservoirs to get better quality of effluents. The MKR is divided into two parts and therefore it is relatively easy to operate in a way that ensures effluent of enough quality to be used for unrestricted irrigation after disinfection.

References

Avnimelech, Y, Wodka M (1988) Accumulation of nutrients in the sediments of Ma'aleh HaKishon reclaimed effluents reservoir. Wat Res 22:1437–1442
Azov Y, Trebugova T (1995) Nitrification processes in stabilization reservoirs. Wat Sci Tech 31:313–319
Eren J (1987) Changes in wastewater quality during long term storage. Water Reuse Symposium N° 4., Denver, Colorado, pp 1291–1296
Juanicó M, Shelef G (1994) Design, operation and performance of stabilization reservoirs for wastewater irrigation in Israel. Wat Res 28:175–186
Juanicó M, Ravid R, Azov Y (1995) Removal of trace metals from wastewater during long-term storage in seasonal reservoirs. Water, Air Soil Poll 82:617–633

Chapter 23

The Negev Desert Reservoirs

Benjamin Teltsch · Anat Maoz · Adam Kanarek · Yossi Azov · Inka Dor

23.1
Introduction

The increased demand for high quality water and the shortage of freshwater resources in Israel led to research on the improvement of the quality of secondary effluent for non-potable uses, especially for agricultural reuse. The best example for the application of this concept is the Dan Region Wastewater Reclamation Project, which provides for the collection, treatment, groundwater recharge, and multi-year storage of the effluent in the aquifer. The wastewater originates from the largest metropolitan area of the country, which includes the city of Tel-Aviv-Jaffa and several other neighboring municipalities. The project serves a total population of about 1.5 million persons, treating about 100 million m^3 per year of municipal wastewater. After storage in the aquifer, the purified effluent is pumped from the aquifer through recovery wells, either directly to the distribution system or into the operational Negev Desert reservoirs which temporarily impound the water before its use for unrestricted agricultural irrigation.

23.2
The Treatment Plant

In the treatment plant the wastewater undergoes mechanical-biological treatment by activated sludge with nitrification-denitrification processes. A good overall removal (above 70% efficiency) is obtained for BOD, BOD_f, TOC, detergents, phenol, Cd, Cn, Pb, Ag, Cu, Fe, Ni and Al. A moderate removal (40% to 70%) is obtained for suspended solids, COD, COD_f, ammonia, N_f, Ba, Mo and Sn. The reduction of total coli-forms, *E. coli* and *S. faecalis* is about four logs, and that of total bacteria is about two logs (Kanarek et al. 1992). The effluent is recharged to a regional groundwater aquifer by means of spreading infiltration basins located near the treatment plant (Azov et al. 1991). The recharge recovery system consists of intermittent one-day flooding and two-day drying leaching basins constructed on sand dunes, controlled passage of the effluent through the unsaturated zone, and its subsequent pumping from the aquifer by recovery wells surrounding the recharge area. A separate zone is created within the regional aquifer, dedicated to effluent prolonged storage lasting from 200 days to several years (Idelovich and Michail 1984; Shelef et al. 1994). The soil and the aquifer serve as a "soil-aquifer treatment" (SAT) system.

23.3
The Soil-Aquifer Treatment (SAT)

The major purification processes occurring in the SAT are: slow sand filtration, chemical precipitation, adsorption, ion exchange, biological degradation, nitrification and denitrification. The purification effect of the SAT which occurs during passage of the effluent through the unsaturated soil zone and the aquifer was evaluated from quality data of the effluent before SAT (RE), and of the reclaimed water after SAT pumped from a representative observation well pumping 100% recharged effluent for a long time (RW). Tables 23.1 and 23.2 summarize the SAT performance (Kanarek et al. 1993).

The relatively high removal efficiencies for a variety of parameters confirm that the SAT sub-system should be considered an integral part of the municipal wastewater treatment process in the Dan Region Project. According to Kanarek et al. (1992, 1993)

Table 23.1. SAT performance – basic water parameters (annual average 1990)

Parameter	Units	Before SAT (RE)	After SAT (RW)	Removal (%)
Suspended solids	$mg\,l^{-1}$	17	0	100
BOD	$mg\,l^{-1}$	20	<0.5	98
BOD f	$mg\,l^{-1}$	3.1	<0.5	84
COD	$mg\,l^{-1}$	69	12.5	82
COD f	$mg\,l^{-1}$	46	12.5	73
TOC	$mg\,l^{-1}$	20	3.3	84
DOC	$mg\,l^{-1}$	13	3.3	75
UV 254 absorbance	$cm^{-1} \times 1\,000$	300	64	79
$KMnO_4$ con-sumption as O_2	$mg\,l^{-1}$	14.1	2.3	84
$KMnO_4$ f con-sumption as O_2	$mg\,l^{-1}$	12.6	2.3	82
Detergents	$mg\,l^{-1}$	0.5	0.08	84
Phenols	$\mu g\,l^{-1}$	8	<2	75
Ammonia as N	$mg\,l^{-1}$	7.6	<0.05	99
Kjeldahl	$mg\,l^{-1}$	11.5	0.6	95
Kjeldahl f	$mg\,l^{-1}$	10.2	0.6	95
Nitrate as N	$mg\,l^{-1}$	3	7.2	–
Nitrite as N	$mg\,l^{-1}$	1.24	0.1	92
Total N	$mg\,l^{-1}$	15.7	7.8	50
Filtered N	$mg\,l^{-1}$	14.4	7.8	46
Phosphorus	$mg\,l^{-1}$	3.4	0.02	99
Alkalinity as $CaCO_3$	$mg\,l^{-1}$	306	300	–
pH	–	7.7	7.9	–

Table 23.2. SAT performance – bacteriological and virological parameters (annual average 1990)

Parameter	Units	Before SAT (RE)	After SAT (RW)
Total bacteria	$N° l^{-1}$	110 000	290
Total coliforms	MPN/100 ml	1 100 000	0
E. coli	MPN/100 ml	130 000	0
Streptococcus faecalis	MPN/100 ml	29 000	0
Enteroviruses	PFU/400 ml	2^a	0

[a] PFU/2001.

the removal of suspended solids (mostly organics) and of *BOD* is virtually complete. The average total and soluble *COD* are reduced from 69 and 46 mg l^{-1} respectively, to 12.5 mg l^{-1}; average *TOC* and *DOC* are reduced from 20 and 13 mg l^{-1} respectively, to 3.3 mg l^{-1}; UV absorbance is reduced from 298 to 64 cm^{-1}× 10^3; concentration of detergents from 0.5 to 0.08 mg l^{-1}; and that of phenols from 8 to <2 µg l^{-1}. Because of the efficient and reliable removal of organics, the SAT system can be regarded as a biological treatment unit. Turbidity decreases from about 17 NTU to 1.5 NTU; total and filtered nitrogen from 15.7 and 14.4 mg l^{-1} respectively to 7.8 mg l^{-1} and ammonia from 7.6 mg l^{-1} to less than 0.05 mg l^{-1}. While nitrogen in the recharge effluent is in the unoxidized forms of ammonia and organic nitrogen (Kjeldahl-N), the residual nitrogen after SAT consists essentially of nitrates. Thus, complete nitrification but only partial denitrification occurs in the soil-aquifer system. Phosphorus removal by SAT is very efficient, from 3.4 mg l^{-1} in the recharge effluent to 0.02 mg l^{-1} in the reclaimed water, which is in the range obtained in natural ground water.

Total coliform bacteria, *E. coli*, *S. faecalis*, and enteroviruses were not detected in the reclaimed water. The salinity of the reclaimed water is acceptable for irrigation of all crops.

The very high quality of the reclaimed water obtained after the SAT system is suitable for a variety of non-potable uses such as unrestricted irrigation (including irrigation of vegetables to be eaten raw and livestock watering), industrial uses, non-potable municipal uses (such as lawn irrigation and toilet flushing) and recreational uses. At present, the only use of this reclaimed water in Israel is unrestricted agricultural irrigation.

23.4
Reservoirs of Reclaimed Water

The water is pumped from the aquifer, then collected and transferred to a 100 km long conveyance main called "The Third Line." There are six open operational reservoirs along the main with volumes between 10 000 to 200 000 m^3 each, with the bottoms lined with geomembranes. The reservoirs hydraulically "float" on the line and are used to compensate for changes in water pressure in the main pipe and to facilitate control of the system during peak demand (floating reservoirs). Retention time of water in the reservoirs is not constant (1–7 days) and depends on operational constrictions. The reclaimed water contains nitrate in concentrations around 5–8 mg N l^{-1} and silica around 11 mg l^{-1}. Therefore, the open reservoirs are a convenient growth habitat for different algae species, resulting in a deterioration of water quality along the line. The problems caused by algae in a wastewater reclamation system are of operational character mainly, due to clogging of filters at the head of the drip irrigation network (Teltsch et al. 1991; Teltsch in this volume). The main difficulties appear in the water supply systems during high peak of water demand – the irrigation season. Reducing the algal development and accumulation in the reservoirs is essential for proper operation of the reclamation system. Research was conducted to understand and to overcome algal growth in the "Third Line" reservoirs (Teltsch and Leventer 1993; Maoz 1995; Azov et al. 1996).

23.5
Growth of Algae in the Reservoirs

Algal Growth Potential (*AGP*) of the reservoir's water was very low (Maoz 1995; Azov et al. 1996). The limiting factor of algal growth is phosphorus, that has a mean concentration as low as 0.02 mg PO_4-P l^{-1}. However, algal accumulation in the reservoirs of the system is much higher than could be determined by the *AGP* measurements. This is due to the "floating" behaviour and the short retention time of these reservoirs with large amounts of new water flowing through the reservoirs every day, allowing attached algae to extract large amounts of phosphorus even if it is found at very low concentrations.

Field observations reveal that most of the algae are benthic forms, attached to the reservoirs' walls and bottoms. In most cases, planktonic algae will not accumulate in the reservoirs due to short retention times and washout of algal cells. Filamentous cyanobacteria and some chlorophycean species form stable biofilms, up to 1.5 cm thick, lining reservoirs' walls and bottoms, while unicellular benthic diatoms develop on the top of these mats. These diatoms are released to the water column by wave action and cause a main clogging nuisance in the filters and drip irrigation network (Teltsch and Leventer 1993; Azov et al. 1996; Teltsch in this volume). To evaluate the algal growth in field conditions, artificial substrates were hung at various depths of one of the "Third Line" reservoirs (Tekuma B Reservoir, 7 m deep, 100 000 m^3 volume, Maoz 1995). The substrates (glass slides) were colonized mainly by benthic diatoms, showing concentrations of up to 22×10^6 cell cm^{-2} in the summer and up to 1.2×10^5 cell cm^{-2} in the winter. Most of the diatom biomass (over 75%) was found at a depth of 5–6 m below the surface during the summer and about 50% of the population at a depth of 4–5 m in the winter. This particular distribution of the algal biomass showing growth limitation at the upper part of the reservoir was caused probably by the frequent lowering of the water level in the reservoir and exposure of the algae on the upper substrates to the direct solar radiation and desiccation.

An operational diatom control method by lowering water level and wall desiccation, based on these findings, is being developed. This can be an alternative algae control method to algicides or biomanipulation by grazing fishes. It should be stressed that the phytotoxicity of certain algicides can cause damage to some agricultural crops (Karnaukhov et al. 1990; Chang et al. 1992), introducing restrictions to water reuse.

23.6
Control of Algae by Grazing Fish

Stocking water reservoirs with fish in order to cope with nuisance organisms and to improve water quality is a common practice in Israel (Leventer 1981; Teltsch and Leventer 1993). Seven different species of fish were stocked by Teltsch and Leventer (1993) in the reservoirs of the "Third Line," each of them with its main food niche (Table 23.3). The fish species, except for the hybrid tilapia and the common carp, do not reproduce in the reservoirs. This fish stocking composition fits the abundant periphytonic development of phyto- and zoobenthos in the reservoirs. The stocking densities and timing were variable and depended on the survival of the fish within the year. The number of fish were usually in the order of thousands of fingerlings per reservoir. The use of the grass carp completely eliminated the macrophytic vegetation and the filamentous algae along the reservoir shore line. The black carp was used as a precaution against colonization of snails in the system.

Field experiments were carried out to find the effect of fish grazing on the attached algal growth on the reservoir's bottom and walls (Azov et al. 1996; Teltsch in this volume). Paving stones were placed in baskets and hung in the water column in duplicates at three different depths: 1, 2 and 3 m below the surface. One of each duplicate was open and the other was covered with a plastic net in order to prevent fish from eating algae accumulated on the paving stones. Analysis of the algae found on the stones indicated that most of them were diatoms, chiefly large *Synedra sp.*, and to a much lesser extent green algae, such as *Scenedesmus*, *Pediastrum* and *Cosmarium*. Data on algae accumulation on the stones were extrapolated into a general algal growth and fish grazing model. The results indicate high levels of attached algal growth in the reservoir, at a rate between 33 to 620 kg d^{-1}. It was found that the fish play an important role in controlling algal growth. The amount of biomass eaten by fish from the reservoir walls and bottom is estimated between 28 to 195 kg d^{-1}. The fish removed between 31% to 84% of the algal biomass, improving considerably the water quality in the reservoirs.

The presence of the filter feeding fish – the silver carp and the periphytonic grazer the hybrid tilapia – resulted in a decrease of the benthonic diatom populations and an increase of the small planktonic chlorophytes which improved the filter clogging capacity of the water.

Table 23.3. Fish species used for the biological control in the "Third Line" reservoirs

Fish	Common name	Feeding habits
Mugil cephalus	Grey mullet	Detritophagic
Oreochromis aureus × *O. niloticus*	Hybrid tilapia	Phytobenthophagic
Mylopharyngodon piceus, *Cyprinus carpio*	Black carp, common carp	Benthophagic
Ctenopharyngodon idella	Grass carp	Macrophytophagic
Hypophthalmichthys molitrix	Silver carp	Phytoplanktophagic
Aristichthys nobilis	Big-head carp	Zooplanktophagic

23.7
Control of Algae by Lowering Water Level and Temporary Desiccation of the Reservoir Walls

Laboratory experiments were conducted in order to evaluate the resistance of various algae to desiccation. The first samples of fresh algal mat collected from the walls of Tekuma B Reservoir were crushed, inoculated on mineral growth medium plates,

Table 23.4. Algal cultures recovered from desiccated mats

Algae in the fresh mat	Days of desiccation						
	1	30	60	90	120	150	180
Cyanophyceae							
Aphanocapsa	████	████	████	████	████	████	████
Aphanothece	████	████	████	████	████	████	████
Chroococcidiopsis	████	████	████	████	████	████	████
Chroococcus	████	████	████	████	████	████	████
Lyngbia	████	████	████	████	████	████	████
Myxocarcina	████	████	████	████	████	████	████
Phormidium	████	████	████	████	████	████	████
Plectonema	████	████	████	████	████	████	████
Schizothrix	████	████	████	████	████	████	████
Scytonema	████	████	████	████	████	████	████
Scytonematopsis	████	████	████	████	████	████	████
Chlorophyceae							
Chlorella	████	████	████	████	████	████	████
Chlorococcum	████	████	████	████	████	████	████
Nannochloris	████	████	████	████			
Scenedesmus	████	████	████	████			
Coelastrum	████	████					
Pediastrum	██						
Dictyospherium	██						
Oocystis	██						
Cosmarium	●						
Ankistrodesmus	●						
Bacillariophyceae							
Gomphonema	0						
Fragillaria	0						
Melosira	0						
Navicula	0						
Surirella	0						
Synedra	0						

and incubated under light. After incubation for 14–20 days the various algae were identified, indicating the composition of the benthic algal community on the reservoir walls, considering that culturing methods may not reveal all the algae species initially present in the sample. Part of the material collected was subjected to desiccation lasting 1–180 days. Already after one day the samples became dry and contracted. Sub-samples of the dry material were taken every day during the first 10 days of desiccation and later after 14, 18, 22, 33, 38, 58, 86, 108 and 180 days. The subsamples were inoculated on the same mineral medium and the algal growth was analysed as above. Table 23.4 summarizes the algae genera recovered from the desiccated mat samples after varying times of desiccation.

Evidently, the cyanobacteria were very resistant to desiccation and all the 11 genera identified survived throughout all the experimental period which lasted half a year. The two green algae *Chlorella* and *Chlorococcum* also succeeded in surviving the entire period of desiccation. *Nannochloris* and *Scenedesmus* survived 108 days, *Coelastrum* 21 days, *Pediastrum* 7 days while *Dictyosphaerium* and *Oocystis* survived 3 days only.

As compared to the above two groups of algae, the diatoms were most sensitive to desiccation and could not recover even after a single day of desiccation leaving empty silica frustules.

These results demonstrate that benthic diatoms can be eliminated from the water by periodical emptying of the reservoirs and temporary drying of the walls and bottoms. However, there are situations in which the indispensable agricultural needs do not allow applying this method, which requires temporary drying of the floating operational reservoirs and excluding them from the distribution system.

References

Azov Y, Juanicó M, Shelef G, Kanarek A, Priel M (1991) Monitoring the quality of secondary effluents reused for unrestricted irrigation after underground storage. Wat Sci Tech 24(9):267–275

Azov Y, Khinich M, Rabkin S, Ben Yosef A, Shelef G (1996) Control of algae in the reservoirs of the "third line". In: Steinberger Y (ed) Preservation of our world in the wake of change, vol VI A/B, ISEEQS Pub., Jerusalem, Israel, pp 707–710

Chang A, Granato T, Page A (1992) A methodology for establishing phytotoxicity criteria for chromium, copper, nickel, and zinc in agricultural land application of municipal sewage sludges. J Env Quality 21(4):521–536

Idelovich E, Michail M (1984) Soil-aquifer treatment: A new approach to an old method of wastewater reuse. JWPCF 56:936–943

Kanarek A, Aharoni A, Michail M, Kogan I, Sherer D (1992) Dan region reclamation project – groundwater recharge with municipal effluents – recharge basins Soreq 1991. Mekorot Water Co. Ltd, Central District, Dan Region Unit, 56 pp

Kanarek A, Aharoni A, Michail M (1993) Municipal wastewater reuse via soil-aquifer treatment for non-potable purposes. Wat Sci Tech 27(7):53–61

Karnaukhov A, Tkachenko V, Shestidesyatnaya N (1990) Adsorption of copper by certain soil types of the Ukraine. Soviet Soil Science 22(5):36–41

Leventer H (1981) Biological control of reservoirs by fish. Bamidgeh, 33(1):3–23

Maoz A (1995) Manipulation of the ecosystem in the third line reservoirs in order to limit algal growth and clogging of the irrigation systems. Msci Thesis, The Hebrew University, 84 pp (in Hebrew)

Shelef G, Azov Y, Kanarek A, Zac Y, Shaw A (1994) The Dan Region sewerage wastewater treatment and reclamation scheme. Wat Sci Tech 30:229–238

Teltsch B, Leventer H (1993) The use of fish for biological control of water supply reservoirs. In: Qian Yi, Hao Jiming, Long Jun (eds) Emerging technologies for environment protection – Preparing for the 21st century, vol 1, International Academic Publishers, Beijing, China, pp 397–405

Teltsch B, Juanicó M, Azov Y, Ben-Harim I, Shelef G (1991) The clogging capacity of reclaimed wastewater: a new quality criterion for drip irrigation. Wat Sci Tech 24(9):123–131

Chapter 24

The Enan Reservoir

Sarig Gafny · Avital Gasith

24.1
Introduction

In regions of water scarcity, where the wet season is short or rainfall is unpredictable, reservoirs are often designed to store large quantities of water for later use, mostly for irrigation and for human use (Mitchell 1974; Fernando 1993; Costa and de Silva 1995; Frempong 1995). Reservoirs are frequently divided into two major groups:

- Mainstream reservoirs with high rate of water exchange (e.g., Søballe and Kimmel 1987)
- Storage impoundments which are characterized by a relatively long retention time and often exhibit thermal stratification (Hannon 1979)

These reservoir groups differ also in the ecological processes affecting the quality of the impounded water (Cook and Carlson 1989).

Water quality of any given water body is determined largely by the quantity (retention time) and quality (nutrients and organic matter loads) of the water entering it, and by internal cycling processes (sedimentation and release of stored nutrients from the sediments; Hannon 1979, Soranno et al. 1997). Abiotic factors, such as the reservoir's depth and shape (Canfield and Bachmann 1981; Cook and Carlson 1989), as well as biological interactions (phytoplankton and/or macrophytes production, food web interactions; Lowe 1979; Carpenter and Kitchell 1993), can also affect water quality. Environmental conditions, such as temperature regime and wind and wave action can also influence nutrient cycling and dynamics (Taylor 1971; Søballe et al. 1992). An inverse relationship may exist between the desired water quality of the reservoir and the nutrients and organic matter concentration. Correspondingly, control and reduction of nutrient concentration are frequently a high priority objective for reservoir management (Cook and Carlson 1989).

Winter-filled reservoirs may exhibit an initial short-term phase of improvement of the water quality associated with sedimentation of silt and of silt-sorbed phosphorous. In addition, algal biomass is usually low because of limiting low temperatures and irradiance, combined with relatively high grazing pressure by zooplankton (Brett and Muller-Navarra 1997). Considerable changes of water quality often occur in reservoirs with a long residence time. Warming up in summer, which enhances biological activity and nutrient cycling, can trigger changes of the algal association. The more edible diatoms and greens which prevail in winter and spring are often replaced in summer by the less edible blue-greens (Taylor 1971; Groeger and Kimmel 1988; Lathrop and Carpenter 1992; Brett and Goldman 1997). Such shifts in algae composition lower the zooplankton grazing pressure and result in the build-up of the algal biomass (Sommer et al. 1986). Irrigation reservoirs that are filled during winter and spring become progressively shallower in summer and may reach a depth in which the stratification is destroyed and hypolimnic nutrients are redistributed throughout the water column (Wunderlich 1971, Søballe et al. 1992). This may trigger massive algal blooms and deterioration of the water quality.

Multi-purpose reservoirs in Israel extend over a total area of about 1 400 ha (Lieberman and Shilo 1989; Sarig 1996). Most of these reservoirs are primarily intended for impounding water for irrigation. Though quality standards for irrigation water are less demanding than for drinking water, relatively high water quality is still a management goal, in order to minimize filter and dripper clogging, and to prevent corrosion of pipes and valves by products of anoxic metabolisms (e.g., H_2S). Reduced nutrient loading and low productivity are therefore desired in irrigation reservoirs. In addition to supplying water for irrigation, many reservoirs in Israel are also used for fish farming. The latter use depends on high water productivity (e.g., Milstain et al. 1994a,b; Sarig 1996). In fish ponds nutrients are commonly added to increase water productivity and to maximize fish yield (e.g., Olah 1986; Diana et al. 1991). One of the principle management goals of multi-purpose reservoirs is to minimize the conflicts emanating from the opposing needs of the different users (Stroud 1966, Jørgensen and Vollenweider 1988).

Enan Reservoir is one of the largest multi-purpose storage impoundment in Israel. The reservoir was constructed in 1983 in northern Israel (Southern Hula Valley, Fig. 24.1), and was primarily designed to intercept low quality runoff that is flowing in the Western Canal of the Jordan River during winter and prevent it from reaching Lake Kinneret. In addition, Enan Reservoir is intended also to supply water for irrigation in the Hula Valley region, for fish farming and (potentially but of a low priority) for recreation.

The purpose of this chapter is to examine the in-reservoir processes that controlled the water quality of Enan Reservoir during the first five years of its operation. Special attention is given to management problems that arise from the needs of different uses of the reservoir and to possible steps to minimize conflicts.

24.2
Methods

24.2.1
Study Site

Enan Reservoir is located in the southern region of the Hula Valley (Fig. 24.1). It is one of the largest and deepest man-made storage impoundments in Israel (750 m max. length, 600 m max. width, 45 ha surface area, 15 m max. depth, 6×10^6 m^3 maximum volume). The rectangularly shaped reservoir is surrounded by a steep embankment. It lacks major embayments (Shoreline Development Index of 1.2) and generally extends in a north-south direction.

The reservoir basin slopes gently toward the south, forming a southern basin that is about one meter deeper than the northern one. At the end of the growing season (September–October) when the water level is lowest, the southern basin remains the area from which the fish are landed. Five perpendicular ramps (R) each at a different

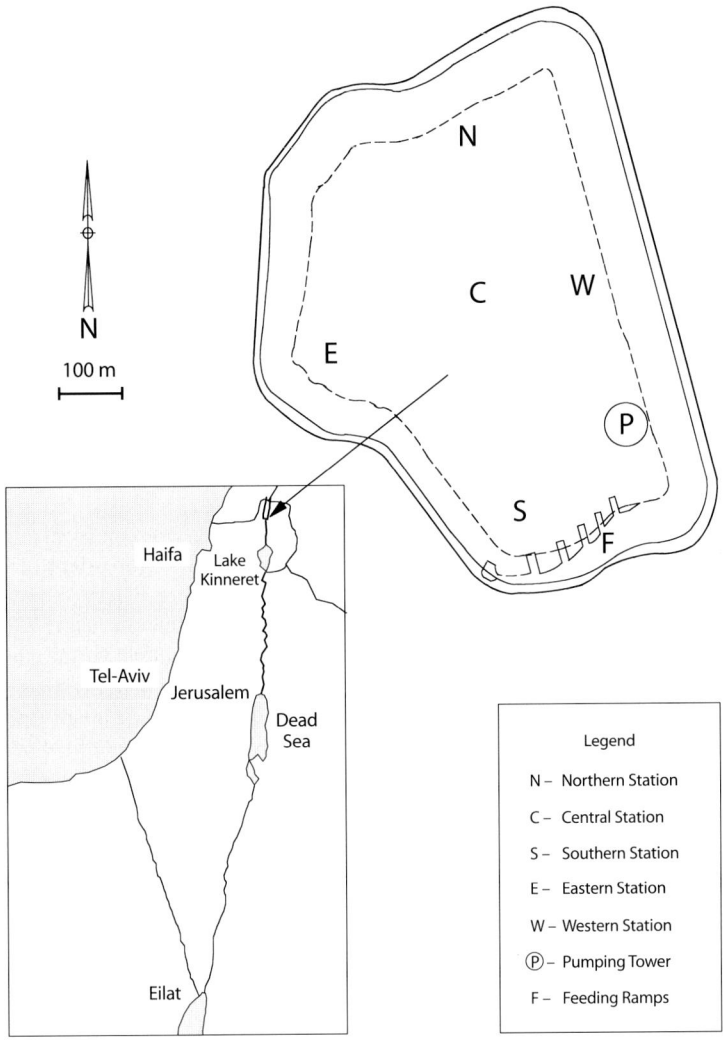

Fig. 24.1. Enan Reservoir in the Hula Valley (Northern Israel). Location of the sampling stations, feeding ramps, feeder and pumping tower is shown

elevation, were constructed along the southern shoreline to allow convenient access for adding fertilizer and fish food (Fig. 24.1). In addition, an automatic fish feeder (F) is located at the south-eastern corner of the reservoir (Fig. 24.1).

Enan Reservoir fills up at the beginning of each winter (December through February), usually in a few major storm events. The water is pumped into the reservoir from the Western Canal of the Jordan River *via* a control tower (P) located on the south-eastern corner of the reservoir (Fig. 24.1). In addition to runoff, the reservoir also receives domestic effluents that occasionally overflow into the Western Canal from wastewater treatment ponds of the town of Qiryat Shemona. Fish-pond and agricultural return water are often added to the reservoir in spring (March–May). During the summer (June–September) water is pumped out of the reservoir into the Western Canal and used for irrigation of fields in the Hula valley. Consequently, the reservoir water level progressively falls and it completely dries out by November.

The reservoir limnology and water quality dynamics were characterized by sampling in 3 main sampling stations located about 250 m apart from each other: S, in the southern basin, C, at the centre, and N, in the northern basin. Samples were also taken occasionally in 3 additional stations: P – pumping tower, E – east and W – west (Fig. 24.1).

24.2.2
Sampling Methods

We studied the limnology of Enan Reservoir from the first year of its operation (June 1983), and for four consecutive years thereafter. The reservoir was sampled twice a month, each time for two consecutive days.

Temperature and dissolved oxygen profiles were taken every 4 h from sunrise to sunset at stations N, C, and S. All three stations were sampled along depth transects (50 cm intervals) using a YSI model 57 Temperature-DO meter. Electric conductivity was measured at each station once a day, at the same depth intervals, using YSI model 33 SCT meter. Light was measured at midday using a Li-Core model LI185A light meter, at 10 cm intervals for the upper 2 m, and at 50 cm intervals at greater depths.

Water samples were collected at 0.5 or 1 m depth intervals using a Van Dorn sampler. Primary productivity was measured at station C using the light-dark bottle method (in quadruplicates, APHA 1985). The samples were incubated for 4–6 h, at their original depth.

The following water quality variables were routinely measured: chloride, pH, turbidity, total suspended solids (TSS), NH_4^+, total phosphorous, biochemical oxygen demand (BOD), chlorophyll a, algae and bacteria (presumptive coliforms, confirmed coliforms, faecal coliforms). Additional water quality variables (including: hardness, alkalinity, H_2S, NO_2, NO_3, Kjeldahl nitrogen, dissolved-P and permanganate demand) were also measured but are not included in this report. Water quality analyses were conducted by the Water Quality Laboratory of the Department of Environmental Engineering and Water Resources at the Technion, and by the Water Quality Laboratory of the Mekorot Water Company.

Statistical analysis of field and laboratory data was performed using a mainframe version of SAS Statistical Package and Systat for Windows (1992)

24.3
Results

24.3.1
Water Temperature and Dissolved Oxygen

During the winter (from reservoir fill-up in December through mid-March), Enan Reservoir was homothermic, with water temperature varying from 12–16° C (Fig. 24.2).

Fig. 24.2. Isotherms of Enan Reservoir (Station C) during the period of December 1983–November 1984

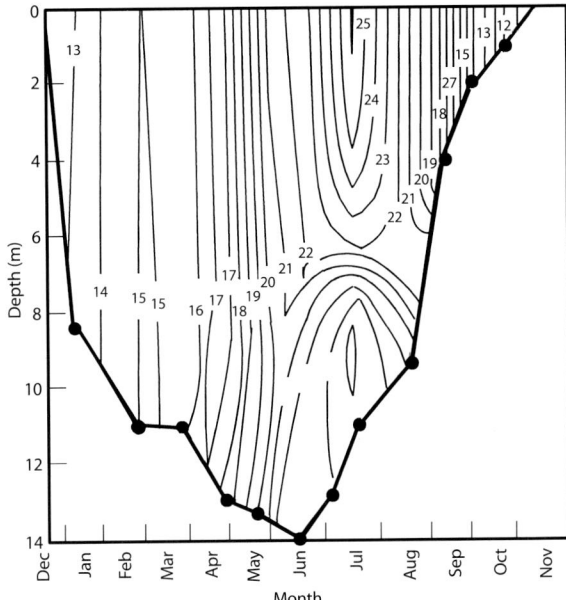

Fig. 24.3. Diurnal change in temperature and dissolved oxygen profiles in Enan Reservoir (Central Station, 11ᵗʰ July 1984)

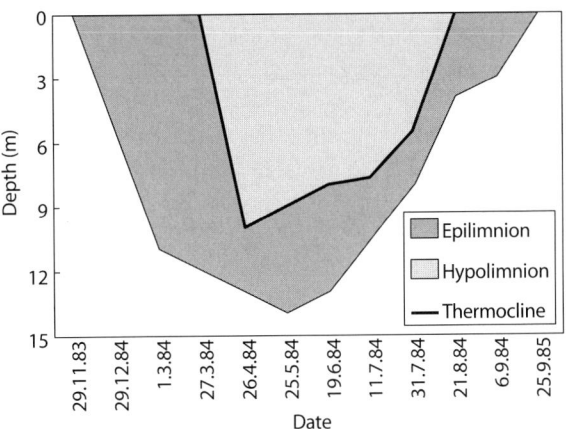

A gradual change in water temperature started in March and became more pronounced from mid-May. Beginning in mid-May, Enan Reservoir underwent daytime thermal stratification, which was destratified at night. This pattern was largely determined by the consistent afternoon north-western wind, which is typical for the Hula Valley in summer. Later in summer (June–August), the thermal stratification stabilised, but during the night the temperature gradient of the thermocline diminished (Fig. 24.3). The withdrawal of water for irrigation increased the epilimnion to hypolimnion volumetric ratio, and by mid-August, when the water depth fell below 5 m, the stratification was destroyed (Fig. 24.4).

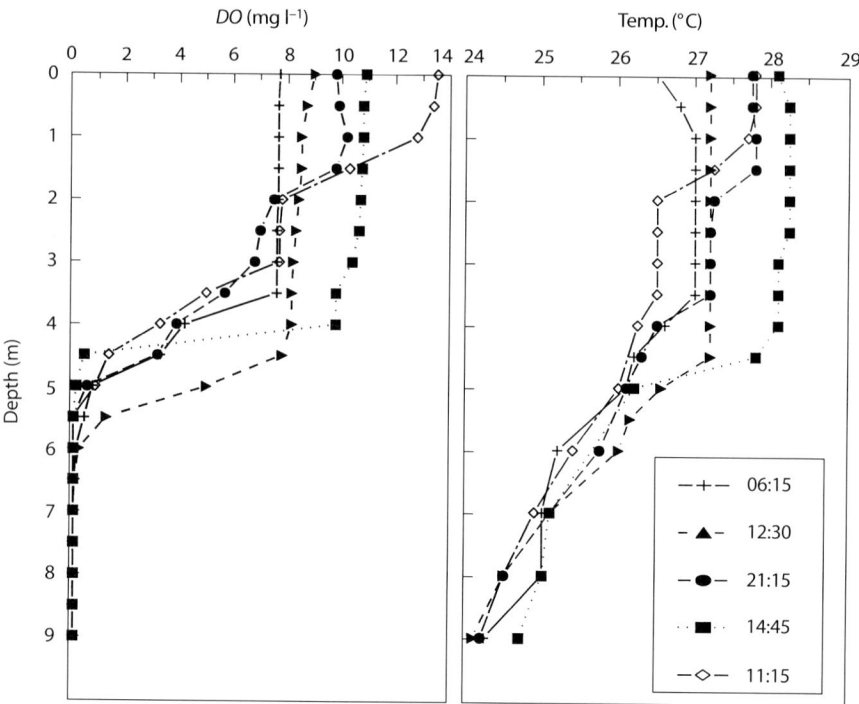

Fig. 24.4. Homothermy and stratification periods in Enan Reservoir (November 1983–September 1984)

During the period of stable stratification the reservoir typically exhibited a daily pattern of change in *the horizontal* distribution of the surface temperature and in the depth and gradient of the thermocline. Until midday, before the north-western wind started to blow, the surface temperature and thermocline depth were similar throughout most of the reservoir, except for slightly warmer surface water in the northern basin (Fig. 24.5a). Soon after the wind began, a horizontal difference of surface temperature developed, and the northern basin became 2° C colder than the southern basin (Fig. 24.5b). Moreover, the thermocline in the northern basin deepened, but the gradient was relatively small (2° C m⁻¹), whereas in the southern basin the thermocline depth remained unchanged, but the gradient increased by about 2.5° C (Fig. 24.5b). As was shown for other reservoirs surrounded by hilly terrain or levees, such phenomenon may be induced by a sudden increase in wind velocity which impacts only part of the water body and causes currents and movement of water layers (Imberger and Parker 1985; Parker and Imberger 1986). In Enan Reservoir, the north-westerly afternoon wind pushed warm surface water from the northern basin southward, causing cooler water to rise from deeper layers in the north, and accentuating the thermocline gradient in the south. Similar temperature difference was not evident between the eastern and western basins.

The distribution of dissolved oxygen (*DO*) in Enan Reservoir was closely associated with the thermal stratification. When the reservoir was homothermic (winter), *DO* was equally distributed from top to bottom (Fig. 24.6). From the beginning of

a WIND – SE: 0 or <0 m s⁻¹ (until 12:10)

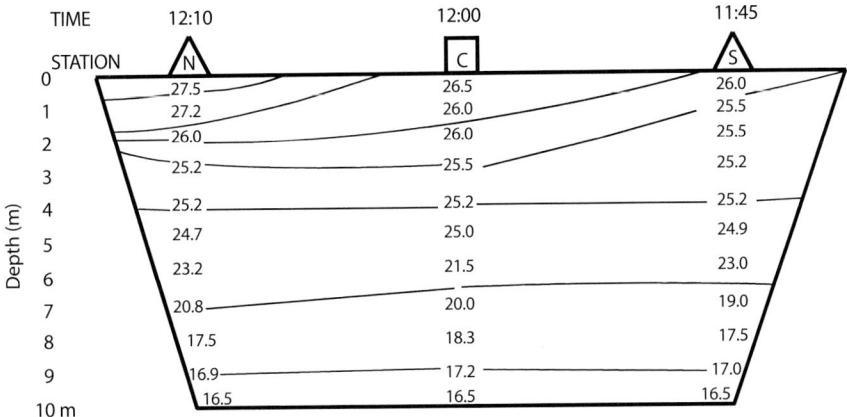

b WIND –NW: >1.5 m s⁻¹ (from 12:10)

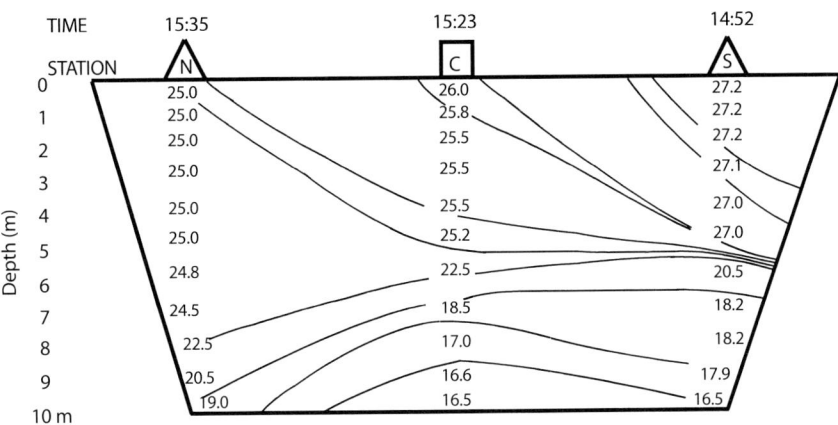

Fig. 24.5. Isotherms based on temperature profiles taken at the Northern (*N*), Central (*C*) and Southern (*S*) stations of Enan Reservoir, July 9, 1985: **a** noon time, before the commencement of the wind, **b** afternoon, about 3 hours of SW wind

summer, (May–June) the epilimnion was often supersaturated (>120% at noon) while the hypolimnion became depleted of oxygen (Fig. 24.6). The diurnal pattern was characterized by a sharp oxycline that was formed in the afternoon (e.g., 14:45, Fig. 24.3) and gradually faded at night (e.g., 21:15, Fig. 24.3) after the wind died out.

The daytime distribution of the *DO* changed horizontally and vertically similar to that observed for the temperature. Until noon, the southern basin in the vicinity of the ramps was relatively oxygen poor (<60% saturation), even near the surface. We attribute this low *DO* to the locally high organic load caused by the addition of chicken manure and fish food in this part of the reservoir. In the afternoon the *DO* conditions

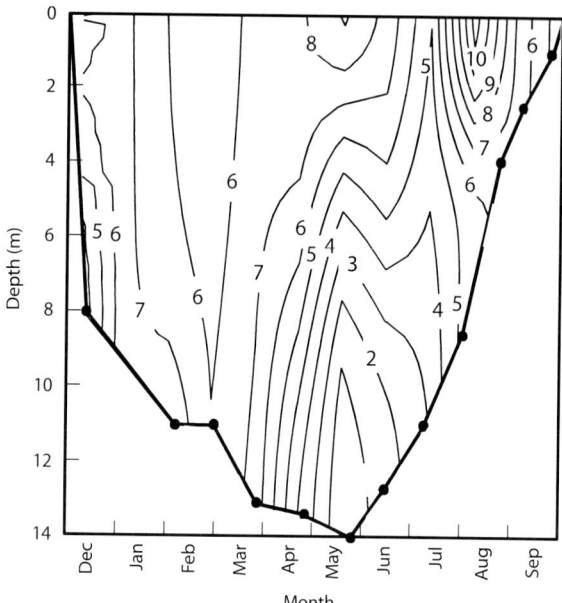

Fig. 24.6. Oxygen isopleth in Enan Reservoir (Station C) during the period of December 1983–November 1984

changed because warm oxygen-rich water from the northern basin was pushed by the wind southward, increasing the oxygen concentration in the southern basin (95% saturation). The north-south surface current caused upwelling of oxygen-poor (65% saturation) deep water to the surface at the northern basin. By the end of the summer, when the water depth was less than 5 m, the *DO* at the surface and in most of the water column of the southern basin remained low (ca. 20% saturation) and the basin was anoxic near the bottom throughout the day.

24.3.2
Phytoplankton and Primary Productivity

Algal abundance was highest in the epilimnion and lowest in the hypolimnion throughout the impoundment period (Fig. 24.7). During winter the green alga *Schroederia* prevailed and the total algal counts were low (<200 cells ml⁻¹). One to three orders of magnitude higher abundance of algae was recorded in March when the free-floating diatom *Cyclotella* predominated. Algal counts declined to pre-bloom level thereafter. In April and May, prior to the formation of stable stratification, the green algae *Platymonas* dominated and blue-greens (e.g., *Microcystis*) appeared in the epilimnion (Fig. 24.7). As summer progressed, the proportion of blue-green algae increased. A more diversified algal community (mostly blue-greens, diatoms, and dinoflagellates) was recorded in midsummer, when the reservoir was relatively shallow (<5 m) and unstratified. The proportion of dinoflagellates (*Peridinium-Glenodinium* complex) progressively increased from 11% in the first year of operation to 54, 63 and 75% in the following years, respectively.

Annual succession of algae in Enan Reservoir conformed with the general pattern of domination of algae that are more edible to zooplankton (e.g., greens) in winter

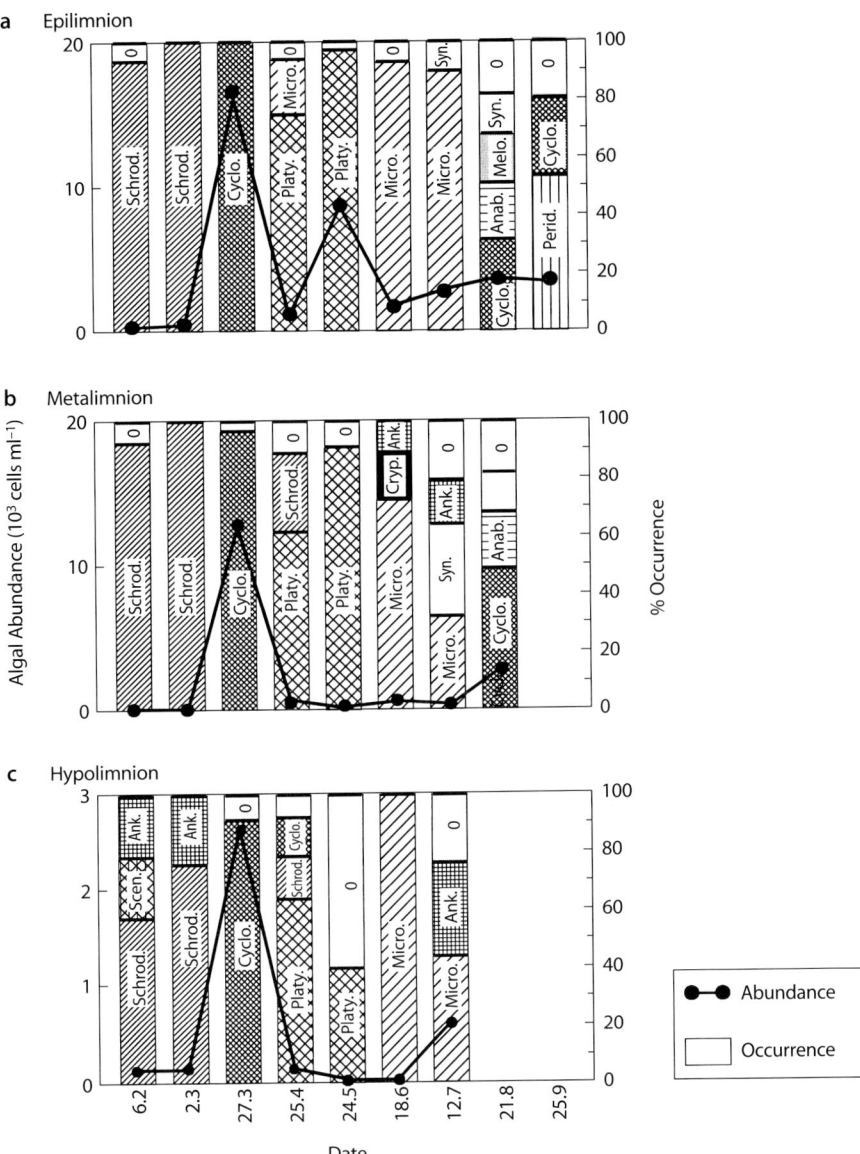

Fig. 24.7. An annual change in total algal abundance (*line*) and relative occurrence of the dominant algae species (*bars*) at the epilimnion, metalimnion and hypolimnion of Enan Reservoir during 1984

and spring and progressive establishment of less edible species such as blue-greens in summer (Brett and Goldman 1977). Correspondingly, higher zooplankton biomass was observed in Enan Reservoir in winter and spring than in summer (Gasith and Gafny, unpublished).

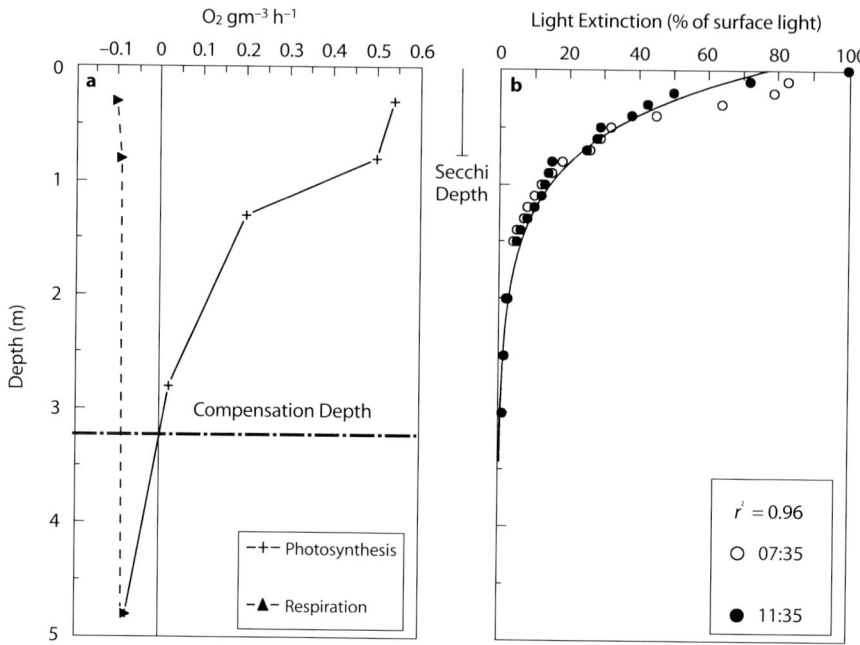

Fig. 24.8. Light attenuation, photosynthesis and respiration rate profiles in Enan Reservoir (Station C, 11th July 1984)

Typical summer depth profiles of light extinction, photosynthesis and respiration rates in Enan Reservoir are shown in Fig. 24.8. Changes of the photosynthetic rate with depth generally followed that of light attenuation, whereas respiration rate in the upper 5 m of the water column remained relatively unchanged. During spring and early summer, photosynthesis in the epilimnion exceeded respiration (P/R >1), and the surface water was supersaturated with oxygen. For example, in July 1984 the highest photosynthetic activity was measured at the surface (0.5 m) layer, where it exceeded respiration by a factor of 5 (Fig. 24.8b). At a depth of 1–1.5 m, photosynthesis was still twice as high as the respiration, and the compensation depth was reached at 3.2 m (Fig. 24.8b). As summer progressed, the compensation depth narrowed and the rate of oxygen depletion in the hypolimnion increased, resulting in a sharp daytime oxycline that separated between the super-saturated epilimnion and the nearly anoxic hypolimnion.

24.3.3
Effect of Impoundment on Water Quality

Improvement of the water quality in Enan Reservoir, relative to that in the western drainage canal of the Jordan River (the source of water for the reservoir), was observed during the initial phase of the impoundment. During the first 3–4 month BOD, suspended solids, turbidity, and NH_4^+ were significantly lower in the reservoir than in the Western Canal (Paired Comparison, Table 24.1). Similar concentrations of total phos-

Table 24.1 Summary statistics (mean and *SD*) for the comparison of selected water quality variables in Enan Reservoir and in the Western Canal of the Jordan River before (winter) and after (summer) thermal stratification

| Variable | Winter (n = 9) | | | | | | Summer (n = 7) | | | | | |
| | Reservoir | | Canal | | | | Reservoir | | Canal | | | |
	X	SD	X	SD	t	P	X	SD	X	SD	t	P
BOD (mg l^{-1})	2.7	1.1	5.9	3.2	3.73	0.009	5.1	3.1	4.7	2.2	0.71	0.51
Turbidity (NTU)	5.8	4.1	20.6	15.2	3.93	0.007	11.9	13.9	14.0	13.9	0.50	0.4
SSL (105°C; mg l^{-1})	6.2	3.0	27.6	19.5	3.56	0.009	12.8	13.9	14.9	14.2	0.58	0.59
NH_3 (mg l^{-1})	0.11	0.14	0.5	0.66	2.28	0.05	0.18	0.13	0.52	0.59	1.46	0.20
TP (mg l^{-1})	0.03	0.05	0.1	0.14	1.87	0.11	0.04	0.07	0.04	0.05	0.27	0.8
Chloroph. a (mg l^{-1})	6.8	6.8	11.9	14.6	1.63	0.18	14.5	12.6	12.1	6.6	0.24	0.82

Table 24.2 Range of total and faecal coliform counts (per 100 ml) in Enan Reservoir and in the Western Canal of the Jordan River during the first two years of operation of the reservoir

	Western Canal		Enan Reservoir	
	---	---	---	---
Year	1983 (n = 17)	1984 (n = 12)	1983 (n = 11)	1984 (n = 26)
Total coliforms	$2 \times 10^3 - 1 \times 10^6$	$5 \times 10^3 - 1.6 \times 10^5$	$2 \times 10^2 - 2 \times 10^4$	$2 \times 10^1 - 2.4 \times 10^3$
Faecal coliforms	$7 \times 10^2 - 3 \times 10^5$	$7 \times 10^3 - 1.1 \times 10^5$	$3.0 - 5 \times 10^2$	$2.0 - 5 \times 10^2$

phorous and chlorophyll a were measured in the drainage canal and the reservoir (Table 24.1). Throughout the impoundment period, bacterial counts (Total and faecal coliform) were lower in the reservoir than in the western canal (Table 24.2).

A typical depth profile of suspended solids, turbidity, and labile organic matter (*BOD*) that was observed during the stratification period is shown in Fig. 24.9. The upper epilimnion layer (ca. 0–3 m) and the lower hypolimnion layer (0.5 m above the bottom), were both relatively rich in suspended particles and correspondingly were most turbid. The *BOD* of the lower hypolimnion was markedly elevated relative to the rest of the water column indicating accumulation of decomposable organic matter near the bottom. The lowest turbidity and concentration of suspended particles was detected in the upper hypolimnion layer (5–8 m) which could thus serve best as the source of water for irrigation with the least clogging problem in the distribution system. However, sharp decline of *DO* in the metalimnion to near anoxic condition, that was detected during the summer, increased the risk of corrosion in the distribution system due to presence of H_2S in the upper hypolimnion.

Pumping water out of the reservoir for irrigation during the summer progressively lowered the water level. When the water level fell below approximately 5 m, the water quality deteriorated and was no more significantly different from that of the water of the drainage canal (Table 24.1).

Fig. 24.9. Depth profiles of water turbidity, suspended solids (*SSL*) biochemical oxygen demand (*BOD*) in Enan Reservoir

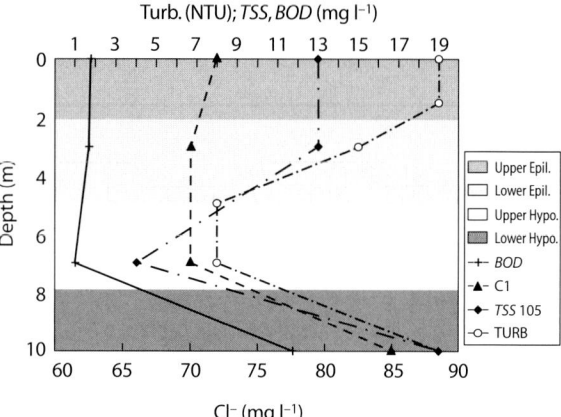

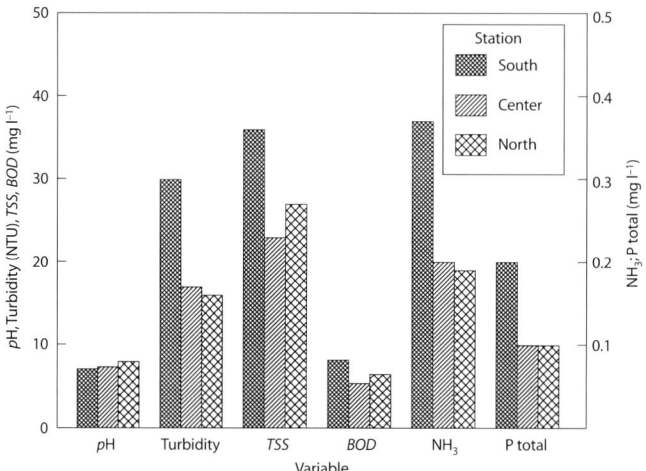

Fig. 24.10. Comparison of selected water quality variables (mean + *SD*) among the southern, central and northern basins of Enan Reservoir (5th September 1985)

The poorest water quality in summer was measured in the southern basin, particularly near the feeders. Except for chlorophyll a concentration, which was uniform throughout the reservoir, the concentrations of suspended solids, NH_4^+, total phosphorus, turbidity, and *BOD* were all highest in the southern basin (Fig. 24.10).

24.3.4
Fish Farming

Fish were introduced into Enan Reservoir as fingerlings, usually in early spring (March–April) and were all well above minimum market size when harvested 7–8 months later (Table 24.3). The common carp (*Cyprinus carpio*) was numerically the most important

Table 24.3 Summary of fish stocking and harvesting records in Enan Reservoir during the first 5 years of its operation

Species	Fish stocked Number	Biomass (kg)	Fish yield Biomass (kg)	Survival (%)	Biomass (kg/ha)
1983					
Common carp	128 000	5 770	88 000	87.1	1 960
Cichlid hybrid	76 000	5 000	26 000	69.1	578
Silver carp hybrid	12 000	384	4 000	20.0	89
Total	216 000	11 154	118 000	77.0	2 622
1984					
Common carp	217 000	5 932	141 500	81.5	3 144
Cichlid hybrid	154 000	7 870	65 670	100.0	1 459
Silver carp hybrid	10 370	550	20 869	>95.0	464
Total	381 370	14 352	228 039	89.5	5 068
1985					
Common carp	288 000	11 320	144 900	64.3	3 220
Cichlid hybrid	176 000	6 150	72 600	>95.0	1 613
Silver carp hybrid	13 000	2 635	35 000	>95.0	778
mullet	11 200	950	0	0	0
Total	488 200	21 055	252 500	76.6	5 611
1986					
Common carp	270 000	17 800	195 000	80.4	4 333
Cichlid hybrid	151 000	7 000	45 760	79.7	1 017
Silver carp hybrid	19 000	560	35 500	>95.0	789
Total	440 000	25 360	278 260	81.0	6 184
1987					
Common carp	187 500	–	165 000	>95.0	3 667
Silver carp hybrid	10 000	–	20 000	>95.0	444
Total	197 500	–	185 000	>95.0	4 111

species followed by a cichlids hybrid (*Oreochromis aureus* × *Oreochromis niloticus*) and a silver carp hybrid (*Hypophthalmichthys molitrix* × *Aristichthys nobilis*) (Table 24.3). Several hundreds of fingerlings of the St. Peter fish *Sarotherodon galilaeus* (Cichlidae) were also introduced into the reservoir.

The number and biomass of the stocked fish increased from year to year and doubled by the third year. Under the largest fish stocking, survival varied from >95%

in silvercarp, 80–100% in cichlids (including in-reservoir reproduction), and 64–80% in the common carp. An attempt to diversify the polyculture by stocking mullet (*Mugil cephalus*), together with the other fish, failed. By the end the growing season, when most of the fish were harvested, the mullet died, probably as a result of their high sensitivity to the high suspended solid concentration and low oxygen conditions that prevailed in the reservoir at the final stage of water drawdown. Since that event, the fishermen modified the harvesting regime. They adopted selective thinning of the population, beginning midway through the growing season. This reduced overcrowding of the fish at the final harvesting stage.

During the first four years, the carp made between 62–75% of the harvested biomass, the cichlid hybrid 17–29% and the silver carp hybrid 3.4–13.8%. The overall fish production ranged between 2 622 and 6 184 kg ha⁻¹, which is similar to that achievable in polyculture in earthen fish ponds (Milstein and Svirsky 1996).

The fish were fed sorghum (supplied daily from the feeding ramps) and protein rich (25%) feed pellets (supplied by an automatic feeder). The reservoir was also supplemented with chicken manure (supplied twice a week from the feeding ramps). In 1983, about 100 metric tons of feed pellets, 140 metric tons of sorghum, and 340 metric tons of chicken manure were introduced into the reservoir during the 7–8 months of the growing season. In 1983 and 1984 the food and manure were provided mainly in late morning hours before the north-westerly wind started to blow. Bottom grab samples indicated that within minutes after it was introduced, most of the chicken manure sank below the thermocline to the oxygen poor water, beyond the reach of the fish. It accumulated on the bottom near the feeding ramps creating an organic rich sediment layer. Moreover, low *DO* conditions that prevailed in the southern basin in the last trimester of the growing season apparently discouraged the fish from approaching the feeding ramps during the morning hours. Consequently, a greater portion of the food settled unconsumed to the bottom, adding to the organic content of the sediment. This further aggravated the water quality conditions in that part of the reservoir. A survey of the bottom sediments at the end of summer (soon after the reservoir was dry) confirmed the existence of organic-rich layer of sediment in the southern basin, where it reached 85% (by weight) in contrast to 5–7% in other parts of the reservoir. The extremely high organic content of the sediment explains the extremely poor water quality conditions that existed in the southern basin at the end of the growing season and probably caused the massive fish-kill in 1985. In 1986 the amount of the chicken manure was reduced by more than 50%, whereas the quantity of sorghum and fish pellets was increased by the same proportion. The operation of the automatic feeder was set for the afternoon hours when the *DO* concentration in the southern basin was highest.

24.4
Discussion

Deep reservoirs often differ from shallow ones in their within-reservoir limnological conditions, such as thermal and chemical stratification (Benson 1968; Lieberman and Shilo 1989). Enan is a warm monomictic reservoir with a short thermal stratification period that is caused by the progressive withdrawal of water, which results in midsummer destratification. Correspondingly, the impoundment period in Enan Reservoir is relatively short (7–8 months). This reduces the efficiency of "self purification" processes (e.g., decomposition of organic matter in the water column and effective sedimentation of inorganic particles) relative to that occurring in deep reservoirs (Cook and Carlson 1989). In contrast, in multi-purpose reservoirs that are used for fish production, water quality may progressively deteriorate because of significant input of allochthonous matter and nutrients, which increase the concentration of dissolved and particulate organic matter and enhance the development of algal biomass (Cook and Carlson 1989). Moreover, it was recently suggested that nutrient excretion by planktivorous fish can account for much of the enhanced algal production in lakes where they are present (Carpenter et al. 1992; Schindler et al. 1993; Schindler et al. 1995). The evacuation rate of the common carp, for example, which is the predominant fish species in Enan Reservoir, is especially high. The carp's turnover rate from food to feces at 26.5° C is 33% of the initial weight of the food consumed per hour (Garcia and Adelman 1985).

Enan Reservoir successfully fulfils its role in intercepting part of the low quality water that flows in the Western Canal of the Jordan River, preventing it from reaching Lake Kinneret. Water quality is one of the major factors that determine the suitability of the impounded water for its other uses. A priori, a conflict exists between the water of high productivity that is desired for fish production and the water with low productivity preferred for irrigation. Comprehensive understanding of the factors that control water quality and of the consequences of their interaction is the basis for prudent management of multi-purpose reservoirs.

The limnology and water quality dynamics of Enan Reservoir are strongly influenced by the south-easterly afternoon wind, which blows daily in late spring and throughout summer. The wind reinforces the thermal stratification daily, and maintains it as long as the reservoir depth exceeds 5 m. A similar pattern of daily change in the intensity of the thermal stratification was reported for other dual-purpose stratified reservoirs in the Jordan Valley (Milstein et al. 1994a; Zoran et al. 1994; Krambeck et al. 1994). If not for the daily wind, the night-time quiescent conditions could result in anoxia and mortality of fish (Milstein et al. 1995a), particularly in the southern basin that is heavily enriched with organic matter. Enan's specific configuration and its position relative to the dominant wind direction (Fig. 24.1) allows for maximal fetch, and therefore, generation of an effective afternoon surface current. The current carries warm and well oxygenated water southward and causes upwelling of colder, less oxygenated water in the northern basin. During the summer, when the reservoir is stratified, the daily shift of the water masses improves the water quality condition in the southern basin where the fish are fed. When the reservoir level falls below 5 m, the wind destroys the stratification and causes resuspension of sedimented organic matter in the southern basin. If not for the daily southward transport of oxygenated water, the southern basin would probably be depleted of oxygen and become unsuitable environment for fish. Resuspension of bottom sediment by the wind also reduces the quality of the water for irrigation. Altogether, a strong daily wind in summer benefits the fish, but at the same time, it reduces sedimentation and increases resuspension, impairing the water quality for irrigation.

The effect of the wind on the reservoir limnology and water quality is superimposed on the impact of organic matter and nutrient enrichment, and on the effect of reservoir drawdown, as the water is withdrawn for irrigation. Increasing the productivity of the water by supplementing fish food with chicken manure is a common practice in fish-pond farming. Studies on fish farming in shallow ponds indicated that adding fertilizers such as chicken manure can increase fish yield (e.g., Hepher 1963; Hepher and Pruginin 1981; Olah 1986; Diana et al. 1991). This was attributed to an increase in algal productivity and to some extent to direct consumption of the added manure (Hepher 1989). In shallow water bodies (<2m), material recycling is highly efficient and loss of organic matter to the sediment is minimal. Likewise, the control of the water quality conditions (e.g., supply of effective aeration) is more feasible in fish ponds than in large, deep water bodies. The experience in Enan Reservoir indicates that in a deep, stratified water body, food and manure that are not immediately consumed are lost to the bottom, beyond the reach of the fish. In situations where the reservoir serves also as a source of water, the water level progressively declines, ultimately reaching a depth in which stratification is no longer possible. At that stage the unconsumed food and organic matter that accumulated on the bottom become resuspended in the water column, causing rapid depletion of oxygen. Moreover, fish and other organisms may suddenly become exposed to toxic levels of products of anaerobic metabolism (i.e., ammonia and hydrogen sulfide) and may die. It appears that the adverse effects associated with the addition of chicken manure on water quality outweigh its potential benefits as an alternative food source and as a source of nutrients that enhance algal productivity. Furthermore, Milstein et al. (1995b) suggested that fertilization is only beneficial to some species, while in others, it leads to a lower growth rate and poorer performance. Therefore, organic load control is one of the management options often recommended for fishery improvement in deep reservoirs (Lee et al. 1984). Overall, optimal fish growth is attained under intermediate organic enrichment (Lee and Jones 1981), whereas the influence of organic enrichment on water used for irrigation is generally negative. In contrast, lowering of the water level negatively influences both fish yield and the quality of the water for irrigation.

In the case of Enan Reservoir, fish production could be improved and the problems limiting the use of the entire impounded volume for irrigation could be minimized, if the marked decline of water quality during the final stage of reservoir drawdown could be avoided. This could be partly achieved by construction of feeding ramps in the northern basin, where high water quality conditions prevail during the morning hours, and thus food consumption by the fish could be enhanced. Furthermore, the accumulation of organic matter in the southern basin and near the pumping tower will be reduced, thus, the adverse effects of low oxygen conditions at the place where the fish are crowded for landing will be minimized, and the concentration of suspended matter in the water pumped for irrigation will be reduced.

Following the second year of operation we recommend a complete abolition of fertilizing with organic manure in Enan Reservoir. This conclusion is supported by the good performance of Enan Reservoir fishery in 1986, after the added manure was cut by more than half. We suggested that in addition to changing of the morning feeding site from the southern to the northern basin, fish production could be improved by increasing the proportion of fish that can efficiently graze on natural food items.

The Cichlids S. galilaeus and O. aureus are examples of efficient planktivorous fish that can benefit from the plentiful plankton that develops in the reservoir, particularly during spring and early summer, and at the same time improve the quality of the irrigation water (Bondari et al. 1983). Despite its relatively low growth rate this fish can attain market size during one growing season.

Altogether, the main management problem of Enan Reservoir is that different reservoir uses are managed separately, each without considering the necessities of the others. For example, water supply managers withdraw water almost without considering fish farming, while fish managers fertilize the water without taking into account water quality requirements. This leads to conflicts between water suppliers and fish farmers and is usually harmful to both uses of the reservoirs.

In conclusion, as suggested by Warona and Cash (1996) multi-purpose reservoirs should be treated as integrated ecosystems. Optimal management of such reservoirs requires prioritizing the uses and identifying the variables that influence the performance of the reservoir in regard to these uses. Assessment of the interactions among these variables will provide the necessary understanding of the limits of system manipulation.

Acknowledgements

The investigation of Enan Reservoir was initiated and coordinated by David Katz of the Mekoroth Water Company, who envisioned the need for scientific knowledge in management of deep-water, multi-purpose reservoirs in Israel. The authors wish to thank Mr. Katz and the staff of Mekorot Water Company at Malacha pumping station, and Central Laboratory at Netofa for their technical assistance and for algal and bacterial analyses. We could not have accomplished the research if not for the assistance and cooperation of Yoav Horin and the fish farming team. We also thank Eli Geffen (Tel-Aviv University) for his help with field sampling. We thank Prof. A. Wachs and Ms. S. Platner (Technion, Haifa) for their part in the water quality analyses. The investigation was supported by the Mekorot Water Company and Tel-Aviv University.

References

APHA (1985) Standard methods for the examination of water and wastewater. American Public Health Association. 16th edn, 1268 pp

Benson, NG (1968) Review of fishery studies on Missouri River main stem reservoirs. Res Rep 71. US Dept. of the Interior. Fish and Wildlife Service. Bureau of Sport Fisheries and Wildlife. Washington, DC, 62 pp

Bondari K, Rhreadgill ED, Bender A (1983) Tilapia culture in conjunction with irrigation and urban farming. In: Fishelson L, Yaron Z (eds) Proceedings of an international symposium on Tilapia in aquaculture. Nazareth, Israel 8–13 May 1983. Tel-Aviv University Pub., Tel-Aviv, Israel, pp 484–493

Brett MT, Goldman CR (1977) Consumer vs. resource control in freshwater pelagic food-webs. Science 275:384–386

Brett MT, Muller-Navarra DC (1997) The role of highly unsaturated fatty acids in aquatic foodweb processes. Freshwater Biology 38:483–499

Canfield DE, Bachmann RW (1981) Predictions of total phosphorus concentrations, chlorophyll a and Secchi depth in natural and artificial lakes. Can J Fish Aquat Sci 38:414–423

Carpenter SR, Kitchell JF (1993) The trophic cascade in lakes. Cambridge University Press, Cambridge, U.K.

Carpenter SR, Cottingham KL, Schindler DE (1992) Biotic feedbacks in lake phosphorus cycles. Trends in Ecol Evol 7:332–336

Cooke GD, Carlson RE (1989) Reservoir management for water quality and THM precursor control. AWWA Res. Foundation and American Water Work Association. Denver, CO., 387 pp

Costa HH, de Silva PK (1995) Limnological research and training in Sri Lanka: State of the art and future needs. In: Gopal B, Wetzels RG (eds) Limnology in developing countries. International Association of Theoretical an Applied Limnology (SIL), New Delhi, India, pp 63–103

Diana JS, Lin CK, Schneeberger PJ (1991) Relationships among nutrient inputs, water nutrient concentrations, primary production, and yield of Oreochromis niloticus in ponds. Aquaculture 92:323–341

Fernando CH (1993) Impact of Sri Lankan reservoirs, their fisheries, management and conservation. In: Erdelen W, Preu C, Ishwaran N, Madduma Bandara CM (eds) Ecology and landscape management in Sri Lanka. Margraf Scientific Books, Weikersheim, pp 351–374

Frempong E (1995) Limnological research and training in Ghana: The past, present and perspectives for future development. In: Gopal B, Wetzels RG (eds) Limnology in developing countries. International Association of Theoretical an Applied Limnology (SIL), New Delhi, India, pp 1–30

Garcia LM, Adelman IR (1985) An in situ estimate of daily food consumption and alimentary canal evacuation rates in the common carp Cyprinus carpio. J Fish Biol 27:487–493

Groeger AW, Kimmel BC (1988) Photosynthetic carbon metabolism by phytoplankton in a nitrogen-limited reservoir. Can J Fish Aquat Sci 45:720–730

Hannon HH (1979) Chemical modifications in reservoir-regulated streams. In: Ward JV, Stanford JW (eds) The ecology of regulated streams. Plenum Press, N.Y., pp 75–91

Hepher B (1963) Ten years of research in fishponds fertilization in Israel. I. The effect of fertilization on fish yields. Bamidgeh 14:29–38

Hepher B (1989) Principle of nutrition. In: Shilo M, Sarig S (eds) Fish culture in warm water systems: problems and trends. CRC Press Inc Boca Raton, FL, pp 146–154

Hepher B, Pruginin Y (1981) Commercial fish farming with special reference to fish culture in Israel. John Wiley and Sons, New York, 261 pp

Imberger J, Parker G (1985) Mixed layer dynamics in a lake exposed to a spatially variable wind field. Limnol Oceanogr 30:473–488

Jørgensen SE, Vollenweider RA (1988) Problems of lakes and reservoirs. In: Jørgensen SE, Vollenweider RA (eds) Guidelines of lake management, vol 1: Principles of lake management. International Lake Environment Committee, UNEP, pp 37–41

Krambeck HJ, Milstein A, Zoran M (1994) Physical limnology of reservoirs in Israel used for crop irrigation and fish farming. Verh Internat Verein Limnol 25:1373–1378

Lathrop RC, Carpenter SR (1992) Zooplankton and their relationship to phytoplankton. In: Kitchell JF (ed) Food web management: a case study of Lake Mendota. Springer-Verlag, N.Y., pp 127–150

Lee GF, Jones RA (1981) Effect of eutrophication of fisheries. Occasional Paper N° 74, Department of Civil Engineering, Texas Tech Univ Lubbock, TX.

Lee GF, Jones RA, Winkler BA, Sweazy RM (1984) Water quality management in the Yellowhouse Canyon lakes – a domestic wastewater reuse project. Proceedings of conference "Water for the 21st Century: Will it be there" Southern Methodist University, Dallas, TX., April 1984

Lieberman OS, Shilo M (1989) Water quality improvement of deep, dual purpose reservoir. Isr J Aquacult, Bamidgeh 41:34–37

Lowe RL (1979) Phytobenthic ecology and regulated streams. In: Ward JV, Stanford JW (eds) The ecology of regulated streams. Plenum Press, N.Y., pp 25–38

Milstein A, Svirsky F (1996) Effect of fish species combinations on water chemistry and plankton composition in earthen fishpond. Aqua Res 27:79–90

Milstein A, Zoran M, Bersadschi D, Krambeck HJ (1994a) Water quality variability in a deep (8 m) reservoir for simultaneous fish farming and field irrigation. Limnologica 24:82–92

Milstein A, Zoran M, Krambeck HJ (1994b) Water quality variability in shallow (4 m) reservoir for simultaneous fish farming and field irrigation. Limnologica 24:71–81

Milstein A, Zoran M, Krambeck HJ (1995a) Seasonal stratification in fish culture and irrigation reservoirs: potential dangers for fish. Aquaculture International 3:116–122

Milstein A, Alkon A, Karplus I, Kochba M, Avnimelech Y (1995b) Combined effects of fertilization rate, manuring and feed pellet application on fish performances and water quality in polyculture ponds. Aquaculture Research 26:55–65

Mitchell DS (1974) Hydrobiological observations on three Rhodesian reservoirs. Freshwat Biol 4:61–72

Olah J (1986) Carp production in manured ponds. In: Billard R, Marcel J (eds) Aquaculture of cyprinids. IRNA Paris, PP 295–303

Parker GJ, Imberger J (1986) Differential mixed-layer deepening in lakes and reservoirs. In: De Deckker P, Williams WD (eds) Limnology in Australia. W Junk Pub, Dordrecht, pp 63–92

Sarig S (1996) The fish culture industry in Israel in 1995. Isr J Aquacult, Bamidgeh 48:158–164

Schindler DE, Kitchell JF, He X, Carpenter SR, Hodgson JR, Cottingham KL (1993) Food web structure and phosphorus cycling in lakes. Trans Am Fish Soc 122:756–772

Schindler DE, Carpenter SR, Cottingham KL, He X, Hodgson JR, Kitchell JF, Soranno PA (1995) Food web structure and littoral zone coupling to pelagic trophic cascades. In: Polis GA, Winemiller KO (eds) Food Webs: Integration of patterns and dynamics. Chapman Hall, N.Y., pp 96–105

Søballe DM, Kimmel BL (1987) A large-scale comparison of factors influencing phytoplankton abundance in rivers, lakes, and impoundments. Ecology 68:1943–1954

Søballe DM, Kimmel BL, Kennedy RH, Gaugush RF (1992) Reservoirs. In: Hackney CT, Adams SM, Martin WH (eds) Biodiversity of the southeastern United States: Aquatic communities. John Wiley & Sons, Inc N.Y., pp 421–474

Soranno PA, Carpenter SR, Lathrop RC (1997) Internal phosphorus loading in Lake Mendota: response to external loads and weather. Can J Fish Aquat Sci (in press)

Stroud RH (1966) American experience in recreational use of artificial waters. In: Lowe-McConnell RH (ed) Man-made lakes. Academic Press, London, pp 189–202

SYSTAT for Windows (1992) Statistic version, 5th edn, Systat Inc Evanston IL, 750 pp

Taylor MP (1971) Phytoplankton productivity response to nutrients correlated with certain environmental factors in six TVA reservoirs. In: Hall GE (ed) Reservoir fisheries and limnology. Am Fish Soc Spec Pub N° 8, Washington, DC, pp 209–231

Warona FJ, Cash KJ (1996) The ecosystem approach to environmental assessment: moving from theory to practice. J Aquatic Ecosystem Health 5:89–97

Wunderlich WO (1971) The dynamics of density-stratified reservoirs. In: Hall GE (ed) Reservoir fisheries and limnology. Am Fish Soc Spec Pub N° 8, Washington, DC, pp 219–231

Zoran M, Milstein A, Krambeck HJ (1994) Limnology of a dual purpose reservoirs in the coastal area and Jordan Valley of Israel. Israeli Journal of Aquaculture Bamidgeh 46:64–75

Index

A

Actinastrum hantzschii 174, 311
Adashim Reservoir 71
adipates 207
Aeromonas 178
AGP, see algal growth potential
Al 221
alcohols 207
algae 51, 115, 129, 139, 173, 233, 247, 289, 309, 364
–, benthic 364
–, control 364
–, counts 372
–, species in reservoirs 175
–, tolerance to pollution 174
algae-bacteria relationships 177
algal growth potential (*AGP*) 364
algicides 241
alkalinity 332
alkyl phenols 207
aluminum sulfate 286
ammonia 173, 312, 332, 372
–, in effluents 160
ammonium 146
Amoeba 188
Anabaena 161
analysis
–, multiple regression a. 101
–, multivariate regression a. 72
–, principal component 265, 275, 339
Anas 185
Ancylostoma 31, 34
Ankistrodesmus 174–177
aquatic animal taxa 183
Araneidae 185
area/volume relationship 66
areas, arid and semiarid 24
Aristichthys nobilis 190, 239
Arthrospira 174
Ascaris 10, 26, 29–34, 44, 45
–, *limbriocoides* 44
Asplanchna 188
ATMLIT 234, 235, 236, 237
ATMMIN 234, 235, 236, 237, 240, 243

B

backwash 236
bacteria 34, 247

–, counts 313
–, heterotrophic 177
–, inhibitors 179
–, nitrifying b. in reservoirs 163
–, slime 233
batch operation 289
Bdelloidea 184
Beggiatoa alba 239
biochemical oxygen demand (*BOD*) 3, 6–9, 37, 40–44, 47, 49–60, 68–75, 78, 86–89, 102, 106–108, 110–112, 126, 127, 150, 154, 160, 163, 171, 175, 191, 199, 205, 212, 214, 229, 284–289, 293–295, 308, 312, 321–324, 327, 335, 336, 339, 344, 358–363, 372, 378–380
–, and algae 175
–, removal 72, 110, 284, 289, 312, 358
biofilms 249, 364
bioflocs 247
biomanipulation 190, 239, 365
bioturbidity 173
birds 185, 290
BOD, see biochemical oxygen demand
Boron 19
Boucher's Law 253
Brachionus 188, 311
–, *angularis* 311
–, *budapestinensis* 312
–, *calyciflorus* 189, 311

C

$CaCO_3$ precipitate 247
Campylobacter 26
Canada 284
Carassius auratus 290
Cerataphyllum 290
chemical oxygen demand (*COD*) 6, 8, 43, 71–74, 110–112, 126–134, 289–291, 308, 339, 361–362
–, removal 72, 110, 289
chemical/floculant treatment 286
chemicals
–, industrial 207
–, organosynthetic in agricultural soils 213
chemoclines 332
Chile 293
China 283, 290
Chironomidae 185, 312
Chironomus 188

Chlamydomonas 51, 59, 174–176, 289, 311, 320, 357
Chlorella 51, 59, 162, 174–180, 189, 191, 242, 264,
      270, 272, 278–281, 289, 309, 310, 318, 319,
      320–322, 325, 326, 357, 367
      –, *pyrenoidosa* 178, 281
      –, *vulgaris* 59, 174, 178–180, 242, 309, 310,
      318–320, 326
chloride 372
chlorine 242
      –, dioxide 251
Chlorobium 289
Chlorococcum 367
chlorophyll 173, 348
chlorophyll a 50, 263, 311, 340, 372
Cholera 29, 31
chromatic coordinates 266
Chromatium 289
ciliates 289
Ciliophora 183
Citrobacter 178
Cladocera 183, 186, 188, 191, 290, 312, 344, 357
clay 235, 247
climatic changes 342
climatic characteristics 329
clogging 233, 364
      –, capacity of effluents 234
      –, of irrigation systems 190
      –, of the drippers 233
Closteridium 51, 59, 174
CO_2 162
coagulation, chemical 258
COD, see chemical oxygen demand
Coelastrum 174, 189, 367
Coleoptera 185, 188, 191
compensation
      –, depth 173
      –, level 308
computer algorithm 86
conductivity, electrical 332, 338, 372
construction 153
continuous-flow
      –, reactor 287, 289
      –, single reservoirs 65
      –, stabilization reservoirs 66
copepoda 183, 186, 191, 238
copper sulfate 241
Corixidae 186, 312
Cosmarium 365
country, arid 305
Cr 221
Cu 221
Culex 185, 190
      –, *pipiens* 190
currents 66, 229
cyanobacteria 238, 364
Cyclopoida 187, 188
Cyclops scutifer 188
Cyclotella 174, 376
Cyprinus carpio 380, 386

D

Dan Region Wastewater Reclamation Project 361

Daphnia 143, 144, 186, 188, 190, 225, 235, 312, 344,
      345
      –, *longispina* 344
      –, *magna* 143, 186, 188, 189, 225, 235, 312
data analysis 81
degradation constants in sediment 147
depth 66, 140, 356, 284, 370
design 61
detergents 74, 205, 207, 290, 340, 358
detritivorous 188
Dictyosphaerium 174, 367
Didinium 188
dissolved organic carbon (*DOC*) 127, 205,
      214–217, 363
dissolved oxygen (*DO*) 40, 58, 59, 79, 131, 174, 308,
      319, 325, 372–375, 379, 382
DO, see dissolved oxygen
DOC, see dissolved organic carbon

E

EC, see electrical conductivity
ecological simulation 80
ecosystem, stability of 188
effluent quality, microbial guidelines for 23
electrical conductivity (*EC*) 329–331, 335,
      341-347
embankment 153, 370
Enterococci 43
epidemiology 28, 34
epilimnion 66, 327, 339, 373
Epiphanes senta 311
Eristalis 185, 188, 191
Escherichia coli 43, 82, 361, 363
esters 207
ethers 207
Euglena 51, 57, 59, 174, 175, 176, 191, 264, 311, 320,
      357
Euglenidae 188
evaporation 66, 116, 331
experimental treatment plant 293

F

FACTOR 327–351
faecal coliforms 43, 74, 101, 131, 141, 359, 372
      –, removal 289
fatty acids 207
ferric chloride 286
fertilization 160
Filinia 188
filter 233
      –, cake 254
      –, simulation of screen f. 234
      –, specific resistance 234
filterability
      –, index 234, 254
      –, test 234
filtration
      –, deep-bed 256
      –, granular media f. 256
      –, mechanical 253
fish 190, 239, 290, 365
      –, farming 370

–, ponds 146
–, reservoirs 336
flocculation 258
food chains 188

G

Gambusia 188, 190
–, *affinis* 190
gas chromatography (GC) 205
GC, *see* gas chromatography
Germany 291
Geta'ot Reservoir 100
Glenodinium 376
Golenkinia 174, 175, 289
golf courses 65, 287, 294
–, irrigation 286
grants and loans 3
grazing 188
green algae 263
ground truth 264

H

Haifa 18, 355
harvesting matrix 94
health
–, requirements 23
–, risks 23
heavy metals 358
helminths 29, 33, 34
–, removal 41
Hemiptera 185, 186, 188, 191
herbivorous 188
Hexarthra 312
high performance liquid chromatography (HPLC) 205
hookworm 31
HPLC, *see* high performance liquid chromatography
humic acid 250, 271
hydraulic
–, age distribution 80, 85, 112
–, short-circuiting 66
Hydrilla 290
hydrological cycle in the reservoirs 61
hypertrophic waters, regulatory potential 324
hypertrophy 305
hypolimnetic accumulation of dissolved solids 333
hypolimnion 66, 327, 339, 373
Hypophthalmichthys molitrix 190, 239, 245

I

India 295
insects, demersal 185
irrigation 291, 305, 355
–, drip i. 233, 247, 364
–, season 181, 356
–, wastewater 23, 285
Italy 292

J

Japan 283

Jeezrael Valley 219, 355
–, Project 18

K

ketones 207
Klebsiella 178

L

lagoons, aerated 18
Lepocinclis 189
Leslie matrix 85, 90, 91, 94, 102
light 160, 372
–, limitation 174
–, penetration 51
lime 286
loading
–, external 340
–, internal 347
–, surface organic l. 69

M

Ma'aleh HaKishon Reservoir 18, 50, 71, 146, 175, 186, 219, 355–360
mass spectrometry (MS) 205
Mauremys caspica *rivulata* 185, 192, 193
MBAS 340, 345
mean residence time (MRT) 68–72, 88, 94, 99, 102, 330, 331
Mesocyclops oqunnus 188, 238
metals, removal of 230
meta-model 113
México 283
Micractinium 51, 59, 174–177
–, *pusillum* 175
Microcyclops minutus 312
Microcystis 190, 319, 320, 376
microorganisms, pathogenic 23
–, in wastewater 26
micropollutants, accumulation of 206
mixotrophic nutrition 177
model 285, 289
–, simulation 113
–, spectral radiance m. 269
–, statistical 110
modelling 150
Moina 188, 312
–, *micrura* 312
monitoring 80, 263, 355
–, program 6, 8
Morocco 294
MRT, *see* mean residence time
MS, *see* mass spectrometry
MS/MS, *see* tandem mass spectrometry
Mugil cephalus 382
multiple regression 111

N

Na'an Reservoir 196, 206, 305–326
Nannochloris 175, 367
Nematoda 184, 188, 191
nitrate 364
Nitrogen 19, 145, 159, 358

Nitrosomonas 164, 166, 167, 171
-, *europaea* 166, 167, 171
Notonectidae 186
nutrients 358
-, balance 148
-, in sediment 145
-, removal 71

O

odours
-, in freshwater, classification 196
-, nuisances 56, 195
olfactory sense 195
Oligochaeta 184, 188, 191
Oocystis 51, 59, 174, 242, 367
operation 61, 243, 306, 328, 356
-, of reservoirs 80, 222
Oreochromis
-, *aureus* 239, 381, 384
-, *niloticus* 239, 381, 385
organic carbon 145
-, loading 78, 106, 141, 284, 327, 383
-, matter 49, 122, 124, 127, 131, 140–142, 263
-, pollutants, degradation mechanisms 214
Oscillatoria 190
Ostracoda 184, 188, 191
outlet 66, 154, 329
-, location 229
oxidants 241, 251
oxygen 289
ozone 242

P

Panagrolaimus 188
PAR, *see* photosynthetic active radiation
PAR, *see* photosynthetically active radiation
particles
-, characterization 247
-, mixture of 248
-, size distribution 250
-, surface composition 248
pathogens 115, 321
-, removal in reservoirs 42
-, survival of 26
Pb 221
PC 275–280
Pediastrum 174, 365, 367
percentage of fresh effluents 68, 87, 220, 243
Percentage of Fresh Effluents (*PFE*) 68, 69, 73,
74, 80, 86–90, 95, 98–102, 129, 131, 142, 220,
330, 336, 338, 340, 347, 360
Peridinium 376
pesticides 205
PFE, *see* Percentage of Fresh Effluents
Phacus 357
Phormidium minnesotense 242
phosphate 173, 312
-, esters 207
phosphorus 20, 145, 364
-, removal 286
photosynthesis 342
-, inhibition by ammonia 160

photosynthetic
-, active radiation (*PAR*) 329
-, bacteria 264, 289
-, reservoirs 264
photosynthetically active radiation (*PAR*) 162,
329, 333, 334
phthalates 207
phytoplankton 357, 376
Planktosphaeria 175
plasticizers 207
Platymonas 376
Pleurochloris pyrenoidosa 242
polivirus 285
polysaccharides 249
pond
-, anaerobic 18, 212, 327
-, controlled discharge p. 284
-, holding p. 285
-, oxidation 59, 198, 312
-, stabilization p. 41, 61, 69, 85, 87, 327, 340
-, waste stabilization 66
Porcellinoides
-, *alchata* 190
-, *coronatus* 190
-, *orientalis* 190
-, *pruinosus* 185, 189
-, *senegallus* 190
potassium 358
precipitation, chemical 247
productivity, primary 308, 372
products, decomposition 200, 201
properties, optical p. of the water constituents
271
protozoa 33
Prymnesium parvum 161, 171
Pseudomonas 178
Pterocles 189
purple bacteria 263

R

radiance 263
-, values 265
rain 116
rate constants 126
raw sewage
-, composition 312
-, quality in Israel 16
reactors
-, non-steady-state r. 61
-, sequential batch 61, 85
-, steady-state flow r. 227
-, with controlled discharge 227
REDOX 340
refractory organic compounds 290
regime, operational 110, 117, 139
-, of stabilization reservoirs 63
regions, arid 66
relationships
-, antagonistic 320
-, competitive 320
remote sensing 263
research and development 3

reservoirs
  –, classification 264
  –, number of 16
  –, physical structure of 117
  –, recreational and landscape wastewater 286
  –, sequential batch 43, 74
  –, typology of the hypertrophic r. 191
residence time 85, 330
respiration 119, 123, 342
  –, rate 50
retention time 284, 289, 363
reuse 13, 23, 61
  –, in agriculture 247
river discharge 284
Rotifera 183, 184, 186–188, 191, 312, 322, 357
rotifers 289
runoff water 370

S

salinity 66, 337
Salmonella 308, 319, 321, 325
  –, faecalis 361, 363
  –, galilaeus 384
salts 13, 19
sampling 48, 81, 101, 196, 220, 307, 329, 339, 372
Sarotherodon galilaeus 381
SAT, see soil aquifer treatment
satellite 263
Scenedesmus 161, 162, 174–180, 235, 242, 289, 311, 325, 365, 367
  –, dimorphus 175, 311
  –, obliqus 161, 178
Schroederia 376
secchi disk 181, 331
sediment 109, 140, 145
  –, traps 148, 221
seepage, losses by 227
Selenastrum 51, 59, 174, 176, 189, 311
  –, minutum 59, 311
sewage, annual flow 15
slimes 249
sludge, activated 361
snails 365
socio-economic research 8
soft ionization techniques 205
soil aquifer treatment (SAT) 16, 18, 362
soil samples 212
solar radiation 116, 329
solvents 205, 207
Spain 287
species
  –, primary aquatic 182
  –, species, secondary aquatic 182
spillway 154
Spirulina 161, 162, 171, 238, 320
  –, platensis 161, 171, 238
Spondylomorum 174
stability 324
standard
  –, California 25, 28, 34
  –, EPA 37

–, WHO 34
Staphylococcus 178
steady-state 85
Stichococcus 175
storage
  –, capacity 355
  –, lagoons 285
stratification 55, 115, 129, 139, 173, 289, 316, 327, 373
succession 191
sulfides 340
sulfur bacteria 239
sulfur, compounds 207
supersaturation 174
surfactants 208
suspended
  –, matter 247
  –, solids 263
Synechococcus 289
Synedra 365

T

Taenia 26, 31, 34, 45, 46
  –, saginata 31, 45, 46
tandem mass spectrometry (MS/MS) 205
tapeworms 31
technology, semi-intensive 18
Tel Aviv 18, 361
temperature 118, 181, 308, 327, 372
  –, gradient, vertical 181
temporary desiccation 366
Tetraedron muticum 311
Tetragnathidae 185
thermoclines 332
Thiocapsa 264, 269–272, 275, 276, 279–281, 289
TOC 126, 127, 131, 136, 138, 361, 363
TON 198, 199
total suspended solids (TSS) 3, 43, 44, 49–54, 58, 59, 71, 72, 173, 256, 284, 287, 308, 339–342, 348, 372
trace
  –, elements 20
  –, metals, base level 224
tracer 335
Trichuris 26, 29, 34, 44
  –, trichuria 44
TSS, see total suspended solids
turbidity 336, 372
typhoid fever 30

U

United States 284
UV absorption 333

V

Vibrio cholerae 30
viruses 34
volume 370
  –, active 65
  –, dead 65
  –, dry 65

W

wastewater treatment 14

wastewater treatment and reuse plan 5
water reclamation projects in Israel 18
water
    –, balance 222
    –, control of quality 350
    –, demand for irrigation 4, 61
    –, discharge curves 64
    –, jet 287
    –, quality 339, 341, 378
    –, scarcity 3, 9, 13

–, supply 5
wind 66, 185, 229, 329, 383
Y
yellow matter 263
Z
zeta potential 249
Zn 221
zooplankton 115, 123, 127, 129, 131, 135, 136, 140,
    142, 174, 183, 193, 233, 238, 239, 249, 308, 311,
    357, 386

Printing: Saladruck, Berlin
Binding: H. Stürtz AG, Würzburg